Studies in Systems, Decision and Control

Volume 279

The series "Studies in Systems, Decision and Control" (SSDC) covers both new developments and advances, as well as the state of the art, in the various areas of broadly perceived systems, decision making and control–quickly, up to date and with a high quality. The intent is to cover the theory, applications, and perspectives on the state of the art and future developments relevant to systems, decision making, control, complex processes and related areas, as embedded in the fields of engineering, computer science, physics, economics, social and life sciences, as well as the paradigms and methodologies behind them. The series contains monographs, textbooks, lecture notes and edited volumes in systems, decision making and control spanning the areas of Cyber-Physical Systems, Autonomous Systems, Sensor Networks, Control Systems, Energy Systems, Automotive Systems, Biological Systems, Vehicular Networking and Connected Vehicles, Aerospace Systems, Automation, Manufacturing, Smart Grids, Nonlinear Systems, Power Systems, Robotics, Social Systems, Economic Systems and other. Of particular value to both the contributors and the readership are the short publication timeframe and the world-wide distribution and exposure which enable both a wide and rapid dissemination of research output.

** Indexing: The books of this series are submitted to ISI, SCOPUS, DBLP, Ulrichs, MathSciNet, Current Mathematical Publications, Mathematical Reviews, Zentralblatt Math: MetaPress and Springerlink.

More information about this series at http://www.springer.com/series/13304

Cengiz Kahraman · Selcuk Cebi
Editors

Customer Oriented Product Design

Intelligent and Fuzzy Techniques

Springer

Editors
Cengiz Kahraman
Department of Industrial Engineering
Faculty of Management
Istanbul Technical University
Macka, Istanbul, Turkey

Selcuk Cebi
Department of Industrial Engineering
Faculty of Mechanical Engineering
Yildiz Technical University
Besiktas, Istanbul, Turkey

ISSN 2198-4182 ISSN 2198-4190 (electronic)
Studies in Systems, Decision and Control
ISBN 978-3-030-42187-8 ISBN 978-3-030-42188-5 (eBook)
https://doi.org/10.1007/978-3-030-42188-5

This Springer imprint is published by the registered company Springer Nature Switzerland AG
The registered company address is: Gewerbestrasse 11, 6330 Cham, Switzerland

We dedicate this book to our Prof. M. Nahit Serarslan who passed away on August 16, 2019

Prof. Dr. Cengiz Kahraman and Prof. Dr. Selcuk Cebi

Prof. Dr. Mehmet Nahit Serarslan was born in Ordu-Gölköy. He graduated from the Management Branch of Mechanical Engineering Department of Istanbul Technical University in 1970. He then became a research assistant in the chair Machinery Tools and Factory Organization. He went to North Carolina State University to

research on operations research and became a research assistant at this university between 1976–1978. He received his Ph.D. from Istanbul Technical University in 1980. He became a Professor in the Industrial Engineering Department of Istanbul Technical University in 1991. Between the years 1991–2019, he had been at several administrative levels: the Head of Industrial Engineering Department, Vice Dean of Management Faculty, etc. His research and interest areas were quality management, statistical process control, probability theory, stochastic processes, and supply chain management. He completed numerous industrial projects successfully in the Turkish Industry. He passed away in Istanbul on August 16, 2019, leaving a lot of friends and students who love him very much behind.

Preface

Today, customers are becoming better organized, well informed, and more demanding. These changes are the reason why companies should develop customer oriented products. Customer oriented product design aims at developing products on the basis of an understanding of customers' or users' needs and requirements. It requires the dissemination of information about customers throughout an organization, formulation of strategies and tactics to satisfy market needs inter-functionally and achievement of a sense of company-wide commitment to these plans. To work in a customer oriented manner, all kinds of customer requirements must be understood.

As the editors of this book, in the first chapter, we present the relations of the customer oriented product design and intelligent and fuzzy techniques. The main steps of customer oriented product design consist of data collection, the definition of customer expectations, integration of customer requirements to design characteristics, implementation of the design and production of the prototype. Under a vague and imprecise environment, the definition of customer expectations and integration of customer requirements to design characteristics require fuzzy and intelligent techniques to be employed. We summarize data collection methods for product design and fuzzy and intelligent design approaches.

The second part is on customer oriented product design using quality function deployment. The first chapter of this part is on the design of noise-canceling earmuffs with quality function deployment (QFD). This chapter aims at defining the parameters that come to the forefront in a noise-canceling earmuff design considering user expectations. QFD method is used to transfer user expectations to the technical specifications of the product. According to the results of the chapter, "Active Adaptation System" and "Passive System with Protective Foam" are the two most important design parameters. The second chapter in this part is on improving the quality of product and process in the manufacturing of particleboard with an integrated QFD approach. It contributes to the relevant literature by means of developing an integrated QFD approach to determine the importance of technical characteristics on the basis of both the customer service level and interests of the firm. The proposed model is applied in a company operating in the wood-based panel industry sector. The third chapter is on fuzzy quality function deployment for smartphone product design. A fuzzy QFD methodology is applied in the smartphone design process to develop the product according to the customer's wishes. To address uncertainties in the subjective assessments of decision-makers, fuzzy sets are integrated with QFD methodology. The fourth section of this part is on an integrated fuzzy QFD methodology for customer oriented multifunctional power bank design. A case study about power banks is illustrated to test the usefulness of the method. There are various power bank models in the market and manufacturers desire to produce the most satisfying power banks for their customers. An integrated fuzzy QFD and fuzzy multi-criteria decision-making (MCDM) method are implemented for customer oriented multifunctional power bank design. The fuzzy analytic hierarchy process (AHP) is applied to determine the customer needs' weights and the fuzzy QFD technique to find the most important design requirements in multifunctional power bank design. The fifth chapter is on the integration of quality function deployment with interval-valued intuitionistic fuzzy (IVIF)-AHP and Kano model for customer oriented product design. An integrated method including QFD, Kano model, and IVIF-AHP is employed for customer oriented product design. Kano model is embedded into a house of quality to define marketing advantage by classifying customer requirements while IVIF-AHP is utilized to determine the importance degrees of customer requirements. The proposed method is applied for laptop design. The sixth and the last chapter of this part is on heterogeneous information integrated QFD model for smart bicycle design. House of Quality (HoQ) of QFD is extended with the 2-Tuple linguistic model to create a flexible environment for decision-makers (DMs). The multi-preference relations technique is used to handle heterogeneous information while calculating customer requirements' importance.

The third part is on customer oriented product design using the axiomatic design method. The first chapter of this part is on an axiomatic design framework of the sustainable product-service systems for the circular economy. An axiomatic design framework is proposed for the sustainable product-service systems so that a complete conceptual design is made based on the environmental, economic, and social dimensions for the products that concern the customer expectations for the circular economy. The second chapter is on developing customer oriented lean production system using axiomatic design and fuzzy value stream mapping. An axiomatic design (AD) based methodology is developed by employing one of the most widely used lean techniques, value stream mapping (VSM). The methodology is designed in three stages which are analyzing the current state, applying the AD method, and designing the future state. The process and lead times are defined by using triangular fuzzy numbers (TFNs) to capture the ambiguity in work-in-process levels in the company. The third chapter is on the website design using Pythagorean fuzzy axiomatic design. Axiomatic design method is used to analyze the design characteristics of banking websites. The website designs of Turkey's leading banks are analyzed.

The fourth part is on customer oriented product design using the Kano model method. The first chapter is on a Kano model in the development of customer oriented products. It tries to facilitate practitioners in future research about the further development of Kano's model and organizations in the future use of the Kano approach to correctly identify customer requirements and channelize their resources in the right direction. The second chapter is on the mobile application design by using a modified fuzzy Kano model. A questionnaire-based Kano Model Method is proposed in this chapter. Since the Kano model questionnaire includes linguistic evaluations, the traditional Kano model is weak to handle uncertainty in linguistic evaluations. Therefore, a modified Fuzzy Kano model is proposed to overcome this difficulty. The third chapter is on the classification of laptop design attributes using the fuzzy Kano model. Design preferences of college students are analyzed by using the fuzzy Kano model, and the results are compared to the results of the traditional Kano model.

The fifth part is on the customer oriented product design using MCDM and heuristics algorithms. The first chapter of this part is on the application of MADM techniques to develop customer oriented products. It presents an approach to apply a multi-attribute decision-making (MADM) method to evaluate the product design alternatives with respect to the customer and designer preferences. Personal data assistant (PDA) selection problem is handled. The alternatives of personal data assistants are evaluated by using different MADM methods. The second chapter presents the application of an advanced fuzzy multi-criteria decision-making method on product development. The authors employ a very recent extension of ordinary fuzzy sets, namely spherical fuzzy sets, in the house of quality development. For the comparison of competitive firms, spherical fuzzy TOPSIS (SF-TOPSIS) is used. The third chapter is on a marketing decision support system (MDSS) for product design based on an outranking approach. A user-friendly and easy-to-learn MDSS is presented. The purpose of this MDSS is to assist a marketer in designing a new product when the customer preferences are not compensatory. The MDSS finds near-optimal solutions for realistically sized problems using evolutionary algorithms. The fourth chapter is on the analysis of a new product development strategy based on a heuristic multi-criteria methodology. The Pythagorean fuzzy-based heuristic approach is proposed as a combination of AHP and ARAS (Additive Ratio Assessment) method. A production system is considered where the manufacturing of a new product is performed in a group decision-making (GDM) setting. The literature review given in the chapter presents the current state of the art and discusses the potential future research trends. The fifth chapter is on the application of heuristics algorithm on product development. It is focused on describing a heuristic algorithm for product development. Heuristic models customer preferences through the disaggregation of global preferences on corn toast products. The sixth chapter, the last chapter of this part, is on the design of a warehouse location involving AS/RS technologies. In the chapter, two different warehouse location design alternatives involving AS/RS technologies are evaluated by using an interval-valued intuitionistic fuzzy analytic hierarchy process (AHP) approach.

The sixth and the last part is on the customer oriented product design using other methods. The first chapter is on the customer oriented product design with conjoint analysis (CA). The factors in the CA are determined as RAM, Graphic Card, Processor Model, Storage Type, Screen Size, and Operating System. The results of the analysis reveal that the computer processor ranks first and other attributes are ranked as RAM, operating system, storage type, graphics card, and screen size, respectively. The second chapter is on assessing customer preferences on a new design of refrigerator using CA. The applicability of CA is experienced in the home appliances industry by designing a new refrigerator. The obtained results are found to be reliable and satisfactory to retain competitive in the market, and the method is suggested as a guide to determine the best design of home appliances considering customer expectations, needs, and perceptions. The third chapter is on the literature review on fuzzy failure mode and effects analysis (FMEA) and its application to the infant car seat design using spherical fuzzy sets. The authors summarize the literature on fuzzy FMEA and propose a spherical fuzzy FMEA model with an illustrative application on infant car seat design. The fourth chapter is on the examination of factors affecting customer oriented product design in the automotive industry. Customer oriented product design for an automobile is examined according to the design concepts and their causal relationships using fuzzy cognitive mapping (FCM) to help the automotive industry. The fifth chapter is on the baby cradle design with Kansei knowledge mining based on rough set theory. To extract more accurate design parameters and Kansei knowledge, rough set rule-based mining is applied for Kansei engineering research to baby cradle design.

We hope that this book will provide a useful resource of ideas, techniques, and methods for additional research on uncertainty modeling in customer oriented product design using intelligent and fuzzy techniques. We are grateful to the referees whose valuable and highly appreciated works contributed to select the high quality of chapters published in this book. I also would like to thank Series Editor, Prof. Janusz Kacprzyk from the Polish Academy of Sciences and Executive Editor Thomas Ditzinger for their supportive roles in this process.

Macka, Turkey Prof. Cengiz Kahraman
Besiktas, Turkey Prof. Selcuk Cebi

Contents

About the Editors

Cengiz Kahraman is a Full Professor at the Industrial Engineering Department of Istanbul Technical University (ITU). His research areas are engineering economics, quality management, statistical decision-making, multi-criteria decision-making, and fuzzy decision-making. He published about 200 journal papers and about 150 conference papers. He is in the editorial boards of many international journals and edited about 10 international books from Springer and Atlantis Press. He guest-edited many special issues of international journals and organized international conferences.

Selcuk Cebi is currently a Professor of Industrial Engineering at Yildiz Technical University. He received his Ph.D. from the Industrial Engineering Program of Istanbul Technical University in 2010 and M.Sc. from the Mechanical Engineering Department of Karadeniz Technical University in 2004. His current research interests are customer Oriented Product Development, Decision Support Systems, Multiple Criteria Decision-Making, and Risk Management. He achieved the publications of international journal papers more than 40 articles having over 2000 total citations with the related research topics.

Part I
Introduction to Customer Oriented Product Design

Customer Oriented Product Design and Intelligence

Selcuk Cebi and Cengiz Kahraman

Abstract The aim of customer oriented product design is to develop products on the basis of an understanding of customers' expectations and requirements. Customer orientation is based on both the experiences of users and customers. This is a crucial issue for the product to be accepted in the market by customers. The main steps of customer oriented product design consist of data collection, definition of customer expectations, integration of customer requirements to design characteristics, implementation of the design and production of the prototype. Under vague and imprecise environment, definition of customer expectations and integration of customer requirements to design characteristics require fuzzy and intelligent techniques to be employed. In this chapter, we summarize data collection methods for product design and fuzzy and intelligent design approaches.

1 Introduction

The first revolution in industrial production was the invention of the steam engine. So, this invention made possible to produce big forces easily for the production and constructed the first step of mass production. In particular, the introduction of computer and automation systems to industry carried this to next step and production quantities rapidly increased while production costs were decreased. The increase in production amounts has created a competitive environment among the companies that want to reach more customers. From World War II to the beginning of 2000s enterprises have tried to reach their customers by adding technological and functional features to their products. However, in the last decade, the price, quality, technological, and functional elements of the products are not adequate to provide sustainable competitive advantage in the competitive environment. Therefore, in recent years, companies have discovered the importance of producing customer-oriented products

S. Cebi (✉)
Department of Industrial Engineering, Yildiz Technical University, Istanbul, Turkey
e-mail: scebi@yildiz.edu.tr

C. Kahraman
Department of Industrial Engineering, Istanbul Technical University, Istanbul, Turkey

C. Kahraman and S. Cebi (eds.), *Customer Oriented Product Design*, Studies in Systems, Decision and Control 279, https://doi.org/10.1007/978-3-030-42188-5_1

that satisfy customer expectations in order to increase their market share. Parallel to industry, researchers have focus on understanding customer expectations while developing products [1–3] since customer expectations have a vital role in providing the competitive advantages for new product development process. In fact, customer expectations define the fundamentals of design requirements. Nowadays, existence of a product on the market for a long time, in other words, success or failure of companies depends on the level of meeting customers' expectations from the product. This has transformed the product-oriented production strategy into a customer-oriented production strategy [4]. In Fig. 1, production development process has been given. The definition and design stages of the production development process (PDP) provide the difference between conventional production development process (C^{PDP}) and customer-oriented product development process (CO^{PDP}). You should collect and analyze the data to obtain an output representing the customer requirements. Then, you should reflect customer requirements into design characteristics and optimize them to catch the expectations of customers.

There are three main components of successful product development. The first one is Definition of Customer Expectations. The second is to determine technical characteristics to satisfy customer expectations and to define innovation ideas on the product. Finally, the third step is to produce the high-quality product that satisfy the features defined first two steps. Among these steps, the first one is a bit more important that the other although all of them are vital for a product life cycle.

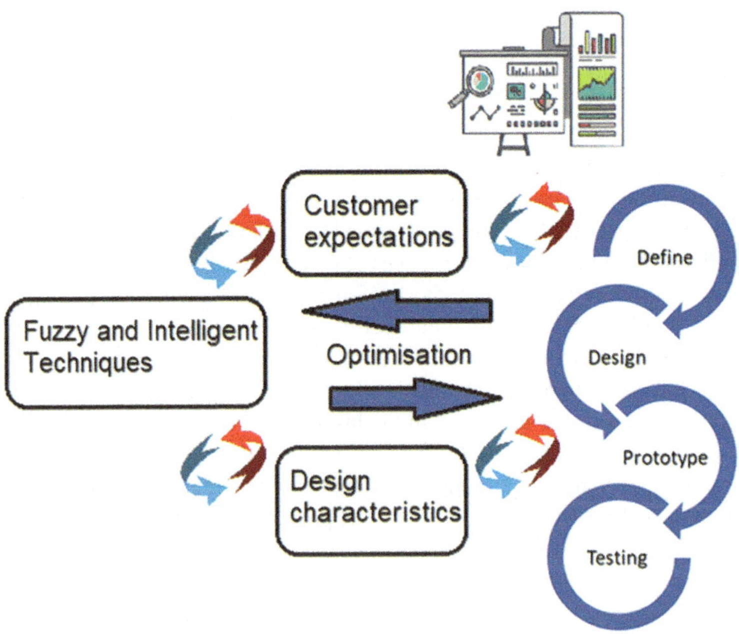

Fig. 1 Product design process

Therefore, Nowadays, many businesses devote considerable resources from product development budgets for incorporating the voice of the customer into the product.

In the scope of this chapter we aim at providing a brief introduction to the techniques which are measuring customer voices and incorporating the voice of the customer into the product. The next section presents data collection methods to understand voice of the customers. In the third section, a brief explanation for the data analysis techniques which are incorporating the voice of the customer into the product has been presented. In Sect. 4, we present the extensions of ordinary fuzzy sets. Section 5 includes the intelligent techniques. Section 6 concludes the chapter.

2 Data Collection Methods

Measuring the Voice of Customer is to understand customers' expectations and requirements on the product. Marketing researchers use different methodologies to ensure what customers' expectations are such as;

- Surveys
- Interviews
- Focus group studies
- Experimental studies (Usability testing)
- Online documents and records.

Each method given above have different strengths and weaknesses [2].

2.1 Surveys

Survey is one of the widely used data collection methods to understand customers' expectations because of their simple structures. The quality and reliability of the collected data are the most important issue to be considered. Otherwise, it causes a waste of time and loss of money. An ideal survey that provides correct and reliability data can be prepared by the following steps;

1. ***Determine Research Objectives***: The main aim of the research should be determined and clearly defined to prepare stunning survey.
2. ***Define potential participants***: If you want perfect outcomes from your research you should ask your research questions to participants relevant to the purpose of the research. At this step, two questions should be answered; the first one is the number of the sample size and the second is the personal type of the participants. For instance, if you want to define design parameters that affect design structures of a mobile phone for elderly people, you should prepare a survey for elderly people. Furthermore, you must consider how you can reach the target population at your survey.

3. ***Prepare an appropriate questionnaire***: The preparation of a questionnaire depends on the methodology that is used for the evaluation of the collected data. There is a strong interdependency between the desired data and the type of questions included in the questionnaire. While a questionnaire is being prepared, the following rules should be considered.

   - Questions should reflect the research objectives
   - Questions should be simple and clear
   - Choices should cover all options
   - The number of questions should be optimized.

4. ***Test and revise the questionnaire***: The prepared questionnaire should be tested whether there are conflicting questions or whether the desired data can be collected.
5. ***Implement the questionnaire***: In order to make data collection easy, online tools will be helpful. In the market, there is a lot of professional questionnaire platform that sends the questionnaire by a link and they provide to reach a broad range of people. The main disadvantage of this option is that you may not be sure that the collected data are reliable. The second option is the paper-based questionnaires. But you need volunteers/professionals to lead surveys. The second method is more expensive than the first one. Furthermore, you pay extra effort and time to transfer collected data to a digital platform. The third way is to use call services. This option is also the most expensive in all. Furthermore, it is impossible to provide any visual object to participants in the questionnaire. However, in some cases, the second and the third one is the most suitable if you conducted a survey just like AHP questionnaires.

The advantages of a questionnaire can be summarized as follows; (i) It can be reached in a large number of participants, (ii) It is cost-effective, (iii) it is easy to visualize and analyze the collected data, and (iv) the identity of participants is generally unknown. On the other hand, the main disadvantage of a questionnaire is that participants may not be clear because of their hidden agenda or they may leave unanswered questions because of their unwillingness. Furthermore, there is no flexibility in the process.

2.2 Interviews

Interviews are a qualitative data collection method that is applied face to face with individuals or groups. Technological tools such as audio recorders and video cameras are utilized to collect data through interviews. This method is categorized as structured, semi-structured and unstructured [5, 6]. The structured interview is a verbal questionnaire that involves predetermined questions and it is directed to participants without any change. On the other hand, the unstructured interview includes an improvisation structure. It means that it does not include any predetermined questions and it is utilized with at most little organization. Finally, the semi-structured interview involves several key questions that restrict the subject to be researched while it allows participants to respond in more detail [5]. Interview methods help researchers to provide to collect detail information and it has a flexible structure in terms of the order of questions. On the other hand, it is hard to analyze the data collected by this method and it takes lots of time to reach in a large number of participants.

2.3 Focus Group Studies

A focus group study is another qualitative data collection method that is utilized by a discussion on a determined topic of an individual group which usually consists of 6–12 people with the assistance of a moderator. Although this method is commonly used in product/marketing research, it is also popular in sociology science as well. The advantage of this method is that it is easy to conduct the survey and it has a flexible structure. On the other hand, it may be difficult to analyze the collected data and the moderator has little control over the groups. Furthermore, the moderator has certain skills to conduct the survey [7, 8].

2.4 Experimental Studies

Experimental studies are another data collection type that is based on the use case scenario. In the scenario, there are several tasks related to the researched product, which should be succeeded by participants. In such studies, the data collection activity is conducted into a laboratory environment or real-life environment [9]. In the literature, since the real-life environment represents the complex requirements of customers, the experimental studies conducted in the real-life environment are more preferred than the studies in the laboratory environment. Thus, the collected data on the user experience the real-life are more reliable and realistic than the results obtained in the laboratory [10]. However, there are two main difficulties for data collection in a real-life environment [11, 12]. (i) it is hard to apply determined data evaluation techniques and (ii) the analysis of the data is complex. On the other hand, the advantage of the data collection in the laboratory environment is that there is full control over the test, and it is easy to measure the feature of the researched product [9, 10]. In the analysis, the collected data are analyses by using objective performance factors such as the success ratio of the participants on the given task in the scenario, error ratio, timing, etc. In the analysis phase, the collected data are analyzed by using objective performance factors such as the success rate, error ratio, timing, etc. when the participants conduct the given task in the scenario. Furthermore, in some studies, the questionnaires are given to participants after an experimental study has been conducted in order to collect detail data on the studied product.

2.5 Online Documents and Records

Nowadays, internet includes big data source on customer complaints, feedbacks, and customer needs. Many firms seek various methods to analyze these data to understand expectations of their customers. For this purpose, some intelligent tools such as machine learning, deep learning, semantic analysis, natural language analysis etc., have been widely utilized to understand customers' expectations and feelings.

3 Customer Oriented Product Design Methods

Customer oriented product design methods is used to integrate customer expectation into design characteristics. These methods are, Quality Function Deployment, Kano Model, Axiomatic Design, Conjoint Analysis, Kansei Engineering, and Multiple Criteria Decision Making Methods.

3.1 Quality Function Deployment

The Quality Function Deployment (QFD) method was proposed to the literature in the late 1960s by Akao [13] to satisfy customers' expectations from a product by matching customers' requirements with technical characteristics for product design. The QFD utilizes a graphical scheme called the house of quality in the design process and the scheme consists of six main sections. The first section presents the customer requirements and importance degrees of them. This section is also knowns as the voice of the customer. The second section includes the technical specifications of the designed product. The third section provides a marketing evaluation of the developed product with respect to competitive products under the defined customer requirements. The fourth section includes the relationships among the customer requirements and technical specifications. The fifth section presents the interdependencies among technical specifications. Finally, the sixth section presents the importance degree of technical specifications based on customer requirements and marketing advantage and contains the technical evaluations of the target.

3.2 Axiomatic Design

The axiomatic design method has been proposed to literature by Suh [14, 15] to present a scientific basis for the design process. The method utilizes two axioms, called independence axiom and information axiom to improve the design of a product, respectively. The independence axiom emphasizes that functional requirements are the minimum set of independent requirements describing the design goals and the independence axiom of these functional requirements has to be fulfilled [14, 15]. The information axiom presents an evaluation tool to select the design which has the smallest information content as the best alternative among the designs when more than one design fulfilling the independence axiom exists [14, 15].

3.3 Kano Model

The Kano model was proposed by Kano [16] to classifying attributes by using a questionnaire containing functional and dysfunctional questions. Functional questions in the questionnaire aim at analyzing the consumer's feelings in the case of fulfillment of an attribute while dysfunctional question examines consumer's feelings in the case of non-fulfillment of an attribute [17]. In the questionnaire, participants are expected to answer both functional and dysfunctional questions by selecting one of the following linguistic terms: "dislike", "live-with", "neutral", "must-be" and "like" [16]. The analysis of the collected data is performed by using an evaluation matrix. Based on the evaluation, attributes are classified as Reverse, Indifferent, Attractive, One-Dimensional and Must-be attributes, respectively [16, 17].

3.4 Conjoint Analysis

The conjoint analysis method which is a statistical-based technique was proposed to analyze the joint effects of customers' preferences related to a set of independent features of the product in the market research [18]. Thus, the importance degree of the features of a product is determined based on consumers' expectations before the design phase of a product. By using a questionnaire, the product alternatives including various combinations of the possible features are presented to customers and they are forced to make a trade-off among features of a product to buy. Then, the importance degree of the product features is determined by using conjoint analysis.

3.5 Kansei Engineering

Kansei Engineering was proposed by Nagamachi [19, 20] in the early 1970s to translate customer's feelings and impressions into product design features. Kansei, a Japanese word, means the psychological feeling and image of customers such as "luxurious, gorgeous and strong" regarding a new product. There are three types of Kansei Engineering (KE) technology called KE Type I, II, and III. Type I provides classification procedures on the new product for the design parameters. Type II is based on computer technologies such as Expert System, Neural Network Model, and Genetic Algorithm etc. Type III includes a mathematical structure.

4 Fuzzy Set Theory

Fuzzy set theory is widely used by integrating into customer oriented product design methods explained in Sect. 3. In the scope of this book, there are lots of applications of these methodologies integrating fuzzy set theory into these methodologies. Fuzzy set theory was proposed to literature in 1965 by Zadeh [21] to handle uncertainties caused by limited knowledge or ambiguity in order to modeling real-life situation. In fuzzy sets, each object in a set is defined by using a membership value between zero and one which represents the membership degree of the object to the set. Based on the imprecision on modeling, the shape of the membership function may also include uncertainties. To cope with this type of problem, extensions of fuzzy set theory have been proposed to literature. There are several extensions of the fuzzy set theory since it proposed to literature. After ordinary fuzzy set theory, Type-2 Fuzzy Sets by Zadeh [22], Interval-Valued Fuzzy Sets (*IVFS*) by Zadeh [22], by Sambuc [23], by Grattan-Guiness [24], and by Jahn [25], Fuzzy Multisets by Yager [26–28], Intuitionistic Fuzzy Sets (*IFS*) by Atanassov [29], Neutrosophic Sets (*NS*) by Smarandache [30–32], Nonstationary Fuzzy Sets (*NFS*) by Garibaldi and Ozen [33–35], Hesitant Fuzzy Sets (*HFS*) by Torra [36], Pythagorean Fuzzy Sets (*PFS*) by Atanassov [37] and by Yager [38], Picture Fuzzy Sets (*PiFS*) by Cuong [39], q-Rung orthopair fuzzy sets (*q-ROFS*) by Yager [40], Spherical Fuzzy Sets (*SFS*) by Kutlu Gundogdu and Kahraman [41], and Fermatean Fuzzy Sets (*FFS*) by Senapati and Yager [42, 43] have been introduced to the literature. In Fig. 2, the historical developments of the fuzzy set theory have been visualized. All of these extensions of fuzzy sets have been proposed to well cope with uncertainties in the modeling of complex problems. Among these methods, *Pythagorean Fuzzy Sets*, *q-Rung orthopair Fuzzy Sets*, and *Fermatean Fuzzy Sets* are based on similar structure and philosophy. In the following, the definitions and main mathematical formulations of the fuzzy sets and its extensions will be presented.

4.1 Ordinary Fuzzy Sets

Fuzzy sets proposed to literature to handle uncertainties in complex problems by Zadeh [21]. The main definition of a fuzzy set can be given as follows [21].

Definition 1 "Let X be a space of points (object), with a generic element of X denoted by x. Thus, $X = \{x\}$. A fuzzy set (class) A in X is characterized by a membership (characteristic) function $f_A(x)$ which associates with each point in X in the interval [0, 1] with the value of $f_A(x)$ at x representing the grade of membership of x in A."

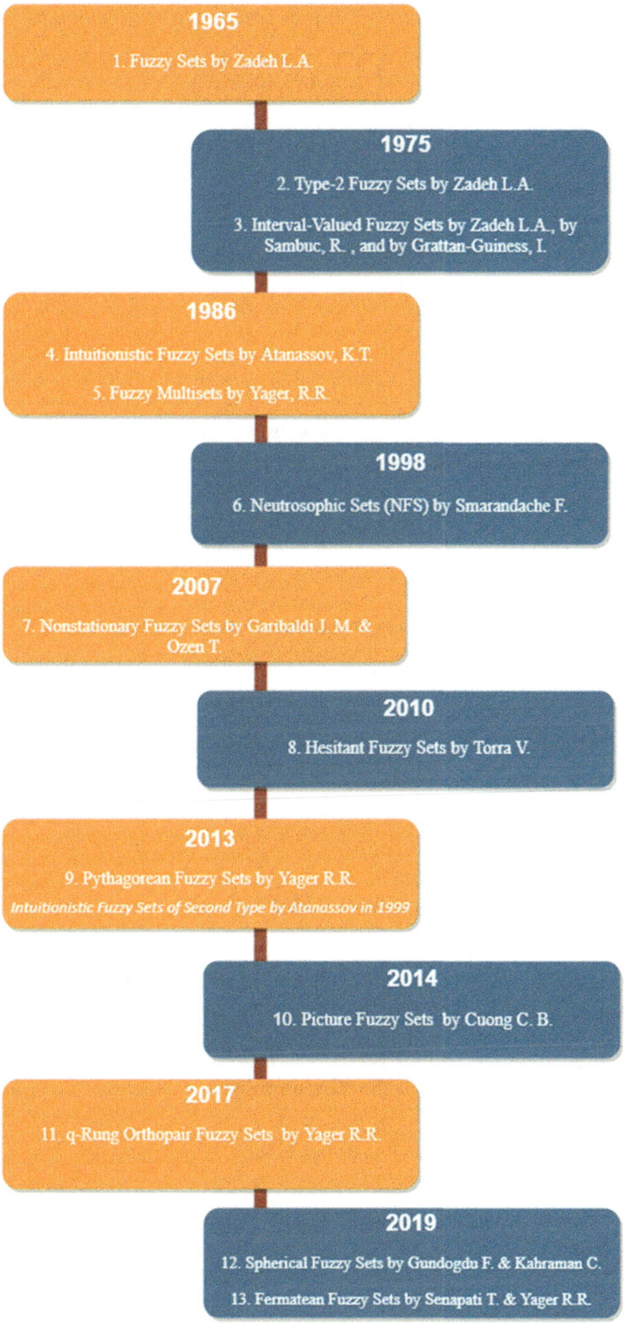

Fig. 2 Chronology of fuzzy set extensions

The fuzzy set defined by a membership degree for an x value is given in Eq. (1)

$$\tilde{A} = \{(x, \mu(x))|x \in X\} \tag{1}$$

where X represents the universe, and $0 \leq \mu(x) \leq 1$ and the non-membership degree is $1 - \mu(x)$ [21].

4.2 *Type-2 Fuzzy Sets*

Type-2 Fuzzy sets proposed by Zadeh in 1975 is defined as follows [22].

Definition 2 "A type-2 fuzzy set $\tilde{\tilde{A}}$ in the universe of discourse X which can be represented by a type-2 membership function $\mu_{\tilde{\tilde{A}}}$ is given as

$$\tilde{\tilde{A}} = \left\{\left((x, u), \mu_{\tilde{\tilde{A}}}(x, u)\right)\middle|\forall x \in X, \forall u \in J_x \subseteq [0, 1], 0 \leq \mu_{\tilde{\tilde{A}}}(x, u) \leq 1\right\} \tag{2}$$

where J_x denotes an interval [0, 1]. Moreover, $\tilde{\tilde{A}}$ represented by:

$$\tilde{\tilde{A}} = \int_{x \in X} \int_{u \in J_x} \mu_{\tilde{\tilde{A}}}(x, u)/(x, u) \tag{3}$$

where $\iint$ denotes union over all admissible x and u."

4.3 *Interval-Valued Fuzzy Sets (IVFS)*

Interval-valued fuzzy (IVFS) sets were proposed to literature in 1975 [22–25]. The definition of the IVFS can be given as follows [22–25].

Definition 3 "An interval-valued fuzzy set (IVFS) A on the universe $X \neq \emptyset$ is a mapping $A: X \rightarrow L([0, 1])$, such that the membership degree of $x \in X$ is given by $A(x) = \left[\underline{A}(x), \bar{A}(x)\right] \in L([0, 1])$, where $\underline{A}: X \rightarrow [0, 1]$ and $\bar{A}: X \rightarrow [0, 1]$ are mappings defining the lower and the upper bound of the membership interval $A(x)$, respectively."

4.4 *Fuzzy Multisets (FM)*

It is the first-time fuzzy multisets were mentioned by Yager [26], and further studied by researchers [27, 28]. Fuzzy multisets (*FM*) are defined as [28]

Definition 4 "Let X be the universe. A fuzzy multiset $\tilde{A}$ over X is characterized by a function $\tilde{A}: X \rightarrow \mathbb{N}^{[0,1]}$. The family of the fuzzy multisets over X is called the fuzzy power multiset over X and is denoted by $\mathcal{FM}(X)$."

4.5 Intuitionistic Fuzzy Sets (IFS)

It is the first time intuitionistic fuzzy sets were proposed to literature by Atanassov [29] as follows;

Definition 5 "Let X be a universe of discourse. An IFS $\tilde{A}$ is an object having the form, $\tilde{A} = \{x, (\mu_{\tilde{A}}(x), v_{\tilde{A}}(x)) | x \in X\}$ where the function $\mu_{\tilde{A}}: X \rightarrow [0, 1]$, $v_{\tilde{A}}(x): X \rightarrow [0, 1]$ and $0 \leq \mu_{\tilde{A}}(x) + v_{\tilde{A}}(x) \leq 1$ are the degree of membership, non-membership of u to $\tilde{A}$, respectively. For any IFS $\tilde{A}$ and $x \in X$, $\pi_{\tilde{A}} = 1 - \mu_{\tilde{A}}(x) - v_{\tilde{A}}(x)$ is called degree of hesitancy of x to $\tilde{A}$."

4.6 Neutrosophic Sets (NFS)

Neutrosophic sets (NFS) have been defined by Smarandache [30–32] as follows;

Definition 6 "Let X be a universe of discourse. Neutrosophic set $\tilde{A}$ in X is an object having the form, $\tilde{A} = \{x, T_{\tilde{A}}(x), I_{\tilde{A}}(x), F_{\tilde{A}}(x) | x \in X\}$ where, $T_{\tilde{A}}$, $I_{\tilde{A}}$, and $F_{\tilde{A}}$ are the truth-membership function, the indeterminacy-membership function, and the falsity-membership function, respectively. The relation among these parameters is given by $0 \leq T_{\tilde{A}}(x) + I_{\tilde{A}}(x) + F_{\tilde{A}}(x) \leq 3$."

4.7 Nonstationary Fuzzy Sets (NFS)

The nonstationary fuzzy sets have been proposed to literature for modeling the variation (over time) of opinion [33, 34]. The definition of the nonstationary fuzzy sets is as follows [35].

Definition 7 "Let A denote a fuzzy set of a universe of discourse X characterized by a membership function μ_A. Let T be a set of time points t_i (possibly infinite) and $f: T \rightarrow \Re$ denote a perturbation function. A nonstationary fuzzy set $\dot{A}$ of the universe of discourse X is characterized by a nonstationary membership function $\mu_{\dot{A}}: T \times X \rightarrow [0, 1]$ that associates with each element (t, x) of $T \times X$ a time-specific variation of $\mu_A(x)$. The nonstationary fuzzy set $\dot{A}$ is denoted by Eq. (4)."

$$\dot{A} = \int_{t \in T} \int_{x \in X} \frac{\mu_{\dot{A}}(t, x)}{x} / t \tag{4}$$

4.8 Hesitant Fuzzy Sets (HFS)

Torra [36] proposed another extension of the ordinary fuzzy sets whose name is hesitant fuzzy set theory in 2010. The definition of fuzzy sets is given as follows [28, 36].

Definition 8 "Let X be the universe. A hesitant fuzzy set $\tilde{A}$ over X is characterized by a function $\tilde{A}$: $P([0, 1])$ where $P([0, 1])$ is the family of all the subsets of the real closed interval [0, 1]. The family of all the hesitant fuzzy sets over X is called the hesitant fuzzy power set over X and is denoted by $\mathcal{FH}(X)$."

4.9 Pythagorean Fuzzy Sets (PFS)

These fuzzy sets were originally proposed by Atanassov in 1999 as *Intuitionistic Fuzzy Sets of Second Type* (*IFS*2) [37]. Later, Yager [38] handled *IFS2* and proposed Pythagorean fuzzy set (*PFS*), which is an extension of intuitionistic fuzzy set (*IFS*) theory. The definition of the Pythagorean fuzzy set is given as follow [37, 38].

Definition 9 "Let X be a universe of discourse. A PFS $\tilde{P}$ is an object having the form, $\tilde{P} = \{x, (\mu_{\tilde{p}}(x), \nu_{\tilde{p}}(x)) | x \in X\}$ where the function $\mu_{\tilde{p}}: X \to [0, 1]$, $\nu_{\tilde{p}}(x): X \to [0, 1]$ are the degree of membership, non-membership of x to P, respectively and $0 \leq \mu_{\tilde{p}}^2(x) + v_{\tilde{p}}^2(x) \leq 1$. For any PFS $\tilde{A}$ and $x \in X$, $\pi_{\tilde{p}} = (1 - \mu_{\tilde{p}}^2(x) - v_{\tilde{p}}^2(x))^{0.5}$ is called degree of hesitancy of x to $\tilde{P}$."

4.10 Picture Fuzzy Sets (PiFS)

The other extension of fuzzy sets and the intuitionistic fuzzy sets is Picture Fuzzy Sets (*PiFS*) recently proposed by Cuong [39]. *PiFS* is used to modeling decisions in situations involving various answer types. The definition of *PiFS* is given as follows [39].

Definition 10 "A picture fuzzy set A on a universe X is an object in the form of $A = \{(x, \mu_A(x), \eta_A(x), \upsilon_A(x) | x \in X)\}$ where $\mu_A(x) \in [0, 1]$ is called the degree of positive membership of x in A, $\eta_A(x) \in [0, 1]$ is called the degree of neutral membership of x in A and $\upsilon_A(x) \in [0, 1]$ is called the degree of negative membership of x in A, and where μ_A, η_A, and υ_A satisfy the following condition:

$$(\forall x \in X) \quad (\mu_A(x) + \eta_A(x) + \upsilon_A(x) \leq 1) \tag{6}$$

Now $(1 - (\mu_A(x) + \eta_A(x) + \upsilon_A(x)))$ is called the degree of refusal membership of x in A."

4.11 q-Rung Orthopair Fuzzy Sets (q-ROFS)

q-Rung orthopair fuzzy sets (q-ROFS) have been proposed to the literature by Yager in 2017 [40]. The definition of the q-ROFS is as follows [40].

Definition 11 "Let X be a universe of discourse. A q-ROFS $\tilde{A}$ is an object having the form, $\tilde{A} = \{\langle x, \mu_{\tilde{A}}(x), \beta_{\tilde{A}}(x)\rangle | x \in X\}$ where the function $\mu_{\tilde{A}}: X \to [0, 1]$, $\beta_{\tilde{A}}(x): X \to [0, 1]$. $\mu_{\tilde{A}}(x)$ and $\beta_{\tilde{A}}(x)$ are the degree of membership and non-membership of x to $\tilde{A}$, respectively and they satisfy the condition $0 \leq \mu_{\tilde{A}}(x)^q + \beta_{\tilde{A}}(x)^q \leq 1$, where $q \geq 1$ for $x \in X$. For any q-ROFS $\tilde{A}$ and $x \in X$ $\pi_{\tilde{A}}(x) = \sqrt[q]{1 - \left(\mu_{\tilde{A}}(x)^q + \beta_{\tilde{A}}(x)^q\right)}$ is called degree of indeterminacy of x to $\tilde{A}$."

4.12 Spherical Fuzzy Sets (PFS)

The concept of Spherical Fuzzy Sets (*SFS*), proposed by Gundogdu and Kahraman [41], another extension of fuzzy sets, provides a larger preference volume for decision makers to assign their judgments about membership, non-membership, and hesitancy degrees. The definition of the sets is as follows [41].

Definition 12 "Let X be a universe of discourse. A SFS $\tilde{S}$ is an object having the form, $\tilde{S} = \{\langle x, (\mu_{\tilde{S}}(x), \nu_{\tilde{S}}(x), \pi_{\tilde{S}}(x))\rangle | x \in X\}$ where the function $\mu_{\tilde{S}}: X \to [0, 1]$, $\nu_{\tilde{S}}(x): X \to [0, 1]$, $\pi_{\tilde{S}}(x): X \to [0, 1]$ and $0 \leq \mu_{\tilde{S}}^2(x) + v_{\tilde{S}}^2(x) + \pi_{\tilde{S}}^2(x) \leq 1$ are the degree of membership, non-membership of x to $\tilde{S}$, respectively. For any SFS $\tilde{S}$ and $x \in X$ $\pi_{\tilde{S}} = (1 - \mu_{\tilde{S}}^2(x) - \nu_{\tilde{S}}^2(x) - \pi_{\tilde{S}}^2(x))^{0.5}$ is called degree of refusal membership of x to $\tilde{S}$."

4.13 Fermatean Fuzzy Sets (FFS)

Senapati and Yager [42, 43] proposed Fermatean fuzzy sets (*FFS*) to cope with uncertainty in decision making.

Definition 13 "Let X be a universe of discourse. An FFS $\tilde{A}$ is an object having the form, $\tilde{A} = \{\langle x, \mu_{\tilde{A}}(x), \beta_{\tilde{A}}(x)\rangle | x \in X\}$ where the function $\mu_{\tilde{A}}: X \to [0, 1]$, $\beta_{\tilde{A}}(x): X \to [0, 1]$. $\mu_{\tilde{A}}(x)$ and $\beta_{\tilde{A}}(x)$ are the degree of membership and non-membership of x to $\tilde{A}$, respectively and they include circumstance $0 \leq \mu_{\tilde{A}}(x)^3 + \beta_{\tilde{A}}(x)^3 \leq 1$ for $x \in X$. For any FFS $\tilde{A}$ and $x \in X$ $\pi_{\tilde{A}}(x) = \sqrt[3]{1 - \left(\mu_{\tilde{A}}(x)^3 + \beta_{\tilde{A}}(x)^3\right)}$ is called degree of indeterminacy of x to $\tilde{A}$."

5 Intelligent Techniques

Intelligent techniques include genetic algorithm, ant colony optimization, simulated annealing method, taboo search method, artificial neural networks, particle swarm optimization, etc.

5.1 Genetic Algorithm

Genetic algorithm (GA) is a random search method that is inspired by the structure of biological evolution and natural genetics to solve both constrained and unconstrained optimization problems. GA uses three rules which are named selection, crossover, and mutation to create the next generation from the current population. The individuals providing the population for the next generation are determined by selection rules. The children for the next generation from two parents are formed by crossover rules. The individual parents are randomly changed to form children concerning mutation rules. The method was developed in the middle of 1970s by Holland [44, 45] to provide a solution to the ill-posed problems such as nonconvex functions, nondifferentiable functions, stochastic, or highly nonlinear functions, multiple local optima, and multiple objectives [46]. Providing an approximate solution in a relatively short time compared with the other metaheuristics is the main advantage of GA [47].

5.2 Ant Colony Optimization

Ant colony (ACO) is a random search method that is inspired by the behavior of ants in finding paths from the nest to solve combinatorial optimization problems [48–50]. The method was proposed to the literature by Dorigo [48]. At first, the ants wander randomly while searching for food sources and they leave pheromones to mark their path. This helps ants to get back home and to find the source the next time. At this stage, there may be more than one path to the source. If the other ants come across one of these paths including these pheromones, they are most likely to follow this path. Each of these ants leaves own pheromones to the path while they bring the food back. In a short time, the shortest path has more pheromones and gets stronger since more ants use the shortest path. Hence, the optimum solution is obtained [48, 51].

5.3 Simulated Annealing Method

It is the first time the method, Simulated Annealing Method (*SAM*), was proposed for optimization problems by Pincus [52], but the pioneer of the well-known form of the method is Kirkpatrick et al. [53] and Cerny [54]. The method was proposed to the literature by being inspired by the physical annealing process in materials science. The process involves heating a material at high temperature and cooling it slowly to compose its crystals' size. This process is implemented as an optimization method by the fact that the random movements of each point are matched with the probability of a dependent variable symbolized as if the temperature of the material. Since the relationship among atomic structure, entropy and temperature during the annealing of a sample of material is based on the Monte Carlo model, it provides the structure of the SAM [55]. Although it is easy to implement for various problems and it gives efficient results, the disadvantage of the method is that it requires initial values of various parameters such as initial temperature, the temperature decrease rule, the temperature stages' duration, etc. [56].

5.4 Tabu Search Method

Tabu search (TS) is another random search algorithm that is used for the solution of combinatorial optimization problems. The method was proposed by Glover [57] and TS is based on neighborhood search procedure. In this procedure, the algorithm iteratively moves from a solution to another solution in the related neighborhood until it reaches any stopping criterion. The method continues a local search until it reaches a local optimum. In this method, Tabu lists are utilized to prevent from cycling back to previously visited points. In this way, the Tabu lists provide a memory recording of the recent history of the search [58].

5.5 Artificial Neural Networks

The method of artificial neural networks (ANNs), also one of the artificial intelligence applications, is an information processing technology that analyzes the existing data by simulating the biological nervous systems and the human brain and provides new information using different learning algorithms. The method was introduced to literature by McCulloch and Pitts in 1943 [59]. In general, ANNs which have been inspired by the interconnection of many biological neurons as in the human brain are formed by connecting artificial neurons of which task simulate the input, processing, and output characteristics of the biological neuron at different effect levels. The most important advantage of ANNs is that it does not require any mathematical model since the method learns from historical data. On the other hand, the method is a black box since it does not provide any model including any relations and patterns among inputs and outputs [60].

5.6 Particle Swarm Optimization

Particle Swarm Optimization (PSO), an optimization algorithm developed by Kennedy and Eberhart [61], was inspired by the movements of the animals moving in the flocks such as birds and fish to simulate the behavior of organisms. In the method, each individual which is searching a solution is called a particle, while the population of particles is called a swarm. The method utilizes a function (fitness function) to understand the closeness of an obtained solution to the optimum solution of the desired problem. The main purpose of the function is to evaluate the suitability of the solution by considering the total value.

6 Conclusion

Understanding the expectations of the end-user from product and predicting the requirements of it are the most important issue to design a long-life cycle product that was demanded by customers all the time. The second step is to integrate customer expectations to the design characteristics of the product. In the literature, many new approaches and strategies have been proposed to identify the relationship between customer requirements and design characteristics. At this point, various methodologies have been utilized to gather data, analyze data, and processing data in the literature. Finally, the quality of the product and customer requirements management comes after these steps. After this chapter, the main objective of this book is to illustrate the first two steps of the customer oriented product design by the applications of fuzzy sets and intelligent techniques.

References

1. Matzler, K., Hinterhuber, H.-H., Bailom, F., Sauerwein, E.. How to delight your customers. J. Prod. Brand. Manag. **5**(2), 6–18 (1996)
2. Lager, T.: The industrial usability of quality function deployment: a literature review and synthesis on a meta-level. R&D Manag. **35**(4), 409–426 (2005)
3. Ulrich, K., Eppinger, S.: Product Design and Development, 5th edn. Mcgraw Hill International edition (2012)
4. Lin, M.-C., Wang, C.-C., Chen, T.-C.: A strategy for managing customer-oriented product design. Concurr. Eng. **14**(3), 231–244 (2006)
5. Gill, P., Stewart, K., Treasure, E., Chadwick, B.: Methods of data collection in qualitative research: interviews and focus groups. Br. Dent. J. **204**, 291–295 (2008). https://doi.org/10.1038/bdj.2008.192
6. McGrath, C., Palmgren, P.J., Liljedahl, M.: Twelve tips for conducting qualitative research interviews. Med. Teach. **41**(9), 1002–1006 (2019). https://doi.org/10.1080/0142159X.2018.1497149
7. Breen, Rosanna L.: A practical guide to focus-group research. J. Geogr. High. Educ. **30**(3), 463–475 (2006). https://doi.org/10.1080/03098260600927575

8. Karen, L., James, A., Ellena, A.: Focus group research: what is it and how can it be used? Can. J. Cardiovasc. Nurs. **24**(1), 16–22 (2014)
9. Nielsen, J.: Heuristic evaluation. Usability Inspection Methods, pp. 25–62. Wiley (1994)
10. Randolph, G.: Use-cases and personas: a case study in light-weight user interaction design for small development projects. Informing Sci. Int. J. Emerg. Transdiscipl. **7**, 105–116 (2004)
11. Zimmermann, G., Vanderheiden, G.: Accessible design and testing in the application development process: considerations for an integrated approach. Univ. Access Inf. Soc. **7**(1–2), 117–128 (2008)
12. Kjeldskov, J., Stage, J.: New techniques for usability evaluation of mobile systems. Int. J. Hum. Comput. Stud. **60**(5–6), 599–620 (2004)
13. Akao, Y.: In: Mazur, G.H. (trans) Quality Function Deployment: Integrating Customer Requirements into Product Design. Cambridge, Productivity Press, MA (1990)
14. Suh, N.P.: The Principles of Design. Oxford Series on Advanced Manufacturing (1990)
15. Suh, H.P.: Axiomatic Design: Advances and Applications MIT-Pappalardo Series in Mechanical Engineering. Oxford University Press, USA (2001)
16. Kano, N.: Attractive quality and must-be quality. Hinshitsu Qual. J. Jpn. Soc. Qual. Control **14**, 39–48 (1984)
17. Mikulić, J., Prebežac, D.: A critical review of techniques for classifying quality attributes in the Kano model. Manag. Serv. Qual. Int. J. **21**(1), 46–66 (2011)
18. Green, P., Rao, V.: Conjoint measurement for quantifying judgmental data. J. Mark. Res. **8**(3), 355–363 (1971). https://doi.org/10.2307/3149575
19. Nagamachi, M.: Kansei Engineering: a new ergonomic consumer-oriented technology for product development. Int. J. Ind. Ergon. **15**, 3–11 (1995)
20. Nagamachi, M.: Kansei Engineering as a powerful consumer-oriented technology for product development. Appl. Ergon. **33**, 273–278 (2002)
21. Zadeh, L.A.: Fuzzy sets. Inf. Control **8**(3), 338–353 (1965)
22. Zadeh, L.A.: The concept of a linguistic variable and its application to approximate reasoning—I. Inf. Sci. **8**(3), 199–249 (1975)
23. Sambuc, R.: Fonctions φ-floues: application a l'aide au diagnostic en pathologie thyroidienne. Ph.D. Thesis, University of Marseille, France (1975)
24. Grattan-Guiness, I.: Fuzzy membership mapped onto interval and many-valued quantities. Z. Math. Logik. Grundladen, Math. **22**, 149–160 (1975)
25. Jahn, K.U.: Intervall-wertige Mengen. Math. Nach. **68**, 115–132 (1975)
26. Yager, R.R.: On the theory of bags. Int. J. Gen. Syst. **13**(1), 23–37 (1986). https://doi.org/10.1080/03081078608934952
27. Miyamoto, S.: Fuzzy multisets and their generalizations. In: Calude, C.S., Păun, G., Rozenberg, G., Salomaa, A. (eds.) Multiset Processing. Lecture Notes in Computer Science, WMC 2000, pp. 225–235. Springer, Berlin, Heidelberg (2000)
28. Riesgo, Á., Alonso, P., Díaz, I., Montes, S.: Basic operations for fuzzy multisets. Int. J. Approx. Reason. **101**, 107–118 (2018)
29. Atanassov, K.T.: Intuitionistic fuzzy sets. Fuzzy Sets Syst. **20**(1), 87–96 (1986)
30. Smarandache, F.: Neutrosophy: Neutrosophic Probability, Set, and Logic: Analytic Synthesis & Synthetic Analysis. American Research Press (1998). ISBN-10: 1879585634
31. Smarandache, F.: Neutrosophy, a new branch of philosophy. Mult. Valued Log. **8**(3), 297–384 (2002)
32. Smarandache, F.: Neutrosophic set—a generalization of the intuitionistic fuzzy set. Int. J. Pure Appl. Math. **24**(3), 287–297 (2005)
33. Ozen, T., Garibaldi, J.M., Musikasuwan, S.: Preliminary investigations into modelling the variation in human decision making. Uncertainty in Knowledge Based System in Perugia, Italy, July 2004
34. Garibaldi, J.M., Ozen, T.: Uncertain fuzzy reasoning: a case study in modelling expert decision making. IEEE Trans. Fuzzy Syst. **15**(1), 16–30 (2007)
35. Garibaldi, J.M., Jaroszewski, M., Musikasuwan, S.: Nonstationary fuzzy sets. IEEE Trans. Fuzzy Syst. **16**(4), 1072–1086 (2008)

36. Torra, V.: Hesitant fuzzy sets. Int. J. Intell. Syst. **25**(6), 529–539 (2010)
37. Atanassov, K.T.: Intuitionistic Fuzzy Sets, Theory and Applications. Springer (1999)
38. Yager, R.R.: Pythagorean fuzzy subsets. In: IFSA World Congress and NAFIPS Annual Meeting (IFSA/NAFIPS), 2013 Joint, pp. 57–61. IEEE (2013)
39. Cuong, C.B.: Picture fuzzy sets. J. Comput. Sci. Cybern. **30**(4), 409–420 (2014)
40. Yager, R.R.: Generalized orthopair fuzzy sets. IEEE Trans. Fuzzy Syst. **25**(5), 1222–1230 (2017)
41. Kutlu Gundogdu, F., Kahraman, C.: Spherical fuzzy sets and spherical fuzzy TOPSIS method. J. Intell. Fuzzy Syst. **36**(1), 337–352 (2019)
42. Senapati, T., Yager, R.R.: Fermatean fuzzy weighted averaging/geometric operators and its application in multi-criteria decision-making methods. Eng. Appl. Artif. Intell. **85**, 112–121 (2019). https://doi.org/10.1016/j.engappai.2019.05.012
43. Senapati, T., Yager, R.R.: Some new operations over Fermatean fuzzy numbers and application of Fermatean fuzzy WPM in multiple criteria decision making. Informatica **30**(2), 391–412 (2019). https://doi.org/10.15388/informatica.2019.211
44. Holland, J.H.: Adaptation in Natural and Artificial Systems. University of Michigan Press, Ann Arbor, MI (1975)
45. Holland, J.H.: Genetic algorithms. Sci. Am. **267**, 66–72 (1992). https://doi.org/10.1038/scientificamerican0792-66
46. Miranda, V., Srinivasan, D., Proença, A.M.: Evolutionary computation in power systems. Electr. Power Syst. Res. **20**(2), 89–98 (1998). https://doi.org/10.1016/S0142-0615(97)00040-9
47. Won, J.R., Park, Y.M.: Economic dispatch solutions with piecewise quadratic cost functions using improved genetic algorithm. Int. J. Electr. Power Energy Syst. **25**(5), 355–361 (2003). https://doi.org/10.1016/S0142-0615(02)00098-4
48. Dorigo, M.: Optimization, learning and natural algorithms. PhD thesis, Dipartimento di Elettronica, Politecnico di Milano, Italy (in Italian) (1992)
49. Dorigo, M., Stützle, T.: Ant Colony Optimization. MIT Press, Cambridge, MA & London, UK (2004)
50. Yang, J., Zhuang, Y.: An improved ant colony optimization algorithm for solving a complex combinatorial optimization problem. Appl. Soft Comput. **10**(2), 653–660 (2010). https://doi.org/10.1016/j.asoc.2009.08.040
51. Macura, W.K.: Ant Colony Algorithm. From MathWorld–A Wolfram Web Resource, Created by Eric W. Weisstein. http://mathworld.wolfram.com/AntColonyAlgorithm.html
52. Pincus, M.: A Monte Carlo method for the approximate solution of certain types of constrained optimization problems. Oper. Res. **18**, 1225–1228 (1970)
53. Kirkpatrick, S., Gelatt, S., Vecchi, M.: Optimization by simulated annealing. Science **220**(4598), 671–680 (1983)
54. Cerny, V.: Thermodynamical approach to the travelling salesman problem. J. Optim. Theory Appl. **45**(1), 41–51 (1985)
55. Özdağoğlu, G.: Atatürk Üniversitesi İktisadi ve İdari Bilimler Dergisi. **22**(2), 357–377 (2010)
56. Jarraya, B., Bouri, A.: Metaheuristic optimization backgrounds: a literature review. Int. J. Contemp. Bus. Stud. **3**(12), 31–44 (2012)
57. Glover, F.: Future paths for integer programming and links to artificial intelligence. Comput. Oper. Res. **13**, 533–549 (1986)
58. Gendreau, M.: An Introduction to Tabu Search. In: Glover, F., Kochenberger, G.A. (eds.) Handbook of Metaheuristics. Kluwer Academic Publishers, New York (2003)
59. McCulloch, W.S., Pitts, W.A.: A logical calculus of the ideas immanent in neural nets. Bull. Math. Biophys. **5**, 115–133 (1943). https://doi.org/10.1007/BF02478259
60. Cebi, S., Kahraman, C., Kaya, I.: Soft computing and computational intelligent techniques in the evaluation of emerging energy technologies. In: Vasant, P., Barsoum, N., Webb, J. (eds.) Innovation in Power, Control, and Optimization: Emerging Energy Technologies, pp. 164–197. IGI Global (2012)
61. Kennedy, J., Eberhart, R.: Particle swarm optimization. In: Proceedings of Conference on Evolutionary Computation (CEC), pp. 1942–1948 (1995)

Part II
Customer Oriented Product Design Using Quality Function Deployment

Design of Noise-Canceling Earmuffs with Quality Function Deployment

Ulas Cinar, Omer Faruk Ugurlu and Selcuk Cebi

Abstract One of the most important responsibilities that employers have to fulfill against the employee is ensuring the safety of employees in the working environment. The personal protective equipment (PPE) is the last resort in case the danger cannot be eliminated despite the required measures taken in the working environment. The user-oriented design of protective equipment is important for effective use in the working environment. Therefore, this study is aimed to define the parameters that come to the forefront in a noise-canceling earmuff design considering user expectations. In the scope of this study, the quality function deployment method, which is widely used in the literature, is used to transfer user expectations to the technical specifications of the product. According to the results of the chapter, "Active Adaptation System" and "Passive System with Protective Foam" are the two most important design parameters.

1 Introduction

Personal protection equipment (PPE) is the latest ring in the chain of measures planned to protect against occupational health and safety risks. If the source of the hazard cannot be exactly eliminated and the collective protection measures to be applied at the source cannot reduce the risk-damage potential to an acceptable level, the use of personal protective equipment is mandatory for the safe operation. The potential for exposure to certain risk factors is very high due to the nature of the work since no intervention at the source is possible. Noise exposure often involves interference planning based on sound blocking, which often interferes with the source

U. Cinar (✉)
Occupational Health and Safety Department, Istanbul Yeni Yuzyil University, Istanbul, Turkey
e-mail: ulas.cinar@yeniyuzyil.edu.tr

O. F. Ugurlu
Mining and Materials Engineering Department, McGill University, Montreal, Canada

S. Cebi
Industrial Engineering Department, Yildiz Technical University, Istanbul, Turkey

C. Kahraman and S. Cebi (eds.), *Customer Oriented Product Design*, Studies in Systems, Decision and Control 279, https://doi.org/10.1007/978-3-030-42188-5_2

and the workflow. Therefore, personal protective equipment is the most preferred protection method for exposure to noise [1].

Noise is generally described as unpleasant, unwanted, disturbing sound and it is a subjective concept that can vary according to the perception of the person. Sound is a physical phenomenon of fluctuations in air pressure through a vibrating source. Physical sound is the pressure changes caused by a vibrating object in the environment. Physiological sound, on the other hand, is the sensation caused by pressure changes in an elastic environment in the hearing organs. The air pressure generated by sound waves is called Sound Pressure, the number of repetitions of vibration in a second is called Frequency, and the distance at which vibrations are spread per unit time is called Sound Speed [2, 3]. The human ear is responsive to the sounds between 20 and 20,000 Hz. Sounds less than 20 Hz are called Infrasound and sounds higher than 20,000 Hz are called Ultrasound Decibel (dB), which is a relative unit, is used to evaluate the sound intensity and it is not a linear unit. 10 dB is 10 times the sound of 1 dB; 20 dB is 100 times that of 1 dB and 40 dB is 10,000 times as strong as 1 dB. The human ear can detect sounds in the 0–140 dB range. A sound of 120 dB gives a feeling of discomfort in the ear, while a pronounced sensation of pain occurs in the 120–130 dB range. Sound intensity of 140 dB and above; may cause severe pain, tear of the eardrum, permanent and temporary hearing loss [4]. Therefore, sound is one of the most important physical pollutants that can cause considerable permanent damage to personal exposure. Nowadays, with the increasing population, developing technology and industrialization, noise becomes an increasing problem [5].

Noise control is provided by two methods, active and passive. The passive method is based on the use of materials that prevent sound transmission in front of the noise source to isolate the sound directly. The active method is based on neutralizing the sound by detecting the external sound waves with a microphone integrated into the headset and producing an inverted wave with the help of an electronic circuit [6]. Earmuffs are passive noise-canceling protectors that completely cover the ear, reduce the sound volume to some extent, but do not completely interrupt speech and hearing. The outer case of the earmuffs, which are produced on the basis of EN 352 standards, is made of hard plastic material that is resistant to falling and impact. Inside the container, materials like a sound-proofing sponge, etc. are used. The cushions surrounding the ear are made of sponge and the contact surface is made of antiallergic material. Earplugs are passive noise-canceling protectors that are used in the ear canal. The earplugs produced according to EN 352-2 standard are made of cotton, polyethylene sponge or silicone rubber for hygiene purposes. The headphones produced according to EN 352-5 standard are active noise-canceling and include an integrated microphone and electronic system in addition to the standards applicable to earplugs [7–10].

Inefficient use of PPE by employees and not adopting them can create occupational safety problems. Preventing these problems can only be possible by selecting the most appropriate PPE by way based on both the personal characteristics of the employees and the conditions of the job. Therefore, the main objective of this study is to design the noise-canceling earmuffs with the most suitable technical specifications for the work conditions by using the associations in "*The House of Quality*" scheme of the "*Quality Function Deployment* (*QFD*)" method. The QFD method is one of the most preferred methods in the industry to improve the system and product quality since it is used effectively to increase customer satisfaction by transferring customer expectations or requirements to product and design parameters [11]. The QFD method covers the operations performed on the matrix called The *House* of Quality where the customer requirements are in the rows and the technical characteristics are in the columns. In the relationship matrix, technical characteristics and customer requirements are associated with weak, medium and strong expressions by assigning numerical values. In the correlation matrix that forms the roof of the house of quality, the extent to which the technical characteristics affect each other in a positive, negative or neutral manner is determined. This association provides an opportunity to see how the improvement of any specification will have an impact on other features and therefore affect the performance of the entire system. As a result of the analysis of matrices, technical characteristics are listed according to their importance and priorities and points of intervention are determined [12].

The rest of this chapter is organized as follows; In Sect. 2, the fundamentals of the QFD method have been proposed. Section 3 presents the customer-oriented design steps of noise cancelling earmuffs. Finally, concluding remarks are presented in Sect. 4.

2 Materials and Methods

In this chapter, the application steps of the both traditional QFD and fuzzy QFD methods will be introduced for the design of noise-canceling earmuffs.

2.1 Traditional Quality Function Deployment Method

The House of Quality of the QFD method consists of six main sections and it is schematically given in Fig. 1. The explanations of the sections indicated with letters are as follows [13]; Section A, is the section where the selection criteria (the characteristics expected of the product) are determined as a result of the customer or expert evaluation. Section B is the section containing the score equivalents of the severity levels corresponding to each feature in Section A. Section C is the section containing the technical specifications for the product. Section D is the section containing the relationship matrix with which the selection criteria and the technical characteristics are associated. Section E is the section containing the correlation matrix with which the technical characteristics are associated. Section F. is the section that includes competitive analyzes. Finally, Section G. is the section that contains the

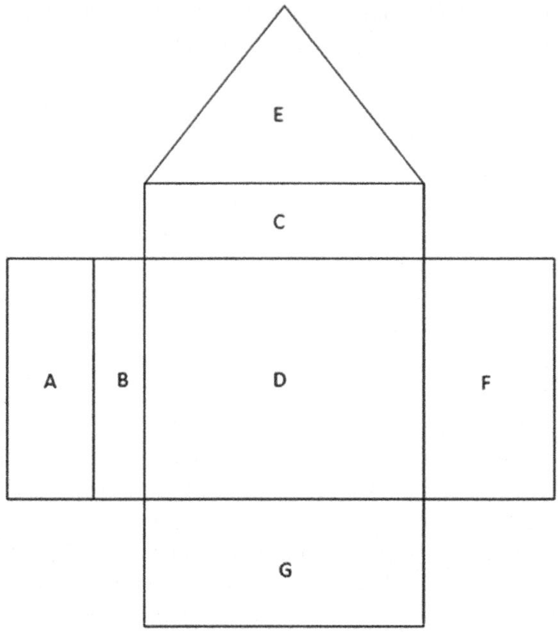

Fig. 1 House of quality

technical evaluations aimed at the target. The application steps of the methodology are given below;

Step 1: First, the selection criteria are determined according to the needs and expectations of the consumer survey, one-to-one interviews, etc. methods. Apart from customer-oriented product design, this method can also be applied for engineering applications, process design, technological improvements, and even managerial applications. Hence, the criteria at this stage can be determined by the evaluation of more than one expert.

Step 2: Determined product expectations are evaluated by the customer or experts and scores are obtained for each feature. These scores usually range from 1 to 5. There is a direct proportion between the scores and the significance level (The higher the score, the higher the significance level).

Step 3: The technical parameters of the analyzed product should be determined in this step.

Step 4: Technical parameters and product expectations are correlated. Scores 1, 3, and 9 are used to make these associations. The scores 9, 3, and 1 represent high relationship, moderate relationship, and mild relationship, respectively.

Step 5: In the expectations, the improvement ratio and the coefficients of advantage expressed by the numbers 1, 1.2 and 1.5 are determined, and absolute weight scores for each customer requirements are obtained by using the following formula.

$$AW = ID \times IR \times A \tag{1}$$

where *AW*, *ID*, *IR*, and *A* represent absolute weight, importance degree, improvement ratio, and marketing advantage, respectively.

Step 6: After the absolute weight scores are determined, the absolute and relative significance levels of the technical properties are calculated by using Eq. (2). For each technical parameter column, the relationship score and the absolute weight score of the respective row are multiplied and the values obtained for all rows in the column are summed to determine the absolute significance of the current technical parameter. After calculating the absolute importance degrees of all technical characteristics, the relative importance degrees are expressed as a percentage.

$$W_j^{TC} = \sum_{i=1}^{n} AW_i^{CR} \times RD_i^j, \quad i = 1, 2, 3, \ldots n \, and \, j = 1, 2, 3, \ldots, m \qquad (2)$$

where W_j^{TC}, AW_i^{CR}, and RD_i^j represent the importance degree of *j*th technical characteristic, importance degree of *i*th customer requirement, and relationship degree between *j*th technical characteristic and *i*th customer requirement, respectively. Furthermore, *n* and *m* represent the number of customer requirements and the number of technical characteristics, respectively.

Step 7: Correlations between technical parameters are determined. The development of technical parameters may affect a different parameter positively or negatively. Therefore, the relations expressed with different symbols such as strongly positive, weakly positive, strongly negative, and weakly negative are indicated in the roof section and evaluations are made according to this association.

Step 8: In this step, the technical parameters are evaluated comparatively with the competing products and the parameters that require improvement with the strengths are interpreted with the determined targets.

2.2 *Fuzzy Quality Function Deployment Method*

The main difference between traditional QFD Method and Fuzzy QFD is based on the linguistic evaluation. Therefore, same steps in QFD method are applied in Fuzzy QFD. The differences between them can be summarized as follow;

The linguistic scale given in Fig. 2 is used to evaluate the importance degree of the customer requirements and to compare performance of competing for the product designs. For the evaluation, the linguistic scale includes *very high* (*VH*), *high* (*H*), *medium* (*M*), *low* (*L*), *and very low* (*VL*) terms.

The linguistic scale given in Fig. 3 is used to evaluate the relationship between customer requirements and technical characteristics. For the evaluation of the relationship, the linguistic scale includes *very strong* (*VS*), *strong* (*S*), and *weak* (*W*) terms.

Furthermore, the linguistic scale given in Fig. 4 is used to evaluate the marketing

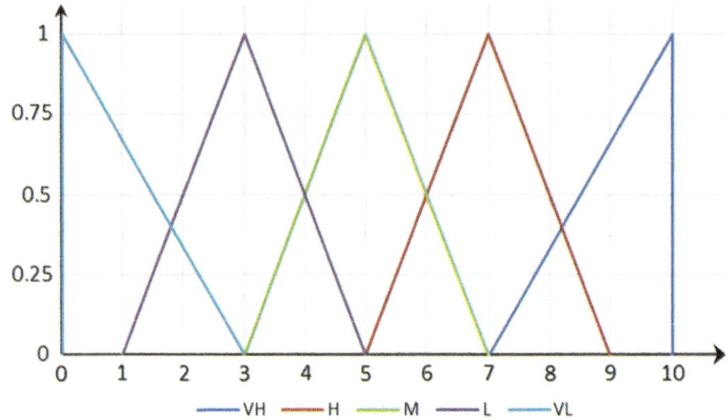

Fig. 2 Membership functions of the linguistic scale for importance degree of customer requirements

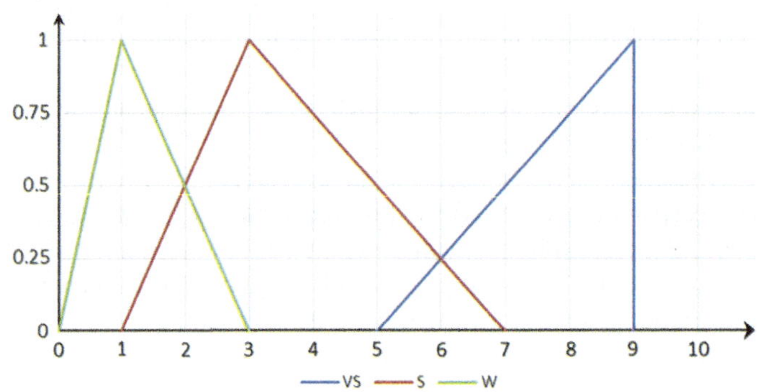

Fig. 3 Membership functions of the linguistic scale for relationship between customer requirements and technical characteristics

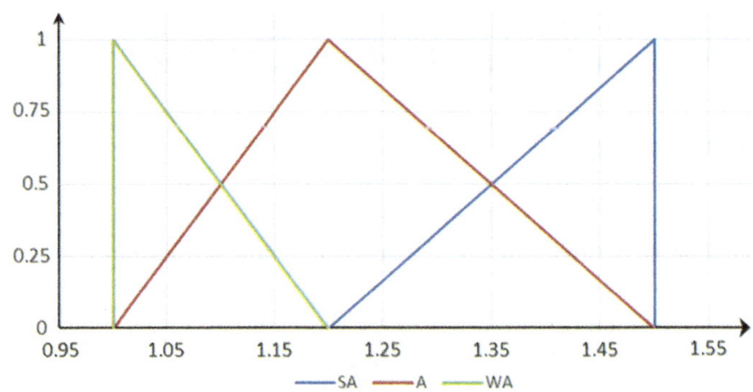

Fig. 4 Membership functions of the linguistic scale for marketing advantage

advantage of the related customer requirements. The linguistic terms given in Fig. 4 includes *strong advantage* (*SA*), *advantage* (*A*), and *weak advantage* (*WA*) terms.

Finally, the absolute fuzzy weight scores ($\tilde{W}$) are obtained using the following formula.

$$\widetilde{W} = \widetilde{ID} \times \widetilde{RD} \times \widetilde{IR} \times \tilde{A} \tag{3}$$

where $\widetilde{ID}$, $\widetilde{IR}$, $\tilde{A}$, and $\widetilde{RD}$ represent fuzzy importance degree of customer requirement, fuzzy improvement ratio, fuzzy marketing advantage, and fuzzy relationship degree between related customer requirement and technical characteristic, respectively. After the relative weights are calculated, the defuzzification is applied to the fuzzy numbers using the following equation [14].

$$\mathrm{w} = \frac{\mathrm{a} + 4\mathrm{b} + \mathrm{c}}{6} \tag{4}$$

where *a*, *b*, and *c* are lower value, middle value, and upper value of a triangular fuzzy number, respectively.

3 Case Study

The pressing area of a metal production plant was chosen as the field of application. In this section, the average sound intensity varies between 100 and 110 dB (A), although it varies in transitions between parts. Sound wall, machine sound insulation, etc. collective protection measures cannot be chosen because they negatively affect the flow of work. Thus, the application of personal protective equipment will be preferred. Selection criteria determined by experts according to the job and the employees are provided in Table 1. The designed earmuff is expected to reduce the

Table 1 Selection criteria

Selection criteria
Compliance with EN 352 Standards
Provides damping of 30 dB (A)
Easy to use
Provides long-term continuous use
Ensures hygiene and comfort standards
Long life
Sensitivity to communication and warning signals should not be lost as much as possible
Lightness
Can be adapted to different noise levels

Table 2 The technical parameters

Impact and chemical resistant outer surface
Antiallergic contact surface
Helmet mountability
Passive system with protective foam
Active noise-canceling system
Replaceable ear cushions
UV protected outer surface
Wireless communication system
Active adaptation system

noise level to below the exposure limit values and not to create compliance problems during inter-departmental transitions. In addition, some parameters have been determined in order to provide ease of use for employees.

EN-352 standards and the characteristics of the earmuffs used in the industry have been analyzed by experts. The technical parameters determined for the noise-canceling earmuffs are given in Table 2.

Impact, a chemical resistant outer surface and antiallergic contact surface are specified in the standards. Helmet mountability, replaceable ear cushions and UV protected outer surface are physical production characteristics of the product. Active noise-canceling system, wireless communication system, and active adaptation system are electronic and special production characteristics and they can be combined with a passive system.

The linguistic relationship between customer expectations and technical characteristics defined by the expert team is presented in Fig. 5. When the conventional QFD steps with the Eqs. (1) and (2) given in Step 5 are applied to the compiled information, the obtained results are given in Fig. 6. Then, Fig. 7 presents the results obtained from the fuzzy calculations with the Eqs. (3) and (4).

According to the conventional calculation; the technical characteristics that come to the forefront in the design of protective earmuffs are "Active Adaptation System", "Passive System with Protective Foam", "Impact and Chemical Resistant Outer Surface" and "Antiallergic Contact Surface", respectively. In the fuzzy calculation, there is no difference in the first 4 parameters compared to the conventional calculation. However, unlike conventional calculation, the "Active Noise-Canceling" parameter is more important than the "Helmet Mountability" parameter.

4 Conclusion

Although personal protective equipment (PPE) is the last resort for occupational health and safety applications, it is the equipment that is necessary for working environments. The use of protective equipment plays an active role in preventing or

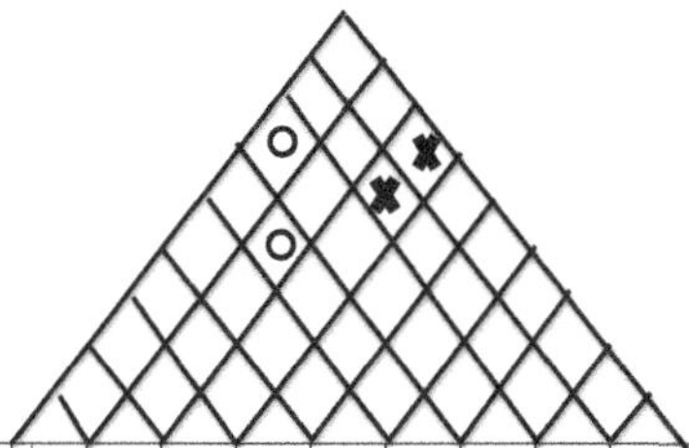

Selection Criterias \ Technical Parameters	Importance Degrees (1-5)	Impact and chemical resistant outer surface	Antiallergic contact surface	Helmet mountability	Passive system with protective foam	Active noise canceling system	Replaceable ear cushions	UV protected outer surface	Wireless communication system	Active adaptation system	Competition Analysis: QFD Applied	Rival 1	Rival 2	Rival 3	Quality Goal	Advantage	Improvement Ratio	Absolute Weight
Compliance with EN 352 Standards	VH	VS	VS								VH	VH	VH	VH	VH	WA		
Provides damping of 30 dB (A)	VH				S	S				S	VH	VH	VH	VH	VH	WA		
Easy to use	H			VS		S				S	H	H	H	H	H	WA		
Provides long-term continuous use	H				VS						M	H	M	VH	H	A		
Ensures hygiene and comfort standards	H		VS	S		W	VS			W	H	H	H	M	H	WA		
Long life	H	VS					S	VS			H	M	H	H	H	WA		
Sensitivity to communication and warning signals should not be lost as much as possible	VH					S			VS	S	VH	L	H	M	VH	A		
Lightness	H			S	S						M	VH	H	VH	H	A		
Can be adapted to different noise levels	VH					S				VS	VH		M		VH	A		
Absolute Importance																		
Relative Importance																		

Correlations
○ weak positive
✖ weak negative

Fig. 5 House of quality including linguistic evaluation

reducing the potential damage for the employees in the accidents. The main reason for the resistance to the use of protective equipment in the working environment is that the equipment is uncomfortable particularly during the operation. In this study, design parameters related to customer-oriented earmuff design were defined and prioritized in order to ensure the effective use of protective equipment in the working environment. In the study, the QFD method, which is one of the most widely used techniques in product design, was used. *Active Adaptation System*, *Passive System with Protective Foam*, *Impact and Chemical Resistant Outer Surface* and *Anti Allergic Contact Surface* are the technical characteristics of the protective earmuff design. In order to increase user satisfaction and better protection, these parameters should be prioritized in the design of earmuffs.

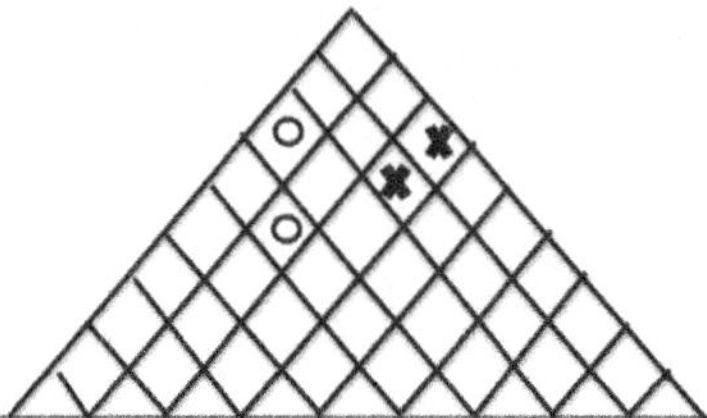

Technical Parameters / Selection Criterias	Importance Degrees (1-5)	Impact and chemical resistant outer surface	Antiallergic contact surface	Helmet mountability	Passive system with protective foam	Active noise canceling system	Replaceable ear cushions	UV protected outer surface	Wireless communication system	Active adaptation system	Competition Analysis							
											QFD Applied	Rival 1	Rival 2	Rival 3	Quality Goal	Advantage	Improvement Ratio	Absolute Weight
Compliance with EN 352 Standards	5	9	9								5	5	5	5	5	1	1.00	5
Provides damping of 30 dB (A)	5				3	3				3	5	5	5	5	5	1	1.00	5
Easy to use	4			9		3				3	4	4	4	4	4	1	1.00	4
Provides long-term continuous use	4				9						3	4	3	5	4	1.2	1.33	6.4
Ensures hygiene and comfort standards	4		9	3		1	9			1	4	4	4	3	4	1	1.00	4
Long life	4	9					3	9			4	3	4	4	4	1	1.00	4
Sensitivity to communication and warning signals should not be lost as much as possible	5					3			9	3	5	2	4	3	5	1.2	1.00	6
Lightness	4			3	3						3	5	4	5	4	1.2	1.33	6.4
Can be adapted to different noise levels	5					3				9	5	0	3	0	5	1.2	1.00	6
Absolute Importance		81.00	81.00	67.20	91.80	67.00	48.00	36.00	54.00	103.00								
Relative Importance		12.88	12.88	10.68	14.59	10.65	7.63	5.72	8.59	16.38								

Fig. 6 Traditional house of quality

The linguistic data were collected and the traditional QFD method was used for processing the linguistic data. In order to understand the uncertainty in the linguistic data, the obtained results were analyzed by using fuzzy QFD. According to fuzzy calculation, only two parameters which are "Active Noise-Canceling" and "Helmet Mountability", changed their places. Since there is no difference between the two calculations in the parameters with the highest importance; the design should be realized by these parameters.

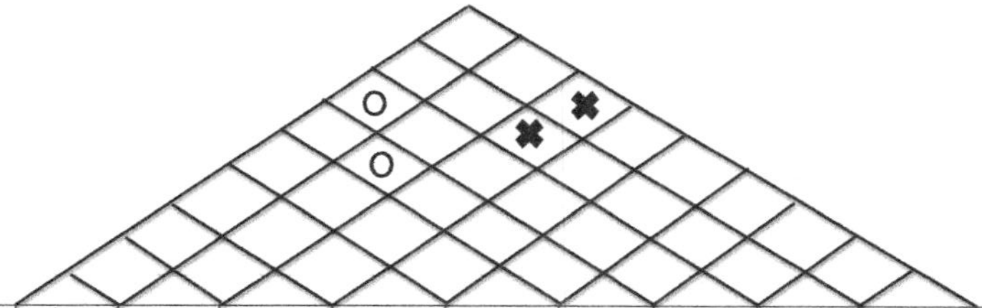

Selection Criterias \ Technical Parameters	Importance Degrees (1-5)	Impact and chemical resistant outer surface	Antiallergic contact surface	Helmet mountability	Passive system with protective foam	Active noise canceling system	Replaceable ear cushions	UV protected outer surface	Wireless communication system	Active adaptation system
Compliance with EN 352 Standards	(7,10,10)	(5,9,9)	(5,9,9)							
Provides damping of 30 dB (A)	(7,10,10)				(1,3,7)	(1,3,7)				(1,3,7)
Easy to use	(5,7,9)			(5,9,9)		(1,3,7)				(1,3,7)
Provides long-term continuous use	(5,7,9)				(5,9,9)					
Ensures hygiene and comfort standards	(5,7,9)		(5,9,9)	(1,3,7)		(0,1,3)	(5,9,9)			(0,1,3)
Long life	(5,7,9)	(5,9,9)					(1,3,7)	(5,9,9)		
Sensitivity to communication and warning signals should not be lost as much as possible	(7,10,10)					(1,3,7)			(5,9,9)	(1,3,7)
Lightness	(5,7,9)			(1,3,7)	(1,3,7)					
Can be adapted to different noise levels	(7,10,10)					(1,3,7)				(5,9,9)
Absolute importance		(60,153,205.2)	(60,153,205.2)	38.33,119.28,294.3	(57,171.12,361.71)	(26,130,402)	(30,84,172.8)	(25,63,97.2)	(35,108,135)	(54,202,432)
Relative importance		(15.57,12.93,8.9)	(15.57,12.93,8.9)	(9.95,10.08,12.77)	(14.79,14.46,15.69)	(6.75,10.99,17.44)	(7.79,7.1,7.5)	(6.49,5.32,4.22)	(9.08,9.13,5.86)	(14.01,17.07,18.74)
Normalization		12.70	12.70	10.51	14.72	11.35	7.28	5.33	8.57	16.84

Selection Criterias	Competition Analysis							
	QFD Applied	Rival 1	Rival 2	Rival 3	Quality Goal	Advantage	Improvement Ratio	Absolute Weight
Compliance with EN 352 Standards	(7,10,10)	(7,10,10)	(7,10,10)	(7,10,10)	(7,10,10)	(1,1,1.2)	(1,1,1)	(7,10,12)
Provides damping of 30 dB (A)	(7,10,10)	(7,10,10)	(7,10,10)	(7,10,10)	(7,10,10)	(1,1,1.2)	(1,1,1)	(7,10,12)
Easy to use	(5,7,9)	(5,7,9)	(5,7,9)	(5,7,9)	(5,7,9)	(1,1,1.2)	(1,1,1)	(5,7,10.8)
Provides long-term continuous use	(3,5,7)	(5,7,9)	(3,5,7)	(7,10,10)	(5,7,9)	(1,1.2,1.5)	(1.67,1.4,1.29)	(8.33,11.76,17.36)
Ensures hygiene and comfort standards	(5,7,9)	(5,7,9)	(5,7,9)	(3,5,7)	(5,7,9)	(1,1,1.2)	(1,1,1)	(5,7,10.8)
Long life	(5,7,9)	(3,5,7)	(5,7,9)	(5,7,9)	(5,7,9)	(1,1,1.2)	(1,1,1)	(5,7,10.8)
Sensitivity to communication and warning signals should not be lost as much as possible	(7,10,10)	(1,3,5)	(5,7,9)	(3,5,7)	(7,10,10)	(1,1.2,1.5)	(1,1,1)	(7,12,15)
Lightness	(3,5,7)	(7,10,10)	(5,7,9)	(7,10,10)	(5,7,9)	(1,1.2,1.5)	(1.67,1.4,1.29)	(8.33,11.76,17.36)
Can be adapted to different noise levels	(7,10,10)		(3,5,7)		(7,10,10)	(1,1.2,1.5)	(1,1,1)	(7,12,15)

Fig. 7 House of quality including triangular fuzzy numbers

References

1. Occupational Safety & Health Council: Guidelines for the Use of Personal Protective Equipment, p. 5 (2001)
2. Kurra, S.: Environmental Noise—Environment Symposium, İzmir, 3–5 June 1982
3. Yücel, M.: Environmental Problems, Cukurova University, No: 150, Adana, 302 s (1995)
4. Güler, Ç., Çobanoğlu, Z.: Noise Environmental Health (vol. 19). Turkey (1994)
5. Fernandez, M.D., Quintana, S., Chavarria, N., Ballesteros, J.A.: Noise exposure of workers of the construction sector. Appl. Acoust. **70**, 753–760 (2009)
6. Elliott, S.J., Nelson, P.A.: Active noise control. IEEE Signal Process. Mag. 12 (1993)
7. EN 352-1. Hearing protectors—general requirements—Part 1: Ear muffs (2002)
8. EN 352-3. Hearing protecters—safety requirements and testing; Part 3: Ear-muffs attached to an industrial safety helmet (2002)
9. EN 352-2. Hearing protectors-safety requirements and testing-Part 2—Ear plugs (2002)
10. EN 352-5. Hearing protectors—safety requirements and testing—Part 5: Active noise reduction ear-muffs (2002)
11. Costa, A.I.A., Dekker, M., Jongen, W.M.F.: Quality function deployment in the food industry: a review. Trends Food Sci. Technol. 306–314 (2001)
12. Bazaati, S., Bayrmi, S., Oral, E.: Quality function deployment and an application of construction sector. Cukurova Univ. J. Eng. **29**(1), 53–61 (2014)
13. Ansari, A., Modarress, B.: Quality function deployment: the role of suppliers. Int. J. Purch. Mater. Manag. **30**(3), 27–35 (1994)
14. Yong, D.: Plant location selection based on fuzzy TOPSIS. Int. J. Adv. Manuf. Technol. **28**, 839–844 (2006)

Improving Quality of Product and Process in the Manufacturing of Particleboard with an Integrated Quality Function Deployment Approach

Yildiz Kose and Emrullah Demirci

Abstract Quality Function Deployment (QFD), which provides customer-oriented product design, is a quality improvement tool that reflects customer requirements (CRs) to the technical characteristics (TCs) of product and production process. The importance rating of technical characteristics in conventional QFD is determined only by customer requirements, without considering company constraints. In determining whether the technical characteristics are effective for the company, QFD is used integrated with Data Envelopment Analysis (DEA). In generating the relations in the House of Quality (HoQ), the linguistic expressions of the experts are converted into fuzzy numbers to provide flexibility to the model and to cope with uncertainty. The Fuzzy DEA-QFD method, which includes customer requirements as well as company constraints, has not been proposed. In this context, this chapter contributes to the relevant literature by means of developing an integrated QFD approach to determine the importance of technical characteristics based on both the customer service level and interests of the firm. The proposed model is applied in a company operating in the wood-based panel industry sector. The importance levels of customer requirements in HoQ are calculated by Goal Programming, which allows for the evaluation of different types of surveys simultaneously. The incorporation of Goal Programming, DEA and Fuzzy Logic to the proposed model shows that the QFD is an interdisciplinary method and its results are reliable owing to the mathematical model.

Y. Kose (✉)
Industrial Engineering Department, Management Faculty, Istanbul Technical University, Istanbul, Turkey
e-mail: kose18@itu.edu.tr

E. Demirci
Industrial Engineering Department, Engineering Faculty, Karadeniz Technical University, Trabzon, Turkey

C. Kahraman and S. Cebi (eds.), *Customer Oriented Product Design*, Studies in Systems, Decision and Control 279, https://doi.org/10.1007/978-3-030-42188-5_3

1 Introduction

In industries, organizations that adapt to changing customer voices faster than their competitors, as well as paying attention to the interests of the company, are more successful than the others. They should be able to monitor customer requests, needs and service levels continuously and update their processes using scientific methods that allow generating customer-oriented systems. In these methods, Quality Function Deployment (QFD), one of the quality improvement tools, is used effectively in decision-making as well as product development. Also, an interdisciplinary system is formed by combining the knowledge and experience of the quality team members from all functional areas within the organization in QFD.

The aim of this study is that customer requirements convert into production processes and final products. In our model that takes into account the company's interest is applied for particleboard with melamine impregnated paper product in a company operating in the wood-based panel industry. No study examined the importance of technical characteristics in the field of the forest industry in terms of both the customer and the firm in the QFD literature. QFD starts with the establishment of the House of Quality. The current chapter constructs HoQ based on three important pillars. The first is the use of Goal Programming on the customer part. Goal Programming can be generally used in the customer part or technical part to determine the importance level of customer requirements and to determine the completion levels or importance ratings of the technical characteristics [10]. Wang and Chin [30] integrated customer requirements gathered in different formats together with the Goal Programming model in the customer part while Chen and Weng [9] determined design parameters in the technical part of HoQ.

The second important point is in the technical part of HoQ. There can have effective design characteristics as long as a company considers its own policy and constraints in product development. In most QFD studies design requirements have been prioritized according to only customer requirements while the number of studies [12, 15, 18, 19, 28, 36] adding factors to CRs such as cost and environmental impact is increasing. The DEA provides a model that allows the calculation of relative importance values in QFD when more factors need to be considered. However, in this study, these factors have been addressed from a more company perspective rather than from a general framework.

The last important point is that there is no study in which Fuzzy DEA and QFD are integrated. In the proposed model, expert opinions with linguistic expressions are used in the relationship matrices in the House of Quality. Since relationships include uncertainty because of the categorical structure of the collected data in conventional QFD, fuzzy sets are integrated with QFD to cope with uncertainties in real case problems. The fuzzy DEA model proposed Wang and Chin [31] is adapted to QFD. Fuzzy DEA model is weighting fuzzy inputs and outputs respectively and then using their expected values to measure the performance of decision-making units.

The rest of this chapter is structured as follows: a literature review related to the topic is given in Sect. 2. The developed approach is proposed in Sect. 3. In Sect. 4, the industrial application of the developed approach is presented. Results and analysis are made in Sect. 5. Finally, concluding remarks are given in Sect. 6.

2 Literature Review

Studies in many different areas such as food, textile, transportation, automobile, electronics, furniture, banking, education and health sectors related to QFD has growing literature. Some of the methods have used together with QFD in these studies are as follows; Analytic Hierarchy Process (AHP), Analytic Network Process (ANP), Kano's Model, DEA, Goal Programming, Failure Mode and Effect Analysis (FMEA) and last but not least Fuzzy Logic.

QFD is the subject of many studies in customer-oriented product development. In the design of Flexible Production System [20], in order to improve the food product [1], for writing instrument design parameters [5], in logistics processes [4], in automobile industry [17], in passenger transport service industry [21], for house electric water heater [26], for small dimensions and low fuel consumption minicars [16] QFD has been used for product developments.

Bevilacqua et al. [2], Bhattacharya et al. [3] and Karsak and Dursun [18] utilized QFD for supplier selection. Also, QFD has been proposed as a multi-criteria decision-making technique in the housing refurbishment contractors selection [14], renewable energy industry [29], in the selection of the design of Panda shared cars [35].

In some studies, only the customer part of the QFD is considered and the importance ratings of customer requirements are calculated. The techniques integrated with the QFD in these calculations are as follows: Analytic Hierarchy Process [22], Analytical Network Process, Fuzzy group decision-making [6], Entropy [7], Goal Programming [30]. There are Conjoint Analysis [11], Markov Chains [34] and Gray System Theory [33] studies that treat customer expectations as a dynamic structure and help to calculate future values.

Mathematical models have been used in most studies in order to provide a quantitative background, to achieve the desired value and to provide results that are more realistic. In order to improve indoor air quality [27], determine quality level writing instruments [9], improve microelectronic packaging design with Kano Model [8] and design of an industrial pincers [25] mathematical model has been used in QFD.

Since QFD is a technique based on linguistic evaluations, it involves uncertainty. Fuzzy set [37] has become an effective tool to deal with ambiguity in the QFD. In most of the above studies, therefore, QFD is often used in combination with Fuzzy Logic to derive the importance of CRs, correlations among CRs and the correlations among TCs [2, 4, 14 16, 17, 20, 26]. In the wood furniture industry Homkhiew et al. [13] and Lin et al. [23] proposed QFD.

To the best knowledge of the authors, Fuzzy DEA-QFD integrated model where technical specifications are considered by the company and the customer has not yet been proposed for particleboard. In this context, this chapter contributes to the relevant literature by means of developing an approach to determine the importance level of TCs in QFD for the first time.

3 Proposed Method

In addition to ensuring the continuity of the quality control and quality improvement studies carried out from the beginning to the end of the production processes, they must be customer-oriented. Quality Function Deployment provides customer focus and converts customer voice into technical characteristics. Since after the start of production the changes in the products or the production process can cause damages for the company, it is therefore important to determine the product to be addressed in the QFD study. Based on this point of view we propose QFD for particleboard with melamine impregnated paper product. Before starting, the team which is generated by experts, clients, and academics should be identified. The installation House of Quality consisting of the customer and technical part is the basic step in the QFD. Our model consists of four main steps which are voice of customer, design characteristics, generating House of Quality and analysis (see Fig. 1).

The House of Quality, which has customer requirements in its row, technical characteristics in the column, relations among the technical characteristics on the roof and relationships between technical characteristics and customer requirements on the main body, consists of 9 steps as seen in Fig. 2. Triangular fuzzy numbers are used to determine the relationship degrees in steps 5 and 8. Competition matrix is not included in the study because the data in the competition matrix could not be collected from the customers accurately and because the focus customer group generally prefers a single company for the particleboard product.

The steps of the proposed model are as follows.

Step 1. Describing customer requirements: If the customer requirements of the product are not collected from the focus customer groups and the QFD team, production of a new product or development of an existing product should not be started.

Step 2. Obtaining importance ratings of CRs: Customers have different technological opportunities, social, political and economic infrastructure. Just as each customer's perception of product characteristics is different, the way they express their preferences on the relative importance ratings of CRs will be different. Therefore, the content of the surveys presented to the customer in order to prioritize customer expectations should also be different such as precise weights, interval weights, ordinal weights, pairwise comparison matrix, etc.

The use of different formats in the rating of customer requirements will provide more accurate results although it is a complex structure. Loss of information may occur if these formats are converted to the same format. These different structures can be combined and solved with the Goal Programming approach without being converted to the same format [30]. There is a difference between the true weights of customer requirements (W_{CRi}, ($i = 1, 2, ..., n$)) and their estimates ($W_{CRi}^{(1)}$). This

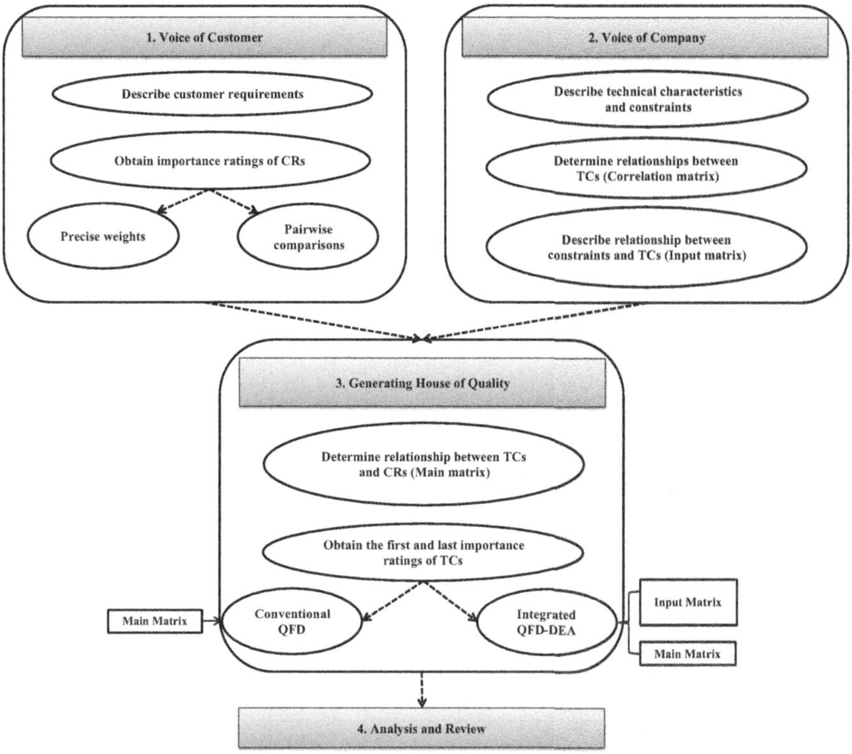

Fig. 1 Steps of the proposed method

difference is defined as the deviation variable. Positive deviation variable ($e_{i+}^{(1)}$) and negative deviation variable ($e_{i-}^{(1)}$) in the first format precise weights; the positive deviation ($E_{+}^{(2)} = (e_{1+}^{(2)}, \ldots, e_{n+}^{(2)})^T$) vector and the negative deviation vector ($E_{-}^{(2)} = (e_{1-}^{(2)}, \ldots, e_{n-}^{(2)})^T$) in the second format pairwise comparison matrix are minimized by the following model.

$$\min \mathrm{J} = \sum_{i=1}^{n} (e_{i+}^{(1)} + e_{i-}^{(1)}) + \sum_{i=1}^{n} (e_{i+}^{(2)} + e_{i-}^{(2)}) \quad (1)$$

Subject to

$$W_{CRi} - e_{i+}^{(1)} + e_{i-}^{(1)} = W_{CRi}^{(1)} \quad (2)$$

$$(A^{(2)} - nI)W_{CRi} - E_{+}^{(2)} + E_{-}^{(2)} = 0 \quad (3)$$

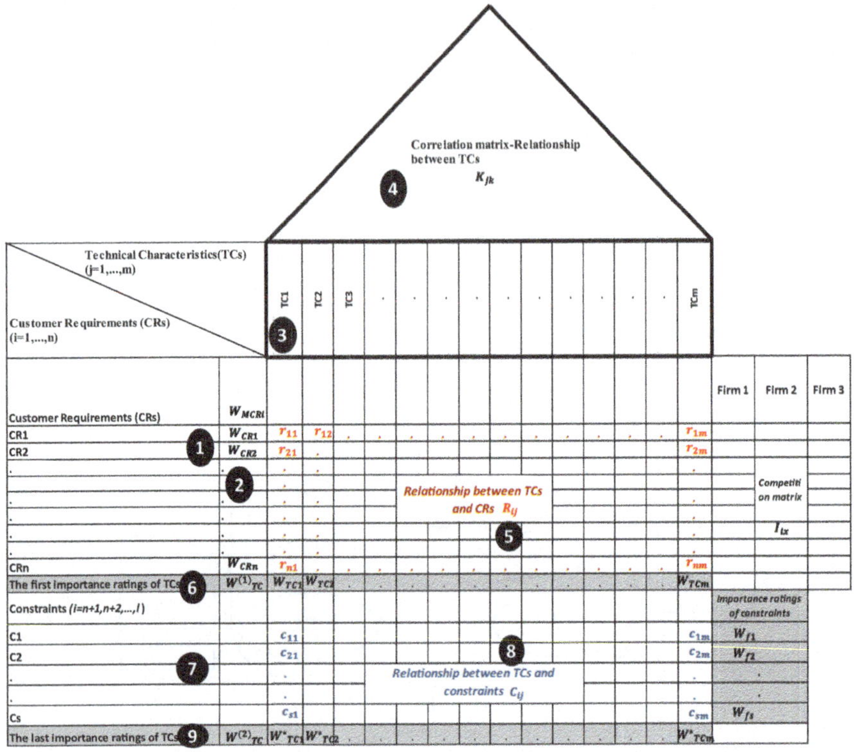

Fig. 2 House of quality for the proposed method

$$\sum_{i=1}^{n} W_{CRi} = 1 \tag{4}$$

$$W_{CRi} \geq 0 \tag{5}$$

In the above model, each objective is treated as a constraint and the desired values are written to the right-hand side. The goal in the first constraint corresponds to the importance of customer estimation for the first format. This constraint ensures that deviations from the true importance ratings determined by all formats are equal to the estimate importance ratings in the first format (Eq. 2). The second constraint ensures that the inconsistency in the second format is zero (Eq. 3). If the comparison matrix $A^{(2)}$ is consistent, then equality ($A^{(2)}W = nW$) is achieved. In the case of inconsistency, the deviation vector is calculated to measure the inconsistency of the pairwise comparison matrix, the unit matrix being "I". It is desirable that sum of importance rating of customer requirements ($i = 1, 2, \ldots, n$) equal to one in Eq. (4) and every variable should be non-negative numbers (Eq. 5).

Step 3. Describing technical characteristics: The aim is to convert the voice of the customer into technical language. As a result of the conversation with experts in the QFD team, the technical characteristics (How?) that will meet the customer requirements (What?) are decided. In Fig. 2, the parameters and indices used in the voice of company steps are given below:

$m = The\, number\, of\, technical\, characteristics\, (TC), (j = 1, 2, \ldots, m)$

$\widetilde{W}_{TCj} = Absolute\, importance\, ratings\, of\, TC\, j$

$\widetilde{W}^{(1)}_{TCj} = The\, first\, importance\, ratings\, of\, TC\, j$

$\widehat{W}^{(2)}_{TCj} = The\, last\, importance\, ratings\, of\, TC\, j$

$\widehat{W}^{(2)}_{TCj} = (W^{L}_{TCj}, W^{M}_{TCj}, W^{U}_{TCj}), lower, medium\, and\, upper\, importance\, ratings$

$l = The\, number\, of\, constraints\, (K), (i = n + 1, n + 2, \ldots, l)$

Step 4. Determining the relationships between technical characteristics: TCs form the column of the House of Quality. There may be negative or positive relationships between them. The scale used to determine the relationship between the technical characteristics is given in Table 1.

Step 5. Determining relationship between TCs and CRs: At this stage, the relationship refers to the degree to which each technical characteristic contributes to the improvement of each customer's expectation. This relationship constitutes the main matrix of the House of Quality and is the most important stage that affects the outcome of the QFD. Therefore, triangular fuzzy numbers are used to get rid of uncertainty in linguistic expressions. The linguistic scale used and the corresponding fuzzy numbers are given in Table 2.

The fuzzy absolute each TCs importance rating is calculated by multiplying the CRs importance ratings determined by Goal Programming and the relationship degrees determined in the main matrix. The equation used to determine the fuzzy absolute importance rating of the technical characteristic j is as follows:

Table 1 Relationship scale between technical characteristics

Linguistic term	Value
High positive	9
Positive	3
Neutral	1
Negative	−3
High negative	−9

Table 2 Linguistic scale for the main matrix

Linguistic terms	Symbol	Fuzzy numbers
Very high	VH	(7, 9, 9)
High	H	(5, 7, 9)
Medium	M	(3, 5, 7)
Low	L	(1, 3, 5)
Very low	VL	(1, 1, 3)

$$\widetilde{W}_{TCj} = \sum_{i=1}^{n} (W^*_{CRi} \times \tilde{r}_{ij}) \tag{6}$$

Variables used in Eq. 6 also shown in Fig. 2 are as follows;
$\hat{r}_{ij} = Fuzzy\,relationships\,degree\,between\,CR\,i\,and\,TC\,j$
$\hat{r}_{ij} = (r^L_{ij}, r^M_{ij}, r^U_{ij}), lower, medium, upper\,relationship\,degree\,of\,\hat{r}_{ij}$
$W^*_{CRi} = The\,last\,importance\,ratings\,of\,CR\,i$
Step 6. Obtaining the first importance ratings of TCs: When determining the first importance ratings of TCs, the correlation matrix, i.e. the relationship matrix between the TCs is also taken into account. Equation 7 is used to determine the fuzzy first importance ratings of TC j [20]. k_{jk} is denoted the relationship degree between TC j and TC k in Eq. 7.

$$\widetilde{W}^{(1)}_{TCj} = \widetilde{W}_{TCj} + \frac{1}{m-1} \sum_{k=1,k\neq j}^{m} (\widetilde{W}_{TCj} \times k_{jk}) \tag{7}$$

Importance ratings of technical characteristics calculated by Eq. 7 are fuzzy numbers. So fuzzy numbers need to be defuzzified and converted to crisp numbers. In this study, the equation developed by Liou and Wang [24] is used (Eq. 8). a, b and c are lower, medium and upper value of the first fuzzy importance ratings of TC j respectively.

$$I^{\alpha}_{T}(\widehat{A}) = \frac{1}{2}\alpha(b+c) + \frac{1}{2}(1-\alpha)(a+b) = \frac{1}{2}[\alpha c + b + (1-\alpha)a] \tag{8}$$

After defuzzification, the importance ratings of TCs should be normalized. The equations used for normalization are as follows ($NW^{(2)}_{TCj}$: $normalized\,value\,TC\,j$);

$$NW^{(2)}_{TCj} = \frac{W^{(2)}_{TCj}}{W^{(2)}_{TCtotal}}, \quad NW^{(1)}_{TCj} = \frac{W^{(1)}_{TCj}}{W^{(1)}_{TCtotal}} \tag{9}$$

Steps 7 and 8. Describing constraints related to TCs and the relationship between constraints and TCs: Companies should not think of technical characteristics in a way that only meets customer needs in order that companies to excel in competition and make a difference in the market but also they should mind many factors or constraints that technical characteristics interact with such as social, economic, technological or politic infrastructure factors. In the scope of the study, the effects of technical characteristics on the environment and the company are taken into consideration and the relationships between these constraints and technical characteristics are evaluated by experts using linguistic scaling. The scale used in the assessment is the same scale in Step 5. $\widetilde{C}_{ij}$ represents fuzzy relation degree of between constraint i and TC j in the HoQ. C^L_{ij}, C^M_{ij} and C^U_{ij} represent the lower, medium and upper value of $\widehat{C}_{ij}$.

Step 9. Obtaining the last importance ratings of TCs: Data Envelopment Analysis is used to determine whether the technical characteristics are effective in terms of both the customer and the company's interests, so their last importance ratings of TCs ($\widehat{W}^{(2)}_{TCj}$) are obtained.

The current chapter utilizes Fuzzy Data Envelopment Analysis because the relationships between TCs and CRs also constraints and TCs are represented by triangular fuzzy numbers. Models of Wang and Chin [31] which allow fuzzy and crisp numbers to be used together, is established. In the model, the decision-making unit (DMU) corresponds to the technical characteristics ($j = 1, 2, \ldots, m$), the constraints that concern the firm are used to as inputs ($i = n + 1, n + 2, \ldots, l$) and customer requirements are used to as outputs ($i = 1, 2, \ldots, n$). The DEA models have been established for the efficiency of technical characteristics. The variables and parameters in the models are as follows:

$\widetilde{r_{ij}} = \left(r^L_{ij}, r^M_{ij}, r^U_{ij}\right)$, *fuzzy output data i of decision making unit j*

$\widetilde{c_{ij}} = \left(c^L_{ij}, c^M_{ij}, c^U_{ij}\right)$, *fuzzy input data i of decision making unit j*

$\widetilde{x_i} = \left(x^L_i, x^M_i, x^U_i\right)$, *fuzzzy weights of output* $i, i = 1, 2, \ldots n$

$\widetilde{y_i} = \left(y^L_i, y^M_i, y^U_i\right)$, *fuzzy weights of input* $i, i = n + 1, n + 2, \ldots, l$

$$\max \theta_0^{best} = \sum_{i=1}^{n} \left(x^L_i r^L_{i0} + 2x^M_i r^M_{i0} + x^U_i r^U_{i0}\right) \tag{10}$$

Subject to

$$\sum_{i=n+1}^{l} \left(y^L_i c^L_{i0} + 2y^M_i c^M_{i0} + y^U_i c^U_{i0}\right) = 1 \tag{11}$$

$$\sum_{i=1}^{n} \left(x^L_i r^L_{ij} + 2x^M_i r^M_{ij} + x^U_i r^U_{ij}\right) - \sum_{i=n+1}^{l} \left(y^L_i c^L_{ij} + 2y^M_i c^M_{ij} + y^U_i c^U_{ij}\right) \le 0, \tag{12}$$

$$x^U_i \ge x^M_i \ge x^L_i \ge 0, i = 1, 2, \ldots, n \tag{13}$$

$$y^U_i \ge y^M_i \ge y^L_i \ge 0, i = n + 1, n + 2, \ldots, l, j = 1, 2, \ldots, m \tag{14}$$

$$\min \theta_0^{worst} = \sum_{i=1}^{n} \left(x^L_i r^L_{i0} + 2x^M_i r^M_{i0} + x^U_i r^U_{i0}\right) \tag{15}$$

Subject to

$$\sum_{i=n+1}^{l} \left(y^L_i c^L_{i0} + 2y^M_i c^M_{i0} + y^U_i c^U_{i0}\right) = 1 \tag{16}$$

$$\sum_{i=1}^{n}\left(x_i^L r_{ij}^L + 2x_i^M r_{ij}^M + x_i^U r_{ij}^U\right) - \sum_{i=n+1}^{l}\left(y_i^L c_{ij}^L + 2y_i^M c_{ij}^M + y_i^U c_{ij}^U\right) \geq 0, \quad (17)$$

$$x_i^U \geq x_i^M \geq x_i^L \geq 0, i = 1, 2, \ldots, n \quad (18)$$

$$y_i^U \geq y_i^M \geq y_i^L \geq 0, i = n+1, n+2, \ldots, l, j = 1, 2, \ldots, m \quad (19)$$

In the solution of models, optimistic and pessimistic efficiency should be considered simultaneously to make an overall assessment of the performance of DMUs. So, Wang et al. [32] proposed geometric average efficiency (Eq. 20).

$$\theta_0^{Geometric} = \sqrt{\theta^{best} \times \theta^{worst}}, j = 1, 2, \ldots m \quad (20)$$

The weight of the outputs, i.e. customer requirements, determined by the Goal Programming should be included in the Fuzzy DEA model. The weights of the outputs are added to the model for each output value with the constraint $x_{i0} = d_i x_{10}$ by the Assurance Region method ($i = 1, 2, \ldots, n$ and $d_1 = 1$).

4 Industrial Application

The proposed QFD technique is applied in a company operating in the forest industry in the Black Sea Region of Turkey. The method is explained to the team which is established under the leadership of senior particleboard business chief in the company. As a result of interviews with customers in the provinces close to the region where the firm is located, 12 customer requirements are determined (see Table 3). In determining these customer requirements, physical, mechanical properties, sales and after-sales attributes of the particleboard are taken into consideration.

Two different questionnaires are prepared by considering customer experiences and their relations with the company and the importance ratings of customer requirements are obtained. In the questionnaires have prepared in the first format (precise weights), customers assigned value to each requirement according to considering their preferences. Since the second format is more time-consuming, made by face-to-face and pairwise comparison matrices are created. The results of both precise weighting and pairwise comparisons, taking the normalize of completed surveys coming from the first group of customers and the average of completed pairwise comparisons coming from the second group, are shown in Tables 3 and 4 respectively.

Precise weighting and pairwise comparison matrices questionnaires are combined with the Goal Programming model in Eqs. (1–5) that aims to minimize deviation variables to get exact importance ratings of CRs. Importance ratings of CRs, results of the models, are given in Table 5.

Table 3 Importance ratings of customer requirements in the first format

Customer requirements	Abbreviation	Weight
Surface smoothness	SurSm	0.107
Same appearance on top and bottom surfaces	Same	0.092
Cutting quality (machinability-workability)	Cutqu	0.115
Homogeneous thickness	HomTh	0.084
Scratch and wear resistance	ScWe	0.099
Suitable resistance	SuRe	0.076
Homogeneous density	HomDe	0.053
Formaldehyde emission	ForEm	0.099
On-time delivery	OnTim	0.069
Communication with customers	ComCus	0.053
Campaign and promotion activities	CamPro	0.061
Economic effects	Eco	0.092

Production stages of the particleboard are examined thoroughly while determined the technical characteristics that meet customer requirements. Technical characteristics that make the greatest contribution to satisfying these requirements are addressed.

Figure 3 shows the correlation matrix of technical characteristics in the roof of the House of Quality. The linguistic scale in Table 1 is used to form the roof matrix.

As for step building the main matrix of the House of Quality, the aim is to highlight those that are significantly related to CRs within the TCs. Linguistic terms are converted to triangular fuzzy numbers in order to avoid uncertainty and to obtain more flexible and consistent results in the main matrix. The main matrix is generated using the scale in Table 2. Once the main matrix has been formed, the absolute importance ratings of each technical characteristic are determined using Eq. 6. Equation 8 is should be used to defuzzification because the absolute importance ratings of TCs are fuzzy values. The α value which is the optimism index in the equation is taken as 0.5. The decision maker can increase the α value to increase optimism. The fuzzy and crisp importance ratings of TCs are given in Table 6.

Hot pressing time, smooth drafting, pressing temperature, raw material blending and laying moisture have the highest absolute importance ratings of the TCs. Closeness to raw material, raw material moisture, dry flake density, closeness to the market and thickness of the flake are the least significant technical characteristics, respectively.

The relationships between the TCs are not considered when calculating the absolute importance ratings of the TCs. Equation 7 calculates the first importance ratings of the TCs, including the correlation matrix. In order to make comparisons between the technical characteristics, the first fuzzy importance ratings are converted to crisp values and are shown in Table 7.

Table 4 Average values of pairwise comparison matrix

No	Kod	Müteri Beklentisi	SurSm	Same	Cutqu	HomTh	ScWe	SuRe	HomDe	ForEm	OnTim	ComCus	CamPro	Eco
1	SurSm	Surface smoothness	**1**	1	0.11	0.14	0.11	0.14	0.11	0.11	0.14	0.14	0.2	0.11
2	Same	Same appearance on top and bottom surfaces	1	**1**	0.11	0.14	0.11	0.14	0.11	0.11	0.14	0.14	0.2	0.11
3	Cutqu	Cutting quality (Machinability-workability)	9	9	**1**	3	1	3	1	1	3	3	5	1
4	HomTh	Homogeneous thickness	7	8	0.33	**1**	0.33	1	0.33	0.2	0.33	1	1	0.33
5	ScWe	Scratch and wear resistance	9	9	1	3	**1**	3	1	1	1	3	5	1
6	SuRe	Suitable resistance	7	8	0.33	1	0.33	**1**	0.33	0.33	1	1	3	0.33
7	HomDe	Homogeneous density	9	9	1	3	1	3	**1**	1	3	5	7	1
8	ForEm	Formaldehyde emission	9	9	1	5	1	3	1	**1**	1	3	5	1
9	OnTim	On-time delivery	8	8	0.33	3	1	1	0.33	1	**1**	3	5	0.33
10	ComCus	Communication with customers	8	8	0.33	1	0.2	1	0.2	0.33	0.33	**1**	3	0.2
11	CamPro	Campaign and promotion activities	5	5	0.2	1	0.14	0.33	0.14	0.2	0.2	0.33	**1**	0.14
12	Eco	Economic effects	9	9	1	3	1	3	1	1	3	5	7	**1**

Table 5 The last importance ratings of customer requirements

Customer requirements	Weights $\left(W^{*}_{CRi}\right)$	Rank
Surface smoothness	0.011	11
Same appearance on top and bottom surfaces	0.005	12
Cutting quality (machinability-workability)	0.145	3
Homogeneous thickness	0.045	9
Scratch and wear resistance	0.129	5
Suitable resistance	0.056	7
Homogeneous density	0.153	2
Formaldehyde emission	0.135	4
On-time delivery	0.092	6
Communication with customers	0.047	8
Campaign and promotion activities	0.027	10
Economic effects	0.155	1
Total	**1**	

The higher importance ratings of technical characteristics, the more they contribute to customer requirements. Laying moisture, smooth drafting, raw material blending and dry flake density are among the most important technical characteristics that contribute to customer satisfaction. Closeness to raw material, amount of glue, closeness to market, qualified staff, hot pressing density are the technical characteristics that make the least contribution to customer satisfaction. Until now the technical characteristics have been evaluated in terms of only customer aspect. In the next stage, it will be investigated whether the technical characteristics are effective or not.

By this time House of Quality has been evaluated as customer-oriented. At this stage, technical characteristics are evaluated by considering firm constraints such as cost, workload, environmental impact, chemical ratio. Each TC is evaluated by experts in terms of these factors. The linguistic scale has used in this evaluation is the same as the scale used to construct the main matrix in the House of Quality (see Table 2). The TCs of high importance are effective as long as they are in line with company policy and interests as well as meeting the CRs. At this stage, a Data Envelopment Analysis (DEA) model which takes into account constraints such as the cost of technical characteristics, the workload per operator, the rate of chemical used and the state of energy consumption, is established. The DEA model should be fuzzy because the relationship between the main matrix, technical characteristics and constraints are expressed in triangular fuzzy numbers. These four constraints are input, customer requirements are output and technical characteristics are the decision-making unit. The optimistic, pessimistic and average efficiency values of the technical characteristics are shown in Table 8 by solving the model Eqs. (10–14) and (15–19) with Lingo 16.0 program. Equation 20 is used for the average efficiency $\widehat{W}^{(2)}_{TCj}$, which are the last importance ratings of the TCs.

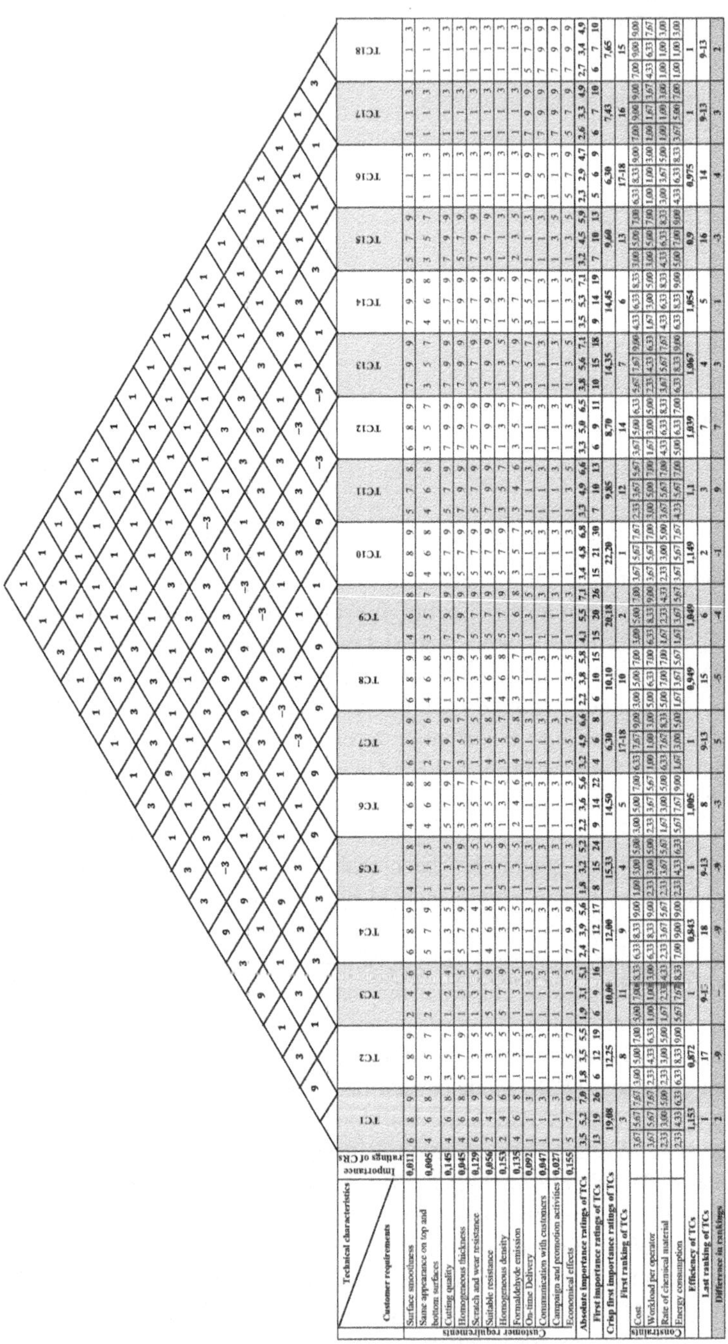

Fig. 3 House of quality for the proposed method

Table 6 The Fuzzy and crisp absolute importance ratings of the TCs

Technical characteristics	$\widetilde{W}_{TCj}$			Crisp value	Rank
1. Raw material blending	3.51	5.18	7.03	5.223	**4**
2. Thickness of flake	1.84	3.51	5.49	3.586	**14**
3. Raw material (wood) moisture input	1.86	3.07	5.07	3.268	**17**
4. Type of chipping machine	2.35	3.89	5.57	3.925	**10**
5. Dry flake density	1.83	3.18	5.18	3.341	**16**
6. Moisture after drying	2.21	3.57	5.57	3.729	**12**
7. Amount of glue	3.20	4.87	6.57	4.875	**8**
8. Homogeneous gluing	2.15	3.82	5.81	3.898	**11**
9. Smooth draft	4.07	5.47	7.09	5.525	**2**
10. Laying moisture	3.45	4.80	6.79	4.961	**5**
11. First pressing density	3.33	4.86	6.64	4.923	**7**
12. Hot pressing density	3.31	4.98	6.48	4.934	**6**
13. Hot pressing time	3.78	5.63	7.11	5.535	**1**
14. Pressing temperature	3.49	5.34	7.12	5.325	**3**
15. Melamine impregnated	3.22	4.49	5.95	4.538	**9**
16. Closeness to raw material	2.27	2.85	4.67	3.160	**18**
17. Closeness to market	2.62	3.26	4.92	3.513	**15**
18. Qualified staff	2.74	3.38	4.92	3.608	**13**

The technical characteristics 1, 3, 5, 7, 9, 10, 11, 12, 14, 17 and 18 are found to be optimism efficient when the fuzzy DEA model is solved. Also in order to make a ranking between the efficiencies of these technical characteristics, a model is established in which pessimistic efficiency values are calculated. The optimistic and pessimistic efficiency rates obtained in the solution of these two models are interpreted in different ways. The final efficiency is calculated by taking the geometric average of the two efficiency values and the final rankings of the technical characteristics are found according to this value.

5 Results and Discussion

The analysis is performed for each method integrated into the House of Quality. The final importance levels of the customer requirements collected in different formats without the need to be converted to the same format and without lack of information or inaccuracy are determined with Goal Programming. As a result of

Table 7 The First importance ratings of the TCs

Technical characteristics	$\widetilde{W}_{TCj}^{(1)}$			Crisp value	Rank
1. Raw material blending	13	19	26	19.08	**3**
2. Thickness of flake	6.3	12	19	12.25	**8**
3. Raw material (wood) moisture input	5.7	9.4	16	10.00	**11**
4. Type of chipping machine	7.2	12	17	12.00	**9**
5. Dry flake density	8.4	15	24	15.33	**4**
6. Moisture after drying	8.6	14	22	14.50	**5**
7. Amount of glue	4.1	6.3	8.5	6.30	**17–18**
8. Homogeneous gluing	5.6	9.9	15	10.10	**10**
9. Smooth draft	15	20	26	20.18	**2**
10. Laying moisture	15	22	30	22.20	**1**
11. First pressing density	6.7	9.7	13	9.85	**12**
12. Hot pressing density	5.8	8.8	11	8.70	**14**
13. Hot pressing time	9.8	15	18	14.35	**7**
14. Pressing temperature	9.5	15	19	14.45	**6**
15. Melamine impregnated	6.8	9.5	13	9.60	**13**
16. Closeness to raw material	4.5	5.7	9.3	6.30	**17–18**
17. Closeness to market	5.5	6.9	10	7.43	**16**
18. Qualified staff	5.8	7.2	10	7.65	**15**

the model, one attributes from the physical, mechanical properties, sales and after-sales attributes of the particleboard constitute the first three places. These are economic effects (CR12), homogeneous density (CR7) and cutting quality (CR3) respectively. The same appearance on top and bottom surfaces (CR2), surface smoothness (CR1), campaign and promotion activities (CR11) are the least important customer expectations.

It is seen that there are differences in the ranking of technical characteristics according to whether taking into account the correlation matrix. The rankings of the technical characteristics 7, 11, 12, 13, 14, 15 and 18 calculated by taking into account the correlation matrix are lower than the order formed without considering the correlation matrix. The fact that these characteristics have a high negative relationship with others leads to a decline in their rankings. Especially, the technical characteristic-the amount of glue, has a negative relationship with the others thus the rank of this TC is the lowest.

The efficiency of the technical characteristics is determined by DEA, taking into account some of the constraints of the technical characteristics. According to the DEA, there are two types of efficiencies, optimistic and pessimistic. The first and the last rankings (average efficiencies) of the technical characteristics are shown in Table 9. According to the table, the dry flake density, the thickness of the flake and the type of chipping machine are technical characteristics with high first importance

Table 8 Efficiencies of TCs under fuzzy weights

Technical characteristics	Optimistic efficiency	Pessimistic efficiency	Average efficiency ($\widehat{W}_{TCj}^{(2)}$)	Rank
1. Raw material blending	1	1.33	**1.153**	1
2. Thickness of flake	0.76	1	**0.872**	17
3. Raw material moisture input	1	1	**1.000**	9–13
4. Type of chipping machine	0.71	1	**0.843**	18
5. Dry flake density	1	1	**1.000**	9–13
6. Moisture after drying	0.99	1.02	**1.005**	8
7. Amount of glue	1	1	**1.000**	9–13
8. Homogeneous gluing	0.9	1	**0.949**	15
9. Smooth draft	1	1.1	**1.049**	6
10. Laying moisture	1	1.32	**1.149**	2
11. First pressing density	1	1.21	**1.100**	3
12. Hot pressing density	1	1.08	**1.039**	7
13. Hot pressing time	0.99	1.15	**1.067**	4
14. Pressing temperature	1	1.11	**1.054**	5
15. Melamine impregnated	0.81	1	**0.900**	16
16. Closeness to raw material	0.95	1	**0.975**	14
17. Closeness to market	1	1	**1.000**	9–13
18. Qualified staff	1	1	**1.000**	9–13

ratings, which increases the level of customer satisfaction. However, considering the constraints, it is seen that the rankings of these parameters are low according to the efficiency values of DEA model in terms of firm interests. This means that some technical characteristics make a high contribution to satisfying customer expectations, while their return to the company is not as expected. Vice versa, there are technical characteristics that have the low value in the first importance ratings but the high value in the efficiency. Some of them are raw material blending, amount of glue, first

Table 9 The First importance ratings and efficiency of the TCs

Technical characteristics	The first importance rating	Rank	Average efficiency	Rank
1. Raw material blending	19.08	3	1.153	1
2. Thickness of flake	12.25	8	0.872	17
3. Raw material (wood) moisture input	10.00	11	1.000	9–12
4. Type of chipping machine	12.00	9	0.843	18
5. Dry flake density	15.33	4	1.000	9-12
6. Moisture after drying	14.50	5	1.005	8
7. Amount of glue	6.30	17–18	1.000	9–12
8. Homogeneous gluing	10.10	10	0.949	15
9. Smooth draft	20.18	2	1.049	6
10. Laying moisture	22.20	1	1.149	2
11. First pressing density	9.85	12	1.100	3
12. Hot pressing density	8.70	14	1.039	7
13. Hot pressing time	14.35	7	1.067	4
14. Pressing temperature	14.45	6	1.054	5
15. Melamine impregnated	9.60	13	0.900	16
16. Closeness to raw material	6.30	17–18	0.975	14
17. Closeness to market	7.43	16	1.000	9–12
18. Qualified staff	7.65	15	1.000	13

pressing density, hot pressing density and hot pressing time. Although these are in line with the interests of the company, they do not contribute sufficiently to customer requirements.

All results for the particleboard with melamine impregnated paper are obtained from the House of Quality in Fig. 3. Customer requirements, technical characteristics that will contribute to the development of customer requirements, relationships between them, constraints affecting technical characteristics for the company are shown together in the House of Quality. Based on this point of view, it provides an overview of customer-oriented product development as well as quality improvement, guiding decision makers in all departments of the firm.

6 Conclusion

Many departments, from production planning to marketing, will utilize from the findings from the House of Quality. Information is provided to the company about priority items and technical requirements related to these priority items and product development in order to increase the customer service level. Also the quality improvement process enables the creation of processes and systems that can combine customer and company requirements.

The use of Goal Programming in the customer part and Fuzzy DEA in the technical part of the study show that Quality Function Deployment is an interdisciplinary method and the results are reliable and realistic due to the mathematical model.

In the House of Quality, linguistic terms are used in the relations section and these linguistic terms are transformed into triangular fuzzy numbers. The conversion of linguistic terms based on human judgments into fuzzy values rather than crisp numbers is for coping with uncertainty.

The specifications in the process of chipping and drying the chips increase the customer satisfaction level but create losses for the company. The fact that the equipment in these processes is quite old and the maintenance cost is high is seen as the reason why the company does not benefit from the technical characteristics here.

With the implementation of Quality Function Deployment, customer requests and needs have been transformed into product specifications for particleboard with melamine impregnated paper. At the same time, the preparation of different surveys according to customer characteristics and the participation of all departments of the company in the implementation of QFD constitute a customer-oriented decision mechanism. At present, the importance of QFD studies emerges at this point, as it is difficult for non-customer-oriented enterprises to survive against their competitors.

In the generation of the House of Quality, it is thought that the expert opinions in the relations matrices have the same weights. It may be more realistic to compare the results by assigning different weights to the expert team consisting of classes such as chef, worker and manager.

In future studies, a QFD model can be developed including the process planning and production planning stages, starting at the supplier and ending at the customer. It can be applied in many different sectors from production to service.

References

1. Benner, M., et al.: Quality function deployment (QFD)—can it be used to develop food products? Food Qual. Prefer. **14**(4), 327–339 (2003). https://doi.org/10.1016/S0950-3293(02)00129-5
2. Bevilacqua, M., Ciarapica, F.E., Giacchetta, G.: A fuzzy-QFD approach to supplier selection. J. Purch. Supply Manag. **12**, 14–27 (2006). https://doi.org/10.1016/j.pursup.2006.02.001
3. Bhattacharya, A., Geraghty, J., Young, P.: Supplier selection paradigm: an integrated hierarchical QFD methodology under multiple-criteria environment. Appl. Soft Comput. J. **10**(4), 1013–1027 (2010). https://doi.org/10.1016/j.asoc.2010.05.025

4. Bottani, E., Rizzi, A.: Strategic management of logistics service: a fuzzy QFD approach. Int. J. Prod. Econ. **103**(2), 585–599 (2006). https://doi.org/10.1016/j.ijpe.2005.11.006
5. Büyüközkan, G., et al.: Determining the importance weights for the design requirements in the house of quality using the fuzzy analytic network approach. Int. J. Intell. Syst. **19**(5), 443–461 (2004). https://doi.org/10.1002/int.20006
6. Büyüközkan, G., Feyzioğlu, O.: Group decision making to better respond customer needs in software development. Comput. Ind. Eng. **48**(2), 427–441 (2005). https://doi.org/10.1016/j.cie.2005.01.007
7. Chan, L.K., et al.: Rating the importance of customer needs in quality function deployment by fuzzy and entropy methods. Int. J. Prod. Res. **37**(11), 2499–2518 (1999). https://doi.org/10.1080/002075499190635
8. Chen, L.H., Ko, W.C.: A fuzzy nonlinear model for quality function deployment considering Kano's concept. Math. Comput. Model. **48**(3), 581–593 (2008). https://doi.org/10.1016/j.mcm.2007.06.029
9. Chen, L.H., Weng, M.C.: A fuzzy model for exploiting quality function deployment. Math. Comput. Model. **38**(5), 559–570 (2003). https://doi.org/10.1016/S0895-7177(03)90027-6
10. Delice, E.K., Güngör, Z.: Determining design requirements in QFD using fuzzy mixed-integer goal programming: application of a decision support system. Int. J. Prod. Res. (2013). https://doi.org/10.1080/00207543.2013.803625
11. Durga Prasad, K., Venkata Subbaiah, K., Narayana Rao, K.: Prioritization of customer needs in house of quality using conjoint analysis. Int. J. Cent. Qual. **4**(2), 145–154 (2010)
12. Fan, W., Xu, M.: The application of QFD based on DEA to improve the TPL quality of service. In: 2010 International Conference on Management and Service Science, MASS 2010 (2010). https://doi.org/10.1109/icmss.2010.5576929
13. Homkhiew, C., Ratanawilai, T., Pochana, K.: Application of a quality function deployment technique to design and develop furniture products. Songklanakarin J. Sci. Technol. **34**(6), 663–668 (2012)
14. Juan, Y.K., et al.: Housing refurbishment contractors selection based on a hybrid fuzzy-QFD approach. Autom. Constr. **18**(2), 139–144 (2009). https://doi.org/10.1016/j.autcon.2008.06.001
15. Kamvysi, K., et al.: Integrating DEAHP and DEANP into the quality function deployment. TQM J. **22**(3), 293–316 (2010). https://doi.org/10.1108/17542731011035532
16. Kang, X., et al.: Integrating evaluation grid method and fuzzy quality function deployment to new product development. Math. Probl. Eng. (2018). https://doi.org/10.1155/2018/2451470
17. Kannan, G.: Implementation of fuzzy quality function deployment in an automobile component to improve the quality characteristics. Qual. Eng. **20**, 321–333 (2008). https://doi.org/10.1080/08982110801982564
18. Karsak, E.E., Dursun, M.: An integrated supplier selection methodology incorporating QFD and DEA with imprecise data. Expert Syst. Appl. **41**(16), 6995–7004 (2014). https://doi.org/10.1016/j.eswa.2014.06.020
19. Karsak, E., Sozer, S., Alptekin, S.: Product planning in quality function deployment using a combined analytic network process and goal programming approach. Comput. Ind. Eng. **44**(1), 171–190 (2003). https://doi.org/10.1016/S0360-8352(02)00191-2
20. Khoo, L.P., Hot, N.C.: Framework of a fuzzy quality function deployment system. Int. J. Prod. Res. **34**(2), 299–311 (1996). https://doi.org/10.1080/00207549608904904
21. Kurtulmuşoğlu, F.B., Pakdil, F., Atalay, K.D.: Quality improvement strategies of highway bus service based on a fuzzy quality function deployment approach. Transp. A: Transp. Sci. **12**(2), 175–202 (2016). https://doi.org/10.1080/23249935.2015.1117535
22. Li, Y.L., Chin, K.S., Luo, X.G.: Determining the final priority ratings of customer requirements in product planning by MDBM and BSC. Expert Syst. Appl. **39**(1), 1243–1255 (2012). https://doi.org/10.1016/j.eswa.2011.07.133
23. Lin, T., et al.: Ergonomics product development of over bed table for bedridden patients. Comput.-Aided Des. Appl. (2016). https://doi.org/10.1080/16864360.2015.1131550

24. Liou, T.S., Wang, M.J.J.: Ranking fuzzy numbers with integral value. Fuzzy Sets Syst. **50**(3), 247–255 (1992). https://doi.org/10.1016/0165-0114(92)90223-Q
25. Luo, X., Tang, J., Kwong, C.K. A QFD-based optimization method for scalable product platform. In: Advances in Product Family and Product Platform Design: Methods and Applications, pp. 141–156 (2014) https://doi.org/10.1007/978-1-4614-7937-6_14
26. Milunovic Koprivica, S., Filipovic, J.: Application of traditional and fuzzy quality function deployment in the product development process. EMJ—Eng. Manag. J. (2018). https://doi.org/10.1080/10429247.2018.1438027
27. Park, T., Kim, K.J.: Determination of an optimal set of design requirements using house of quality. J. Oper. Manag. **16**, 569–581 (1998)
28. Ramanathan, R., Yunfeng, J.: Incorporating cost and environmental factors in quality function deployment using data envelopment analysis. Omega **37**(3), 711–723 (2009). https://doi.org/10.1016/j.omega.2007.12.003
29. Schillo, R.S., Isabelle, D.A., Shakiba, A.: Linking advanced biofuels policies with stakeholder interests: a method building on quality function deployment. Energy Policy **100**, 126–137 (2017). https://doi.org/10.1016/j.enpol.2016.09.056
30. Wang, Y.M., Chin, K.S.: A linear goal programming approach to determining the relative importance weights of customer requirements in quality function deployment. Inf. Sci. **181**(24), 5523–5533 (2011). https://doi.org/10.1016/j.ins.2011.08.016
31. Wang, Y.M., Chin, K.S.: Fuzzy data envelopment analysis: a fuzzy expected value approach. Expert Syst. Appl. **38**(9), 11678–11685 (2011). https://doi.org/10.1016/j.eswa.2011.03.049
32. Wang, Y.M., Chin, K.S., Yang, J.B.: Measuring the performances of decision-making units using geometric average efficiency. J. Oper. Res. Soc. **58**(7), 929–937 (2007). https://doi.org/10.1057/palgrave.jors.2602205
33. Wu, H.H., Liao, A.Y.H., Wang, P.C.: Using grey theory in quality function deployment to analyse dynamic customer requirements. Int. J. Adv. Manuf. Technol. **25**(11–12), 1241–1247 (2005). https://doi.org/10.1007/s00170-003-1948-8
34. Wu, H.H., Shieh, J.I.: Using a Markov chain model in quality function deployment to analyse customer requirements. Int. J. Adv. Manuf. Technol. **30**(1–2), 141–146 (2006). https://doi.org/10.1007/s00170-005-0023-z
35. Wu, X., Liao, H.: An approach to quality function deployment based on probabilistic linguistic term sets and ORESTE method for multi-expert multi-criteria decision making. Inf. Fusion (2018). https://doi.org/10.1016/j.inffus.2017.11.008
36. Zhang, X.: User selection for collaboration in product development based on QFD and DEA approach. J. Intell. Manuf. **30**(5), 2231–2243 (2019). https://doi.org/10.1007/s10845-017-1386-3
37. Zadeh, L.A.: Fuzzy Logic. IEEE Comput. **21**(4), 83–93 (1988). https://doi.org/10.1109/2.53

Fuzzy Quality Function Deployment Method for Smart Phone Product Design

Sukran Seker

Abstract In today's rapidly-changing environment, gaining competitive advantages of organizations is provided by meeting customer needs and expectations. Quality function deployment (QFD) is a systematic method that transforms customer needs (customer's voice) into engineering features for a product or service. In this study, Fuzzy-QFD methodology is applied in the smartphone design process to develop the product according to the customer's wishes. To address uncertainties in the subjective assessments of decision makers, fuzzy sets are integrated with QFD methodology. Thus, the presented method is performed to meet the expectations of the product/service in the best way. This is provided by converting customer voice (needs) into technical requirements and determining the systematic relationship between them. The results of the analysis prove the applicability and effectiveness of the Fuzzy QFD methodology in new product design.

1 Introduction

In today's fast-growing and rapidly-changing environment, new product development (NPD) which satisfies customer requirements (CRs) and differentiates the company among competitors is a major factor for success and profitability of a company [1]. Some tools such as IT technology, computer-aided design, engineering and manufacturing (CAD/CAE/CAM) and advanced prototyping technology are performed to enhance product design, reduce lead time and the production cost [2].

However, an extensive number of product design and development methods such as Reverse engineering, Value engineering, Taguchi method, and Quality function deployment (QFD) have been performed and implemented in numerous areas. The first three methods give emphasis to product functions mostly and the less to customer requirements and production operations [3]. On the other hand, QFD is known as a methodology to evaluate features or functions that affect the quality of products or services for satisfying customer expectations. Accordingly, in QFD methodology,

S. Seker (✉)
Department of Industrial Engineering, Yildiz Technical University, Istanbul, Turkey
e-mail: sseker@yildiz.edu.tr

C. Kahraman and S. Cebi (eds.), *Customer Oriented Product Design*, Studies in Systems, Decision and Control 279, https://doi.org/10.1007/978-3-030-42188-5_4

the voice of the customer (VOC) is translated into detailed engineering specifications to design products that fulfill product and service quality characteristics [1].

A characteristic of QFD process consists of four main stages include determination of CRs, the determination of engineering characteristics (ECs) which are actions of the company to meet CRs, determination of process for the customer satisfaction and the process control. During a new product design, in order to meet CRs, QFD team organize the first two stages named as product planning and part deployment of QFD process to maximize customer satisfaction. The house of quality (HOQ) is an important part of QFD methodology, including information on "WHAT", "HOW", relationship between "WHAT" and "HOW", and, the relationship between the "HOW" factors themselves [4].

While constructing the HOQ, prioritization of customer desires and preferences plays an important role since new product design should provide customer satisfaction and service quality to meet CRs [5]. As a result, the main aim of QFD methodology is to maximize customer satisfaction meeting CRs. To obtain CRs for service or product, brainstorming and tree analogy are two effective ways.

However, in most cases, importance degree of CRs, sales point, relationship between CRs and ECs, technical difficulty to change or maintain design requirements cannot be evaluated by either crisp values or random procedures in traditional QFD process. To handle subjectivity and ambiguity in the transformation of customer evaluations for a company's brand and the competitor performances, linguistic variables and corresponding fuzzy numbers can be used. In this regard, fuzzy set theory (FST) has been used to model uncertain information rely on human judgments [6].

In this chapter, the systematic procedure of QFD methodology with respect to FST, is performed in product design. Smart phone is designed and developed using Fuzzy QFD methodology. The aim is to understand Fuzzy QFD application well in new product development by translating CRs ("WHATs") into appropriate ECs ("HOWs") at every phase from research through production design and development to manufacture.

The organization of this chapter is arranged as follows: In the following section, a brief literature review on QFD, fuzzy QFD applications are introduced. In Sect. 3, some preliminaries used throughout this chapter are presented. The application of Fuzzy QFD is performed for smart phone design. Conclusion and discussions about study are presented in the last section.

2 Literature Review

QFD was first developed by Akao in the late 1960s [7]. Since then, QFD was employed at the Kobe shipyards of Mitsubishi Heavy Industries Ltd., Toyota, Ford Motor Company, General Motors, Chrysler, Digital Equipment, Hewlett-Packard, AT&T, Procter and Gamble, and Baxter Healthcare etc. [6]. QFD has been widely used as a promising tool to develop new products that satisfy customer needs. Literature provides numerous publications on QFD in various fields [5, 8–12].

However, Fuzzy QFD was used to deal with the impreciseness in customer voice. Chen and Weng [13] formulated a fuzzy model to obtain design requirements for customer satisfaction using QFD methodology. Liu [14] proposed Fuzzy QFD methodology performing mathematical programming approach to rank design requirements for product development. Chen and Ko [15] suggested a fuzzy nonlinear model for QFD considering Kano's concept for new product planning. Chen and Weng [16] proposed fuzzy goal programming models to determine design requirements in QFD process. Kahraman et al. [17] presented a fuzzy optimization model for identifying product's technical requirement which could be used during design process in QFD. Lin et al. [18] proposed a model to evaluate environmental necessities of sustainable products with linguistic terms using Fuzzy QFD and Analytic Network Process (ANP). Fung et al. [19] proposed an asymmetric fuzzy linear regression model to evaluate the functional associations for product planning using QFD procedure. Buyukozkan et al. [20] suggested a new fuzzy group decision making approach to meet CRs in product development with QFD. Proposed method is demonstrated with an example which concern the development of the hatch door of a car. Zaim et al. [21] presented ANP with fuzzy logic to prioritize technical characteristics of a product (or a service) while applying QFD. Younesi and Roghanian [22] proposed an integrated QFD for Environment (QFDE), fuzzy Decision Making Trial and Evaluation Laboratory (DEMATEL) and Fuzzy Analytic Network Process (FANP) are introduced for sustainable product design. Yang et al. [23] suggested Fuzzy QFD to develop new product design using former production design. Hsuan et al. [24] presented a new product planning model to meet customer needs using Fuzzy goal programing and QFD. Vinodh et al. [25] proposed fuzzy QFD approach to rank relevant CRs, sustainability parameters and sustainability initiatives for the consumer electronics product. Wu et al. [26] proposed QFD based on DEMATEL technique and VIKOR method for product planning and development under the hesitant fuzzy environment. In the following sections, fuzzy QFD method is used to design a new smart phone.

3 Proposed Method

In the following, some basic concepts, with respect to FST and QFD methodology which will be used throughout the chapter are presented.

3.1 Fuzzy Set Theory

FST is suggested by Zadeh [27]. FST is used to deal with imprecise and uncertain information. A FST refers to a fuzzy number which is between zero and one. A fuzzy set $\widetilde{A}$ identified on R must have the specifications given as [28–30].

(a) $\mu\widetilde{A}(x) = 0$ for all $x \in (-\infty, c] \cup [d, \infty)$, where $c < d$.

(b) $\mu_{\tilde{A}}(x)$ is hardly increasing on [c, a] and hardly decreasing on [b, d] for $c \le a \le b \le d$.
(c) $\mu_{\tilde{A}}(x) = 1$ for all $x \in [a, b]$ ensured $a \le b$.

There are various applications of fuzzy numbers such as Triangular Fuzzy Number (TFN) and Trapezoidal Fuzzy Number (TrFN) in literature. To quantify subjectivity of the CRs and prioritize them, TFNs are used in this study. A TFN can be illustrated by a vector (a_1, a_2, a_3) that represents the lower bound, most likely, and upper bound values.

While a fuzzy number $(\widetilde{A})$ is shown as TFN $\widetilde{A} = (a_1, a_2, a_3)$, the membership function of fuzzy number $\widetilde{A}$ is illustrated by Eq. (1):

$$\mu_{\tilde{A}}(x) = \begin{cases} 0, & if\, x \le a_1 or\, x \ge a_2 \\ \frac{x-a_1}{a_2-a_1}, & if\, a_1 \le x \le a_2 \\ \frac{a_3-x}{a_3-a_2}, & i\, a_2 \le x \le a_3 \end{cases} \quad (1)$$

where $x, a_1, a_2, a_3 \in R$.

Linguistic expressions are represented with words or sentences. Linguistic terms and corresponding triangular values used in this study are shown in Table 1.

Let $\widetilde{A} = (a_1, a_2, a_3)$ and $\widetilde{B} = (b_1, b_2, b_3)$ be two TFNs. The three arithmetic operations utilized in the present work for TFNs are:

$$\widetilde{A} \oplus \widetilde{B} = (a_1 + b_1, a_2 + b_2, a_3 + b_3) \quad (2)$$

$$\widetilde{A} \ominus \widetilde{B} = (a_1 - b_3, a_2 - b_2, a_3 - b_1) \quad (3)$$

$$\widetilde{A} \otimes \widetilde{B} = (a_1 \times b_1, a_2 \times b_2, a_3 \times b_3) \quad (4)$$

Once two TFNs are summed, again a TFN is obtained. Likewise, when one TFN is subtracted from another TFN, a TFN is obtained again.

Assume $\widetilde{A}_i = (a_i, b_i, c_i)$ and $\widetilde{B}_i = (d_i, e_i, f_i)$ where a_i, b_i, c_i and d_i, e_i, f_i are all positive numbers where $i = 1, 2, \ldots, n$.

Table 1 Linguistic expressions and corresponding fuzzy numbers

Linguistic assessment variables	Characteristic function of fuzzy numbers
Very high (VH)	(7, 9, 10)
High (H)	(5, 7, 9)
Medium (M)	(3, 5, 7)
Low (L)	(1, 3, 5)
Very low (VL)	(0, 1, 3)

$$\widetilde{A}_1 \otimes \widetilde{B}_1 \oplus \widetilde{A}_2 \otimes \widetilde{B}_2 \oplus \ldots \oplus \widetilde{A}_n \otimes \widetilde{B}_n = \left(\sum_{i=1}^{n} a_i d_i, \sum_{i=1}^{n} b_i e_i, \sum_{i=1}^{n} c_i f_i \right) \quad (5)$$

3.2 *Quality Function Deployment (QFD)*

QFD is a very important planning tool for development of new products by taking into account CRs. QFD performs several matrices to represent relationships between success factors of the company for customer satisfaction. These matrices are constructed based on HOQ, which is key planning part of QFD. HOQ is implemented by a multidisciplinary product design and development team to identify CRs and ECs [31].

HOQ begins with the determination of CRs which means 'WHAT' in HOQ. The assignment of priorities is obtained based on VOC. Then, the design requirements or ECs which means 'HOW' are determined with respect to the CRs. The relationships among vertical column of WHATs (CRs) and horizontal row of HOWs (ECs) are established [32]. Using the symbols given in Fig. 2 the roof of the matrix is built considering correlation of HOWs. Later, the weights of the ECs are determined and located in HOQ. The last step is competitive analysis. To apply the competitive analysis, the decision makers (DMs) evaluate the performance ratings of the company's product and its competitor's product for each ECs by means of linguistic variables.

Figure 1 represents the typical chart of the HoQ [33, 34].

- Area B shows the relative importance of CRs.
- Area C represents ECs or technical specifications, which means 'how' the product or services have to be made.
- Area D represents relationship matrix, which is the core element of the HOQ.
- Area E shows weight of the ECs.
- Area F represents correlation matrix, which expressed how ECs are correlated to each other.
- Area G shows target values of ECs with respect to competitors.

In the following, basic steps of QFD are explained:

Step 1. Determine CRs. The CRs of product design which form list of 'WHATs' are identified. This step is critical as it requires obtaining and expressing what the customer expect from a new product.
Step 2. Identify the ECs. The technical criteria which form list of external variables or 'HOWs' are identified with respect to product design.
Step 3. Determine the relative importance of the 'WHATs'. In this step, the relative importance of CRs are determined.
Step 4. Obtain the "WHAT"–"HOW" correlation matrix and constructing the HOQ. The correlation matrix between CRs and ECs are determined.
Step 5. Determine the weight of the 'HOWs'. The weights of ECs are determined.

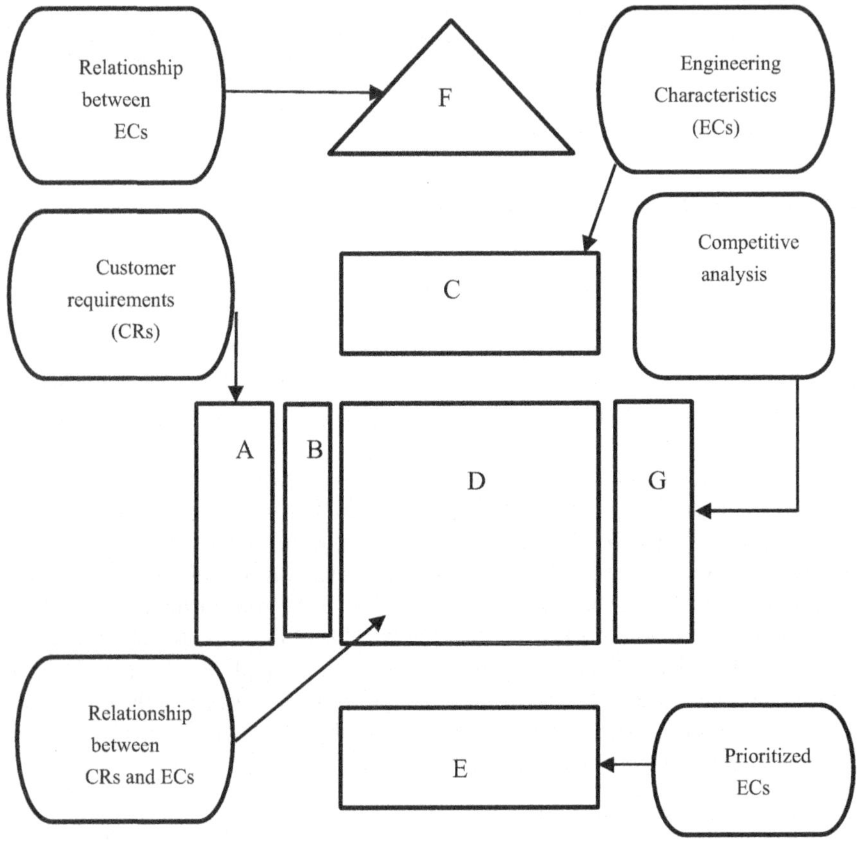

Fig. 1 House of quality (HoQ) [33, 34]

Step 6. Prepare the matrix for correlating the 'HOWs'. The correlation between ECs are determined and constructed in the roof of HOQ.
Step 7. Determine each potential competitor's impact on the factors considered HOWs. The designed product is compared with potential competitors with respect to ECs.
Step 8. Determine the target values for the ECs. Overall priorities and performance values of ECs are determined.

4 Fuzzy-QFD Methodology

Since QFD methodology has been successfully adopted in the new product development, to handle the vague nature of product development processes, FST is integrated with QFD process in this chapter. In the following, the implementation of

Fuzzy QFD methodology based on Bevilacqua's Fuzzy QFD approach is presented for smart phone design [35].

Step 1. Identify the CRs and ECs to satisfy CRs. CR and ECs to satisfy CRs are identified by DMs with respect to the manufacturing and performance of the product/service.
Step 2. Use linguistic variables to weight the 'WHATs' and translate into fuzzy numbers. To determine the degree of importance (or weight) of each 'WHAT', the linguistic variables and corresponding TFNs given in Table 1 are used. The aggregated weight of each WHAT' attribute is calculated using Eq. (6).

$$\text{WEIGHT}_{WHAT} = \{w_i, where\, i = 1, \ldots, k\}$$

$$w_i = \frac{1}{n} \otimes (w_{i1} \oplus w_{i2} \oplus \ldots \oplus w_{in}) \tag{6}$$

where k is the number of 'WHAT's, n is number of DMs.

Step 3. Construct the "HOW"–"WHAT" relationship matrix and weight the 'HOWs'. Each DM express their judgment on the impact of each 'HOW' on each 'WHAT' using linguistic variables as given in Table 1. The evaluation results establish the relationships matrix. The RATING is the average correlation score in the matrix of the 'HOW'–'WHAT'.

r_{ij} is the TFN assigned by DM and shows an aggregate correlation score between ith 'what' and the jth 'how'.

$$RATING = \left\{r_{ij}, where\, i = 1, \ldots, k\, and\; j = 1, \ldots, m\right\}$$

$$r_{ij} = \frac{1}{n} \otimes \left\{r_{ij1} \oplus r_{ij2} \oplus \ldots r_{ijn}\right\} \tag{7}$$

The weight of each 'HOW' attribute is calculated using Eq. (8).

$$\text{WEIGHT}_{HOW} = \left\{W_j, where\; j = 1, \ldots, m\right\}$$

$$W_j = \frac{1}{k} \otimes \left[\left(r_{j1} \otimes w_1\right) \oplus \ldots \oplus \left(r_{jk} \otimes w_k\right)\right] \tag{8}$$

where k is the number of 'WHAT's, m is number of 'HOW's, n is number of DMs.

Step 4. Develop 'HOWs' correlation matrix. The correlations between the 'HOWs' attributes are determined and placed in the roof of HOQ using symbols as shown in Fig. 2.

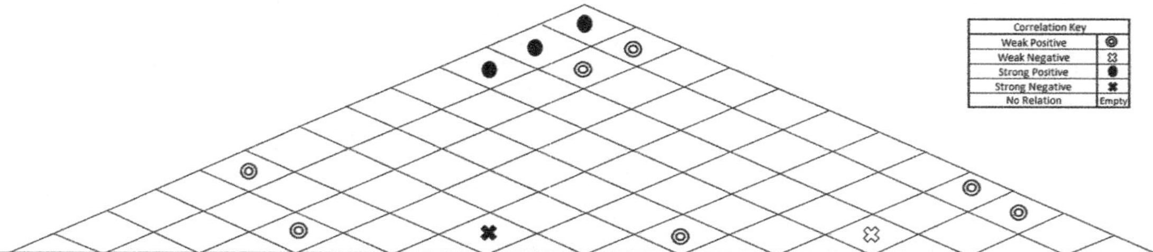

Customer Requirements (WHATs)	WHATs (weight)			Hardware (HW) design			Software(SW) design			High MP camera resolution			LED flash		
Ease of Use	5	7	8.67	6.33	8.33	9.67	5.67	7.67	9.33				0.33		
High camera resolution	4.33	6.333	8.33	3.67	5.67	7.67	3.67	5.67	7.67	7	9	10	6.33	8.33	9.67
Long lasting battery life	5.67	7.667	9.33	5.67	7.67	9.33	2.33	4.33	6.33	1			0.33		
Durable screen	7	9	10	4.67											
Network performance	6.33	8.333	9.67	1.67			5	7	9						
Fast operating system	7	9	10	2.33			7	9	10						
High internal memory	3.67	5.667	7.67	6.33	8.33	9.67	1.67								
Variety of applications	3.67	5.667	7.67	2.33			5.67	7.67	9.33						

Customer Requirements (WHATs)	High battery capacity			Fast charging battery			Tempered glass screen protector			Material (dust and water resistant)		
Ease of Use	5.67	7.67	9.33	5	7	9	4.33	6.33	8.33	5	7	9
High camera resolution												
Long lasting battery life	7	9	10	5.67	7.67	9.33						
Durable screen							7	9	10	7	9	10
Network performance	0.33											
Fast operating system												
High internal memory												
Variety of applications												

Customer Requirements (WHATs)	Network technology supported by device			High RAM			Latest OS			Connection capabilities such as (Bluetooth, WIFI, A/GPS)		
Ease of Use	5.67	7.67	9.33	5	7	9	5.67	7.67	9.33	3.67	5.67	7.67
High camera resolution				0.33			5.67	7.67	9			
Long lasting battery life	0.33			0.33			0.33			0.33		
Durable screen												
Network performance	6.33	8.33	9.67	2.67			7	9	10	6.33	8.33	9.67
Fast operating system	3.33			2.33			7	9	10	3.33		
High internal memory				7	9	10	3.33					
Variety of applications	3.33			5	7	8.67	7	9	10	3.33		

W_1			W_2			W_3			W_4			W_5			W_6		
α	β	γ	α	β	γ	α	β	γ	α	β	γ	α	β	γ	α	β	γ
21.4	25	38.6	20.6	38.2	57.8	4.5	7.13	10.4	3.88	6.6	10.1	8.76	15.3	21.8	7.14	13.5	20.6

W_7			W_8			W_9			W_{10}			W_{11}			W_{12}		
α	β	γ	α	β	γ	α	β	γ	α	β	γ	α	β	γ	α	β	γ
8.83	15.7	21.5	9.25	16.3	22.3	13.2	15.4	21.8	13.2	17.5	27.6	23.3	38.7	53.7	12	13.6	20

Fig. 2 The HOQ of smart phone design

Step 5. Determine the strength of each competitor on the ECs attributes. Each DM compare the various competitors in terms of attributes of ECs using linguistic variables and corresponding fuzzy numbers as given in Table 1. The rating of competitors with respect to ECs are calculated using Eq. (9).

$$COMPETITOR_{RATING} = \{CR_{hj}, where\, h = 1, \ldots, p, j = 1, \ldots, m\}$$

$$CR_{hj} = \frac{1}{n} \otimes \{cr_{hj1} \oplus cr_{hj2} \oplus \ldots cr_{hjn}\} \tag{9}$$

where m is the number of attributes 'HOWs', p is the number of competitors, n is the number of DMs. cr_{hjn} shows the TFN denoted by the nth DM for hth competitor with respect to jth EC attribute. CR_{hj} is identified by TFN and shown with $CR_{hj} = (CR_{hj\alpha}, CR_{hj\beta}, CR_{hj\gamma})$.

Step 6. Rank the competitors. The competitor strength in terms of meeting ECs is obtained using Eq. (10). To rank the degree to which each competitor satisfies a given requirement, fuzzy suitability index (FSI) is calculated.

$$FSI = \{FSI_h\, where\, h = 1, \ldots, p\}$$

$$FSI_h = \frac{1}{m} \otimes [(AR_{h1} \otimes W_1) \otimes \ldots \otimes (AR_{hm} \otimes W_m)] \tag{10}$$

FSI_h scores which are TFNs for each competitor are defined by $FSI_h = (FSI_{h\alpha}, FSI_{h\beta}, FSI_{h\gamma})$ and the calculations are as follows:

$$FSI_{h\alpha} = \frac{1}{m} \sum_{j=1}^{m} SR_{hj\alpha} \cdot W_{j\alpha} \tag{11}$$

$$FSI_{h\beta} = \frac{1}{m} \sum_{j=1}^{m} SR_{hj\beta} \cdot W_{j\beta} \tag{12}$$

$$FSI_{h\gamma} = \frac{1}{m} \sum_{j=1}^{m} SR_{hj\gamma} \cdot W_{j\gamma} \tag{13}$$

5 Illustrative Example

In this section, to illustrate applicability of Fuzzy QFD methodology in production design, a smart phone is designed and developed. CRs of a new smart phone is identified by conducting a market survey. ECs are specified by a multidisciplinary team. The steps of fuzzy QFD are implemented as follows:

Step 1. Determine CRs and ECs to satisfy CRs for developing smart phone. CRs and ECs to satisfy CRs are determined by conducting a market survey. The CRs and ECs for smart phone design are identified as given in Table 2.
Step 2. Weigh the WHATs. CRs are evaluated using linguistic scale and then converted into fuzzy numbers as given Table 1. The weights for each WHAT given by DMs are shown in Table 3. After determining ECs to satisfy CRs, HOQ for the smart phone is built as given in Fig. 2. The aggregated weight of each CRs is calculated using Eq. (6). The weights acquired by aggregating the opinion expressed by each DM is illustrated in Fig. 2.
Step 3. Construct relationship matrix and determine weight of 'HOWs'. The correlation between CRs and ECs to design smart phone is constructed by DMs as a relationship matrix using linguistic variables as given in Table 1. The judgments expressed by the three DMs are shown in Table 4. The aggregated correlation score between CRs and ECs in the 'HOW'–'WHAT' matrix is calculated using Eq. (7).

Table 2 Identified WHATs and HOWs for smart phone design

CRs (WHATs)	ECs (HOWs)
Ease of use (EU)	Hardware (HW) design
High camera resolution (HCR)	Software (SW) design
Long lasting battery life (LLF)	High MP camera resolution
Durable screen (DS)	LED flash
Network performance (calling, being called, data) (NP)	High battery capacity
Fast operating system (FOS)	Fast charging battery
High internal memory (HIM)	Tempered glass screen protector
Variety of applications (VA)	Material (dust and water resistant)
	Network technology supported by device
	High RAM
	Latest OS
	Connection capabilities such as Bluetooth, WiFi, A/GPS

Table 3 The weights for each WHAT given by DMs

WHAT's	DM1	DM2	DM3
Ease of use	H	VH	M
High camera resolution	H	M	H
Long lasting battery life	VH	H	H
Durable screen	VH	VH	VH
Network performance (calling, being called, data)	VH	H	VH
Fast operating system	VH	VH	VH
High internal memory	H	M	M
Variety of applications	M	H	M

The weight of each 'HOW' attribute is determined using Eq. (8). The results are seen in Fig. 2.
Step 4. Develop correlations matrix between ECs. The correlation matrix between ECs are built in the roof of HOQ using the symbols as shown in Fig. 2.
Step 5. Determine strength of manufacturers with respect to ECs. To design a new smart phone, three manufacturers are evaluated in terms of meeting ECs. In the same way as before, the linguistic variables are expressed via TFNs as given in Table 1. Once DM's judgments for the various manufacturers are determined with respect to each ECs, DMs' evaluations for each manufacturer are aggregated using Eq. (9).
Step 6. Rank the manufacturers. Manufacturer analysis can assist the DMs to identify the strength, weakness, market position of the manufacturer's product. FSI_h is calculated for each manufacturer using Eqs. (11–13). This index represents the degree to which each manufacturer fulfills the customer wants and needs. The results are seen in Table 5. To rank each manufacturer according to their score, FSI_h should be defuzzified. Yager (1981) proposed the defuzzify procedure to provide simplicity in ranking. The crisp value *(CV)* of any TFN (a, b, c) is computed using Eq. (14).

$$CV = \frac{a + 2b + c}{4} \tag{14}$$

Using the fuzzy ranking principle, after defuzzification the ranking order for the manufacturers is obtained as in Table 6.

The results obtained from the case study shows that ECs that have large influence on the CRs such as "Network performance", "Fast operating system" and "High internal memory" in the design process of smart phone. The final ranking of ECs in the light of CRs of the new smart phone production design and development represents that: "Latest OS" come the first, "Software (SW) design" come the second position, the ECs of "Hardware (HW) design" come the third position and finally the ECs of "LED flash" in the last position. Considering crisp ECs, the manufacturers are ranked as M3 $\succ$ M1 $\succ$ M4 $\succ$ M2. Then, Manufacturer 3 is evaluated as the best manufacturer which meets ECs among the rest.

Table 4 The opinions expressed by the three DMs between WHATs and HOWs

WHATs	Hardware (HW) design			Software (SW) design			High MP camera resolution			LED flash			High battery capacity			Fast charging battery		
	DM1	DM2	DM3	DM1	DM2	DM3	DM1	DM2	DM3	DM1	DM2	DM3	DM1	DM2	DM3	DM1	DM2	DM3
EU	VH	H	VH	H	H	VH						L	H	VH	H	H	H	H
HCR	M	M	H	H	H	L	VH	VH	VH	VH	VH	H						
LLF	VH	H	H	M	M	L			M			L	VH	VH	VH	H	H	VH
DS	VH	VH																
NP			H	H	H	H									L			
FOS			VH	VH	VH	VH												
HIM	H	VH	VH			H												
VA			VH	H	H	VH												

WHATs	Tempered glass screen protector			Material (dust and water resistant)			Network technology supported by device			High RAM			Latest OS			Connetion capabilities such as (Bluetooth, Wi-Fi, A/GPS)		
	DM1	DM2	DDM3	DM1	DM2	DM3	DM1	DM2	DM3	DM1	DM2	DM3	DM1	DM2	DM3	DM1	DM2	DM3
EU	H	H	M	H	H	H	H	VH	H	H	H	H	H	H	VH	H	H	L
HCR			VL						VL			L	VH	VH	M			
LLF									L			L			L			L
DS	VH	VH	VH	VH	VH	VH												
NP							VH	H	VH	M	H		VH	VH	VH	VH	VH	H
FOS							H	H				VH	H	H	VH	H	H	
HIM										VH	VH	VH	H	H				
VA						H	H		H	VH	M	VH	VH	VH	H	H		

Table 5 The *FSI* for each manufacturer

FSI			
Manufacturer	α	β	γ
M1	53.839	114	214
M2	48.313	106	204
M3	58.859	121	220
M4	51.412	108	206

Table 6 The ranking order of manufacturers

Manufacturer	Score	Ranking
M1	124.12	2
M2	116.03	4
M3	130.07	1
M4	118.65	3

6 Conclusion

QFD has been widely applied in various fields due to its effectiveness particularly with the support of fuzzy logic. In this chapter, to illustrate the applicability of fuzzy QFD methodology in the product design successfully, a new smart phone design is conducted by performing fuzzy QFD methodology. To do this, the identified CRs for different customer profiles are used to translate them into ECs that can support product or service from idea to detailed designs and ultimately manufacturing process. As a result, the case study validates that fuzzy QFD is a powerful tool to maximize customer satisfaction for the development of new products. Thus, fuzzy QFD approach can provide an effective procedure for improving a new product design characteristics, enhancing customer satisfaction and gaining competitive advantage in the marketplace. However, different fuzzy extensions can be integrated with QFD method to cope with uncertainty better in further studies.

References

1. Jeddi, K., Haghshenas, M.: Design of new product specification with fuzzy AHD & QFD (Case Study: Digital scales Production Company). Int. Acad. J. Bus. Manag. **2**(10), 38–44 (2015). ISSN 2454-2768
2. Kazımıerska, M., Krawczyk, M.: New product development (Npd) process—an example of industrial sector. Manag. Syst. Prod. Eng. **25**(4), 246–250
3. Liu, H.: Product design and selection using fuzzy QFD and fuzzy MCDM approaches. Appl. Math. Model. **35**(2011), 482–496 (2011)
4. Chen, L., Ko, W.: Fuzzy approaches to quality function deployment for new product design. Fuzzy Sets Syst. **160**(2009), 2620–2639 (2009)
5. West, B.M., Olhager, J.: The house of flexibility: using the QFD approach to deploy manufacturing flexibility. Int. J. Oper. Prod. Manag. **22**, 50–79 (2002)

6. Karsak, E.: Fuzzy multiple objective programming framework to prioritize design requirements in quality function deployment. Comput. Ind. Eng. **47**(2004), 149–163 (2004)
7. Akao, Y.: Quality Function Deployment: Integrating Customer Requirements into Product Design. Productivity Press, Cambridge, MA (1990)
8. Bouchereau, V., Rowlands, H.: Quality function deployment: the unused tool. Eng. Manag. J. 45–52 (2000)
9. Tang, J., Fung, R.Y.K., Baodong, X., Wang, D.: A new approach to quality function deployment planning with financial consideration. Comput. Oper. Res. **29**, 1447–1463 (2002)
10. Nahm, Y.E., Ishikawa, H., Inoue, M.: New rating methods to prioritize customer requirements in QFD with incomplete customer preferences. Int. J. Adv. Manuf. Technol. **65**(9–12), 1587–1604 (2013). https://doi.org/10.1007/s00170-012-4282-1
11. Lam, J.S.L.: Designing a sustainable maritime supply chain: a hybrid QFD–ANP approach. Transp. Res. Part E Logist. Transp. Rev. **2015**(78), 70–81 (2015). https://doi.org/10.1016/j.tre.2014.10.003
12. Lombardi, M., Fargnoli, M.: Prioritization of hazards by means of a QFD-based procedure. Int. J. Saf. Secur. Eng. **8**(2), 342–353 (2018). https://doi.org/10.2495/safe-v8-n2-342-353
13. Chen, L.H., Weng, M.C.: A fuzzy model for exploiting quality function deployment. Math. Comput. Model. **38**, 559–570 (2003)
14. Liu, S.: Rating design requirements in fuzzy quality function deployment via a mathematical programming approach. Int. J. Prod. Res. **43**(3) (2005)
15. Chen, L.H., Ko, W.C.: A fuzzy nonlinear model for quality function deployment considering Kano's concept. Math. Comput. Model. **48**(3–4), 581–593 (2008)
16. Chen, L.H., Weng, M.C.: An evaluation approach to engineering design in QFD processes using fuzzy goal programming models. Eur. J. Oper. Res. **172**, 230–248 (2006)
17. Kahraman, C., Ertay, T., Buyükozkan, G.: A fuzzy optimization model for QFD planning process using analytic network approach. Eur. J. Oper. Res. **171**(2), 390–411 (2006)
18. Lin, L.Z., Chen, W.C., Chang, T.J.: Using FQFD to analyze island accommodation management in fuzzy linguistic preferences. Expert Syst. Appl. **38**(6), 7738–7745 (2011)
19. Fung, R.K., Chen, Y., Tang, J.: Estimating the functional relationships for quality function deployment under uncertainties. Fuzzy Sets Syst. **157**, 98–120 (2006)
20. Buyukozkan, G., Feyzioglu, O., Ruan, D.: Fuzzy group decision making to multiple preference formats in quality function deployment. Comput. Ind. **58**, 392–402 (2007)
21. Zaim, S., Sevkli, M., Akdağ, H., Demirel, O., Yayla, Y., Delen, D.: Use of ANP weighted crisp and fuzzy QFD for product development. Expert. Syst. Appl. **41**(9), 4464–4474 (2014)
22. Younesi, M., Roghanian, E.: A framework for sustainable product design: a hybrid fuzzy approach based on quality function deployment for environment. J. Clean. Prod. **108**(Part A), 385–394 (2015)
23. Yang, S., Ong, S.K., Nee, A.Y.C.: Design for remanufacturing—a fuzzy-QFD approach. In: Nee, A., Song, B., Ong, S.K. (eds.) Re-engineering Manufacturing for Sustainability. Springer, Singapore (2013)
24. Hsuan, C., Chang, K., Ting, Y.: Approach based on fuzzy goal programing and quality function deployment for new product planning. Eur. J. Oper. Res. **259**(2), 654–663
25. Vinodh, S., Manjunatheshwara, K.J., Karthik Sundaram, S., Kirthivasan, V.: Application of fuzzy quality function deployment for sustainable design of consumer electronics products: a case study. Clean Technol. Environ. Policy **19**(4), 1021–1030 (2017)
26. Wu, S., Liu, H., Wang, L.: Hesitant fuzzy integrated MCDM approach for quality function deployment: a case study in electric vehicle. Int. J. Prod. Res. **55**(15), 4436–4449 (2017). https://doi.org/10.1080/00207543.2016.1259670
27. Zadeh, L.A.: Fuzzy sets. Inform. Control **8**(3), 338–353 (1965)
28. Wang, L.X.: A Course in Fuzzy Systems and Control. Prentice-Hall International Inc. (1997)
29. Lee, K.: First Course on Fuzzy Theory and Applications. Springer (2005). ISBN 978-3-540-32366-2
30. Gani, A., Assarudean, S.: A new operation on triangular fuzzy number for solving fuzzy linear programming problem. Appl. Math. Sci. **6**(11), 525–532 (2012)

31. Temponi, C., Yen, J., Tiao, W.: House of quality: a fuzzy logic-based requirements analysis. Eur. J. Oper. Res. **117**(1999), 340–354 (1999)
32. Vinodh, S., Chintha, S.K.: Application of fuzzy QFD for enabling sustainability. Int. J. Sustain. Eng. **4**(4), 313–322 (2011). https://doi.org/10.1080/19397038.2010.536853
33. Kusiak, A.: Qualıty Functıon Deployment And Process Models, Engineering Design: Products, Processes, and Systems (Chapter 6), p. 1999. Academic Press, San Diego, CA (1999)
34. Ding, J.: Applying fuzzy quality function deployment (QFD) to identify solutions of service delivery system for port of Kaohsiung. Qual. Quant. **43**, 553–570 (2009). https://doi.org/10.1007/s11135-007-9138-7
35. Bevilacqua, M., Ciarapica, F., Giacchetta, G.: A fuzzy-QFD approach to supplier selection. J. Purch. Supply Manag. **12**(2006), 14–27 (2006)

An Integrated Fuzzy QFD Methodology for Customer Oriented Multifunctional Power Bank Design

Gülçin Büyüközkan, Merve Güler and Esin Mukul

Abstract Due to the intense global competition and rapid advances in technology, companies are forced to continually invest into the new product development (NPD). The main reason for NPD is offering new value to customers. In this context, customer-oriented product development is crucial. The purpose of the article is to propose a new method for determining and prioritizing the most important design requirements for customer-oriented product design. Quality Function Deployment (QFD) tool is selected for this purpose; since it efficiently considers customer needs. Fuzzy logic is applied thanks to its ability to represent uncertain information. It enables decision makers (DMs) to assess different elements with linguistic term sets. A case study about power banks is illustrated to test the usefulness of the method. Power banks are saviors when one has no access to power outlets, and are gradually entering our lives as accessories. However, there are various power bank models in the market and manufacturers desire to produce the most satisfying power banks for their customers. Therefore, in this chapter, an integrated fuzzy QFD and fuzzy multi criteria decision making (MCDM) method are implemented for customer-oriented multifunctional power bank design. We applied the fuzzy Analytic Hierarchy Process (AHP) technique to determine the customer needs' weights and the fuzzy QFD technique to find the most important design requirements in multifunctional power bank design. The results of the case study are shared in the final section of the chapter.

1 Introduction

Companies invest in new product development (NPD) to compete and to maintain their existence in the market. NPD is crucial since products and services are the lifeblood for companies and societies. The key point is developing products that create enhanced value for customers. Therefore, customer-oriented design is an essential objective of companies. Quality Function Deployment (QFD) is a capable method that aims to improve products according to customer needs by translating them into

G. Büyüközkan (✉) · M. Güler · E. Mukul
Department of Industrial Engineering, Galatasaray University, 34349 Istanbul, Turkey
e-mail: gbuyukozkan@gsu.edu.tr

C. Kahraman and S. Cebi (eds.), *Customer Oriented Product Design*, Studies in Systems, Decision and Control 279, https://doi.org/10.1007/978-3-030-42188-5_5

suitable technical requirements. QFD is widely used in various decision-making fields [1].

Connectedness has become a basic need since people always desire to stay connected anytime and anywhere. Portable electronic devices (e.g. smartphones, tablets, and smart watches) are widely used in our daily lives. According to International Data Corporation, 1.4 billion smartphones were sold in 2018 [2]. With the increased number of new smartphones, high energy consumption due to device charging access to power outlets can sometimes be limited [3, 4]. Thus, power banks provide an alternative solution to this problem [5].

Power banks have become an important accessory and various power bank models are introduced day-by-day [6]. A power bank can be defined as a portable charger with a charging circuit, discharging circuit, and rechargeable batteries [4]. In the design of power banks, the most customer-satisfying product design should be the priority of companies. Different customer needs have to be considered (e.g. safety, capacity, and power). Here, the problem is to decide on the most important technical components that meet the highest range of the customer needs. So, selecting technical components is a crucial step for production and design processes [7].

In this chapter, it is aimed to introduce a new integrated method for identifying and prioritizing the most important design conditions for customer-oriented product design. To meet this purpose, the QFD's first phase (i.e. House of Quality-HOQ) is implemented. To achieve an efficient HOQ, this tool is combined with a fuzzy multi-criteria decision making (MCDM) technique. Moreover, fuzzy logic is preferred since it considers vague information and enables decision makers (DMs) to present their evaluations with linguistic term sets. MCDM methodology is applied since the problem consists of many different components. Fuzzy Analytic Hierarchy Process (AHP) technique is employed to prioritize customer needs. This algorithm is then applied for determining the most important technical elements in a power bank design.

The chapter is structured in the following order. First, Sect. 2 provides an overview of the state of the art on power bank and QFD and fuzzy MCDM tools. Section 3 describes the research method and Sect. 4 illustrates the utilization of this method on a case study involving multifunctional power bank design. Finally, concluding remarks and future directions are summarized.

2 Literature Survey

2.1 Power Bank

Portable electronic devices have become an indispensable part of our lives, since people desire to be connected anywhere, any time. New smartphones drain the battery faster with their large screens and advanced features; people have to frequently charge their devices. Power banks are saviors when charging in power outlets is not possible and are increasingly accepted as important accessories in daily life.

Many studies are conducted about power banks recently. Liu et al. [3] conducted a survey to analyze the main problems of traditional charging and to investigate the acceptance of portable power bank sharing. To draw attention to reducing environmental impacts, dos Reis Benatto et al. [8] compared the environmental impacts of solar powered and regular power banks. In another study, Palczynska [9] measured electromagnetic emissions of a portable power bank. Thakur et al. [10] fabricated a prototype that harvests electrical energy from clean resources of nature.

Some of these studies investigate technical configurations of devices. For instance, Moo et al. [11] proposed a power bank configured for buck-boost type battery modules. Another paper by Hou et al. [12] explored battery power module configurations. Also, Wu et al. [13] presented a battery power bank that equalizes charging, regulates voltage, and realizes fault tolerance. In this study, [4] the interoperability capabilities between a power bank and computer USB ports are examined.

Several studies have developed novel systems. Xie et al. [14] designed a self-charged power bank that harvests energy from humans to sustain recharging. Haselton [2] produced a self-powered power bank with an integrated hybrid nano-generation unit that harvests ambient mechanical energy. He [5] built an Arduino system to allow the measurement of portable power banks in laboratories.

This literature survey on power banks shows that power bank design and configuration have been investigated in several studies [2, 4, 5, 11–14]. Nevertheless, no previous study analyzed power bank design with QFD methodology. This literature gap of using analytical tools (e.g. House of Quality (HOQ)) in the design of power banks prompted this chapter to develop this method.

2.2 QFD and Fuzzy MCDM

QFD is a customer-oriented method for maximizing customers' satisfaction. The HOQ matrix, which is part of the QFD, is a process that facilitates structured product development and improvement. This way, customer needs are converted into technical characteristics. Considering the complex profile of this problem, many evaluations should be taken into account by experts for deciding on a suitable solution. However, it is difficult to decide on the most appropriate solution when information is of uncertain nature [15–17]. For this reason, fuzzy logic is combined with QFD to handle such uncertainty.

One of the first studies using fuzzy logic principles in the QFD method was made by Khoo and Ho [15]. In this study, it was stated that the importance of linguistic variables in the QFD process was neglected and then the probability theory and fuzzy arithmetic based QFD applications were developed. The studies involving fuzzy MCDM and QFD methods are given in Table 1.

Table 1 Literature survey of fuzzy MCDM and QFD

Year	Author	Integrated techniques	Application area
2019	Liu et al. [18]	Fuzzy QUALIFLEX	Robot development
2018	Aboutorab et al. [19]	Fuzzy BWM	Supplier development
2018	Suen et al. [6]	Fuzzy COPRAS	Material selection for a detergent manufacturer in Turkey
2018	Liu et al. [20]	Fuzzy BWM, fuzzy MULTIMOORA	Selecting suppliers of fresh green products
2018	Peng et al. [21]	HFLTS	Hydraulic retarder project
2018	Tian et al. [22]	Fuzzy MULTIMOORA	Performance evaluation for smart bike-sharing programs in China
2018	Tsai et al. [23]	Fuzzy AHP, fuzzy DEMATEL	Logistics center service quality
2018	Van et al. [24]	Fuzzy TOPSIS	Green supplier evaluation and selection
2018	Yu et al. [25]	Fuzzy Choquet integral	Design of steering wheel for electric vehicles
2017	Akbaş and Bilgen [26]	Fuzzy ANP, fuzzy AHP	Ideal gas fuel selection in Turkey
2017	Alkhatib [27]	Fuzzy DEMATEL	Strategic logistics outsourcing
2017	Hsu et al. [28]	Fuzzy Delphi, fuzzy AHP, fuzzy TOPSIS	KPIs for sustainability of SMEs
2017	Pamucar et al. [29]	Interval rough DEMATEL, MAINCA, ANP	Bidder evaluation in the public procurement procedure
2017	Roy et al. [30]	Fuzzy AHP	Non-traditional machining (NTM) processes in manufacturing
2016	Arsenyan and Büyüközkan [31]	Fuzzy AD	Information planning in collaborative product development
2016	Ignatius et al. [32]	Fuzzy ANP	Green building evaluation in Malaysia
2015	Anusha et al. [33]	Fuzzy DEMATEL, fuzzy TOPSIS	Evaluation of social networking sites in India
2015	Büyüközkan and Çiftçi [34]	Multi-preference relations	Portable entertainment and game System design
2015	Büyüközkan and Güleryüz [35]	Multi-preference relations	IT planning

(continued)

Table 1 (continued)

Year	Author	Integrated techniques	Application area
2015	Wang [36]	Fuzzy DEMATEL, fuzzy AHP	Performance measurement of Business Intelligence (BI) systems
2015	Karsak and Dursun [37]	OWA, 2-tuple linguistic approach	Supplier evaluation and selection
2015	Dursun and Karsak [38]	FWA (fuzzy weighted average)	Selection of medical suppliers
2014	Li et al. [39]	Intuitionistic fuzzy TOPSIS	Knowledge management system selection from a user point of view
2014	Onut and Tosun [40]	Fuzzy ANP, fuzzy TOPSIS	Supplier selection in the banking sector, Turkey
2013	Ayağ et al. [41]	Multi-objective mathematical programming	Supply chain management strategies for the dairy industry
2013	Büyüközkan and Çiftçi [42]	Multi-preference relations	Sustainable supply chain
2013	Dursun and Karsak [43]	FWA (fuzzy weighted average)	Supplier selection
2013	Wey and Chiu [44]	Fuzzy ANP	Urban design of pedestrian space in New Taipei City
2012	Büyüközkan and Çiftçi [45]	Multi-preference relations	Collaborative software development process
2012	Wang and Chen [46]	Fuzzy Delphi, fuzzy DEMATEL	Collaborative product design and optimal selection of module mix
2011	Büyüközkan and Berkol [47]	ANP, Zero-one goal programming	Designing a sustainable supply chain
2011	Liu [48]	Fuzzy AHP	Product design and selection
2010	Bhattacharya et al. [49]	Fuzzy AHP	Supplier selection
2008	Feyzioğlu and Büyüközkan [50]	Fuzzy Choquet integral	New product development
2007	Büyüközkan et al. [51]	Multi-preference relations	Development of the hatch door of a car
2006	Kahraman et al. [16]	Fuzzy ANP, fuzzy AHP	Quality improvement for PVC window and door systems company
2005	Büyüközkan and Feyzioğlu [52]	Multi-preference relations	Software development

(continued)

Table 1 (continued)

Year	Author	Integrated techniques	Application area
2005	Ertay et al. [53]	Fuzzy ANP, fuzzy AHP	Product planning stage of a car design
1999	Chan et al. [54]	Fuzzy and entropy methods	Determination of relative importance ratings of customer needs in the automotive industry

According to Table 1, there is a gap in the literature for these methodologies since there exist any study that integrates fuzzy MCDM methods and QFD for multifunctional power bank as an application area. In this study, literature gap is filled by presenting an integrated fuzzy QFD methodology that consists of new criteria and alternatives for multifunctional power bank.

3 The Research Methodology

In this chapter, the aim is to introduce a new methodology for identifying and prioritizing the most important design conditions for customer-oriented product design. In this context, the study proposes an integrated QFD methodology.

QFD's HOQ matrix method is a critical tool for product development. In this context, customer needs and technical requirements for multifunctional power bank are determined.

Secondly, the complex profile of the evaluation needs to be taken into account by experts for deciding on a suitable solution. However, it is difficult to decide on the most appropriate solution when information possess an uncertain nature. Therefore, the method is supported by the fuzzy set theory. This theory is preferred since it considers vague information and enables DMs to voice their evaluations with linguistic term sets.

In this chapter, fuzzy AHP technique is deployed to calculate the weights of customer needs. The relation of customer needs to technological requirements in the HOQ matrix is evaluated. Furthermore, the prioritization of technological requirements are obtained. Its flowchart is provided in Fig. 1.

3.1 Quality Function Deployment

QFD is an important planning, development, communication and design technique that enables the transformation of customers' needs into product or service characteristics in all functional components [55]. It is a decision-making method that

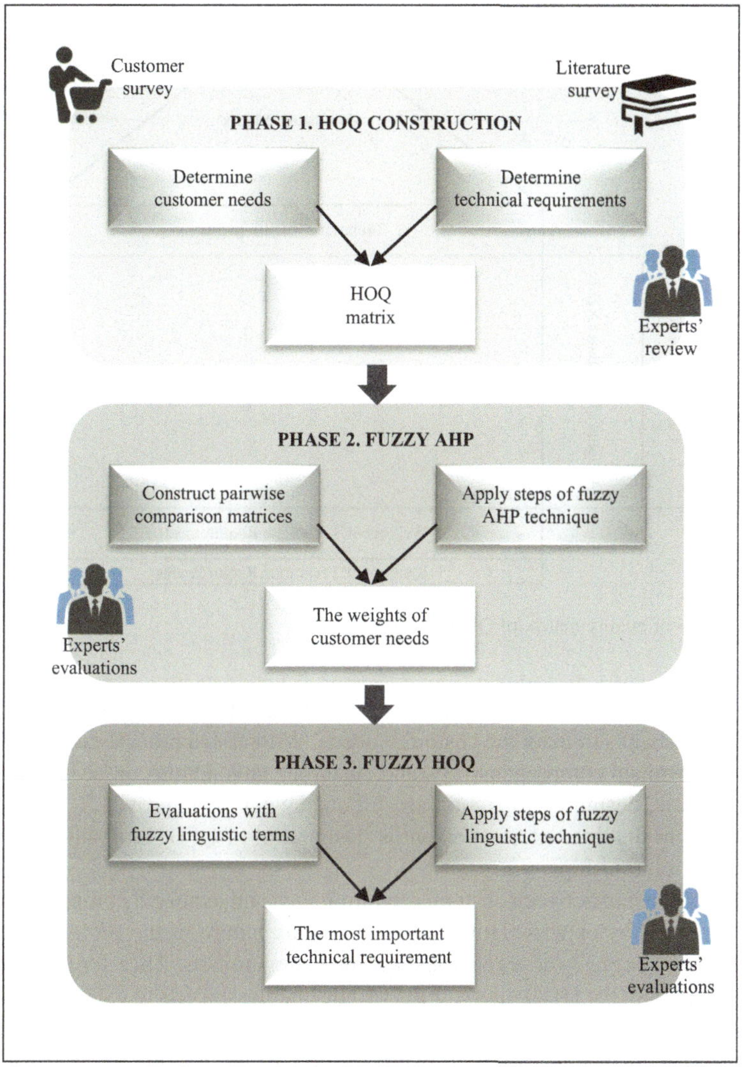

Fig. 1 The flowchart of the integrated methodology

can be employed for designing enhanced products and services which address customers' expectations. This approach aims to translate customer needs into appropriate technical or technological requirements.

HOQ is the most frequently used matrix in the QFD process (Fig. 2). This matrix is conducted by a team of multidisciplinary experts to translate customer needs gathered from market research and benchmark data into technical requirements that are sought when designing a new product or service [56].

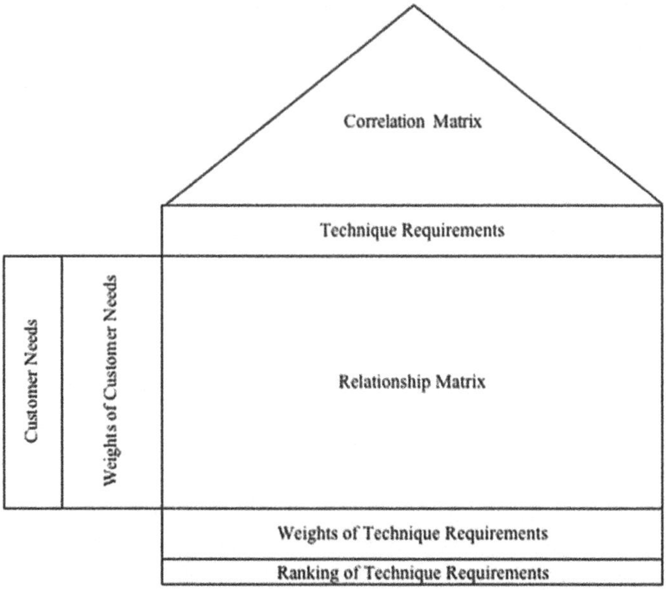

Fig. 2 House of quality matrix of QFD

HOQ matrix of QFD is used to evaluate the technical requirements for multifunctional power banks to meet the customer needs. With this approach, customer needs can be determined comprehensively, and customer satisfaction can be increased. At the same time, design optimization and efficiency can be improved. It is a systematic algorithm that is used as an essential design tool of QFD. It organizes data and establishes relationships.

The relationship between customer requirements and technical components needs to be evaluated by experts. However, the complex nature of the problem makes it challenging to express these relationships by using numbers. Therefore, fuzzy logic is integrated with the HOQ matrix. Fuzzy logic enables experts to express their ideas with linguistic term sets. For instance, experts may evaluate CN1 and TR1 by using a linguistic term (e.g. strongly more important). Moreover, the uncertainty in the evaluation process can be reflected in fuzzy QFD methodology.

3.2 *Fuzzy Set Theory*

The fuzzy set theory is introduced by [57] to simulate human cognitive processes in terms of their vagueness and imprecision. This mathematical theory is applied to decision problems to incorporate vague information. In this theory, an element has a membership degree in a fuzzy set [58]. A triangular fuzzy number is denoted as

Table 2 Table of linguistic scale [60]

Linguistic term	Abb.	TFN
Equal	E	(1, 1, 1)
Equally important	EI	(0.5, 1, 1.5)
Weakly more important	WMI	(1, 1.5, 2)
Strongly more important	SMI	(1.5, 2, 2.5)
Very strongly more important	VSMI	(2, 2.5, 3)
Absolutely more important	AMI	(2.5, 3, 3.5)

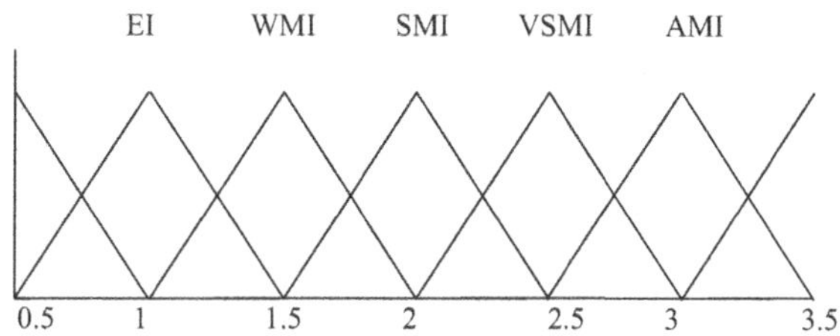

Fig. 3 Table of seven boundaries with its semantics [60]

$\widetilde{A} = (l, m, u)$ where $l \leq m \leq u$. The membership function of $\widetilde{A}$ is given as the following [59]:

$$\mu_{\widetilde{A}}(x) = \begin{cases} \frac{(x-l)}{(m-l)}\, if\ l \leq x \leq m, \\ \frac{(u-x)}{(u-m)}\, if\ m \leq x \leq u, \\ 0\, otherwise. \end{cases} \tag{1}$$

Table 2 and Fig. 3 shows the triangular fuzzy numbers (TFNs) used in this study.

Let $\widetilde{A_1} = (l_1, m_1, u_1)$ and $\widetilde{A_2} = (l_2, m_2, u_2)$ be two triangular fuzzy numbers. Arithmetic operations on fuzzy numbers are defined as:

$$\widetilde{A_1} \oplus \widetilde{A_2} = (l_1 + l_2, m_1 + m_2, u_1 + u_2) \tag{2}$$

$$\widetilde{A_1} \ominus \widetilde{A_2} = (l_1 - u_2, m_1 - m_2, u_1 - l_2) \tag{3}$$

$$c \otimes \widetilde{A_1} = \begin{cases} (cl_1, cm_1, cu_1), c > 0 \\ (cu_1, cm_1, cl_1), c < 0 \end{cases} where\ c\ is\ a\ scalar \tag{4}$$

$$\widetilde{A_1} \otimes \widetilde{A_2} = (l_1 * l_2, m_1 * m_2, u_1 * u_2), if\ l_1 \geq 0, l_2 \geq 0 \tag{5}$$

$$\widetilde{A_1} \oslash \widetilde{A_2} = \left(\frac{l_1}{l_2}, \frac{m_1}{m_2}, \frac{u_1}{u_2}\right), if\ l_1 \geq 0, l_2 > 0 \tag{6}$$

3.3 Fuzzy AHP Method

AHP is pioneered by Saaty [61] as a decision-making methodology that determines the priorities among different criteria. AHP is a powerful technique since it validates consistency and it has a simple concept since it conducts pairwise comparisons. However, AHP is criticized for its inability for taking into account vague information. Therefore, AHP is extended in a fuzzy environment to deal with the perception uncertainty of a decision maker [62–64]. Therefore, Fuzzy AHP method facilitates the decision-making process in uncertain situations. The steps of this method are provided in the following steps [65]:

Step 1: Build the fuzzy comparison matrices by using TFNs in Table 2.
Step 2: Construct α-cut matrices. α-cut incorporates decision makers' confidence over his/her preferences. μ denotes the index of optimism that represents the satisfaction for the judgment matrix. The optimality index is defined as [66]:

$$\tilde{a}_{ij}^{\alpha} = \mu \tilde{a}_{ijl}^{\alpha} + (1 - \mu)\tilde{a}_{ijl}^{\alpha} \quad \forall \alpha \in [0, 1] \tag{7}$$

Step 3: Normalize the matrix and calculate the consistency ratio (*CR*) using:

$$CI = \frac{\lambda \max - n}{n - 1} \tag{8}$$

$$CR = \frac{CI}{RI} \tag{9}$$

where CI stands for the consistency index, λ_{max} is the biggest eigenvector of the matrix, n represents the number of criteria and RI is the random index. *CR* should be less than 0.1 for consistent results.

Step 4: Apply steps 1–3 for each sub-criterion to calculate local weights. Then, calculate final weights, by the multiplication of the weights of sub-criteria by those f the main criteria.

4 Case Study

4.1 Background

With the perpetual changes in customer demands and needs, technology and competition, the market dynamics, and the speed of technological changes enterprises need to come up with new product designs to gain market share and increase their competitiveness.

In this study, it is requested to develop a multifunctional power bank as a product. Today, the majority of people use mobile phones and/or tablets, and the biggest problem of users is to charge their batteries. One of the suitable solutions to this problem is the use of a power bank. Since mobile devices can be charged anywhere, power banks have become very favored products. In this chapter, customer needs and technical requirements of power bank are determined, and fuzzy methods are used for the evaluation of these components. Together with this methodology, it is determined which technical requirement should be emphasized in the improvement of the power bank.

4.2 Proposed House of Quality Model

In this study, it is requested to develop a multifunctional power bank as a product. Firstly, a survey was made to 87 people in order to determine customer needs and experts' opinions and literature review was taken into consideration. In this survey, they were asked to specify gender and age ranges and thus it is aimed to determine the expectations of different customer groups. All other questions are related to power banks. As a result, customer needs were determined as in Table 3.

Performance (CN1): This main criterion includes the criteria relevant to the performance of the power bank such as long-term use, number of charge times, ergonomic, durability and lightness. All of these features are very critical for customers. They want to a long-lasting and comfortable use the power bank. The number of maximum charge cycles is one of the other important points for customers. The number of charge times is related to the performance of the power bank [2–5, 12, 67].

Usability (CN2): The product should be easy to use by the customer to meet the customer needs. This main criterion includes criteria such as ease of use, ease of cleaning, portability, environmental sensitive material, and elegant view/design. Especially the fact that the power bank is portable is very important for the customers since they carry this device with them at any time [6, 8, 14, 24, 65].

Peripherals (CN3): To design the power bank more attractive for customers, there should be some additional features such as led lights, USB socket terminal, cable and adapter, flash lighting and sound system. With the combination of these features, the power bank becomes multifunctional. The addition of lighting is particularly preferred by customers [2, 4, 5, 8, 12, 26, 67].

Our review of the state of the art and experts' opinions provide a basis for the technical requirements given in Table 4.

Table 3 Customer needs of power bank

Customer needs		References
Performance (CN1)	Long-term use (CN11)	[3, 4, 67]
	Number of charge times (CN12)	[3–5, 11, 12]
	Ergonomic (CN13)	[5, 14, 55, 65]
	Durability (CN14)	[2, 10, 67]
	Lightness (CN15)	[2, 8, 10, 14]
Usability (CN2)	Ease of use (CN21)	[5, 10, 14, 65, 68]
	Ease of cleaning (CN22)	[8, 67, 68]
	Portability (CN23)	[2, 8, 14, 67]
	Environmental sensitive material (CN24)	[8, 6, 14, 24]
	Elegant view/design (CN25)	[5, 14, 65]
Peripherals (CN3)	Led lights (CN31)	[14, 26, 67]
	USB socket terminal (CN32)	[2, 4, 5, 67]
	Cable and adapter (CN33)	[2, 8, 10, 67]
	Flash lighting (CN34)	[4, 8, 67]
	Sound system (CN35)	[4, 5, 12, 67]

4.3 Calculation of Customer Needs Weights

Fuzzy AHP method is used to prioritize customer needs. First, pairwise comparison matrices are constructed by using the verbal expressions given in Table 2 voiced by the DMs. Table 5 provides the DMs' linguistic assessments for the main criteria.

Then, the α-cut matrices are constructed by using (7) and by taking the optimality index as 0.5. Table 6 shows the α-cut matrix for the main criteria.

The *CI* and *CR* values for each evaluation are calculated with (8) and (9). The *CR*s are calculated as follows. $CR_{main} = 0.008$; $CR_{CN1} = 0.019$; $CR_{CN2} = 0.007$; $CR_{CN3} =$ 0.019. Steps 3–4 are replied to find the local sub-criteria weights. The final priorities are calculated by multiplying the sub-criteria weights with the main criteria weights. Table 7 illustrates the priorities of customer needs found with the fuzzy AHP method.

Finally, the most important customer need is found as long-term use (CN11); the second one is environmentally sensitive material (CN24), and the third one is elegant View/Design (CN25) by the fuzzy AHP technique.

Table 4 Technical requirements of power bank

Technical requirements	References
Adding the appropriate solar panel to the device (TR1)	[8, 26, 28]
High capacity battery (TR2)	[3, 5, 11, 13, 65, 67]
Integrating USB input into the device (TR3)	[2, 4, 14, 67]
Placement of keys (TR4)	[5, 55, 65, 68]
Adding the speaker (TR5)	[2, 5, 8, 14]
Production of cable and adapter (TR6)	[2, 8, 10, 67]
Durable material (TR7)	[2, 10, 67]
Integrating flash input into the device (TR8)	[5, 6, 14, 67]
Lightweight material (TR9)	[2, 8, 14, 65, 67]
Adding the appropriate Bluetooth system to the device (TR10)	[2, 5, 14, 26]
Converting solar energy into electrical energy (TR11)	[8, 10, 28]
Integrating LED light (TR12)	[14, 26, 67]
Easy to use design (TR13)	[5, 10, 14, 65]
Production in different colors (customization) (TR14)	[5, 14, 65, 67]

Table 5 Main criteria evaluation

	Linguistic terms			Triangular fuzzy numbers		
	CN1	CN2	CN3	CN1	CN2	CN3
CN1	–	EI	SMI	1	(0.5, 1, 1.5)	(1.5, 2, 2.5)
CN2		–	WMI	(0.67, 1, 2)	1	(1, 1.5, 2)
CN3			–	(0.4, 0.5, 0.67)	(0.5, 0.67, 1)	1

Table 6 Main criteria α-cut matrix

	CN1	CN2	CN3
CN1	(1, 1)	(0.5, 1.5)	(1.5, 2.5)
CN2	(0.67, 2)	(1, 1)	(1, 2)
CN3	(0.4, 0.67)	(0.5, 1)	(1, 1)

4.4 Evaluation of Technical Requirements of Power Bank

DMs made a collective decision and formed the relationship matrix. For each actor group such as customer needs, technique requirements, weights of customer needs and the matrix relating customer needs to technical requirements are placed in the House of Quality, as Table 8.

DMs evaluate the relationship of customer needs to technical requirements by using the linguistic scale in Table 2. Using the triangular fuzzy numbers in Table 2, linguistic expressions are transformed into fuzzy numbers.

Table 7 Weights of customer needs

Criteria	Weights	Sub-criteria	Local weights	Weights	Ranking
CN1	0.386	CN11	0.379	0.146	1
		CN12	0.213	0.082	5
		CN13	0.149	0.057	7
		CN14	0.115	0.044	12
		CN15	0.144	0.055	8
CN2	0.387	CN21	0.151	0.059	6
		CN22	0.120	0.046	11
		CN23	0.138	0.053	9
		CN24	0.312	0.121	2
		CN25	0.279	0.108	3
CN3	0.227	CN31	0.159	0.036	13
		CN32	0.105	0.024	15
		CN33	0.143	0.032	14
		CN34	0.373	0.085	4
		CN35	0.220	0.050	10

Triangular fuzzy numbers are multiplied by weights of customer needs to obtain the importance degree of technical requirements as fuzzy numbers. Then, these triangular fuzzy numbers are transformed into real numbers with the following equation:

$$BNPi = \frac{[(u_i - l_i) + (m_i - l_i)]}{3} + l_i \quad (10)$$

These defuzzified values are normalized and the final importance degree of technique requirements are shown as in Table 8.

As a result of the evaluation, the priority order of the technical requirements is as follows: adding the appropriate solar panel to the device (TR1), integrating LED light (TR12), high capacity battery (TR2), production of cable and adapter (TR6) and durable material (TR7). According to these results, adding the appropriate solar panel to the device is the most appropriate technique requirement for improvement multifunctional power bank.

5 Conclusion

Recent advances in technology force companies to develop new products to survive in the market. The NPD process consists of several stages which take a long time and bring high costs. In this process, it is very decisive to analyze the product, to determine what customers' needs are. Furthermore, defining the technical requirements

Table 8 HOQ matrix for power bank

		Wi	Technical requirement														
			TR1	TR2	TR3	TR4	TR5	TR6	TR7	TR8	TR9	TR10	TR11	TR12	TR13	TR14	
Customer needs	CN11	0.146	SMI	SMI				E	E					AMI			
	CN12	0.082		AMI				SMI									
	CN13	0.057			EI		SMI			SMI			E		SMI		
	CN14	0.044							E		EI				WMI		
	CN15	0.055									E						
	CN21	0.059											AMI		E		
	CN22	0.046			AMI												
	CN23	0.053													AMI		
	CN24	0.121	SMI						AMI								
	CN25	0.108	WMI		SMI	E		AMI				E				AMI	
	CN31	0.036	VSMI	SMI				EI	E			AMI		AMI			
	CN32	0.024	SMI	WMI				E	EI					VSMI			
	CN33	0.032	WMI	SMI				EI	EI					AMI			
	CN34	0.085	SMI	WMI				E	E	E				VSMI			
	CN35	0.050				E	AMI										Sum
			1.052	0.839	0.412	0.158	0.265	0.812	0.730	0.200	0.100	0.216	0.233	0.916	0.400	0.324	6.658
		Imp.	0.158	0.126	0.062	0.024	0.040	0.122	0.110	0.030	0.015	0.033	0.035	0.138	0.060	0.049	1.000
		Rank	1	3	6	13	9	4	5	12	14	11	10	2	7	8	

accurately is also essential for the company. In this context, QFD is an important tool that provides these features.

The objective of this chapter is to introduce a new method for selecting the most important technical components in multifunctional power bank design. Therefore, QFD is combined with fuzzy logic and MCDM techniques. Fuzzy logic considers uncertain information and enables DMs to assess elements with linguistic term sets. Fuzzy AHP technique is applied to find customer needs' weights and the fuzzy QFD tool is applied to find the most important technical requirements.

According to the obtained result, adding an appropriate solar panel to the device is the most suitable technical requirement for improving a multifunctional power bank. The addition of this technical requirement, enables power banks to be charged if it is needed.

In future studies, the number of customers and technological requirements can be increased and the same problem can be solved using aggregation operators for group decision making to aggregate DMs' evaluations. As a second perspective, this method can be coupled with hesitant fuzzy linguistic term sets.

Acknowledgements The authors would like to thank the industrial experts for their support, as well as acknowledge the grants of Galatasaray University Research Fund (Projects Numbers: 19.402.001, 19.402.003 and 19.402.006).

References

1. Bottani, E., Rizzi, A.: Strategic management of logistics service: a fuzzy QFD approach. Int. J. Prod. Econ. **103**(2), 585–599 (2006)
2. Haselton, B.: CNBC Tech. Apple isn't alone: 2018 was the 'worst year ever' for smartphone shipments, IDC says. https://www.cnbc.com/2019/02/01/idc-2018-smartphone-shipments-worst-year-ever.html. Accessed March 2019
3. Liu, L., Tang, W., Chen, B., Deng, C., Zhong, W., Cao, X., Wang, Z.: A self-powered portable power bank based on a hybridized nanogenerator. Adv. Mater. Technol. **3**, 1700209 (2018)
4. Minghao, C., Yang, Y., Yuwen, H.: Portable power bank, dare to ask where the road is–Survey of the Capital University student charging treasure. In: 2017 7th International Conference on Social Network, Communication and Education (SNCE 2017). Atlantis Press (2017)
5. He, F.: USB port and power delivery: an overview of USB port interoperability. In: 2015 IEEE Symposium on Product Compliance Engineering (ISPCE), pp. 1–5. IEEE (2015)
6. Suen, C.Y.B., Chan, K.T., Hung, T.K., Lee, C.C.: Remote monitoring on capacity of portable power bank in testing laboratories. In: IECON 2017-43rd Annual Conference of the IEEE Industrial Electronics Society, pp. 4734-4739. IEEE (2017)
7. Buyukozkan, G., Ertay, T., Kahraman, C., Ruan, D.: Determining the importance weights for the design requirements in the house of quality using the fuzzy analytic network approach. Int. J. Intell. Syst. **19**(5), 443–461 (2004)
8. dos Reis Benatto, G.A., Espinosa, N., Krebs, F.C.: Life-cycle assessment of solar charger with integrated organic photovoltaics. Adv. Eng. Mater. **19**(8), 1700124 (2017)
9. Palczynska, B.: Radiated emissions measurements of a portable power bank in a GTEM cell. In: 2017 IEEE International Conference on Environment and Electrical Engineering and 2017 IEEE Industrial and Commercial Power Systems Europe (EEEIC/I&CPS Europe), pp 1–6. IEEE (2017)

10. Thakur, P., Kool, A., Hoque, N.A., Bagchi, B., Khatun, F., Biswas, P., Das, S.: Superior performances of in situ synthesized ZnO/PVDF thin film based self-poled piezoelectric nanogenerator and self-charged photo-power bank with high durability. Nano Energy **44**, 456–467 (2018)
11. Moo, C.S., Wu, T.H., Hou, C.H., Hsieh, Y.C.: Balanced discharging of power bank with buck-boost battery power modules. In: 2014 International Power Electronics Conference (IPEC-Hiroshima 2014-ECCE ASIA), pp. 1796–1800. IEEE (2014)
12. Hou, C.H., Yen, C.T., Wu, T.H., Moo, C.S.: A battery power bank of serial battery power modules with buck-boost converters. In: 2013 IEEE 10th International Conference on Power Electronics and Drive Systems (PEDS), pp. 211–216. IEEE (2013)
13. Wu, T.H., Moo, C.S., Hou, C.H.: A battery power bank with series-connected buck–boost-type battery power modules. Energies **10**(5), 650 (2017)
14. Xie, L., Li, J., Cai, S., Li, X.: Design and experiments of a self-charged power bank by harvesting sustainable human motion. Adv. Mech. Eng. **8**(5), 1687814016651371 (2016)
15. Khoo, L.P., Ho, N.C.: Framework of a fuzzy quality function deployment system. Int. J. Prod. Res. **34**(2), 299–311 (1996)
16. Kahraman, C., Ertay, T., Büyüközkan, G.: A fuzzy optimization model for QFD planning process using analytic network approach. Eur. J. Oper. Res. **171**(2), 390–411 (2006)
17. Bevilacqua, M., Ciarapica, F.E., Giacchetta, G.: A fuzzy-QFD approach to supplier selection. J. Purch. Supply Manag. **12**(1), 14–27 (2006)
18. Liu, H.C., Quan, M.Y., Shi, H., Guo, C.: An integrated MCDM method for robot selection under interval-valued Pythagorean uncertain linguistic environment. Int. J. Intell. Syst. **34**(2), 188–214 (2019)
19. Aboutorab, H., Saberi, M., Asadabadi, M.R., Hussain, O., Chang, E.: ZBWM: the Z-number extension of best worst method and its application for supplier development. Expert Syst. Appl. **107**, 115–125 (2018)
20. Liu, A., Xiao, Y., Ji, X., Wang, K., Tsai, S.B., Lu, H., Wang, J.: A novel two-stage integrated model for supplier selection of green fresh product. Sustainability **10**(7), 2371 (2018)
21. Peng, J.G., Xia, G., Sun, B.Q., Wang, S.J.: Systematical decision-making approach for quality function deployment based on uncertain linguistic term sets. Int. J. Prod. Res. **56**(18), 6183–6200 (2018)
22. Tian, Z.P., Wang, J.Q., Wang, J., Zhang, H.Y.: A multi-phase QFD-based hybrid fuzzy MCDM approach for performance evaluation: a case of smart bike-sharing programs in Changsha. J. Clean. Prod. **171**, 1068–1083 (2018)
23. Tsai, J.Y., Ding, J.F., Liang, G.S., Ye, K.D.: Use of a hybrid MCDM method to evaluate key solutions influencing service quality at a port logistics center in Taiwan. Brodogradnja: Teorija i praksa brodogradnje i pomorske tehnike **69**(1), 89–105 (2018)
24. Van, L., Yu, V., Dat, L., Dung, C., Chou, S.Y., Loc, N.: New integrated quality function deployment approach based on interval neutrosophic set for green supplier evaluation and selection. Sustainability **10**(3), 838 (2018)
25. Yu, L., Wang, L., Bao, Y.: Technical attributes ratings in fuzzy QFD by integrating interval-valued intuitionistic fuzzy sets and Choquet integral. Soft. Comput. **22**(6), 2015–2024 (2018)
26. Akbaş, H., Bilgen, B.: An integrated fuzzy QFD and TOPSIS methodology for choosing the ideal gas fuel at WWTPs. Energy **125**, 484–497 (2017)
27. Alkhatib, S.F.: Strategic logistics outsourcing: upstream-downstream supply chain comparison. J. Glob. Oper. Strat. Sourc. **10**(3), 309–333 (2017)
28. Hsu, C.H., Chang, A.Y., Luo, W.: Identifying key performance factors for sustainability development of SMEs—integrating QFD and fuzzy MADM methods. J. Clean. Prod. **161**, 629–645 (2017)
29. Pamučar, D., Mihajlović, M., Obradović, R., Atanasković, P.: Novel approach to group multi-criteria decision making based on interval rough numbers: hybrid DEMATEL-ANP-MAIRCA model. Expert Syst. Appl. **88**, 58–80 (2017)
30. Roy, M.K., Ray, A., Pradhan, B.B.: Non-traditional machining process selection-an integrated approach. Int. J. Qual. Res. **11**(1) (2017)

31. Arsenyan, J., Büyüközkan, G.: An integrated fuzzy approach for information technology planning in collaborative product development. Int. J. Prod. Res. **54**(11), 3149–3169 (2016)
32. Ignatius, J., Rahman, A., Yazdani, M., Šaparauskas, J., Haron, S.H.: An integrated fuzzy ANP–QFD approach for green building assessment. J. Civ. Eng. Manag. **22**(4), 551–563 (2016)
33. Anusha, C., Parappilly, J.J., Sangaiah, A.K.: A new MCDM approach integrating QFD, dematel with TOPSIS for exploring the effect of social network usage on academic performance. IIOAB J. **6**(4), 32–42 (2015)
34. Büyüközkan, G., Çifçi, G.: An extended quality function deployment incorporating fuzzy logic and GDM under different preference structures. Int. J. Comput. Intell. Syst. **8**(3), 438–454 (2015)
35. Büyüközkan, G., Güleryüz, S.: Extending fuzzy QFD methodology with GDM approaches: an application for IT planning in collaborative product development. Int. J. Fuzzy Syst. **17**(4), 544–558 (2015)
36. Wang, C.H.: Using quality function deployment to conduct vendor assessment and supplier recommendation for business-intelligence systems. Comput. Ind. Eng. **84**, 24–31 (2015)
37. Karsak, E.E., Dursun, M.: An integrated fuzzy MCDM approach for supplier evaluation and selection. Comput. Ind. Eng. **82**, 82–93 (2015)
38. Dursun, M., Karsak, E.E.: Fuzzy decision approach based on QFD and FWA for selection of medical suppliers. Proc. World Congr. Eng. **2**, 1–3 (2015)
39. Li, M., Jin, L., Wang, J.: A new MCDM method combining QFD with TOPSIS for knowledge management system selection from the user's perspective in intuitionistic fuzzy environment. Appl. Soft Comput. **21**, 28–37 (2014)
40. Onut, S., Tosun, S.: An integrated methodology for supplier selection under the presence of vagueness: a case in banking sector, Turkey. J. Appl. Math. (2014)
41. Ayağ, Z., Samanlioglu, F., Büyüközkan, G.: A fuzzy QFD approach to determine supply chain management strategies in the dairy industry. J. Intell. Manuf. **24**(6), 1111–1122 (2013)
42. Büyüközkan, G., Çifçi, G.: An integrated QFD framework with multiple formatted and incomplete preferences: a sustainable supply chain application. Appl. Soft Comput. **13**(9), 3931–3941 (2013)
43. Dursun, M., Karsak, E.E.: A QFD-based fuzzy MCDM approach for supplier selection. Appl. Math. Model. **37**(8), 5864–5875 (2013)
44. Wey, W.M., Chiu, Y.H.: Assessing the walkability of pedestrian environment under the transit-oriented development. Habitat Int. **38**, 106–118 (2013)
45. Büyüközkan, G., Çifçi, G.: A new incomplete preference relations based approach to quality function deployment. Inf. Sci. **206**, 30–41 (2012)
46. Wang, C.H., Chen, J.N.: Using quality function deployment for collaborative product design and optimal selection of module mix. Comput. Ind. Eng. **63**(4), 1030–1037 (2012)
47. Büyüközkan, G., Berkol, Ç.: Designing a sustainable supply chain using an integrated analytic network process and goal programming approach in quality function deployment. Expert Syst. Appl. **38**(11), 13731–13748 (2011)
48. Liu, H.T.: Product design and selection using fuzzy QFD and fuzzy MCDM approaches. Appl. Math. Model. **35**(1), 482–496 (2011)
49. Bhattacharya, A., Geraghty, J., Young, P.: Supplier selection paradigm: An integrated hierarchical QFD methodology under multiple-criteria environment. Appl. Soft Comput. **10**(4), 1013–1027 (2010)
50. Feyzioğlu, O., Büyüközkan, G.: An integrated group decision-making approach for new product development. Int. J. Comput. Integr. Manuf. **21**(4), 366–375 (2008)
51. Büyüközkan, G., Feyzioğlu, O., Ruan, D.: Fuzzy group decision-making to multiple preference formats in quality function deployment. Comput. Ind. **58**(5), 392–402 (2007)
52. Büyüközkan, G., Feyzioğlu, O.: Group decision making to better respond customer needs in software development. Comput. Ind. Eng. **48**(2), 427–441 (2005)
53. Ertay, T., Büyüközkan, G., Kahraman, C., Ruan, D.: Quality function deployment implementation based on analytic network process with linguistic data: an application in automotive industry. J. Intell. Fuzzy Syst. **16**(3), 221–232 (2005)

54. Chan, L.K., Kao, H.P., Wu, M.L.: Rating the importance of customer needs in quality function deployment by fuzzy and entropy methods. Int. J. Prod. Res. **37**(11), 2499–2518 (1999)
55. Akao, Y., King, B., Mazur, G.H.: Quality Function Deployment: Integrating Customer Requirements into Product Design, vol. 21. Productivity Press, Cambridge, MA (1990)
56. Hauser, J.R., Clausing, D.: The House of Quality, pp. 1–13. Harvard Business Review (1988)
57. Zadeh, L.A.: Fuzzy sets. Inf. Control **8**, 338–353 (1965)
58. Ross, T.J.: Fuzzy Logic with Engineering Applications. Wiley (2005)
59. Wu, H., Tzeng, G., Chen, Y.: A fuzzy MCDM approach for evaluating banking performance based on balanced scorecard. Expert Syst. Appl. **36**(6), 10135–10147 (2009)
60. Kulak, O., Kahraman, C.: Fuzzy multi-attribute selection among transportation companies using axiomatic design and analytic hierarchy process. Inf. Sci. **170**, 191–210 (2005)
61. Saaty, T.L.: The Analytic Hierarchy Process. McGraw-Hill, New York (1980)
62. Büyüközkan, G.: Determining the mobile commerce user requirements using an analytic approach. Comput. Stand. Interfaces **31**(1), 144–152 (2009)
63. Büyüközkan, G., Çifçi, G., Güleryüz, S.: Strategic analysis of healthcare service quality using fuzzy AHP methodology. Expert Syst. Appl. **38**(8), 9407–9424 (2011)
64. Büyüközkan, G.: Multi-criteria decision making for e-marketplace selection. Internet Res. **14**(2), 139–154 (2004)
65. Büyüközkan, G., Çifçi, G.: A combined fuzzy AHP and fuzzy TOPSIS based strategic analysis of electronic service quality in healthcare industry. Expert Syst. Appl. **39**(3), 2341–2354 (2012)
66. Lee, A.R.: Application of modified fuzzy AHP method to analyze bolting sequence of structural joints. UMI Dissertation Service, A Bell & Howell Company (1999)
67. Gergin, Z., Akbas, M., Akture, A.O., Yurt, M.F.: Kalite Fonksiyonu Göçerimi Yöntemiyle Müşterinin Marka Algısının Arttırılması: Taşınabilir Şarj Cihazları Üreten Bir Firmada Uygulama (2018)
68. Zhu, Z., Johguchi, K., Mattausch, H.J., Koide, T., Hironaka, T.: Low power bank-based multi-port SRAM design due to bank standby mode. In: The 2004 47th Midwest Symposium on Circuits and Systems, MWSCAS'0 (2004)

Integration of Quality Function Deployment with IVIF-AHP and Kano Model for Customer Oriented Product Design

Merve Ozalp, Deniz Kucukbas, Esra Ilbahar and Selcuk Cebi

Abstract Nowadays, meeting the needs and expectations of customers has become more important to ensure the sustainability of the products in the market. Product designers utilize various methods and approaches to achieve this goal. The most favorable methods for product design are Quality Function Deployment (QFD) and Kano Model. QFD protects products against design errors and reduces customer dissatisfaction by matching customer expectations with technical characteristics of products. Kano Model categorizes product attributes with respect to customer expectations by examining the customer satisfaction levels in the cases of fulfillment and non-fulfillment of attributes. Therefore, in this chapter, an integrated method including QFD, Kano Model and interval-valued intuitionistic fuzzy analytical hierarchy process (IVIF-AHP) is employed for customer oriented product design. Kano Model is embedded into house of quality to define marketing advantage by classifying customer requirements while IVIF-AHP is utilized to determine the importance degrees of customer requirements. The proposed method is applied for laptop design. It is shown that processor, graphic card and battery are selected to be the most important technical characteristics.

1 Introduction

Customer expectations are constantly changing in today's market environment and designing products with only technical excellence does not seem to be sufficient for the success of the company. It is important to meet the changing customer needs and to match the technical parameters of the product with the customer requirements. The products need to be differentiated from similar ones on the market with various features.

Understanding new expectations of customers has a huge role in keeping products on the market. Quality Function Deployment (QFD) was proposed to the literature to match customer expectations with technical characteristics for product design. Since

M. Ozalp (✉) · D. Kucukbas · E. Ilbahar · S. Cebi
Department of Industrial Engineering, Yildiz Technical University, Istanbul, Turkey
e-mail: mervozlp@gmail.com

C. Kahraman and S. Cebi (eds.), *Customer Oriented Product Design*, Studies in Systems, Decision and Control 279, https://doi.org/10.1007/978-3-030-42188-5_6

QFD reduces product design time, it helps us quickly market products at a lower cost. It prevents customer dissatisfaction arising from design errors and gives certainty to design decisions.

QFD basically has six components. These are customer requirements, planning matrix, technical characteristics, relationship matrix between technical characteristics & customer requirements, and the roof showing the relationship between technical characteristics and targets. These six components are represented with a graphical tool called "*House of Quality*" which allows comparison and evaluation of customer requirements with respect to company capabilities and product characteristics. In the literature, there are many applications of QFD varying from product development to evaluation of supplier reliability. In this chapter, QFD is utilized to illustrate a customer-oriented product design strategy and indicate its effectiveness in product design process.

The rest of this chapter is organized as follows; In Sect. 2, the fundamentals of QFD method is presented. Then, a comprehensive literature review on studies including product development applications of QFD is provided. An illustrative example is given in Sect. 3 to demonstrate how QFD is applied. Findings are presented in Sect. 4. Finally, concluding remarks on QFD and its utilization in product development are given in Sect. 5.

2 Literature Review

QFD is the most popular customer-oriented product design method in the literature. There are various applications of QFD in product development. Zadry et al. [16] used QFD to design an ergonomic spinal board. Wood et al. [17] employed QFD to determine the constituent elements of the green hospital. Chen and Chen [27] utilized QFD to design a remote-controlled racing car. Wu et al. [39] adopted QFD to identify customer needs and to determine priorities in bicycle design. Ju and Sohn [23] utilized QFD to develop robotic products. Alba-Elias et al. [26] employed QFD to obtain an overall design of the automatic tablet device. Homkhiew et al. [40] analyzed furniture development via QFD. Režek Jambrak et al. [12] used QFD to analyze apple beverages treated with high-power ultrasound in order to convert customer requirements into design targets and major quality assurance points. Kowalska et al. [5] employed QFD to identify consumer preferences about sponge-fatty cake. Fátima Cardoso et al. [22] developed an organic fruit jelly by QFD. Furthermore, researchers have integrated some methods into QFD structure to improve its analytical capability. In this section, various studies published in the international journals indexed in Web of Science are investigated in order to analyze the techniques integrated with QFD method. A detailed literature review study was conducted by Sunday Ayoola Oke [41] on the topic of QFD in 2013. In the study, it is emphasized that the integration of QFD with other techniques such as neurofuzzy, fuzzy concept, genetic algorithm, mathematical programming, multiple criteria decision tools improves its analysis power. Therefore, in the scope of this chapter, in order to see how this trend is progressing, the studies published in Web of Science between the years 2012 and 2018 are examined. Particularly, the studies including applications of QFD in product design have come to the forefront.

In the literature, one of the most used methods along with QFD structure is Theory of Inventive Problem Solving (TRIZ). Frizziero et al. [1] used QFD and TRIZ method in open molds design. The QFD method was used to evaluate the quality of the design process and the TRIZ method was utilized to solve critical problems that might affect the design. In the study of Wang on smartphones, TRIZ and QFD were also used together [14]. In the study, TRIZ was preferred to address conflicts between technical attributes. Chen and Huang [20] employed TRIZ to solve the contradictions between the technical requirements in the QFD relationship matrix for Notebook PC design. Similarly, Zhang et al. [33] used TRIZ and QFD to present a new stove design. Melemez et al. [36] used QFD, Analytical Hierarchy Process (AHP) proposed by Saaty [44], and TRIZ for forestry trailer design. In the study, AHP was used to select the best concept. QFD was adopted to convert customer requirements into quality characteristics. Then, TRIZ was used to solve negative correlations between quality characteristics. In the study of Vinodh et al. [30], "*environmentally conscious quality function deployment (ECQFD)*", TRIZ and AHP were utilized for automotive component design. Customer requirements were determined using ECQFD and they were converted into technical characteristics. After TRIZ was used to associate design options with innovative alternatives, AHP was employed to select the best design.

Another method which is generally integrated with QFD is Kano Model. Gangurde et al. [2] utilized QFD and Kano Model method for mobile phone design. In the study, they used Kano Model to determine customer requirements while QFD method was employed to identify priorities for customer requirements and technical characteristics. Similarly, Ji et al. [29] used QFD and Kano Model for notebook design. Kang et al. [3] employed QFD, fuzzy AHP, Evaluation Grid Method (EGM) and Kano Model for the design of mini cars. Fuzzy QFD was used to convert customer requirements into technical requirements while fuzzy Kano Model combined with fuzzy AHP was used to determine the priority of requirements. EGM was used to build qualitative relations among customer requirements and design characteristics. He et al. [10] proposed a new Kano Model approach which is called as importance-frequency Kano (IF-Kano) model and the proposed approach was integrated into QFD for the design of an elevator. In order to classify customer requirements by considering the interaction and importance between frequencies, IF Kano Model was preferred. The results of the IF Kano Model was integrated into the QFD model and then, outputs of QFD were used as inputs to non-linear programming model. The multi-population adaptive genetic algorithm was employed to solve this model.

In some studies, QFD has been integrated with multiple criteria decision-making methods. In the study of Wu et al. [15], QFD, DEMATEL and VIKOR methods were employed in electric vehicle design. QFD was used to define engineering characteristics while the hesitant fuzzy DEMATEL was utilized to analyze the interrelationships among customer requirements and their weights. Then the hesitant fuzzy VIKOR was adopted to prioritize engineering characteristics. Wang and Chen [42] used fuzzy

DEMATEL, fuzzy Delphi, integer programming, QFD and ANP for design of digital water camera. Zaim et al. [32] employed QFD and ANP for the design of new equipment to squeeze the polyethylene pipes, and consequently to stop the gas flow without damaging the pipes. Nixon et al. [37] used Linear Fresnel Reflector, AHP and QFD to develop solar thermal collector. Cebi et al. [48] utilized QFD, fuzzy sets theory, and DEMATEL methods in order to evaluate design parameters of the vessel engine room.

QFD and fuzzy sets theory have been preferred by researchers to handle uncertainties in design process [3, 15, 42, 49, etc.]. For instance, Koprivica and Filipovic [4] used these methods to improve the design quality of the boiler. Vinodh et al. [13] utilized integration of fuzzy sets theory with QFD to provide sustainable electronics products. Zhong et al. [34] used QFD for the car design. Li et al. [6] used Interval Linguistic Information (ILI) and QFD in their study on a practical product improvement of a turbine engine. ILI was adopted to express uncertain assessments in the product information. Wu and Ho [25] used QFD and fuzzy sets for green mobile phone design. The reason for using fuzzy logic is to handle the subjectivity in voice of customers. Wu and Liao [8] used QFD and Probabilistic Linguistic-Organisation Rangement EtAynhése de Données Relationnelles (PL-ORESTE) for the evaluation of Panda shared cars. To obtain fuzzy weights of design parameters, QFD was used to extend the probability linguistic term. The PL-ORESTE method was developed to obtain preferences and relationships between alternatives. Yuen [31] used fuzzy QFD, Fuzzy Cognitive Network Process, and The Fuzzy Aggregative Grading Clustering for development of cloud software product. The reason for using fuzzy QFD is to address uncertainty in expert assessment.

The methods integrated with QFD are not limited with the literature given above. There are also some methods that are rarely used with QFD. For instance, Murali et al. [18] used QFD and importance performance analysis for water purifier. Choi et al. [21] employed QFD and usability evaluation method in their study for smart car design. Neyses and Sandberg [24] used QFD and multivariate data analysis for electric guitar design. Han and Shin [28] utilized QFD, bibliometric analysis, and trend impact analysis for carbon nanotube biosensor technology. In the study of Du et al. [35], QFD and axiomatic design were used in machine tool remanufacturing. The reason for preferring QFD was to standardize and optimize the redesign process while axiomatic design was used for redesign of used products.

Based on the comprehensive literature review on applications of QFD, the most used techniques integrated with QFD for product development were presented in Table 1. According to Table 1, fuzzy sets theory is quite popular in order to consider uncertainties in design process. The most commonly utilized methods integrated with QFD are Kano Model, AHP, and TRIZ. TRIZ is generally used to solve problems related to conflicting technical characteristics. AHP is utilized to define relationships among customer requirements and their importance degrees while DEMATEL and ANP techniques are employed to measure interdependencies among technical characteristics and their importance degrees. MCDM techniques are used for the performance evaluation of design alternatives.

Table 1 The techniques and methods integrated with QFD

Article	Fuzzy	Crisp	QFD	AHP	TRIZ	Kano	Others
[1]		✓	✓		✓		
[2]		✓	✓			✓	
[3]	✓		✓	✓		✓	Evaluation Grid Method (EGM)
[4]	✓		✓				
[5]		✓	✓				
[6]		✓	✓				Interval Linguistic Information
[7]			✓				Artificial Neural Networks
[8]		✓	✓				ORESTE
[10]		✓	✓			✓	
[12]		✓	✓				
[13]	✓		✓				
[14]	✓		✓		✓		
[15]		✓	✓				DEMATEL, VIKOR
[16]		✓	✓				
[17]		✓	✓				
[18]		✓	✓				
[20]		✓	✓		✓		
[21]		✓	✓				Usability Evaluation Method
[22]		✓	✓				
[23]		✓	✓				
[24]		✓	✓				
[25]		✓	✓				Fuzzy Logic
[26]		✓	✓				
[27]		✓	✓				
[28]		✓	✓				
[29]		✓	✓			✓	
[30]		✓	✓	✓	✓		
[31]	✓		✓				
[32]		✓	✓				ANP
[33]		✓	✓		✓		
[34]	✓		✓				
[35]		✓	✓				Axiomatic Design
[36]		✓	✓	✓	✓		
[37]		✓	✓	✓			
[39]		✓	✓				
[40]		✓	✓				
[42]		✓	✓				Delphi, Integer Programming, ANP, DEMATEL

3 Quality Function Deployment (QFD)

QFD is one of the techniques considering customer requirements to provide customer satisfaction. After World War II, QFD method was proposed to provide successful product design since the quality and innovation had become more important in product development. The method was implemented by some well-known companies such as Mitsubishi Heavy Industries Ltd., Kobe shipyards, and Toyota at the beginning of the 1970s. Although QFD was implemented by companies, it did not stand out in the academic literature in the 1970s. QFD, which was just a tool until the beginning of the 1980s, started to be used in the USA by Ford Motor Company and the Cambridge Corporation [46] as a result of the article written by Kogure and Akao [52]. QFD has an important role in customer-oriented product design since the requirements of product design and expectations of customers have to be well understood and the importance degrees of these expectations for the customer need to be evaluated correctly. For this purpose, QFD presents all required information for the product design stage by using the House of Quality (HoQ) given in Fig. 1 [11].

QFD method consists of nine steps as given in Fig. 2 and these steps marked in Fig. 1 are as follows [9]:

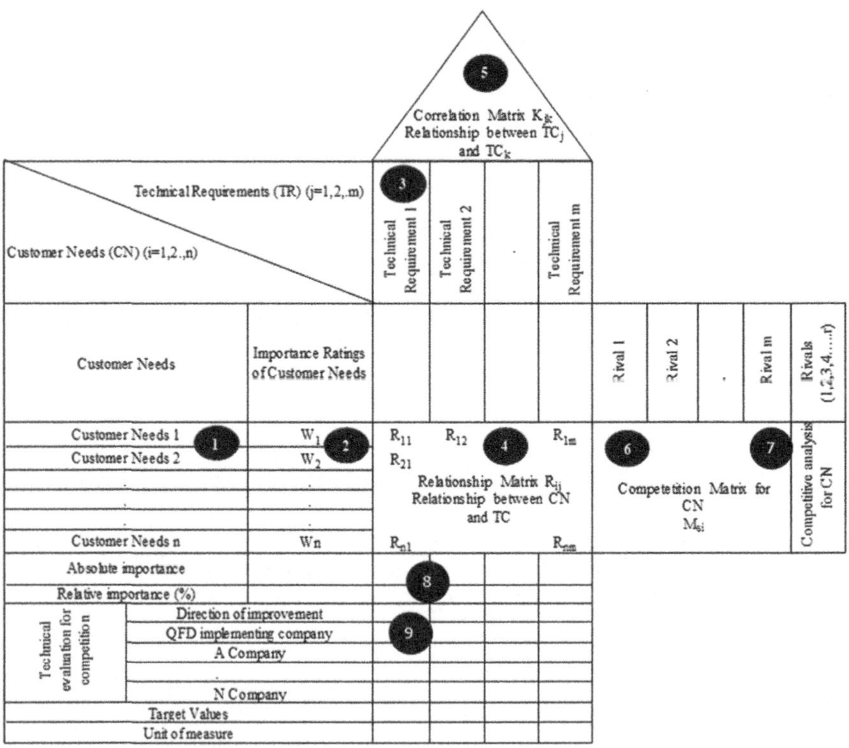

Fig. 1 The house of quality [9]

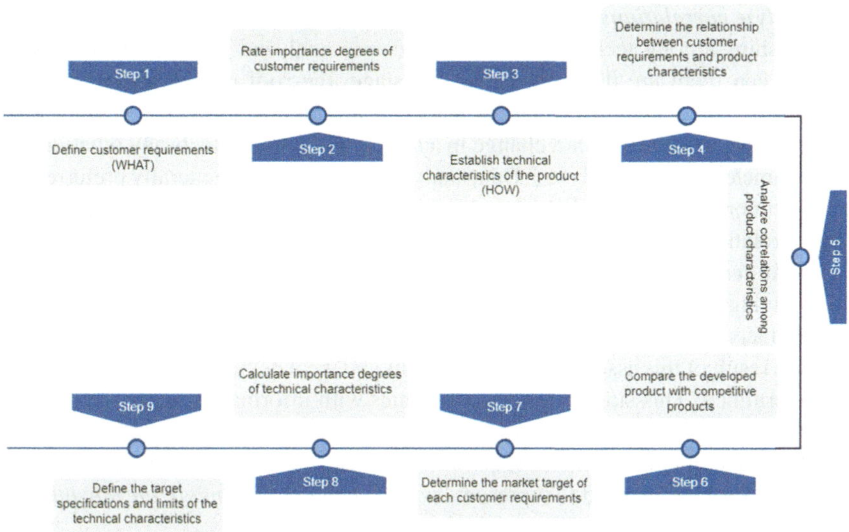

Fig. 2 Steps of quality function deployment

1. ***Define customer requirements (What)***: This is the stage of meeting customer requirements, known as the voice of customer (VOC) [45]. Customer needs and expectations are the features that customers want to have in the designed product. These needs are collected by using some methods such as surveys with clients, focus group interviews, and internet blog searches. After the customer needs are determined, customers are asked to prioritize these needs and expectations.
2. ***Rate importance degrees of customer requirements***: As the number of requirements collected from customers is high, they have to be graded to identify the important ones. Customers can express their ratings using a scale of 5-7-9 for each requirement or a scale from 1 to 10 (10 = extremely important, 1 = least important) [46]. In addition to these methods, there are many studies using fuzzy QFD method in the literature.
3. ***Establish technical characteristics of the product (How)***: At this stage, design parameters to meet customer needs are determined. Technical requirements should be in the manufacturer's control and should be measurable. Measurability of technical requirements is important for understanding whether customer needs are met [43].
4. ***Analyze relationships between customer requirements and technical characteristics (What and How)***: At this stage, the relationships between customer requirements and design parameters are determined. Various scales are used to determine this relationship. For this purpose, 9, 3, 1, and 0 values are used for "*strong relationship*", "*moderate relationship*", "*poor relationship*", and "*no relationship*", respectively [4].

5. ***Analyze correlations among the technical characteristics***: In this step, the relationships among technical characteristics are analyzed and interdependencies between them are determined. At this stage, the roof of the quality house is formed. There may be positive or negative correlations between design parameters. Therefore, whether a change in a design parameter affects any other design parameter must be analyzed. Data analysis techniques are generally preferred for other components of QFD while brainstorming might be sufficient to determine the relationships between technical characteristics [47].
6. ***Make technical comparisons and competitive assessments***: At this stage, the planning matrix of HoQ is created. Customers are asked to evaluate competitors' products for designated needs. Likert scale is generally used in this evaluation. As a result of this assessment, the strengths and weaknesses of competitors can be determined. This stage provides companies with information about the market.
7. ***Determine market target for each customer requirement***: In this step, the market advantage of each customer requirement is determined. For this purpose, 1, 1.2, and 1.5 values are used to represent "*no any advantage*", "*moderate advantage*", and "*strong advantage*", respectively.
8. ***Calculate importance degrees of technical characteristics***: In this step, information embedded into HoQ is utilized to obtain importance degrees of design characteristic. For this purpose, the following equation is utilized:

$$W_j^{TC} = \sum\nolimits_{i=1}^{n} W_i^{CR} \times R_i^j \times IR_i \times MA_i,\ \mathrm{i} = 1, 2, 3, \ldots, \mathrm{n};\ \mathrm{j} = 1, 2, 3, \ldots, \mathrm{m} \quad (1)$$

   where W_j^{TC}, W_i^{CR}, R_i^j, IR_i, and MA_i represents the importance degree of *j*th technical characteristic, importance degree of *i*th customer requirement, improvement ratio for *i*th customer requirement, and marketing advantage for *i*th customer requirement, respectively while *n* and *m* represent the number of customer requirements and number of technical characteristics, respectively.
9. ***Define the targets specifications and limits of technical characteristic***: In this step, the importance of each design parameter is calculated and the technical matrix of HoQ is created. This matrix includes the direction of improvement, standard units of measurements and values, and relative importance values of technical requirements. The direction of improvement indicates the direction in which the technical requirement should be developed. Improvement directions are expressed by the following symbols [43]; "*more, longer better* (▲)", "*Less, the shorter the better* (▼)", and "*Target value is better* (●)".

4 Proposed Method

In this section, we will indicate the distinction of our proposed method from the traditional QFD. In the proposed method, importance degrees of customer requirements will be calculated by using interval-valued intuitionistic fuzzy (IVIF) AHP proposed by Wu et al. [49]. IVIF-AHP is preferred because it is better able to represent vagueness in human judgments. Moreover, we will utilize Kano Model [50] to determine marketing advantage for each customer requirement since it is hard to identify marketing advantage of a customer requirement based on qualitative methods. In general, they are determined based on experts' judgements and experiences. There are also some studies using Kano Model to determine customer requirements in the literature. However, it is the first-time marketing advantage has been determined based on Kano Model, which is able to classify attributes based on the effect they have on customer satisfaction.

5 Case Study: The Design of Laptop

The laptops have become one of the preferred products in the electronics market, and a large part of the market is composed of college students [51]. For this reason, a Kano questionnaire is used to determine the priorities of college students for laptops. 82 respondents participated in the questionnaire and the detailed information about the questionnaire is given in Part 4 Chap. 3.

1. ***Define customer requirements (What)***: Customer requirements are identified based on the Kano questionnaire as high processor capability (*CR1*), light design (*CR2*), high storage capacity (*CR3*), high resolution (*CR4*), and long battery life (*CR5*).
2. ***Rate importance degrees of customer requirements***: There may be uncertainties in determining the relationship between customer requirements and technical characteristics or determining the importance degrees of customer requirements.
   Cases in which fuzzy logic is needed for QFD method [4]:
   - Linguistic expressions in the information on the product
   - Estimated values obtained from missing or uncertain information
   - Ease of reasoning under uncertainty
   - Ease of use and understanding of products with multiple inputs and outputs.

   Therefore, in this study, weights of customer requirements are determined by using IVIF-AHP. Table 2 shows the aggregated decision matrix used for IVIF-AHP and Table 3 demonstrates the weights of customer requirements obtained from it.
3. ***Establish technical characteristics of the product (How)***: Technical characteristics are identified as processor, RAM, hard drive, graphic card, battery, and material type and quality.

Table 2 Aggregated decision matrix for IVIF-AHP

Aggregated matrix	CR1	CR2	CR3	CR4	CR5
CR1	E	H	H	H	E
CR2	L	E	H	E	L
CR3	L	L	E	E	VL
CR4	L	E	E	E	VL
CR5	E	H	VH	VH	E

Table 3 Weights of customer requirements obtained from IVIF-AHP

Customer requirements	Weights
CR1	0.23
CR2	0.19
CR3	0.16
CR4	0.17
CR5	0.26

4. ***Analyze relationships between customer requirements and technical characteristics (What and How)***: The relationships between customer requirements and technical characteristics are provided in Fig. 3 by using the scale which has the choices, "*strong relationship*", "*moderate relationship*", "*poor relationship*", and "*no relationship*".
5. ***Analyze correlations among the technical characteristics:*** The relationships among technical characteristics are demonstrated in Fig. 3.
6. ***Make technical comparisons and competitive assessments***: In this study, competitive assessment is not performed since the main aim of this application is to provide importance degrees of design characteristics in laptop design.
7. ***Determine market target for each customer requirement***: The marketing advantage of each customer requirement is identified according to the results of Kano Model. When the customer requirement is classified as "Must-Be", "One-Dimensional" and "Attractive", the marketing advantage takes the value of 1, 1.2, and 1.5, respectively.
8. ***Calculate importance degrees of technical characteristics***: Importance degrees of design characteristics are obtained and presented in Fig. 3.
9. ***Define the targets specifications and limits of technical characteristics***: Since this study does not include a competitive assessment phase, targets specifications and limits of technical characteristics are not provided in Fig. 3.

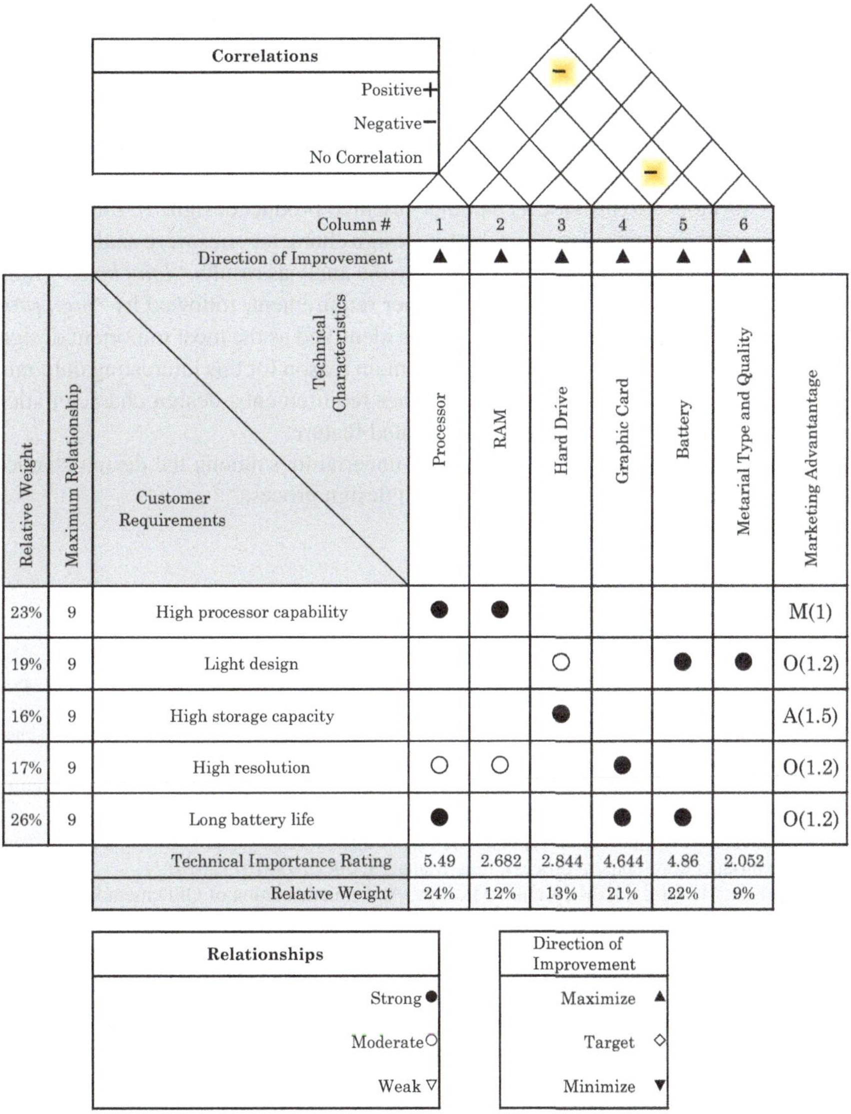

Fig. 3 The house of quality for laptop design

6 Conclusion

The main objective of this chapter is to illustrate the customer-oriented product design strategy while introducing QFD method. For this purpose, a detailed literature review on customer-oriented product design between the years 2012 and 2018 is provided.

As a result of the literature review, the following results are reached: (*i*) In the literature, the number of studies using QFD integrated with other methods has increased in recent years. (*ii*) Among multiple criteria decision-making techniques, AHP and TRIZ are often preferred for integration with QFD. (*iii*) Fuzzy QFD was used in the studies, which have uncertainty in determining HoQ relations. Furthermore, in this chapter, an integrated method that consists of the QFD, Kano Model and IVIF-AHP was developed to provide a customer-oriented product design. To illustrate the application of the proposed method, laptop design characteristics were analyzed considering customer expectations. According to the analysis results, "*long battery life*" is selected to be the most important customer requirement, followed by "*high processor capability*". Moreover, "*processor*" is identified as the most important design characteristics, followed by the battery. The main reason for this interesting outcome is rooted in relationships among the customer requirements, design characteristics and also the marketing advantage of the related feature.

For further study, interdependencies and uncertainties among the design characteristics might be taken into consideration in design process.

References

1. Frizziero, L., Francia, D., Donnici, G., Liverani, A., Caligiana, G.: Sustainable design of open molds with QFD and TRIZ combination. J. Ind. Prod. Eng. **35**(1), 21–31 (2018)
2. Gangurde, S.R., Patil, S.S.: Benchmark product features using the Kano-QFD approach: a case study. Benchmarking: Int. J. **25**(2), 450–470 (2018)
3. Kang, X., Yang, M., Wu, Y., Ni, B.: Integrating evaluation grid method and fuzzy quality function deployment to new product development. Math. Probl. Eng., Hindawi (2018)
4. Koprivica, S.M., Filipovic, J.: Application of traditional and fuzzy quality function deployment in the product development process. Eng. Manag. J. **30**(2), 98–107 (2018)
5. Kowalska, M., Pazdzior, M., Krzton-Maziopa, A.: Implementation of QFD method in quality analysis of confectionery products. J. Intell. Manuf. **29**, 439–447 (2018)
6. Li, Y.L., Du, Y.F., Chin, K.S.: Determining the importance ratings of customer requirements in quality function deployment based on interval linguistic information. Int. J. Prod. Res. **56**(14), 4692–4708 (2018)
7. Patil, C.K., Husain, M., Halegowda, N.W.: Study of quality function deployment model based on artificial neural network with optimization techniques. J. Adv. Manuf. Syst. **17**, 119–136 (2018)
8. Wu, X., Liao, H.: An approach to quality function deployment based on probabilistic linguistic term sets and ORESTE method for multi expert multi-criteria decision making. Inf. Fusion **43**, 13–26 (2018)
9. Dehe, B., Bamford, D.: Quality Function Deployment and operational design decisions – a healthcare infrastructure development case study. Prod. Plan. Control Manag. Oper. **28**(14), 1177–1192 (2017)
10. He, L., Song, W., Wu, Z., Xu, Z., Zheng, M., Ming, X.: Quantification and integration of an improved Kano model into QFD based on multi-population adaptive genetic algorithm. Comput. Ind. Eng. **114**, 183–194 (2017)
11. Köse, Y., Demirci, E.: Fuzzy quality function deployment approach for selection of mineral water manufacturer companies. Turk. J. Occup./Environ. Med. Saf. **1**(3), 465–478 (2017)

12. Režek Jambrak, A., Šimunek, M., Grbeš, F., Mandura, A., Djekic, I.: Analysis of apple beverages treated with high-power ultrasound: a quality function deployment approach. J. Sci. Food Agric. **98**, 2258–2266 (2017)
13. Vinodh, S., Manjunatheshwara, K.J., Sundaram, S.K., Kirthivasan, V.: Application of fuzzy quality function deployment for sustainable design of consumer electronics products: a case study. Clean Technol. Environ. Policy **19**, 1021–1030 (2017)
14. Wang, C.H.: Incorporating the concept of systematic innovation into quality function deployment for developing multi-functional smart phones. Comput. Ind. Eng. **107**, 367–375 (2017)
15. Wu, S.M., Liu, H.C., Wang, L.E.: Hesitant fuzzy integrated MCDM approach for quality function deployment: a case study in electric vehicle. Int. J. Prod. Res. **55**(15), 4436–4449 (2017)
16. Zadry, H.R., Susanti, L., Rahmayanti, D.: Ergonomics intervention on an alternative design of a spinal board. Int. J. Occup. Saf. Ergon. **23**(3), 393–403 (2017)
17. Wood, L.C., Wang, C., Abdul-Rahman, H., Abdul-Nasir, N.S.J.: Green hospital design: integrating quality function deployment and end-user demands. J. Clean. Prod. **112**, 903–913 (2016)
18. Murali, S., Pugazhendhi, S., Muralidharan, C.: Integration of IPA and QFD to assess the service quality and to identify after sales service strategies to improve customer satisfaction – a case study. Prod. Plan. Control **27**(5), 394–407 (2016)
19. Kecek, G., Yüksel, R.: Analitik hiyerarşi süreci (ahp) ve promethee teknikleriyle akıllı telefon seçimi. Sos. Bilim. Derg. **49**, 46–63 (2016)
20. Chen, C.H., Huang, C.Y.: The synergy of QFD and TRIZ for solving EMC problems in electrical products – a case study for the notebook PC. J. Ind. Prod. Eng. **32**(5), 311–330 (2015)
21. Choi, I.K., Kim, W.S., Lee, D., Kwon, D.S.: A weighted qfd-based usability evaluation method for elderly in smart cars. Int. J. Hum.-Comput. Interact. **31**(10), 703–716 (2015)
22. Fátima Cardoso, J.D., Filho, N.C., Miguel, P.A.C.: Application of quality function deployment for the development of an organic product. Food Qual. Prefer., Braz. **40**, 180–190 (2015)
23. Ju, Y., Sohn, S.Y.: Patent-based QFD framework development for identification of emerging technologies and related business models: a case of robot technology in Korea. Technol. Forecast. Soc. Chang., Korea, 44–64 (2015)
24. Neyses, B., Sandberg, D.: A new methodology to select hardwood species for wooden products. Wood Mat. Sci. Eng. **10**(4), 344–352 (2015)
25. Wu, Y.H., Ho, C.C.: Integration of green quality function deployment and fuzzy theory: a case study on green mobile phone design. J. Clean. Prod. **108**, 271–280 (2015)
26. Alba-Elías, F., Guerra-Sánchez-De-La-Nieta, Á., Las-Heras-Casas, J., González-Marcos, A., Alfonso-Cendón, J., Castejón-Limas, M.: A new device for dosing additives in the food industry using quality function deployment. J. Food Process. Eng., Spain (2014)
27. Chen, L.H., Chen, C.N.: Normalisation models for prioritising design requirements for quality function deployment processes. Int. J. Prod. Res. **52**(2), 299–313 (2014)
28. Han, K., Shin, J.: A systematic way of identifying and forecasting technological reverse salients using QFD, bibliometrics, and trend impact analysis: a carbon nanotube biosensor case. Technovation **34**, 559–570 (2014)
29. Ji, P., Jin, J., Wang, T., Chen, Y.: Quantification and integration of Kano's model into QFD for optimising product design. Int. J. Prod. Res. **52**(21), 6335–6348 (2014)
30. Vinodh, S., Kamala, V., Jayakrishna, K.: Integration of ECQFD, TRIZ and AHP for innovative and sustainable product development. Appl. Math. Model. **38**, 2758–2770 (2014)
31. Yuen, K.K.F.: A hybrid fuzzy quality function deployment framework using cognitive network process and aggregative grading clustering: an application to cloud software product development. Neurocomputing, China **142**, 95–106 (2014)
32. Zaim, S., Sevkli, M., Camgöz-Akdağ, H., Demirel, O.F., Yayla, A.Y., Delen, D.: Use of ANP weighted crisp and fuzzy QFD for product development. Expert Syst. Appl. **41**, 4464–4474 (2014)

33. Zhang, F., Yang, M., Liu, W.: Using integrated quality function deployment and theory of innovation problem solving approach for ergonomic product design. Comput. Ind. Eng., China **76**, 60–74 (2014)
34. Zhong, S., Zhou, J., Chen, Y.: Determination of target values of engineering characteristics in QFD using a fuzzy chance-constrained modelling approach. Neurocomputing **142**, 125–135 (2014)
35. Du, Y., Cao, H., Chen, X., Wang, B.: Reuse-oriented redesign method of used products based on axiomatic design theory and QFD. J. Clean. Prod. **39**, 79–86 (2013)
36. Melemez, K., Gironimo, G.D., Esposito, G., Lanzotti, A.: Concept design in virtual reality of a forestry trailer using a QFD-TRIZ based approach. Turk. J. Agric. For., Turk. 789–801 (2013)
37. Nixon, J.D., Dey, P.K., Davies, P.K.: Design of a novel solar thermal collector using a multi-criteria decision-making methodology. J. Clean. Prod. **59**, 150–159 (2013)
38. Özçifçi, V., Arsu, T.: Lojistik servis sağlayicisi seçiminde ahp uygulaması. Sos. Ve Beşeri Bilim. Derg. **5**(1) (2013)
39. Wu, C.T., Pan, T.S., Shao, M.H., Wu, C.S.: An extensive QFD and evaluation procedure for innovative design. Math. Probl. Eng. (2013)
40. Homkhiew, C., Ratanawilai, T., Pochana, K.: Application of a quality function deployment technique to design and develop furniture products. Songklanakarin J. Sci. Technol. **34**(6), 663–668 (2012)
41. Oke, S.A.: Manufacturing quality function deployment: literature review and future trends. Eng. J. **17**(3), 79–103 (2013)
42. Wang, C.H., Chen, J.N.: Using quality function deployment for collobrative product design and optimal selection of module mix. Comput. Ind. Eng. **63**, 1030–1037 (2012)
43. Uppalanchi, A.: Application of quality function deployment in new product and service development. Masters' thesis. Missouri University of Science and Technology (2010)
44. Saaty, T.L.: Decision making with the analytic hierarchy process. Int. J. Serv. Sci. **1**(1), 83–98 (2008)
45. Gonza´lez, M.E., Quesada, G., Bahill, A.T.: Improving product design using quality function deployment: the school furniture case in developing countries. Qual. Eng. **16**(1), 47–58 (2003)
46. Karsak, E.E., Sozer, S., Alptekin, S.E.: Product planning in quality function deployment using a combined analytic network process and goal programming approach. Comput. Ind. Eng. **44**, 171–190 (2002)
47. Verma, R., Maher, T., Pullman, M.: Effective product and process development using quality function deployment. Integr. Prod. Process. Dev. 339–354 (1998)
48. Cebi, S., Ozkok, M., Demirci, E.: Evaluation of design parameters for vessel engine room by using a modified QFD technique. J. Mult.-Valued Log. Soft Comput. **23**(5–6), 559–587 (2014)
49. Wu, J., Huang, H.B., Cao, Q.W.: Research on AHP with interval-valued intuitionistic fuzzy sets and its application in multi-criteria decision making problems. Appl. Math. Model. **37**(24), 9898–9906 (2013)
50. Kano, N.: Attractive quality and must-be quality. Hinshitsu (Qual., J. Jpn. Soc. Qual. Control) **14**, 39–48 (1984)
51. Shiau, C.S., Tseng, I.H., Heutchy, A.W., Michalek, J.: Design optimization of a laptop computer using aggregate and mixed logit demand models with consumer survey data. In: ASME 2007 International Design Engineering Technical Conferences and Computers and Information in Engineering Conference (pp. 175–185). American Society of Mechanical Engineers Digital Collection (2007)
52. Kogure, M., Akao, Y.: Quality function deployment and CWQC in Japan. Qual. Prog. 10/1983, **16**(10), 25–29 (1983)

Heterogeneous Information Integrated QFD for Smart Bicycle Design

Gülçin Büyüközkan, Deniz Uztürk and Öykü Ilıcak

Abstract The developing technology started to define a new set of functions and potentials for the products. Customers that keep up with the developing technologies expect to experience cutting-edge technologies in their products. Amid these expectations, the market has become a very competitive place where everybody wants to strike the customer's attention with high technology. On the other hand, non-customer oriented high technology often intimidates end-users, and that decreases the preferability of the product. Motivated by the critical importance of customer-oriented design, this contribution presents a smart product design methodology with heterogeneous information. The non-homogeneity of the information comes from its variant decision-making group as the potential customers and experts. Quality Function Deployment (QFD), which is a very well known, powerful tool for product design is chosen for this methodology. House of Quality (HoQ) of QFD is extended with the 2-Tuple linguistic model to create a flexible environment for decision makers (DMs). Furthermore, multi-preference relations technique is used to handle heterogeneous information while calculating the customer requirements' importance. A case study about the smart bicycle design is presented with the results to test the plausibility of the suggested methodology at the end of the contribution.

1 Introduction

Customer oriented product design is a very crucial tool to gain a competitive advantage in the sector. Today, customers' satisfaction has a fundamental role while creating the market success. Product design, which is a procedure to create a new product, has become a very efficient and successful process if the customer's voice is well heard [1]. However, the customer requirements are variant, and they need to be well investigated in order to achieve a successful product design. Today, with high

G. Büyüközkan (✉) · Ö. Ilıcak
Department of Industrial Engineering, Galatasaray University, 34349 Istanbul, Turkey
e-mail: gbuyukozkan@gsu.edu.tr

D. Uztürk
Department of Management, Galatasaray University, 34349 Istanbul, Turkey

C. Kahraman and S. Cebi (eds.), *Customer Oriented Product Design*, Studies in Systems, Decision and Control 279, https://doi.org/10.1007/978-3-030-42188-5_7

customization requests, it is a very critical challenge to reflect the diversified requirements assigned by customers. In addition, the customers may not be able to provide a homogeneous information because of the diversified segments. Hence, aggregating the non-homogeneous information, which is a complex phase, becomes essential during decision making. In this study, motivated by the diversity of customer segments, a customer-oriented product design process is proposed with heterogeneous information.

To handle the multi-formatted data, first of all, multi-preference relations technique is suggested for criteria weighting process. Afterward, Quality Function Deployment (QFD) which is a frequently used technique that converts customer needs (CNs) into design requirements (DRs) [2] is proposed for the design process. The main purpose of QFD is to listen to the voice of the customer to develop a new product or to improve an existing product. QFD is an approach that takes into consideration many DMs opinions in the prioritization of DRs. Therefore, in the essence it can be said that QFD is a group decision-making (GDM) approach. House of Quality (HoQ) of QFD is suggested for a smart bicycle design methodology for this contribution.

However, HoQ can be empowered by its 2-Tuple Linguistic Model [3] extension for linguistic GDM applications where decision-makers (DMs) assess the criteria with linguistic variables. Since, multiple DMs are involved in decision-making process, the data provided by DMs may not be homogeneous regarding their level of knowledge. So, 2-Tuple extension of HoQ provides the calculations with multi-granular linguistic data with minimum loss of information. The extension of HoQ with 2-Tuple linguistic model enable HoQ to deal with linguistic data from various levels. Different levels create multi-granular data and 2-Tuple gives the possibility to deal with this multi-granular data.

Different linguistic label sets can be assigned to different DM according to their level of knowledge. Thus, DMs make their assessment with the linguistic sets that they are comfortable to use. This flexible environment forms more robust and effective decision-making environment. So, the main reason to integrate QFD with 2-Tuple is to create the flexibility to DMs and have calculations with linguistic variables without any loss of information.

Furthermore, GDM where more than one DM is involved, is often preferred to overcome the DMs' assessments' subjectivity from prejudices, preferences, and limited experiences [4]. Moreover, in GDM process, DMs may have different backgrounds, experiences, and ideas and may want to give their evaluations in different formats according to their preferences. They may give their evaluations in many different formats such as numerically, linguistically etc. depending on their preferences [5].

In this study, an extended 2-Tuple HoQ methodology for smart bicycle design is proposed with multi-preference relations technique, which is a GDM approach, that handles heterogeneous information.

The remainder of the study is organized as follows: Sect. 2 provides the literature review about the methodologies. Section 3 gives computational steps of the applied methodology. Section 4 presents the case study in smart bicycle design. Section 5 provides the obtained results and discussion while Sect. 6 concludes the study and provides future research directions.

2 Literature Review

In this section, a review about the suggested topic and techniques are given with details. First a literature review about the QFD applications for product design is given. After that, multi-preference tool in QFD applications on new product design is presented. Later Finally, the review about the 2-Tuple integrated QFD tool are presented.

QFD Applications for Product Design

QFD is a technique used to design products and services for customer satisfaction. The aim of this study, product design, will be discussed by modeling with the QFD technique.

In this section, QFD studies on product design are reviewed and seventeen different works are summarized in Table 1.

In the studies indicated in Table 1, we can say that QFD is frequently used in the field of product development/product design and is integrated with many different techniques. In this study, QFD technique was expanded with 2-tuple and used with multi-preference relations technique. The place of these techniques in the literature with QFD is described below.

Multi-Preference Relations Integrated QFD

Multi-preference relations allow DMs, who have different educational/cultural background or different perspectives, to present their preferences in various ways. This increases the flexibility in the GDM process by applying different formats of judgments.

When we examine the literature, there are many studies that combine multi-preference relations technique with QFD methodology since it includes a decision process involving many DMs.

In this section, studies on multi-preference relations technique with QFD methodology is proposed and due to the main focus of this study, studies about "new product design/development" are presented. Seven different works on new product design/development are gathered from this literature review and their details are given in Table 2.

As it is seen from Table 1, there exists no study which combines QFD with the technique of multi-preference relations on product design in recent years. In addition, there are no such studies that involve multi-preference relations integrated QFD methodology for smart bicycle design. Therefore, the plausibility of the multi-preference integrated QFD in product design field is shown by these studies in Table 2.

Table 1 Studies about QFD applications on product design

Year and reference	Methodology	Subject	Type
2009 [6]	QFD, analytic network process (ANP), Kano model	Customer-oriented web design	Case study
2009 [7]	Fuzzy QFD, fuzzy failure mode and effects analysis (FMEA)	Semiconductor packing design	Illustrative example
2011 [8]	QFD, fuzzy Delphi, ANP, Interpretive structural modeling (ISM)	New product development	Case study
2012 [9]	QFD	Pipe choking equipment improvement	Case study
2012 [10]	QFD, fuzzy Delphi, fuzzy DEMATEL, linear integer programming	Development of various types of sport and water digital cameras	Case study
2013 [11]	QFD, fuzzy preference relations, GDM	Writing instrument assessment	Illustrative example
2014 [12]	QFD, fuzzy cognitive network process	Cloud software product development	Illustrative example
2015 [13]	QFD	Biometric identification system design	Case study
2015 [14]	QFD, ANP	Design and application of management patterns	Case study
2015 [15]	QFD, Fuzzy ANP, FMEA	Design of a green and low-carbon product	Case study
2016 [16]	QFD, preference ordering method	Respiratory mask development	Illustrative example
2017 [17]	Fuzzy QFD, design structure matrix (DSM)	Development of a new type of automotive part	Case study
2018 [18]	QFD, design of experiments (DOE) method	Development of new food products	Case study
2018 [19]	QFD, fuzzy COPRAS, 2-tuple	Material selection	Case study
2018 [20]	QFD, ANP	Cloud product design	Case study
2019 [21]	QFD, proportional hesitant fuzzy linguistic term set, best-worst method	QFD model proposition for product design problems	Illustrative example
2019 [22]	QFD, data envelopment analysis	User selection for mobile phone development	Illustrative example

Table 2 Studies on multi-preference relations with QFD about product design

Year and reference	Methodology	Subject	Type
2005 [23]	Multi-preference relations, Fuzzy GDM, QFD	Software development	Numerical example
2007 [24]	Multi-preference relations, Fuzzy GDM, QFD	Hatch door development of a car	Numerical example
2012 [25]	GDM, multi-format preference analyses in QFD	Personal digital assistant product development	Case study
2012 [26]	QFD, Multi-preference, Nonlinear programming	Pencil design	Numerical example
2013 [27]	QFD, incomplete preference, multi-preference	Sustainable SCM development	Case study
2015 [28]	Fuzzy GDM in QFD with multiple and incomplete preference relations	Game systems development	Case study
2015 [4]	Fuzzy GDM in QFD with multiple and incomplete preference relations	Software development	Case study

In addition, in this study, the QFD methodology is extended with 2-Tuple linguistic model and the literature review based on 2-Tuple Integrated QFD approach is given in the following section.

2-Tuple Integrated QFD

In this contribution, the main focus is to create a product design process with heterogeneous information. HoQ of QFD is selected as the main tool to create customer-oriented product. It is extended with 2-Tuple linguistic model to overcome the multi-granularity in linguistic evaluations.

In the literature, "2-Tuple Linguistic Model" integrated QFD applications are searched in Web of Science portal to see the trending subjects studied with the suggested techniques. Sixteen different works are gathered from this literature research and their details are given in Table 3.

As it is seen from the Table 3, 2-Tuple integrated QFD applications are mostly used for multi-criteria decision making (MCDM) selection problems. In this chapter, the product design subject is selected as an application area like in [29, 31, 44]; however the methodology is expanded with multi-preference technique while calculating CNs' importance. The preliminaries for the proposed methodology and the detailed steps of the techniques will be given in the next section.

Table 3 2-Tuple integrated QFD studies

Year and reference	Subject	Type
2008 [29]	CNs' importance calculations for product design/development	Illustrative example
2012 [30]	Supplier selection	Numerical example
2012 [31]	Product design under multi-granularity linguistic environment	Numerical example
2013 [32]	Group linguistic variable to construct kano model oriented QFD	Numerical example
2013 [33]	Determining customer requirement weights by TFMF and TLR	Numerical example
2015 [34]	Failure prioritizing model development	Numerical example
2015 [35]	Supplier evaluation and selection	Case study
2016 [36]	Third party logistics supplier selection	Numerical example
2016 [37]	Group decision making model proposition	Numerical example
2017 [38]	Determining importance ratings of patients' requirements	Case study
2017 [39]	Warehouse design and selection	Case study
2017 [40]	Supplier evaluation	Case study
2017 [41]	Fuzzy Kano's model to understand the requirements	Case study
2018 [42]	Solution selection in crowdsourcing contests	Illustrative example
2018 [43]	Emergency Routes Evaluation	Numerical example
2018 [44]	Product redesign evaluation	Case study

3 Methodology

In this section, the computational steps of the proposed methodology are provided. The structure of the proposed methodology is illustrated in Fig. 1. First the basic preliminaries are defined for 2-Tuple linguistic model. Later, the details about the calculations are given respectively.

Preliminaries for 2-Tuple Linguistic Model

The 2-Tuple fuzzy linguistic representation model shows linguistic information with a so-called 2-Tuple form of (s, α), where 's' is a linguistic label and 'α' is a numerical value representing this symbolic translation's value [45].

Let $S = \{S_0 \ldots, Sg\}$ be a linguistic term set and $\beta \in [0, g]$ a value supporting the result of a symbolic aggregation operation, then the 2-Tuple that expresses the equivalent information to β is obtained with the following function [45]:

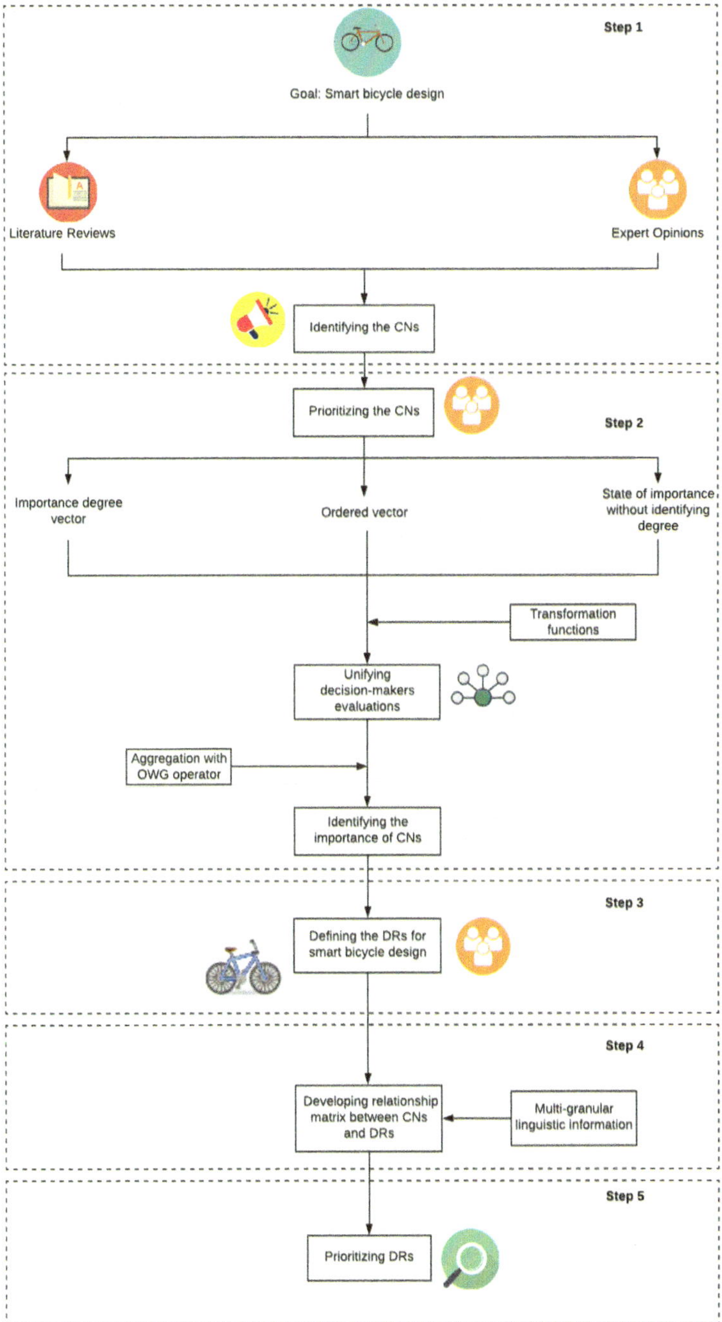

Fig. 1 Framework of the proposed methodology

$$\Delta : [0, g] \rightarrow \mathcal{S} \times [-0.5, 0.5)$$
$$\Delta(\beta) = \begin{cases} Si \; i = round(\beta) \\ \alpha \;\; \alpha \in [-0.5, 0.5) \end{cases} \quad (1)$$

Let $\mathcal{S} = \{S_0 \ldots, Sg\}$ be a linguistic term set and $(s_i, \alpha) \in \bar{S}$ be a 2-Tuple linguistic value and the reciprocal function is defined as [3]:

$$\Delta^{-1} : \;\; \bar{\mathcal{S}} \rightarrow [0, g]$$
$$\Delta^{-1}(s_i, \alpha) = i + \alpha = \beta \quad (2)$$

where β is the equivalent numerical value which belongs to $[0, g]$.

The linguistic term set can be transformed into 2-Tuple linguistic value by adding zero as $s_i \in S \Rightarrow (s_i, 0) \in \bar{\mathcal{S}}$ [3].

The aggregation of multiple values into one form is the crucial step for decision making, in this contribution, in order to reflect the importance of DMs into aggregated form, *Weighted Aggregation Operator (WAO)* has been chosen to use, *WAO* relation is computed as [3].

$$\vec{x} = \left(\frac{\sum_{i=1}^{n} \Delta^{-1}(r_{ij}, \alpha_{ij}) \times \Delta^{-1}(w_i, \alpha_i)}{\sum_{i=1}^{n} \Delta^{-1}(w_i, \alpha_i)} \right) = \Delta\left(\frac{\sum_{i=1}^{n} \beta_i \times w_i}{\sum_{i=0}^{n} w_i} \right) \quad (3)$$

Multi-granular linguistic information can be managed in different ways, here we propose the use of *Linguistic Hierarchies (LH)*.

A *LH* is the union of all levels *t* where each level *t* corresponds to a linguistic term set symmetrically distributed with an odd granularity. The linguistic levels selected for this study are: S^3 *(Low-Medium-High)* and S^5 *(Very Low-Low-Medium-High-Very High)*.

The transformation function to translate linguistic term set with granularity *n(t)* to a linguistic term set having granularity *n(t')* is as follows [3]:

$$TF_{t'}^{t} = \left(S_i^{n(t)}, \alpha^{n(t)}\right) = \Delta\left(\frac{\Delta^{-1}(\left(S_i^{n(t)}, \alpha^{n(t)}\right) \times \left(n(t') - 1\right)}{n(t) - 1} \right) \quad (4)$$

Quality Function Deployment

QFD is a very powerful tool which is usually used for service or product design cases [2]. It is mostly used with crisp numbers such as 1-3-9 (O-Δ-□) where you can assess while creating a stronger contrast between the stronger and the weakest relation between CNs and DRs [46]. QFD is based on a design tool called HoQ. The shape of the assessment tool looks like basic house, so it is called the house of the quality requirements for a product or a service.

The foundation of the HoQ is the belief that products should be designed while marketing people, design engineers, and manufacturing staff must work closely

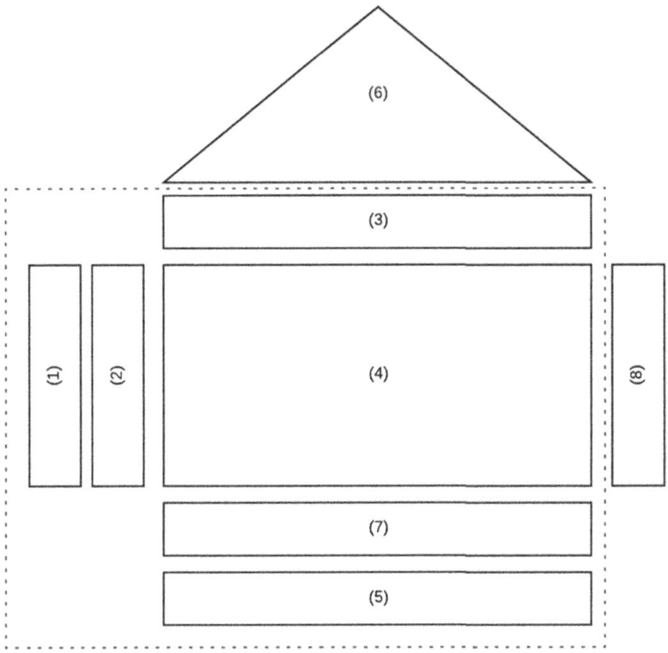

Fig. 2 Constituents of HoQ of QFD

together. The HoQ is a conceptual guideline that gives the means for inter-functional planning and communications.

In the construction of HoQ, group of DMs are involved to every step of the process, and they are required to give their opinions about the importance of CNs, correlations between DRs, and relationships between CNs and DRs. Figure 2 shows the parts of the HoQ, the definitions and properties of each part are given as follows [47]:

(1) *Voice of the customer.*

The first step in a QFD project is to identify who the customers are (target audience). Then the DM team, gathers information from customers on the requirements they request for the product or service. That information will produce a data for CN section for the HoQ.

(2) *Customer Importance Ratings.*

After completing defining customers voice as in (1), a relative importance for CNs are assigned for each one. To obtain the importance rates, different methodologies can be used such as MCDM techniques (AHP, ANP…etc.), linguistic approaches (2-Tuple, fuzzy logic…etc.), surveys and hybrid tools (linguistic information integrated MCDM tools). Once the importance rates are gathered, the first and the most important stage that lead the design is completed.

(3) *Technical Descriptors - "Voice of the Engineer".*

Technical descriptors may exist that your organization is already using to determine product or service specification, however new measurements can be added to ensure that your product is satisfying CNs. These are the engineering/design requirements for a product or a service.

(4) *Relationship Matrix.*

The relationship matrix is the area where the relationship between CNs and the company's ability is determined by the DMs. The team asks the question, "what is the strength of the relationship between the technical descriptors and the CNs?" Relationships can either be weak, moderate, or strong and carry a numeric value of 1, 3 or 9. Furthermore the relationship matrix can be extended with linguistic sets assigned to each DM [4].

(5) *Priorities of DRs.*

This is the results obtained from the HoQ steps. It gives the overall ranking of DRs according to the CNs.

(6) *Correlation Matrix.*

This is the part where all the relations between DRs are examined. Negative relations between DRs must be well examined to reduce the physical contradictions later.

(7) *Target Values for Technical Descriptors.*

At this stage, the QFD team begins to establish target values for each DR. Target values represent "how much" for a DR are needed to be achieved. This stage creates an essential and robust roadmap to create a customer-oriented product.

(8) *Customer Rating of the Competition.*

Understanding how customers rate the competition can be a tremendous competitive advantage. In this step asking the customers about the rating of the product or service in relation to the competition, is a good idea.

In this study, the parts indicated by the dashed lined square are selected to be used in the product design phase. Correlation of DRs and the competitive analysis parts are ignored for the suggested methodology.

Proposed Methodology

Step 1—Identifying the CNs: In this step CNs are identified, which can be presented as "the voice of customer". These requirements are identified by benefiting from literature researches.

Step 2—Prioritizing the CNs: In this step, the importance of CNs is determined by expert opinions. DMs can give their opinions in different formats. For this reason, in order to collect the different formats of opinions under a single group decision, multi-preference relations technique was used, and the steps are explained below.

Step 2.1—Unifying DMs evaluations: Assessments taken from individuals were unified to convene under a common group decision. Some of the most commonly used preference formats can be given as preference ordering vectors, importance degree vectors, linguistic preference relations, selected subsets, fuzzy preference relations, multiplicative preference relations and utility functions [4, 23, 24, 28, 48, 49].

In this study, individual assessments are gathered in different format as follows:

An importance degree vector ($u_1, \ldots, u_N$) where $u_i \in [0, 1]$ $i = 1, \ldots, N$. Closer to 1 means more important for u_i. With the formula below we can turn it to relevance of relative importance:

$$z_{ij} = u_i/u_i \, for \, all \, 1 \leq i \neq j \leq N \tag{5}$$

An ordered vector $(o(1), \ldots, o(N))$. Here, $o(i)$ represents the importance ranking of CN $_i$, where the most important is 1 and the least important is N. With the formula below we can turn it to relevance of relative importance:

$$\begin{aligned} z_{ij} &= 9u_i - u_j \, for \, all \, 1 \leq i \neq j \leq N \\ where & \\ u_i &= (N - o(i))/(N - 1) \end{aligned} \tag{6}$$

An importance of CNs but not stating the rating clearly. In this case, if the criteria i is more important than criteria j, $z_{ij} = 9$ and $z_{ji} = 1/9$. If nothing mentioned, then $z_{ij} = 1$.

Step 2.2—Collecting the evaluations: Each assessment is aggregated to define a common group decision in this step. With this step, dominant opinions of DMs are reflected. The *order weighted geometric (OWG)* operator is defined by the following formula:

$$\Phi^G\left\{\left(\bar{w}^{k1}, p_{ij}^{k1}\right), \ldots \left(\bar{w}_k^{kL}, p_{ij}^{kL_k}\right)\right\} = \prod_{1=1}^{L_k}\left(p_{ij}^{k[l]}\right) \tag{7}$$

Here, $\{1, \ldots, L_k\} \rightarrow \{1, \ldots., L_k\}$ is a permutation such that $\bar{w}\, kl \geq \bar{w}\, k[l + 1]$, $l = \{1, \ldots., L_{k-1}\}$, so $\bar{w}\, k1$ is the lth largest value in the set $\bar{w}k1, \ldots, \bar{w}\, kL_k$). Proportional quantifiers, such as "most", "as many as possible" etc., can be represented by fuzzy subsets of the unit interval [0, 1]. When the ratio t is suitable with the purpose of the quantifier it demonstrates then for any $t \, \varepsilon \, [0, 1]$, $Q(t)$ indicates the degree. For a non-decreasing relative quantifier, Q, the weights can be acquired with the formula below:

$$W_k = Q(k/K) - (Q(k - 1)/K) \, k = 1, \ldots., K \tag{8}$$

where $Q(t)$ is described as;

$$Q(t) = \begin{cases} 0, & if \, t < s \\ \frac{t-s}{v-s}, & if \, s \leq t \leq v \\ 1, & if \, t \geq v \end{cases} \tag{9}$$

Some of the relative quantitative determinants can be given as; "most" (0.3, 0.8), "at least half" (0, 0.5) and "as many as possible" (0.5, 1). The fuzzy quantifier Q is symbolized by Φ_Q^G. For this reason, the whole multiplicative relative importance is acquired as follows:

$$p_{ij}^k = \Phi_Q^G\left(p_{ij}^{k1}, p_{ij}^{k2}, \ldots, p_{ij}^{kL_k}\right), \quad 1 \le i \ne j \le N \tag{10}$$

Step 2.3—Identifying the importance of CNs: To define importance weights of CNs, the evaluations of the DM groups aggregated in the matrix P_k which acquired from the Eq. (6), must be utilized. Next, the importance of one factor against other in a fuzzy majority sense will be measured. Benefiting from *OWG* operator, Φ_Q^G, given in [50] we have;

$$QGID_i^k = 1/2\left(1 + \log_9 \Phi_Q^G\left(p_{ij}^k : j = 1, \ldots, N\right)\right) \; \textit{for}\, \text{all}\, i = 1, \ldots\ldots, N \tag{11}$$

After normalization, the importance degrees in percentage for the group k is given below:

$$QGID_i^k = QGID_i^k / \sum_i QGID_i^k \tag{12}$$

Step 3—Defining the DRs: Identifying DRs is a critical step for the product design. This is the essential step where the physical components of the product are defined. Decision-making group and/or in-depth research about the subject can help to define and develop DRs for the product.

Step 4—Developing relationship matrix between CNs and DRs: The main focus of this step is to create an objective and flexible decision-making environment to the decision-making group. The experts express their opinion about the relations between each CN-DR pair. The relations will provide necessary information while calculating the DRs priorities. In this suggested methodology, unlike ordinary QFD applications, experts express their opinions in linguistic variables. Different granulated linguistic sets are assigned to each expert accordingly their level of knowledge about the subject [51].

Step 5—Prioritizing DRs: This step is the result of the HoQ calculations of QFD. Prioritizations of DRs created a roadmap to follow while creating the product according to CNs. Engineering requirements are ranked according to their importance at this stage and 2-Tuple integrated QFD provides advantageous and easy computational steps for this stage while dealing with linguistic data.

Finally, by using the target and ranking values for each DR, a product design can be achieved with non-homogeneous information.

The next section presents the applied case study for a smart bicycle design to test the plausibility of the suggested techniques.

4 Case Study: Smart Bicycle Design

Nowadays, smart design is very topical subject in every sector. Smarter systems are starting to become indispensable in every part of our lives. They facilitate our lives and they facilitate the complex systems in different sectors. For example, manufacturing sector where machine learning facilitates the provision of demand, construction sector where smarter buildings creates more comfortable and secure places for residents, and transportation sector where new solutions are developed to ease the complex transportation routes in urban areas, etc.

Today, smart design is a necessity; but it becomes successful when the customer's needs are well reflected in the design. Motivated by this necessity, in this case study a smart bicycle design subject is selected. Due to the population increase in urban areas, the concept of traffic/congestion becomes unbearable for the residents. For this reason, bicycles have become a preferred transportation tool. Smarter solutions for the bikes have started to be developed in order to provide comfort and easy driving experience to the cyclists.

Depending on the customer's needs and preferences, bicycles may vary. It is, therefore, essential to design bicycles that matches different customer profiles. Since bicycles are the products with a high number of customer complaints, it will be a crucial issue to consider customers' will especially in designing phase.

In this study, the subject of smart bicycle design is discussed by using an extended 2-Tuple QFD methodology with multi-preference relations technique.

First, customer needs, namely CNs for smart bicycle design are addressed. A detailed literature review was carried out to determine CNs for smart bicycle design and then thirteen CNs were taken into consideration. This section constitutes the first step of the proposed approach.

Then, the importance of thirteen CNs was revealed. For this step, a decision-making group of four people (three expert and one group of possible customers) is gathered for the evaluations.

At this stage, as mentioned before an in-depth literature review and research is conducted to generate CNs. Later, a survey is conducted with a large group of possible bicycle customers. They are asked to assign an importance degree (1–5) to each detected CN. The result of the survey was the normalized as an importance percentage for each CN. Moreover, three experts evaluated the importance of each CN by their preferred assessment style. As a result of these four different-formatted evaluations obtained from DMs, they are combined with multi-preference relations technique and the prioritization of CNs was demonstrated.

Following these steps, design requirements namely DRs for the design of smart bicycles were determined and these requirements were prioritized as a result of three expert assessments with 2-tuple integrated QFD technique. All these steps are explained in detail as follows:

Step 1—Identifying the CNs: In this step, a comprehensive literature review was made, and CNs have been created as follows by focusing on the question of "What does a customer expect from a bicycle?".

CN1—Fitness into most places (public transport): For customers, it is definitely an important feature for bicycles to fit into most places, moreover, folding bicycles are much more preferred than the others.

CN2—Durability: Durability refers to the longevity of the bicycle. When customers buy a bike, they want to be able to use it for at least 5 years.

CN3—Safety: This emphasizes the importance of safety of the bicycle. When there is no safe place or a garage for a bicycle, a creative locking system is absolutely necessary.

CN4—Technologic features: Technological features such as night illumination, ringing and reflector for use in darkness, bring many customers closer to cycling and increase interest in the product. The reflector is especially important since the vehicles sometimes do not notice the bikes during the night trips.

CN5—Cleanness while using: When the cyclist passes through muddy water or dusty areas, the rear wheel can contaminate the user. Therefore, users want to travel without any contamination.

CN6—Lightness: It is one of the must-have features and makes it easier to carry the bicycle.

CN7—Tirelessly and fast drive in all terrain: Using a bicycle during uphill can be tiresome. For this reason, a bicycle that can be used without fatigue is important for customers.

CN8—Driving comfort: In terms of health, it is very important to be comfortable when riding a bicycle. An uncomfortable position can cause back pain in the user, and in this case the cycling may never preferred again.

CN9—Driving quality: Even in cars, this feature is widely common. It is an important criterion in every vehicle.

CN10—Easy maintenance: When a bike fails, the customer wants to repair it quickly. Therefore, it needs some simple equipment.

CN11—Solid braking system: It is impossible to think of a bike without a good braking system since it directly saves life. The customer's expectation is to stop completely after a short distance from the moment the brake is pressed.

CN12—Portable features: For those who place great importance on carrying goods, this feature is an important customer expectation. These can be the first-aid kit, the main drink, the transport of water etc.

CN13—Solid tire: Tire strength is an expectation that a customer will pay attention directly.

Step 2—Prioritizing the CNs: After the identification of CNs for smart bicycle design, CNs are prioritized by using multi-preference relations technique.

Step 2.1—Unifying DMs evaluations:

First DM says CN8 is important than CN2 and CN7; CN7 is more important than CN2 and CN4; CN2 is more important than CN4 and CN13.

Second DM provides an ordering vector of {8, 3, 5, 7, 11, 10, 1, 2, 6, 13, 4, 12, 9}. Here, 1 means that the CN7 is the most important and 13 means that the CN10 is the least important requirement. The ordering vector is transformed into a relative importance relation by using Eq. (6) as $u_1 = (13-8)/(13-1) = 0.42$ and $u_2 = (13-3)/(13-1) = 0.83$ $P^2_{12} = 9u_i - u_j = 9^{0.42-0.83} = 0.40$.

Third DM gives an importance degree vector as {0.6, 0.9, 0.8, 0.8, 0.3, 0.4, 0.7, 1, 0.8, 0.2, 0.9, 0.4, 0.6}. Here, CN2 and CN11 are evaluated as most important requirements with the importance degree of 0.9 which is closer to 1, and CN10 is evaluated as the least important one with the importance degree of 0.2. The importance degree vector is transformed into a relative importance relation using Eq. (5) as $P^{3}_{12} = 0.6/0.9 = 0.67$.

Finally, the fourth DM also gives an importance degree vector (as a result of the survey conducted within the possible customers) as {0.7, 1, 0.9, 0.8, 0.5, 0.5, 1, 1, 0.9, 0.3, 0.9, 0.3, 0.6}. Here, CN2, CN7 and CN8 are evaluated as most important requirements with the importance degree of 1 and CN10 and CN12 are evaluated as the least important requirements with the importance degree of 0.3. The importance degree vector is transformed into a relative importance relation using Eq. (5) as $P^{4}_{12} = 0.7/1 = 0.7$.

As an instance, P3, the evaluation matrix of the third DM is given as follows (Table 4).

Step 2.2—Collecting the evaluations: The matrices P1–P4 are collected with the help of Eqs. (7)–(10) and the *OWG* operator is used with the fuzzy linguistic quantifier "at most—(0.3, 0.8)". The weighing vector is calculated as (0, 0.4, 0.5, 1). Using Eq. (7), the relation matrix is found as follows (Table 5).

Step 2.3—Identifying the importance of CNs: For obtaining priorities from the evaluation matrix, Eqs. (7)–(10) are applied once more to compute the weighing vector (0, 0, 0, 0.2, 0.15, 0.15, 0.15, 0.15, 0.15, 0.15, 0.6, 0, 0) linked to the fuzzy linguistic quantifier "at most". Thereafter, using Eqs. (11)–(12), the aggregated importance values of are calculated as 0.50, 0.64, 0.59, 0.56, 0.37, 0.42, 0.65, 0.67, 0.57, 0.28, 0.62, 0.37, 0.48. After normalization, the final ranking of CNs are given as follows (Table 6).

Step 3—Defining the DRs: The DRs are defined with the help of three experts and a researched conducted about the smart bicycle design. The details of the DRs are provided as in Table 7.

Step 4—Developing relationship matrix between CNs and DRs: Each DM is provided their opinion about the relations between CN and DR for a smart bicycle design with linguistic variables S^3*(Low-Medium-High)* and S^5 *(Very Low-Low-Medium-High-Very High)*. First DM provided the assessments in S^3 and other two experts gave their opinion in S^5.

The Table 8 gives the evaluation of the third experts as an example and later Table 9 presents the aggregated linguistic assessments for the ultimate aggregated relation matrix.

First a unification phase is applied by using *LH* (see Sect. 3) (Eq. (4)), afterward an aggregation process is applied by applying Eq. (3) with the weight vector (0.2, 0.4, 0.4) for DMs.

Step 5—Prioritizing DRs: The prioritization as in Table 10, is obtained by applying the following relation:

Table 4 Evaluation matrix of the third DM

	CN1	CN2	CN3	CN4	CN5	CN6	CN7	CN8	CN9	CN10	CN11	CN12	CN13
CN1	1.00	0.67	0.75	0.75	2.00	1.50	0.86	0.60	0.75	3.00	0.67	1.50	1.00
CN2	1.50	1.00	1.13	1.13	3.00	2.25	1.29	0.90	1.13	4.50	1.00	2.25	1.50
CN3	1.33	0.89	1.00	1.00	2.67	2.00	1.14	0.80	1.00	4.00	0.89	2.00	1.33
CN4	1.33	0.89	1.00	1.00	2.67	2.00	1.14	0.80	1.00	4.00	0.89	2.00	1.33
CN5	0.50	0.33	0.38	0.38	1.00	0.75	0.43	0.30	0.38	1.50	0.33	0.75	0.50
CN6	0.67	0.44	0.50	0.50	1.33	1.00	0.57	0.40	0.50	2.00	0.44	1.00	0.67
CN7	1.17	0.78	0.88	0.88	2.33	1.75	1.00	0.70	0.88	3.50	0.78	1.75	1.17
CN8	1.67	1.11	1.25	1.25	3.33	2.50	1.43	1.00	1.25	5.00	1.11	2.50	1.67
CN9	1.33	0.89	1.00	1.00	2.67	2.00	1.14	0.80	1.00	4.00	0.89	2.00	1.33
CN10	0.33	0.22	0.25	0.25	0.67	0.50	0.29	0.20	0.25	1.00	0.22	0.50	0.33
CN11	1.50	1.00	1.13	1.13	3.00	2.25	1.29	0.90	1.13	4.50	1.00	2.25	1.50
CN12	0.67	0.44	0.50	0.50	1.33	1.00	0.57	0.40	0.50	2.00	0.44	1.00	0.67
CN13	1.00	0.67	0.75	0.75	2.00	1.50	0.86	0.60	0.75	3.00	0.67	1.50	1.00

Table 5 Relation matrix

	CN1	CN2	CN3	CN4	CN5	CN6	CN7	CN8	CN9	CN10	CN11	CN12	CN13
CN1	1.00	0.55	0.68	0.79	1.82	1.47	0.54	0.48	0.73	2.72	0.59	1.79	1.09
CN2	1.83	1.00	1.24	1.45	3.34	2.68	0.98	0.88	1.34	4.98	1.09	3.27	2.00
CN3	1.48	0.81	1.00	1.17	2.69	2.16	0.79	0.71	1.08	4.01	0.88	2.64	1.61
CN4	1.26	0.69	0.85	1.00	2.29	1.85	0.67	0.61	0.92	3.42	0.75	2.25	1.38
CN5	0.55	0.30	0.37	0.44	1.00	0.80	0.29	0.26	0.40	1.49	0.33	0.98	0.60
CN6	0.68	0.37	0.46	0.54	1.24	1.00	0.36	0.33	0.50	1.85	0.41	1.22	0.75
CN7	1.87	1.02	1.27	1.48	3.41	2.74	1.00	0.90	1.36	5.08	1.11	3.34	2.04
CN8	2.08	1.13	1.41	1.65	3.78	3.04	1.11	1.00	1.51	5.64	1.23	3.71	2.27
CN9	1.37	0.75	0.93	1.09	2.50	2.01	0.73	0.66	1.00	3.73	0.81	2.45	1.50
CN10	0.37	0.20	0.25	0.29	0.67	0.54	0.20	0.18	0.27	1.00	0.22	0.66	0.40
CN11	1.68	0.92	1.14	1.34	3.07	2.47	0.90	0.81	1.23	4.58	1.00	3.01	1.84
CN12	0.56	0.31	0.38	0.44	1.02	0.82	0.30	0.27	0.41	1.52	0.33	1.00	0.61
CN13	0.92	0.50	0.62	0.73	1.67	1.34	0.49	0.44	0.67	2.49	0.54	1.64	1.00

Table 6 Final ranking of CNs

	Importance	Normalized	Rank		Importance	Normalized	Rank
CN1	0.51	0.075	8	CN8	0.67	0.100	1
CN2	0.63	0.095	3	CN9	0.57	0.086	6
CN3	0.58	0.088	5	CN10	0.28	0.041	13
CN4	0.55	0.083	7	CN11	0.62	0.093	4
CN5	0.39	0.055	12	CN12	0.37	0.055	11
CN6	0.41	0.062	10	CN13	0.48	0.072	9
CN7	0.66	0.096	2				

Table 7 Detected DRs for a smart bicycle design

	Design requirements		Design requirements
DR1	Mobile phone stabilizer	DR11	Practical partitions
DR2	Geometry of bicycle frame	DR12	Speed increasing design
DR3	Gear quality	DR13	Light and sound features
DR4	Shock absorber/suspension	DR14	Position and number of bicycle fender
DR5	Disc brake	DR15	Multiple saddle
DR6	Pedal choice	DR16	Bicycle fork
DR7	Right tire and tire width	DR17	Bicycle frame material
DR8	Handlebar width and material	DR18	Lock system and traceability
DR9	Chain quality	DR19	Foldable design
DR10	Saddle comfort		

$$\left(v_j, \alpha_j\right) = 1/m \sum_{i=1}^{m} \Delta^{-1}(r_i, \alpha_i) \times \Delta^{-1}\left(s_{ij}, \alpha_{ij}\right) \tag{13}$$

where, m is the number of CNs, $\left(v_j, \alpha_j\right)$ is the importance of DRs as a result, (r_i, α_i) is the weights of each CN and $\left(s_{ij}, \alpha_{ij}\right)$ is the values in relationship matrix for *ith* CN and *jth* DR.

The importance is measure for DRs on the total impact on CNs. DR with the highest importance critical points to consider during the designing phase.

5 Results and Discussion

According to the applied case study, the most three important DRs for a smart bicycle design are:

1. DR13 Light and sound features
2. DR7 Right tire and tire width

Table 8 Evaluation of the third expert as an example

	DR1	DR2	DR3	DR4	DR5	DR6	DR7	DR8	DR9	DR10	DR11	DR12	DR13	DR14	DR15	DR16	DR17	DR18	DR19
C1																			VH
C2		VH		H			M										H		
C3					M	M					VL		H			M	L	VH	
C4	VH										VL		VH					H	
C5														VH					
C6		H							H								VH		VL
C7			VH			M	H			VL		VH							
C8	H			H		M		H		VH	M		M	H	M	L			
C9		VL		VH												VH			
C10								M				H		VL					
C11					VH														
C12	L										VH				VH				
C13							H												

Table 9 Ultimate aggregated relationship matrix for smart bicycle design

	DR1	DR2	DR3	DR4	DR5	DR6	DR7	DR8	DR9	DR10	DR11	DR12	DR13	DR14	DR15	DR16	DR17	DR18	DR19
C1																			(L, 0.31)
C2		(L, 0.31)		(L, 0.05)			(L, − 0.34)										(L, 0.05)		
C3					(L, − 0.34)	(L, − 0.34)					(VL,0)		(L, 0.05)			(L, − 0.34)	(VL, 0.26)	(L, 0.31)	
C4	(L, 0.31)										(VL,0)		(L, 0.31)					(L, 0.05)	
C5														(L, 0.31)					
C6		(L, 0.05)							(L, 0.05)								(L, 0.31)		(VL,0)
C7			(L, 0.31)			(L, − 0.34)	(L, 0.05)			(VL,0)		(L, 0.31)							
C8	(L, − 0.08)			(L, 0.05)		(L, − 0.34)		(L, 0.05)		(L, 0.31)	(L, − 0.34)		(L, − 0.34)	(L, 0.05)	(L, − 0.34)	(VL, 0.26)			
C9		(VL,0)		(L, 0.31)												(L, 0.31)			
C10								(L, − 0.34)				(L, 0.05)		(VL,0)					
C11					(L, 0.31)														
C12	(VL, 0.26)									(L, 0.31)				(L, 0.31)					
C13							(L, 0.05)												

Table 10 Rating of the DRs

Rank	DR	2-Tuple value	Rank	DR	2-Tuple value
1	DR13	(L, −0.31)	9	DR12	(L, −0.45)
2	DR7	(L, −0.36)	10	DR1	(L, −0.46)
3	DR4	(L, −0.38)	11	DR6	(VL, −0.47)
4	DR17	(L, −0.41)	12	DR16	(VL, 0.47)
5	DR5	(L, −0.42)	13	DR2	(VL, 0.46)
6	DR15	(L, −0.43)	14	DR8	(VL, 0.38)
6	DR11	(L, −0.43)	15	DR3	(VL, 0.34)
7	DR14	(L, −0.44)	16	DR10	(VL, 0.33)
8	DR18	(L, −0.45)	17	DR9	(VL, 0.3)
			18	DR19	(VL, 0.3)

3. DR4 Shock absorber/suspension
4. DR17 Bicycle frame material
5. DR5 Disc brake.

As it is seen from Table 9, DRs detected as the most essential ones are the requirements related to the drive and comfort quality, since the customer group and the experts assigned driving comfort as the most important criteria while designing smart bicycle. So, the fundamental issue while developing a smart bike is to create a comfortable driving experience. Moreover, the targets are assigned to each DR with experts and they are given in Table 11. According to the targets and the rating of DRs a smart bicycle design proposition is made.

Since we consider the bicycle to be more functional in terms of using electrical, we believe that an electric bicycle will meet the above specifications and become a more suitable product for customers.

Table 11 Targets assigned for each DR

DR	Target	DR	Target
DR1	Yes	DR11	First aid/maintenance bag
DR2	Double triangle	DR12	Electric motor
DR3	Twelve speed SPRAM 5 × 5	DR13	Yes
DR4	Double suspension	DR14	Two (back and front)
DR5	Yes	DR15	Two saddles
DR6	Locked pedal	DR16	Solid fork
DR7	2.1 inc and more	DR17	Carbon
DR8	Soft- 60 cm	DR18	Micro-GPS
DR9	SHIMANO chain	DR19	Yes
DR10	Adjustable saddle		

Design proposal: *Carbon Body Electrical Bicycles Connected to The Internet.*

This bike has an e-tron electric motor and is ultra-light thanks to its carbon body. The weight of the carbon fiber body is only 1.6 kg. Bicycles, each with a weight of 600 grams, electric motors and batteries do not exceed 21 kilograms. The E-bike has a 3 HP electric motor that produces 250 Nm of torque. The bike's electric motor can be charged in 2.5 h. The e-bike is available in three driving modes and can only be operated with pedal power. This allows the bike to reach a speed of 80 km. Another innovation is that the bicycle can be controlled via the smartphone. Anti-stealing, automobile immobilizer technology used in bicycles, the user's smart phone via Wi-Fi allows communication. When the e-bike communicates with the smartphone, the immobilizer is disengaged. Also includes a USB port and headphone jack.

As a result;

1. Thanks to the electric motor speed increasing design feature is satisfied.
2. Complete carbon body feature allows the customer expectations about the driving quality are provided.
3. USB input, headphone jack and WI-FI connection to smartphones provides the customer expectations about the light and sound features are obtained.
4. Micro-GPS add-in augments the traceability and the driving comfort.

6 Conclusion and Perspectives

In today's competitive world, product design is known to have a key position in new product development. Businesses develop products to serve many customers with different needs. The more suitable the designed products to the needs of the customers, the greater the acceptability. Therefore, a customer-oriented approach is crucial in product design processes.

In product design processes, customers' expectations from the product must be correctly analyzed. The QFD approach, which is known as a customer-oriented systematic method aimed at guaranteeing the general quality in the process of product development/design, is used frequently in the process of understanding CNs and converting them into technical requirements.

QFD is also a GDM approach since it is an approach involving many DMs. Considering the GDM approach, the DMs' evaluations, in terms of their opinions or educational levels etc., can be in different formats. Some DMs may want to present a linguistic assessment, and some may wish to assess the criteria numerically. Therefore, in this study, multi-preference relations technique was used in the evaluation of CNs in QFD. In addition, in the evaluation of technical requirements, the QFD was extended with 2-tuple linguistic model to overcome the multi-granularity in linguistic evaluations. Smart bicycle design has been discussed in this study and the most important technical requirement with the proposed approach is determined as "Light and sound features".

The contributions of this study can be summarized as follows:

In this study, CNs and technical requirements were evaluated for smart bicycle design and the results were interpreted.

A customer-oriented product design approach was presented by taking the evaluations from a large group of possible bicycle customers, and three experts' opinions.

A more flexible GDM process has been introduced by combining individual evaluations in different formats.

The QFD approach is used with multi-preference relations technique and extended with 2-tuple linguistic information.

For future works, this study can be extended by taking into consideration of incomplete information of DMs by using incomplete preference relations technique to reduce the uncertain nature of GDM. Also, in future research, some other types of fuzzy data could be adopted in QFD such as intuitionistic fuzzy sets. Besides, the proposed approach can be extended by applying different aggregation operators such as ordered weighted averaging (OWA), majority additive OWA (MA-OWA) or induced ordered weighted geometric (IOWG) operators, and different GDM approaches may be proposed to find the consensus of assessments in QFD.

Acknowledgements The authors would like to thank the industrial experts for their support, as well as acknowledge the grants of Galatasaray University Research Fund (Projects Numbers: 19.402.001, 19.402.003 and 19.402.006).

References

1. Xu, Y., Chen, G., Zheng, J.: An integrated solution—KAGFM for mass customization in customer-oriented product design under cloud manufacturing environment. Int. J. Adv. Manuf. Technol. **84**, 85–101 (2016). https://doi.org/10.1007/s00170-015-8074-2
2. Akao, Y.: An Introduction to Quality Function Deployment. Productivity Press, Cambridge, Massachusetts (1990)
3. Martínez, L., Rodriguez, R.M., Herrera, F.: The 2-tuple Linguistic Model. Springer International Publishing, Cham (2015)
4. Büyüközkan, G., Güleryüz, S.: Extending fuzzy QFD methodology with GDM approaches: an application for IT planning in collaborative product development. Int. J. Fuzzy Syst. **17**, 544–558 (2015). https://doi.org/10.1007/s40815-015-0065-9
5. Büyüközkan, G., Ilıcak, Ö.: Integrated SWOT analysis with multiple preference relations: selection of strategic factors for social media. Kybernetes **48**, 451–470 (2019)
6. Iranmanesh, H., Tabrizi, B.H.: An integrated framework for customer-oriented web design using QFD, Kano model and ANP. In: 2009 International Conference on Computers & Industrial Engineering, pp. 1674–1679. IEEE, Troyes, France (2009)
7. Chen, L.-H., Ko, W.-C.: Fuzzy linear programming models for new product design using QFD with FMEA. Appl. Math. Model. **33**, 633–647 (2009). https://doi.org/10.1016/j.apm.2007.11.029
8. Lee, A.H.I., Lin, C.-Y.: An integrated fuzzy QFD framework for new product development. Flex. Serv. Manuf. J. **23**, 26–47 (2011). https://doi.org/10.1007/s10696-011-9076-5
9. Camgoz-Akdag, H., Zaim, S., Acar, M.F., et al.: Product improvement with quality function deployment (QFD) technique. Adv. Mater. Res. **445**, 1058–1063 (2012). https://doi.org/10.4028/www.scientific.net/AMR.445.1058

10. Wang, C.-H., Chen, J.-N.: Using quality function deployment for collaborative product design and optimal selection of module mix. Comput. Ind. Eng. **63**, 1030–1037 (2012). https://doi.org/10.1016/j.cie.2012.06.014
11. Chen, L.-H., Ko, W.-C., Tseng, C.-Y.: Fuzzy approaches for constructing house of quality in QFD and its applications: a group decision-making method. IEEE Trans. Eng. Manag. **60**, 77–87 (2013). https://doi.org/10.1109/TEM.2012.2204063
12. Yuen, K.K.F.: A hybrid fuzzy quality function deployment framework using cognitive network process and aggregative grading clustering: an application to cloud software product development. Neurocomputing **142**, 95–106 (2014). https://doi.org/10.1016/j.neucom.2014.03.045
13. Ionica, A.C., Leba, M.: QFD integrated in new product development—biometric identification system case study. Procedia Econ. Finance **23**, 986–991 (2015). https://doi.org/10.1016/S2212-5671(15)00454-2
14. Aghdam, M.M., Mahdavi, I., Shirazi, B., Vahidi, J.: House of quality improvement by new design requirements generation. In: 2015 International Conference on Industrial Engineering and Operations Management (IEOM), pp 1–9. IEEE, Dubai (2015)
15. Lin, C.-Y., Lee, A.H.I., Kang, H.-Y.: An integrated new product development framework—an application on green and low-carbon products. Int. J. Syst. Sci. **46**, 733–753 (2015). https://doi.org/10.1080/00207721.2013.798447
16. Zheng, P., Xu, X., Xie, S.Q.: A weighted preference graph approach to analyze incomplete customer preference information in QFD product planning. In: 2016 IEEE International Conference on Industrial Engineering and Engineering Management (IEEM), pp 1070–1074. IEEE, Bali, Indonesia (2016)
17. Liu, A., Hu, H., Zhang, X., Lei, D.: Novel two-phase approach for process optimization of customer collaborative design based on fuzzy-QFD and DSM. IEEE Trans. Eng. Manag. **64**, 193–207 (2017). https://doi.org/10.1109/TEM.2017.2651052
18. Ahmadabadi, H.Z., Zamzam, F., Meybodi, F.R. et al.: Development of a new sesame product using QFD and DOE methods: a case study of sesame product in yazd. Montenegrin J. Econ. **14**, 27–44 (2018). https://doi.org/10.14254/1800-5845/2018.14-1.2
19. Dursun, M., Arslan, Ö.: An integrated decision framework for material selection procedure: a case study in a detergent manufacturer. Symmetry **10**, 657 (2018). https://doi.org/10.3390/sym10110657
20. Alptekin, S.E., Alptekin, G.I.: A fuzzy quality function deployment approach for differentiating cloud products. Int. J. Comput. Intell. Syst. **11**, 1041 (2018). https://doi.org/10.2991/ijcis.11.1.79
21. Huang, J., You, X.-Y., Liu, H.-C., Si, S.-L.: New approach for quality function deployment based on proportional hesitant fuzzy linguistic term sets and prospect theory. Int. J. Prod. Res. **57**, 1283–1299 (2019). https://doi.org/10.1080/00207543.2018.1470343
22. Zhang, X.: User selection for collaboration in product development based on QFD and DEA approach. J. Intell. Manuf. **30**, 2231–2243 (2019). https://doi.org/10.1007/s10845-017-1386-3
23. Büyüközkan, G., Feyzioğlu, O.: Group decision making to better respond customer needs in software development. Comput. Ind. Eng. **48**, 427–441 (2005)
24. Büyüközkan, G., Feyzioğlu, O., Ruan, D.: Fuzzy group decision-making to multiple preference formats in quality function deployment. Comput. Ind. **58**, 392–402 (2007)
25. Li, Y.-L., Tang, J.-F., Chin, K.-S., et al.: On integrating multiple type preferences into competitive analyses of customer requirements in product planning. Int. J. Prod. Econ. **139**, 168–179 (2012). https://doi.org/10.1016/j.ijpe.2012.03.031
26. Wang, Y., Zhang, Z., Koh, C.-K., et al.: Passivity enforcement for descriptor systems via matrix pencil perturbation. IEEE Trans. Comput.-Aided Des. Integr. Circuits Syst. **31**, 532–545 (2012)
27. Büyüközkan, G., Çifçi, G.: An integrated QFD framework with multiple formatted and incomplete preferences: a sustainable supply chain application. Appl. Soft Comput. **13**, 3931–3941 (2013)
28. Büyüközkan, G., Çifçi, G.: An extended quality function deployment incorporating fuzzy logic and GDM under different preference structures. Int. J. Comput. Intell. Syst. **8**, 438–454 (2015). https://doi.org/10.1080/18756891.2015.1017379

29. Li, Z., Gao, Q., Zhang, D., Liu, G.: A New Method of Rating the Importance of Customer Needs in Quality Function Deployment. IEEE, New York (2008)
30. Dursun, M., Karsak, E.E.: Supplier selection using an integrated decision making approach based on QFD and 2-tuple fuzzy representation. World Congress on Engineering and Computer Science, WCECS 2012, vol. Ii, pp. 1309–1315. Int Assoc Engineers-Iaeng, Hong Kong (2012)
31. Li, M.: The extension of quality function deployment based on 2-tuple linguistic representation model for product design under multigranularity linguistic environment. Math. Probl. Eng. 989284. (2012). https://doi.org/10.1155/2012/989284
32. Wang, S.-Y.: Applying the superior identification group linguistic variable to construct kano model oriented quality function deployment. Technol. Econ. Dev. Econ. **19**, S304–S325 (2013). https://doi.org/10.3846/20294913.2013.880082
33. Ai, Q., Shu, T., Liu, Q., et al.: A method for determining customer requirement weights based on TFMF and TLR. Enterp. Inf. Syst. **7**, 569–580 (2013). https://doi.org/10.1080/17517575.2012.763190
34. Mi, C., Qiang, Y., Liu, S., et al.: An integrated failure prioritizing model of complex equipment. J. Grey. Syst. **27**, 39–50 (2015)
35. Karsak, E.E., Dursun, M.: An integrated fuzzy MCDM approach for supplier evaluation and selection. Comput. Ind. Eng. **82**, 82–93 (2015). https://doi.org/10.1016/j.cie.2015.01.019
36. Sheng, Z., Yan, L.: The quality function deployment and 2-tuple linguistic based approach to third party logistics supplier selection. Sichuan Univ Press, Chengdu (2016)
37. Wang, Z.-L., You, J.-X., Liu, H.-C.: Uncertain quality function deployment using a hybrid group decision making model. Symmetry-Basel **8**, 119 (2016). https://doi.org/10.3390/sym8110119
38. Li, X., He, Z.: Determining importance ratings of patients' requirements with multi-granular linguistic evaluation information. Int. J. Prod. Res. **55**, 4110–4122 (2017). https://doi.org/10.1080/00207543.2016.1253890
39. Buyukozkan, G., Uzturk, D.: Combined QFD TOPSIS approach with 2-tuple linguistic information for warehouse selection. In: 2017 IEEE International Conference on Fuzzy Systems (Fuzz-IEEE). IEEE, New York (2017)
40. Liu, M., Gao, Q.: Supplier evaluation in TSC based on fuzzy linguistic term sets and QFD. In: 2017 29th Chinese Control and Decision Conference (CCDC), pp 4671–4675. IEEE, New York (2017)
41. He, L., Ming, X., Li, M., et al.: Understanding customer requirements through quantitative analysis of an improved fuzzy Kano's model. Proc. Inst. Mech. Eng., Part B: J. Eng. Manuf. **231**, 699–712 (2017). https://doi.org/10.1177/0954405415598894
42. Zhang, X., Su, J.: An integrated QFD and 2-tuple linguistic method for solution selection in crowdsourcing contests for innovative tasks. J. Intell. Fuzzy Syst. **35**, 6329–6342 (2018). https://doi.org/10.3233/JIFS-181122
43. Mei, Y., Liang, Y., Tu, Y.: A multi-granularity 2-tuple QFD method and application to emergency routes evaluation. Symmetry-Basel **10**, 484 (2018). https://doi.org/10.3390/sym10100484
44. Mi, C., Chen, Y., Zhou, Z., Lin, C.-T.: Product redesign evaluation: An improved quality function deployment model based on failure modes and effects analysis and 2-tuple linguistic. Adv. Mech. Eng. **10**, 1687814018811227 (2018). https://doi.org/10.1177/1687814018811227
45. Herrera, F., Martínez, L.: A 2-tuple fuzzy linguistic representation model for computing with words. IEEE Trans. Fuzzy Syst. **8**, 746–752 (2000)
46. Akao, Y., Mazur, G.H.: The leading edge in QFD: past, present and future. Int. J. Qual. Reliab. Manag. **20**, 20–35 (2003)
47. Akao, Y.: QFD: integrating customer requirements into product design. Camb MA (1990)
48. Zhang, B., Dong, Y., Xu, Y.: Maximum expert consensus models with linear cost function and aggregation operators. Comput. Ind. Eng. **66**, 147–157 (2013). https://doi.org/10.1016/j.cie.2013.06.001
49. Xia, M., Xu, Z., Chen, N.: Some hesitant fuzzy aggregation operators with their application in group decision making. Group Decis. Negot. **22**, 259–279 (2013). https://doi.org/10.1007/s10726-011-9261-7

50. Herrera, F., Herrera-Viedma, E., Chiclana, F.: Multiperson decision-making based on multiplicative preference relations. Eur. J. Oper. Res. **129**, 372–385 (2001). https://doi.org/10.1016/S0377-2217(99)00197-6
51. Herrera, F., Herrera-Viedma, E., Martínez, L.: A fusion approach for managing multi-granularity linguistic term sets in decision making. Fuzzy Sets Syst. **114**, 43–58 (2000). https://doi.org/10.1016/S0165-0114(98)00093-1

Part III
Customer Oriented Product Design Using Axiomatic Design Method

An Axiomatic Design Framework of Sustainable Product-Service Systems for Circular Economies

Alperen Bal and Sule Itir Satoglu

Abstract In recent years, sustainable product design process gained notable attentions and evolved having Triple Bottom Line (TBL) accounting perspective. TBL requires integrating environmental, social as well as economic factors at the design phase. Making a product sustainable is important from the customers' point of view, because today customers show tendency not to prefer non-sustainable products. Besides, customers today are more and more interested in accessing to product usage or functions instead of its ownership. So, in this study, an Axiomatic Design Framework is proposed for the Sustainable Product-Service Systems, for the first time in the literature, so that a complete conceptual design is made based on the environmental, economic and social dimensions for the products that concerns the customer expectations for the circular economy. Besides, a washing machine design is discussed based on the proposed sustainable product service-system framework.

1 Introduction

Sustainability is defined as the ability of meeting the needs of the present without compromising the ability of future generations to meet their needs. Traditionally, firms are focused on economic incentives. However, considering merely the economic objectives is not sufficient. The triple bottom-line approach of sustainability suggests that the economic, environmental and social aspects must be considered during the strategic, tactical and operational decisions [23].

In recent years, sustainable product design process gained notable attentions and evolved having Triple Bottom Line (TBL) accounting perspective. TBL requires integrating environmental factors (e.g., End of Life recovery strategies, energy use, water consumption), social factors (e.g., health and safety issues) as well as economic factors at the design phase. Making a product sustainable is important from the

A. Bal
Faculty of Engineering, Yalova University, Yalova, Turkey

S. I. Satoglu (✉)
Faculty of Management, Istanbul Technical University, Besiktas, Istanbul, Turkey
e-mail: onbaslis@itu.edu.tr

C. Kahraman and S. Cebi (eds.), *Customer Oriented Product Design*, Studies in Systems, Decision and Control 279, https://doi.org/10.1007/978-3-030-42188-5_8

customers' point of view, because today customers show tendency not to prefer non-sustainable products [29]. Results of the Green Brand Survey which is one the largest survey with 9000 consumers in the Australia, Brazil, China, France, Germany, India, UK and US indicates that 50% of consumers have concerns on environment and expect environmental pillar should be included in the products [12].

Bovea and Belis [6] comprehensively present environmental aspects into the design process. Used products should be rehabilitated using different strategies. Implementing End of Life (EoL) strategies ensures a closed-loop approach to material flow [22]. EoL product recovery includes reuse, remanufacture and/or recycling after product use. These operations reduce required raw material usage and energy consumption in production. For instance, in the apparel industry virgin polyester is intensively used. One of the largest apparel and footwear manufacturer Adidas set their goal to get rid of virgin polyester by 2024 [32]. Thus, all footwear produced by Adidas will be having recycled polyester. Reducing virgin material provides significant advantages such as using less energy for production while preventing EoL products/materials going into the nature. In addition, use of raw materials is crucial in many sectors such as Electrical and Electronics Equipment (EEE) production. Especially, rarely find elements used in EEE (e.g., Indium, Palladium, Samarium, Neodymium, Gallium, Cobalt, Platinum) are highly critical due to economic importance and supply risk [8]. Recycling of waste EEE products must be planned in a multi-period and multi-product basis considering the economic, environmental and sustainable targets, based on a TBL perspective [4].

On the other hand, social pillar requires designing healthier products for the community. Hebrok [20] studied the social aspects of product-life. However, the social aspects of sustainability attracted relatively less attention on sustainable product design studies.

To address the environmental and social concerns mentioned above, the concept of Circular Economy has recently gained importance. Product characteristics should be considered during the design phase to reach the Circular Economy. Circular Economy concept enables new business models in support of the technological advances. Besides, customers today are more and more interested in accessing to product usage or functions instead of its ownership [37] Product Service System (PSS) concept typically makes manufacturer to retain product ownership but to provide the customer the right to use it for a fee. In a classical sell model, the Original Equipment Manufacturer (OEM) delivers the product to the vendor and customer buys the product. However, in a PSS, the OEM holds the possession of it. Thus, products can be pursued by the OEM throughout their lifecycle. Three main categories of PSS are identified by Tukker [36]. First one is use-oriented PSS. The product is still sold to customer but some additional services, such as maintenance contracts, are added. Second one is use-oriented PSS where the product ownership is retained by the provider, and the product is offered through several forms, e.g. renting, leasing, pay-per-use or sharing. Third one is result-oriented PSS. In this category, the provider agrees in principle on a result, with no pre-determined product involved. Catering services and pay-per-use contract are typical examples of this category.

PSS enables manufacturers to collect end-of-life (EoL) products more easily compared with the traditional sell model. Besides, products can be used more than once in their lifecycle by applying reuse and remanufacture. Thus, repaired/refurbished products can have second lifecycle to go into the market again. Besides, since the ownership belongs to the manufacturer, no purchasing cost but a probably lower leasing cost incurs to the customer.

To achieve sustainable product design, different tools are available such as Quality Function Deployment (QFD), Life Cycle Assessment (LCA), Theory of Inventive Problem Solving (TRIZ) and Axiomatic Design (AD). Ahmad et al. [1] reviewed these tools and applications for sustainable product design. The authors mentioned that sustainable product design methods that consider all of the environmental, economic and social domains are limited, but some of these are merely considered in most of the past studies. New approaches and new business models to product design is required considering TBL perspective. In this context, this chapter presents a sustainable product design framework using AD. Figure 1 explains how the proposed product design framework pave the way for a closed loop material flow.

The raw materials extracted from the nature are purchased by the component manufacturers and these are converted into the parts/components of the product, by means of the manufacturing processes. Then the purchased components are processed and/or assembled according to the product design and hence the manufacturing of the products are completed. The products are either sold or leased to the customers

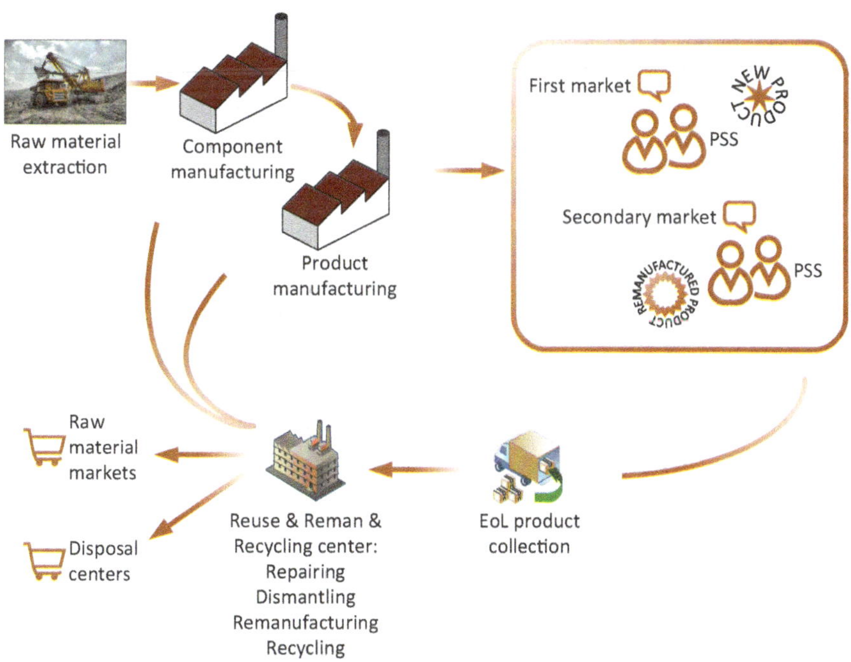

Fig. 1 Closed-loop material flow for sustainably designed products via PSS business model

as brand-new products. Otherwise, remanufactured products are either sold or leased to the customers for the reuse. In a Product Service System (PSS), the manufacturer retains the product ownership and provides the customer the right to use it at the expense of a fee. By means of the PSS, the EoL products can be better kept track of and collected from the users.

These EoL products are later sent to the Reuse, Remanufacturing and Recycling centers where it is decided whether the product can be reused after some refurbishment/repair, and the components can be dismantled for remanufacturing and reuse or for recycling. Ultimately, the material and components that cannot be repaired and reused are recycled. The cycled materials are sold to the raw material markets and the material that could not be recycled are disposed.

However, the product design significantly affects disassembly and material recovery from the EoL products, and also the manufacturing costs of the products. Several past studies suggested that the majority of the manufacturing costs can be determined at the product design stage [38]. Besides, during the design stage, the selection of the environment-friendly materials [28] and components are strongly desired. The novel methodologies that focus on Design-for-Disassembly [19, 21] are also strongly required to be implemented at the product design stage, as the products' bill-of-materials become more complex that may cause difficulty in disassembly and component or material recovery unless the product design is not performed properly.

In this study, an Axiomatic Design (AD) methodology has been used for the sustainable product-service system design, because AD provides a scientific basis to the design studies. As it instructs the designer to determine the functional requirements of the design based on the customer expectations, and the design parameters that helps fulfill these requirements, the trial-and-error type non-value added activities are eliminated [34]. Coupled designs are also prevented by means of the Information Axiom of the AD.

To the best of authors' knowledge, this is the first study that proposes an axiomatic design framework for a sustainable product-service system design. Besides, any past study that developed an axiomatic model for the sustainable product designs could not be encountered. These show the unique aspect of our study.

The chapter is organized as follows: In the next section the fundamentals of the Axiomatic Design methodology proposed by Suh are explained, and the proposed Axiomatic Design methodology for the Sustainable Product-Service System is explained in detail. In the Application Section, the proposed framework is applied on a white goods product design case. Finally, the conclusion and future work of the study are presented.

2 Methodology

In this section, first the fundamental of the Axiomatic Design methodology proposed by Suh [34] is explained. Later, the framework of Sustainable Product Service System developed based on the Axiomatic Design principles is explained, in detail.

2.1 Fundamentals of Axiomatic Design Methodology

Traditional design methodologies are usually based on a trial-and-error approach such that first an initial or pilot design is made and it is intended to be improved incrementally. In other words, "Design has depended on intuitive and innate reasoning rather than on rigorous scientific study [34]". However, Axiomatic Design Methodology provides a scientific basis to all design activities. It suggests the designer to systematically consider what the design must achieve functionally, and how these requirements will be achieved.

In the Axiomatic Design methodology there are important concepts that needs to be comprehended at the beginning. The *functional requirements* (*FR*s) are the minimum set of independent requirements of the design that need to be satisfied. In other words, these show *what* functions the design must be able to carry out. These must be defined in the *Functional Domain* according to the *customer expectations*. The *design parameters* (*DP*s) are the physical means of satisfying the functional requirements [34]. In other words, these show *how* the functional requirements can be achieved. However, there are limitations on the acceptable solutions that are called *constraints* that need to be considered during the design endeavor.

There are Independence and Information Axioms of Axiomatic Design. Axiom 1 that is the Independence Axiom stipulates that the independence of the functional requirements of a specific design must be achieved. Axiom 2 that is the Information Axiom orders selection of the design that requires the minimum information content [34]. In other words, the simplest design must be selected among the alternatives. The Independence Axiom stipulates that the functions of a design must be independent from each other, but the physical parts are allowed to be dependent, and this does not violate the Independence Axiom [34].

The relationship among the FRs and DPs are shown in Eq. (1). Here, *FR* and *DP* represent the vectors of the functional requirements and that of the design parameters, respectively. $[A]$ represent the design matrix based on which we can decide whether the Independence Axiom is satisfied. So, the design matrix explicitly shows the relationship among the FRs and DPs.

$$\{FR\} = [A]\{DP\} \tag{1}$$

The Design matrix is shown in Eq. (2) for a case of three FRs and DPs.

$$[A] = \begin{bmatrix} A_{11} & A_{12} & A_{13} \\ A_{21} & A_{22} & A_{23} \\ A_{31} & A_{32} & A_{33} \end{bmatrix} \tag{2}$$

There are three types of a product or system design in terms of independence of the functional requirements. In an uncoupled design that is ideal, the design matrix has elements on the diagonal and other elements are equal to zero. Equation (3) shows an example of an uncoupled design matrix.

$$[A] = \begin{bmatrix} A_{11} & 0 & 0 \\ 0 & A_{22} & 0 \\ 0 & 0 & A_{33} \end{bmatrix} \tag{3}$$

In a decoupled design, the matrix must be either upper diagonal or lower diagonal. A decoupled design is still an acceptable one in terms of the Independence Axiom. Equation (4) shows an example of the decoupled design matrix.

$$[A] = \begin{bmatrix} A_{11} & 0 & 0 \\ A_{21} & A_{22} & 0 \\ A_{31} & A_{32} & A_{33} \end{bmatrix} \tag{4}$$

However, in a coupled design, several the functional requirements affect each other, and the design matrix has elements at the upper and bottom of the diagonal. A coupled design is not intended and violates the Independence Axiom [34]. An example of a coupled design is shown in Eq. (5).

$$[A] = \begin{bmatrix} A_{11} & 0 & A_{13} \\ A_{21} & A_{22} & A_{23} \\ 0 & A_{32} & A_{33} \end{bmatrix} \tag{5}$$

After this short summary of the fundamentals of the Axiomatic Design Methodology, the pioneer studies that employed the Axiomatic Design principles during the systems design will be explained. Kulak et al. [26] reviewed Axiomatic Design literature and concluded that most of the past studies the Information Axiom to select the most suitable design alternative among the candidates, but the design studies on the basis of Independence Axiom are limited.

Axiomatic Design was used for the conceptual design of the software systems [25], manufacturing systems [33], flexible manufacturing systems [2], cellular manufacturing systems [27] and the hybrid cellular manufacturing systems [15], on the basis of the Independence Axiom. Cebi and Kahraman [10] extended the Independence Axiom of AD into the fuzzy Information Axiom, and proposed that the relationship between the FRs and DPs can be uncertain and should be defined in terms of fuzzy membership functions. The authors also applied of the fuzzy Independence Axiom for the automobile seat design in a subsequent study [9]. The most recent AD study for a manufacturing system design belongs to Delaram and Valilai [13] where the authors developed an Axiomatic framework for the computer integrated manufacturing systems.

The work of Stiassnie and Shpitalni [31] is a pioneer that considered the environmental issues during the Axiomatic Design of the products. Zhang and Li [39] attempted to estimate the total energy content of the product through its whole lifecycle including the raw material extraction, manufacturing, assembly, use, disassembly and recycling, but the authors did not provide any axiomatic design framework. Later,

Du et al. [14] developed an Axiomatic Design and Quality Function Deployment methodology for redesigning the used products, for re-use purposes.

2.2 *Proposed Sustainable Product Service System Framework*

In this sub-section, we explain the Axiomatic Design Framework for Sustainable Product-Service Systems. At first, the highest level FR is defined as follows: FR: Design the Product for Sustainability.

Then, we zigzag to the physical domain and determine the associated DP as follows: DP: Sustainable Product Design Methodology. Later, by returning to the Functional domain, the highest level FR s decomposed into the following FRs, as shown in Table 1.

After returning to the physical domain, the DPs to satisfy the above mentioned design parameters are determined as shown in Table 2.

Figure 2 illustrates the decomposition of the highest level FR and DP, and the relationships among the DPs and FRs. In other words, the DPs that affect several FRs are shown with the arrows in this figure. DP1 Customer-Product Survey must be

Table 1 The decomposition of the highest level FR

FRs-explanation
FR1: Determine the customer expectations
FR2: Learn the laws and patents about the product design
FR3: Design the product with environmental consciousness
FR4: Select environment friendly manufacturing processes
FR5: Minimize the material and manufacturing costs of the product
FR6: Optimize the energy consumption and performance of the product
FR7: Minimize the maintenance and usage costs of the product after sales/lease

Table 2 The decomposition of the highest level DP

DPs-explanation
DP1: Customer-product survey
DP2: Legislation and patent search about the specific product type
DP3: Environment friendly product design
DP4: Green production processes
DP5: Cost optimization of the product design
DP6: Energy and performance efficient product design
DP7: Customer-oriented product-service system model

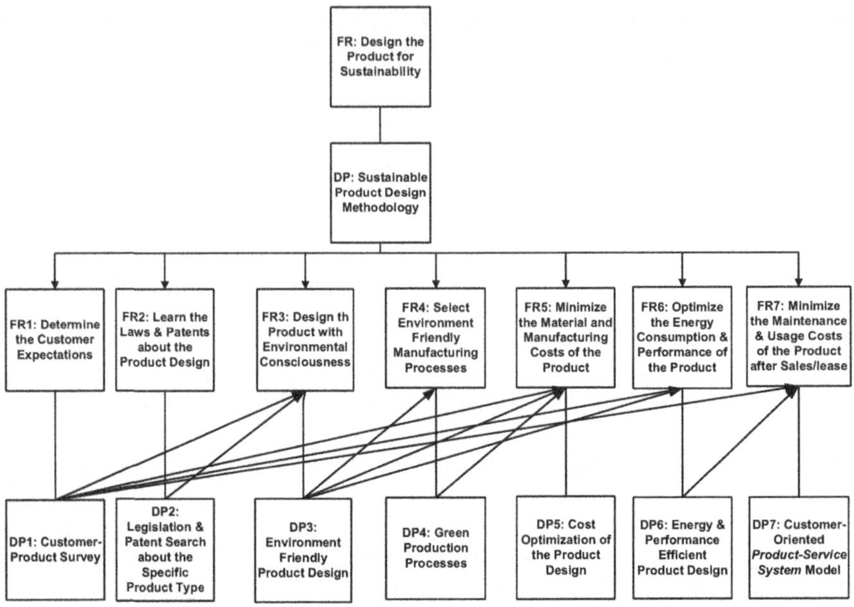

Fig. 2 The decomposition of the highest level FR and DP

conducted so that the customer expectations are well comprehended and considered during the product design. So, this DP1 affects FR3, since many customers expect environmentally-conscious product design; and affects FR5 as the product-design must be cost-effective for an economically sustainable product design. DP1 also affects FR6, since there is a trade-off between the energy consumption and performance of the product while the product is used by the customers, and the customers expect both an energy-effective and high-performance product design. Besides, DP1 also affects DP7, since Product Service Systems provide maintenance services to the customer cost-effectively, and recovery of the EoL products are better managed by means of the PSSs.

Besides, there are legal regulations that stipulate recovery or recycling of the EoL products at a target percentage on an annual basis for the manufacturers [4, 24]. So, these legal regulations mentioned in DP2 affect the FR3: Design the Product with Environmental Consciousness. Besides, during the design stage, some patents filed before by the competitors have to be considered during the design. The Environment-Friendly Product Design (DP3) affects the Selection of Environment-Friendly Manufacturing processes (FR4), and Optimization of the Energy Consumption and Performance of the Product (FR6) during the usage stage. This is due to the fact that most of the sustainability influence stems from the product design [1]. However, the manufacturing processes and systems can be improved in terms of energy-efficiency and material wastes, based on lean manufacturing tools and techniques [5]. The energy-efficient and green production processes (DP4) help to decrease the associated material and manufacturing costs (FR5), as the energy consumption will be lower. Energy and Performance Efficient Product Design (DP6) enhances minimization of the maintenance and usage costs of the product (FR7), as the trade-off between the energy consumption and the performance of the product is optimized not to cause excess energy and maintenance costs and reach an acceptable level of performance. Finally, Customer-Oriented Product Service System Model (DP7) should be employed for the minimization of the maintenance and usage costs of the product after sales/lease. New business models where the product is leased to the customer and the maintenance costs are covered by the product-owner help satisfy the customer needs.

The design matrix (A1) between these seven FRs and DPs are presented below, and shows that the design is decoupled and acceptable:

$$A1 = \begin{vmatrix} X & 0 & 0 & 0 & 0 & 0 & 0 \\ 0 & X & 0 & 0 & 0 & 0 & 0 \\ X & X & X & 0 & 0 & 0 & 0 \\ 0 & 0 & X & X & 0 & 0 & 0 \\ X & 0 & X & 0 & X & 0 & 0 \\ X & 0 & X & 0 & 0 & X & 0 \\ X & 0 & 0 & 0 & 0 & X & X \end{vmatrix}$$

When we decompose FR3, we obtain the following FRs and the associated DPs are also denoted below, in Table 3.

As Ljungberg [28] suggested, during a sustainable product design, the materials of the products must be selected based on their environmental impacts and eco-efficiency, and some guidelines are suggested to the product developers. So, Environmentally-Safe Materials and Components need to be Selected (FR31) and an Easy-to-recycle product must be designed (FR32). To achieve these functional requirements, Environment Hazard Analysis of the Raw Materials and Components must be conducted (DP31) and a Recyclable Product Design (DP32) must be developed. As the product components and materials are more widely recycled or recovered, the amount of material disposed to the nature is minimized, that shows the importance of the DP32. The associated design matrix *A2* is presented below, and this

Table 3 The decomposition of the FR4 and DP4

FRs-explanation
FR31: Use environment-safe raw materials and components
FR32: Design the product easy-to-recycle
DP31: Environment hazard analysis of the raw materials and components
DP32: Recyclable product design

implies a decoupled design. DP31 (Environment Hazard Analysis of the Raw Materials and Components) affects both FR31 (Use Environmentally-Safe Raw Materials and Components) and FR32 (Design an Easy-to-Recycle Product).

The associated design matrix *A2* is as follows, and it is a decoupled design:

$$A2 = \begin{vmatrix} \boldsymbol{X} & \mathbf{0} \\ \boldsymbol{X} & \mathbf{1} \end{vmatrix}$$

The decomposition of FR4 results in four new FRs, and the associated DPs are determined as shown in Table 4.

In order to select the Environmental Friendly Manufacturing Processes (FR4), we need to first Determine the alternative Manufacturing Processes for the Product Design (FR41). Hence, many alternative manufacturing processes can be generated. Later, among many alternatives, the Processes with minimum energy consumption must be selected (FR42) by means of an Energy Efficiency Analysis of the Manufacturing Processes (DP42). Baysan et al. [5] explained the energy-efficiency analysis of the manufacturing processes by means of a Value-stream Mapping and energy consumption analysis methodology. After ensuring the energy effectiveness, the Processes with minimum environmental hazardous waste release must be selected (FR43) by means of an Environmental Hazard Analysis of the Process Wastes (DP43). Finally, Occupationally-Safe Manufacturing Processes must be selected (FR44), since human or social dimension of the manufacturing processes must be considered for a sustainable product design. This can be reached by an Occupational Risk Assessment Scheme for the Manufacturing Processes (DP44). Authors would like to mention that during the process selection stages, the Information Axiom of the Axiomatic Design [34] should be employed, so as to determine the most suitable one. There are many applications of the Information Axiom for decision making,

Table 4 The decomposition of the FR4 and DP4

FRs/DPs-explanation
FR41: Determine alternative manufacturing processes for the product design
FR42: Select the processes with minimum energy consumption
FR43: Select the process with minimum hazardous waste release
FR44: Select occupationally-safe manufacturing processes
DP41: Manufacturing process alternatives for the product design
DP42: Energy-efficiency analysis of the manufacturing processes
DP43: Environmental hazard analysis of process wastes
DP44: Occupational risk assessment scheme for the manufacturing processes

such as production system parameter selection [16], robot arm selection [3], 3-D printer selection [11], among many others.

The associated design matrix *A3* that implies a decoupled design is as follows:

$$A3 = \begin{vmatrix} X & 0 & 0 & 0 \\ X & X & 0 & 0 \\ X & X & X & 0 \\ X & 0 & X & X \end{vmatrix}$$

3 Application

A washing machine has been selected to apply proposed sustainable product service system oriented AD methodology. Washing machine is a typically standardized product with respect to product characteristics and manufacturing process. A machine weights between 70–100 kg and contains on average more than 20 different materials encompassing steel, cast iron, concrete, different kinds of plastics, aluminum, glass, printed circuit boards and polypropylene [30]. Components of the washing machine include electric motor, drum, pump for water drainage, heating elements, printed circuit board for the control panel and cables. User interface and top panel are made from plastics. Glass and plastics are used for the loading cover. Cabinet and drum are fabricated from stainless steel. The analyzed washing machine is produced by a famous Turkish appliance manufacturer for mid-market.

Highest level functional domain requires optimizing cost parameters together with environmental and social concerns. FR1 is related to understanding customer expectations. Because of changing technology and evolving business models, customer expectations can change. One of the most suitable way of understanding this change is to conduct surveys. Therefore, current trends in the market can be understood.

At the production phase, it is seen that most of the design characteristics are under protection by patents. Also, manufacturers to obey sector related regulations [17]. FR2 requires following recent regulations for appliances and being aware of patents. For example, a unique S shaped side panel design to prevent noise is under protection by a patent (see Fig. 3).

Environment friendly material use is very important to make a product sustainable. Recyclable or natural material such as wooden material can be used for the upper panel of the washing machine. On the other hand, manufacturing processes of the washing machine should also be environment friendly. In addition to lead time, total energy and water consumption should be calculated. Sustainable Value Stream Mapping (Sus-VSM) is an appropriate tool to assess environmental impact along with lead time of the production process [18]. Today, energy consumption of appliances is strictly regulated. Annual energy consumption is targeted below 175 kwh for the redesigned washing machine which is stated by European Union energy

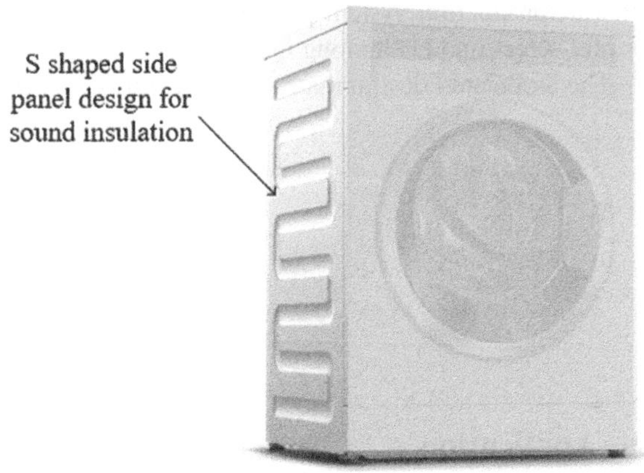

Fig. 3 Patented sound insulation design

label [7]. Another feature of the product should include less water usage and better cleaning performance thanks to sensors for FR6 and FR7. Customer-Oriented *Product-Service System* Model (DP7) requires redesign the product to offer it as a service. Considering, a washing machine is used 175 cycle/year and has 11 years' lifetime [35], looking the idle time of washing machine one can see most of the time the product is idle. This persuades people only to pay whenever they use the service. Thus, use-oriented PSS model can be implemented. Also, the condition of the product must be tracked by the provider to sustain maintenance and to manage collection process of retired items. Thus, IoT sensors are added to the washing machine to make it a connected device.

3.1 Decomposition of Functional Requirement 3: Design the Product with Environmental Consciousness

Environmental concerns require environment friendly product design (DP3). The decomposition of FR3two new FRs and the associated DPs are denoted below in Fig. 4.

Environment friendly raw material and component use is important in manufacturing and use phase as well as recycling. Various components of the washing machine are suitable for remanufacturing and recycling. Also, legislations enforce companies to recover EoL products [17]. In the washing machine; drum, water pump, electric motor, capacitor and electronic board are the key components for disassembly. There are three EoL strategies for retired product: reuse, remanufacture, and recycle. If the product condition is fairly good, they can be resold to secondary markets after refurbishments. Otherwise, remanufacturing option is evaluated. Electric motor and water pump are suitable components for remanufacturing. Recycling is the last option for product recovery. Instead of simply shredding the washing machine, recycling process requires improvement. Design for easier disassembly significantly reduce

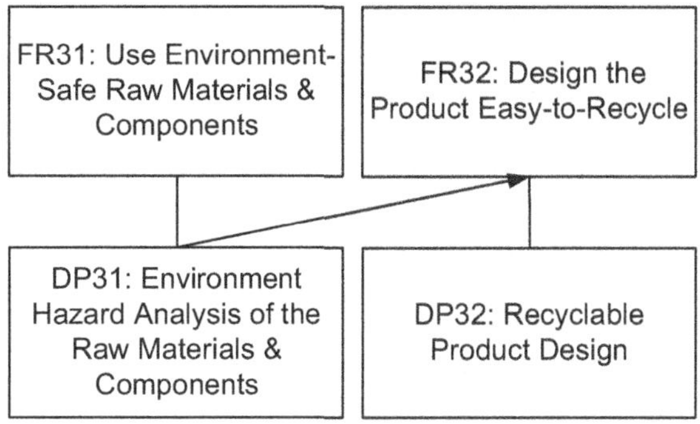

Fig. 4 Decomposition of FR3

recovery time and increase the potential for benefitting more material or component from the EoL product.

On the other hand, recovery of critical raw materials also become a critical issue for manufacturers. As the very essential part of electronics, the printed circuit boards include these elements such as neodimium, samarium and terbium. Therefore, easy to dismantle electronic board is evaluated.

3.2 Decomposition of Functional Requirement 4: Select Environment Friendly Manufacturing Processes

The decomposition of FR4 results in four new FRs, and the associated DPs are determined in the Fig. 5. Use of Sus-VSM [18] is proposed to achieve green production process (DP4) by assessing environmental impact, energy consumption, cost, and waste. Regarding the associated Design Parameters with decomposition of FR4, operational cost items of the manufacturing process needs to be identified. In this

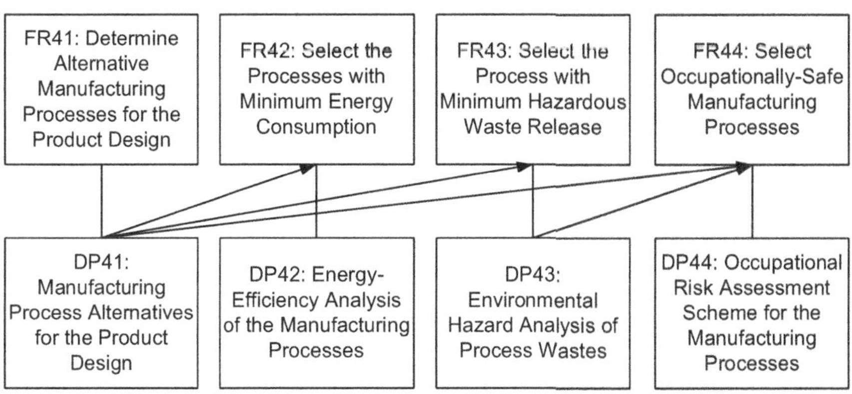

Fig. 5 Decomposition of FR4

respect, labor cost, cost for use of energy, cost of consumables, maintenance cost, indirect labor cost needs to be assessed.

Energy efficiency of the manufacturing processes is very critical to achieve sustainable manufacturing. Therefore, energy consumption of sub-processes, energy consumption for transportation into/out of the line, ratio of renewable energy used, and energy consumption on maintaining facility environment needs to be calculated to achieve DP42. DP43 requires calculating greenhouse gas emission from energy consumption of the line, total water consumption, mass of restricted disposals, mass of disposed chips and scraps, mass of disposed consumables, consumables reuse ratio. DP44 ensures more humane work environment. Thus, injury rate, mist/dust level, noise level, physical load index, health-related absenteeism rate needs to be calculated. DP42, DP43 and DP44 creates Sus-VSM, and future state of it enables applying manufacturing process alternatives for the product design (DP41).

4 Conclusion

This chapter presents an Axiomatic Design framework for sustainable product-service system design that exploits the emerging business models. The model explores the customer expectations at the very beginning stage, and then considers them through the design stages. Since sustainable product design stipulates us to consider economic, environmental and social dimensions, the proposed framework also developed on the basis of these three aspects. Besides, Product-Service System models that supports the EoL product returns and decreases maintenance costs of the customers, are considered in the model. The proposed framework is exemplified with a washing machine product design case based on sustainability concerns, and the validity of the proposed model is shown.

Nevertheless, this research has limitations. The results related to the case is only for one type of product and cannot be generalized. In other words, different product types have different customer expectations and design constraints that need to be considered. For future research, different product types can be considered for the Sustainable Product Service Systems design. Besides, new business models to facilitate the sustainability of the Product Service Systems can be developed, according to the changing needs of the customers.

References

1. Ahmad, S., Wong, K.Y., Tseng, M.L., Wong, W.P.: Sustainable product design and development: a review of tools, applications and research prospects. Resour. Conserv. Recycl. **132**, 49–61 (2018)
2. Babic, B.: Axiomatic design of flexible manufacturing systems. Int. J. Prod. Res. **37**(5), 1159–1173 (1999)

3. Bahadir, M.C., Satoglu, S.I.: A novel robot arm selection methodology based on axiomatic design principles. Int. J. Adv. Manuf. Technol. **71**(9–12), 2043–2057 (2014)
4. Bal, A., Satoglu, S.I.: A goal programming model for sustainable reverse logistics operations planning and an application. J. Clean. Prod. **201**, 1081–1091 (2018)
5. Baysan, S., Kabadurmus, O., Cevikcan, E., Satoglu, S.I., Durmusoglu, M.B.: A simulation-based methodology for the analysis of the effect of lean tools on energy efficiency: an application in power distribution industry. J. Clean. Prod. **211**, 895–908 (2019)
6. Bovea, M.A.D., Pérez-Belis, V.: A taxonomy of ecodesign tools for integrating environmental requirements into the product design process. J. Clean. Prod. **20**(1), 61–71 (2012)
7. Bressanelli, G., Perona, M., Saccani, N.: Assessing the impacts of circular economy: a framework and an application to the washing machine industry. Int. J. Manag. Decis. Mak. 1–27 (2019)
8. Campadello, L., Deubzer, O., Langle, A., Amadei, A., Arienti, G., Sala, M.: Upgrading regulations and standards to enable recycling of CRM from WEEE. Solutions for Critical Raw materials—a European Expert Network (SCRREEN). Report no: D8.2 (2019)
9. Cebi, S., Kahraman, C.: Determining design characteristics of automobile seats based on fuzzy axiomatic design. Int. J. Comput. Intell. Syst. **3**(1), 43–55 (2010)
10. Cebi, S., Kahraman, C.: Extension of axiomatic design principles under fuzzy environment. Expert Syst. Appl. **37**(3), 2682–2689 (2010)
11. Cebi, S., Kahraman, C.: A new weighted fuzzy information axiom method in production research. J. Enterp. Inf. Manag. **32**(1), 170–190 (2019)
12. Chen, C., Zhu, J., Yu, J.Y., Noori, H.: A new methodology for evaluating sustainable product design performance with two-stage network data envelopment analysis. Eur. J. Oper. Res. **221**(2), 348–359 (2012)
13. Delaram, J., Valilai, O.F.: An architectural view to computer integrated manufacturing systems based on axiomatic design theory. Comput. Ind. **100**, 96–114 (2018)
14. Du, Y., Cao, H., Chen, X., Wang, B.: Reuse-oriented redesign method of used products based on axiomatic design theory and QFD. J. Clean. Prod. **39**, 79–86 (2013)
15. Durmusoglu, M.B., Satoglu, S.I.: Axiomatic design of hybrid manufacturing systems in erratic demand conditions. Int. J. Prod. Res. **49**(17), 5231–5261 (2011)
16. Ertay, T., Satoğlu, S.I.: System parameter selection with information axiom for the new product introduction to the hybrid manufacturing systems under dual-resource constraint. Int. J. Prod. Res. **50**(7), 1825–1839 (2012)
17. European Commission: Directive 2012/19/EU of the European Parliament and of the Council of 4 July 2012 on waste electrical and electronic equipment (WEEE). OJEU L 197/38 of 24.07.2012 (2012)
18. Faulkner, W., Badurdeen, F.: Sustainable value stream mapping (Sus-VSM): methodology to visualize and assess manufacturing sustainability performance. J. Clean. Prod. **85**, 8–18 (2014)
19. Favi, C., Marconi, M., Germani, M., Mandolini, M.: A design for disassembly tool oriented to mechatronic product de-manufacturing and recycling. Adv. Eng. Inform. **39**, 62–79 (2019)
20. Hebrok, M.: Design for longevity: taking both the material and social aspects of product-life into account. J. Des. Res. **12**(3), 204–220 (2014)
21. Huang, C.C., Liang, W.Y., Yi, S.R.: Cloud-based design for disassembly to create environmentally friendly products. J. Intell. Manuf. **28**(5), 1203–1218 (2017)
22. Jawahir, I.S., Dillon, O.W., Rouch, K.E., Joshi, K.J., Venkatachalam, A., Jaafar, I.H.: Total life-cycle considerations in product design for sustainability: a framework for comprehensive evaluation. In: Proceedings of the 10th International Research/Expert Conference, pp. 1–10, Barcelona, Spain (2006, September)
23. Jeurissen, R.: John Elkington, Cannibals with forks: the triple bottom line of 21st century business. J. Bus. Ethics **23**(2), 229–231 (2000)
24. Kapukaya, E.N., Bal, A., Satoglu, S.I.: A bi-objective model for sustainable logistics and operations planning of WEEE recovery. An Int. J. Optim. Control.: Theor. Appl. (IJOCTA) **9**(2), 89–99 (2019)

25. Kim, S.J., Suh, N.P., Kim, S.G.: Design of software systems based on axiomatic design. Robot. Comput.-Integr. Manuf. **8**(4), 243–255 (1991)
26. Kulak, O., Cebi, S., Kahraman, C.: Applications of axiomatic design principles: a literature review. Expert Syst. Appl. **37**(9), 6705–6717 (2010)
27. Kulak, O., Durmusoglu, M. B., Tufekci, S.: A complete cellular manufacturing system design methodology based on axiomatic design principles. Comput. Ind. Eng. **48**(4), 765–787 (2005)
28. Ljungberg, L.Y.: Materials selection and design for development of sustainable products. Mater. Des. **28**(2), 466–479 (2007)
29. Rasmi, S.A.B., Kazan, C., Türkay, M.: A multi-criteria decision analysis to include environmental, social, and cultural issues in the sustainable aggregate production plans. Comput. Ind. Eng. **132**, 348–360 (2019)
30. Rüdenauer, I., Gensch, C.-O., Quack, D.: Eco-efficiency analysis of washing machines
31. Stiassnie, E., Shpitalni, M.: Incorporating lifecycle considerations in axiomatic design. CIRP Ann. **56**(1), 1–4 (2007)
32. Storbeck, O.: Adidas vows to use only recycled plastics by 2024. The Financial Times Limited 2019. https://www.ft.com/content/73ca70d8-84e1-11e8-96dd-fa565ec55929 (2018). Accessed 30 July 2019
33. Suh, N.P., Cochran, D.S., Lima, P.C.: Manufacturing system design. CIRP Ann. **47**(2), 627–639 (1998)
34. Suh, N.P.: Axiomatic Design: Advances and Applications. Oxford University Press (2001)
35. Tasaki, T., Hashimoto, S., Moriguchi, Y.: A quantitative method to evaluate the level of material use in lease/reuse systems of electrical and electronic equipment. J. Clean. Prod. **14**(17), 1519–1528 (2006)
36. Tukker, A.: Eight types of product–service system: eight ways to sustainability? Experiences from SusProNet. Bus. Strategy Environ. **13**(4), 246–260 (2004)
37. Tukker, A.: Product services for a resource-efficient and circular economy—a review. J. Clean. Prod. **97**, 76–91 (2015)
38. Ulrich, K.T., Pearson, S.: Assessing the importance of design through product archaeology. Manag. Sci. **44**(3), 352–369 (1998)
39. Zhang, H.C., Li, H.: An energy factor based systematic approach to energy-saving product design. CIRP Ann. **59**(1), 183–186 (2010)

Developing a Customer Oriented Lean Production System Using Axiomatic Design and Fuzzy Value Stream Mapping

Ömer Faruk Yılmaz, Gökhan Özçelik and Fatma Betül Yeni

Abstract This chapter proposes a comprehensive methodology to design a customer-oriented production system. To this end, an axiomatic design (AD) based methodology is developed by employing one of the most widely used lean techniques, value stream mapping (VSM). The methodology is designed in three stages: (i) analyzing the current state, (ii) applying the AD method, and (iii) designing the future state. To consider the inherent vagueness of the processes, a fuzzy approach is embedded into the VSM within the context of the proposed methodology, in the first and third steps. To show the applicability of the proposed methodology, a real case study from a water-meter producer is conducted. The lead time is an indicator to show how fast a company responds to customer demands. Therefore, it is employed as a metric to compare current and future states. The process and lead times are defined by using triangular fuzzy numbers (TFNs) to capture the ambiguity in work-in-process (WIP) levels in the company. Computational results show the effectiveness of the proposed methodology in terms of comparison metric, i.e. manufacturing lead-time. This study provides a guideline for academicians and researchers and aims to be a stepping stone for future studies.

1 Introduction

In today's competitive market environment, companies place more emphasis on designing a customer-oriented production system to react to changes in expectations [6]. Customer expectations are highly variable due to the high product assortments and small-volume orders. This makes it necessary for a company to redesign its traditional production system. The design process starts with analyzing the current state of a production system to determine requirements to be transformed. By doing so, methods that can be used for design are determined. After carefully examining the system, a comprehensive roadmap should be employed to obtain a customer-oriented

Ö. F. Yılmaz · G. Özçelik (✉) · F. B. Yeni
Industrial Engineering Department, Engineering Faculty, Karadeniz Technical University, Trabzon, Turkey
e-mail: gozcelik@ktu.edu.tr

C. Kahraman and S. Cebi (eds.), *Customer Oriented Product Design*, Studies in Systems, Decision and Control 279, https://doi.org/10.1007/978-3-030-42188-5_9

production system. AD is an effective method to construct a roadmap to design a system from beginning to end. AD is proposed by Suh [35] and since then it has been used in many studies for the product or system design [5, 25, 31]. The AD provides a systematic guideline for design activities with a theoretical foundation based on logical thought processes [10]. On one hand, AD paves the way for designing a complete production system. On the other hand, it allows using lean manufacturing techniques while designing the system, which is important to respond to customers within shorter lead time by eliminating wastes [7].

In this study, a methodology based on AD and lean techniques are preferred to be used, because a customer-oriented production system design can be achieved with high product quality, shorter lead time, and high response rate [43]. That being the case, first, the current state of the production system should be analyzed by considering the uncertainties in the system, such as process durations and WIP levels. For this reason, the fuzzy value stream mapping (FVSM) technique is employed to show the points that need to be improved in the production system. Then, design parameters (DP), functional requirements (FR), and their relations are determined in the second step of the methodology, i.e. AD step. Lastly, the FVSM technique is utilized again to visualize the future state of the production system.

A real-life case study from a company producing water-meter is provided to show how the proposed methodology is implemented. The improvement achieved by the implementation of the methodology is analyzed in terms of the lead time reduction and the computational results validate the effectiveness of it.

The noteworthy contributions of this study are outlined from different perspectives as follows.

- From theoretical perspective, the main contribution of this study is to present a comprehensive methodology based on AD and lean tools to obtain a customer-oriented lean production system.
- From practical perspective, a real case study is provided to show the applicability of the proposed methodology.
- As for the managerial point of view, this study provides several managerial insights related to AD and FVSM based on the research data taken from a company producing water-meters. In this manner, it intends to be a stepping stone for the researchers and practitioners in the industry.

The remainder of this study is organized as follows. A comprehensive literature review for the AD applications on lean manufacturing is presented in Sect. 2. The detailed explanations and the proposed methodology are provided in Sect. 3. A real-life case study and computational results are given in Sect. 4. The implemented improvements and the comparisons of current and future systems are discussed in Sect. 5. Concluding remarks and future research directions are given in Sect. 6.

2 Literature Review

In the academic literature, the AD concept is well studied in many scopes such as product design, production system, decision making, software design. In this study, the AD studies handling for designing production systems are reviewed and classified in terms of the type of production system.

Duda et al. [9], Babic [1], Molina and Sanchez [27] studied the AD for flexible production systems; Luo et al. [24], Bi et al. [3], Farid [12], Farid and Ribeiro [13], Puik et al. [28] and Farid [14] studied the AD for cellular and reconfigurable production systems; Durmuşoğlu and Satoğlu [10], Ertay and Satoglu [11], Bahadır and Satoğlu [2] studied the AD for hybrid and automated production systems; Franco and Batocchio [15], Vinodh [40], Matt [25], Weber et al. [42], Holzner et al. [19] studied the AD for holonic/agile/changeable production systems; Mokhtar and Houshmand [26], Han et al. [17], Taha et al. [38] studied the AD for virtual and computer integrated production systems.

Due to the concept of this study, the remarkable AD studies regarding the lean production system are explained in detail. For the review of the literature, "lean manufacturing" and "axiomatic design" words are considered together using the "title, abstract, keywords" search in the Scopus database. All the 29 relevant proceedings and articles in English between 1999 and 2018 are selected and shown in Fig. 1. This reveals that there are a few studies considering the lean and AD principles together in the literature.

Duda et al. [9] presented a framework depending on the principles of a lean production system and AD to design an automotive component production system. Cochran et al. [8] employed the methodology composed of lean management and AD principles for improving the competitiveness of a company. Houshmand and

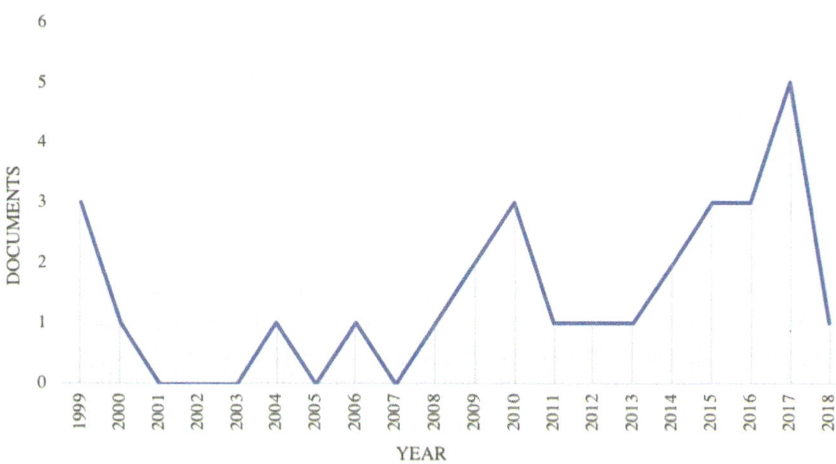

Fig. 1 Studies considering AD and lean manufacturing by years

Jamshidnezhad [20] studied axiomatic modeling for a lean production system considering process variables and developed a hierarchical structure to model the design process of a lean production system, consisting of FRs and DPs. Jadeja et al. [21] applied the just-in-time concept through the AD principles to a company. Jadeja et al. [22] introduced a systematic structured approach combining of lean production principles and Six Sigma through the instrument of the key elements of AD templates for small-scale industry. Shirwaiker and Okudan [34] investigated the appropriateness of Theory of Inventive Problem Solving (TRIZ) and AD for lean design applications enunciated a close correlation between these two tools and lean design metrics by evaluating TRIZ and AD for their contributions to leanness. Vinodh and Aravindraj [41] developed a conceptual model for designing a lean production system based on the axiomatic modeling approach used by Houshmand and Jamshidnezhad [20]. They also conducted a study regarding designing a lean production system of an Indian rotary switches manufacturing organization. Rauch et al. [29] presented an AD-based approach to deduce a catalog of guidelines to the design of Lean Product Development processes. Giambalvo et al. [16] proposed a decision support tool to help designers or project managers select the best methodology for their specific problems through the reviewing of several topics including six sigma, systems thinking, axiomatic design and lean manufacturing, etc. Shah et al. [33] presented a comprehensive guide for manufacturing enterprises to achieve long-term business sustainability by addressing several issues including lean and axiomatic design concepts too.

Our extensive survey of existing studies in the literature shows that a few studies have been conducted to develop a lean-based methodology with AD principles. Therefore, a comprehensive methodology is proposed to handle the design requirements of a customer-oriented production system by considering the vagueness inherent of processes through fuzzy techniques. From this perspective, a theoretical contribution is provided to the existing literature. On the other hand, to show the applicability of the methodology, it is validated by a real case study from a company producing water-meters. Besides, several managerial insights are presented within the context of this study to contribute to the literature from the managerial perspective.

3 Methodology

In this section, a lean methodology is proposed combining the concepts of AD and FVSM to design a customer-oriented production system. The method which allows the system to update itself continuously handles the current state of the system by using FVSM first and constructs the hierarchical structure according to the requirements of the system. Then it helps to design the future state of the system by using FVSM for comparison with the current state. The comparison is made in terms of WIP and thus fuzzy lead times. In the study, triangular fuzzy numbers are used in the method since there are uncertainties about the process and lead times. Therefore, the FVSM is preferred to reveal the current and future state by employing TFNs. The

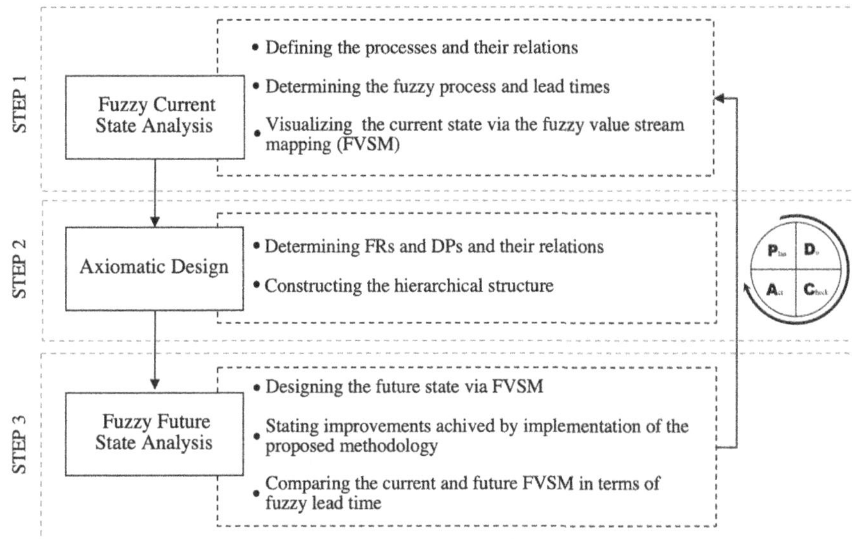

Fig. 2 Schematic representation of the proposed methodology

proposed method is comprised of three main steps: (i) fuzzy current state analysis, (ii) axiomatic design concept, and (iii) fuzzy future state analysis. The proposed methodology is constructed based on the PDCA cycle (plan-do-control-act) which is applied for continuous improvement to enhance the system performance. The method is explained step by step and its schematic representation is presented in Fig. 2.

Step 1: *Fuzzy Current State Analysis*

The method starts with defining the processes and their relations. After these definitions, process and lead times are determined by consulting with practitioners. The last operation of the current system analysis is to draw the FVSM that visualizes the current state of the system. In the literature, Rother and Shook [30] presented a VSM technique for the first time in the existing literature. Notoriously, VSM is one of the best lean tools to map the processes of a production system, and visualize the non-value adding processes, waiting times, the WIP in the systems to enhance effectiveness [32]. In the study, the reason for using FVSM instead of the VSM technique is the uncertainties in the WIP, process and lead times. To this end, the WIP, process and lead times are defined as TFNs, depicted in Fig. 3, to cope with the vagueness inherent of processes. Describing the time intervals spent to produce a product in the production processes through TFNs is the main idea of the FVSM. The usage of TFNs is suitable in the context of VSM to reflect the current state in a detailed manner. Process and lead times can be readily determined using TFNs. For this procedure, the studies of Hapke et al. [18] and Braglia et al. [4] can be followed.

TFNs are defined as $\widetilde{A} = (a, b, c)$ where "a" corresponds to the smallest likely value (lower bound), "b" the most likely value, and "c" the largest likely value (upper bound) for a fuzzy event [39]. Here, the fuzzy set $\widetilde{A}$ is defined as

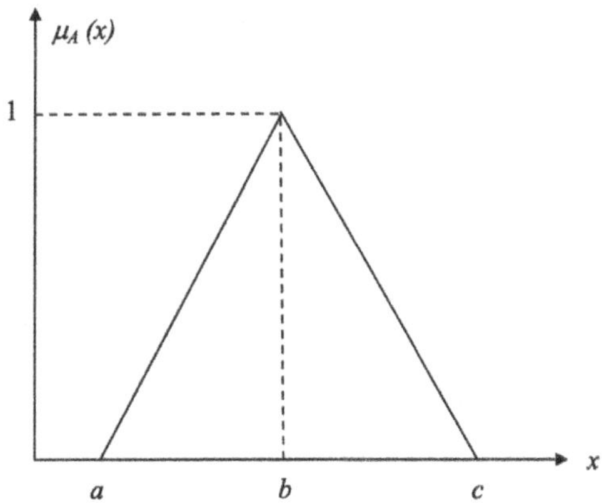

$$\mu_A(x) = \begin{cases} 0 & for\, x < a \\ \dfrac{x-a}{b-a} & for\, a < x < b \\ \dfrac{c-x}{c-b} & for\, b < x < c \\ 0 & for\, x > c \end{cases}$$

Fig. 3 Representation of triangular fuzzy numbers

$\widetilde{A} = \{(x, \mu_A(x)) : x \in A,\ \mu_A(x) \in [0, 1]\}$ where the x belongs to the classical set A and the membership function $\mu_A(x)$ belongs to the interval [0, 1]. For more information about the uncertainties in VSM, the study of Braglia et al. [4] can be followed.

Step 2: *Axiomatic Design*

In this step, a scientific design framework is constructed to boost the system performance within the lean philosophy following the AD principles. The AD method, presented by Suh [35], ensures that realizing of any problems that may occur before the design and also finding of suggestions regarding the system. The main philosophy of AD relates to involving an interplay between "what we want to achieve" and "how we satisfy the need (i.e., the what)" [36]. The important issue about the design with axioms is that the designer fulfills the issues listed below [37]:

- Understanding customer needs,
- Identifying problems to meet the needs,
- Creating solutions through synthesis,
- Using the analysis method to optimize the solution,
- Checking whether the resulting design solution meets customer needs.

There are two axioms in AD methodology. The first one is the independence axiom that necessitates independent FRs. It distinguishes between acceptable or unacceptable design. The second one is the information axiom and minimizes the information content to select the optimal design solutions among the ones that satisfy 1st axiom [1]. In the AD methodology, the high-level FR is decomposed to the next level hierarchically. The decomposition of requirements is conducted in the function domain, in a similar way the corresponding DPs which fulfill the FRs are determined in the physical domain [10]. The relationship between FRs and DPs are formulated as follows [23]:

Fig. 4 **a** Uncoupled matrix **b** decoupled matrix **c** coupled matrix

(a) $$[A]=\begin{bmatrix} X & 0 & 0 \\ 0 & X & 0 \\ 0 & 0 & X \end{bmatrix}$$ **(b)** $$[A]=\begin{bmatrix} X & 0 & 0 \\ X & X & 0 \\ X & X & X \end{bmatrix}$$ **(c)** $$[A]=\begin{bmatrix} X & 0 & X \\ X & X & 0 \\ X & 0 & X \end{bmatrix}$$

$$\{FR\} = [A]\{DP\} \tag{1}$$

where $\{FR\}$ is the functional requirement vector, $\{DP\}$ is the design parameter vector and $[A]$ is design matrix. Each element A_{ij} of the $[A]$ relates a component of the $\{FR\}$ to a component of $\{DP\}$ (Suh 1992). In the AD, the desired design (diagonal matrix) is an uncoupled one in which there is no interaction between the FRs (see Fig. 4a). Also, triangular matrix is acceptable depending on information axiom. $[A]$ of a decoupled design can be either lower or upper triangular (see Fig. 4b). However, a coupled matrix is unacceptable since the changes in a DP may affect all FRs (see Fig. 4c).

The decomposition of FRs and DPs must be ongoing as far as no other decomposition is required. The designer can make a decision regarding the level of the hierarchy and must be able to zigzag between the domains back and forth to decompose the FRs and DPs. So, the design hierarchies are obtained for the FRs and DPs [10].

Step 3: *Fuzzy Future State Analysis*

In the final step, the future state of the system is designed via FVSM, mentioned in Step 1. Improvements in the system will be seen following the suggestions determined in the AD step thanks to illustrating the future state of the system using FVSM. Besides, the current and future FVSM are compared in terms of total fuzzy lead time.

4 Case Study

In this section, a real case study from a water-meter producer is presented to show the applicability of the proposed methodology. The water-meters produced in the company vary according to their diameter and product family, known as KDM-5, is considered for the implementation of the methodology. The application of the proposed methodology is expressed step by step.

Step 1: The processes are defined as below, and the relations among them are depicted through the FVSM:

1. Core process: Cores are obtained by using the raw materials and then send to the casting process.
2. Casting process: This process is carried out on semi-finished products taken from the core process.
3. Cut and sandblasting process: In this process, semi-finished products taken from the casting process are cleaned.

4. Painting process: In this process, semi-finished products taken from the previous process are painted.
5. Subtractive manufacturing process: In this process, drilling and clamping operations are carried out.
6. Assembly process: In this process, the assembly operations on water-meters are conducted.
7. Polymer process: In this process, polymer operations are conducted, and semi-finished products are sent to assembly process based on request.
8. Packaging process: In this process, packaging operations are conducted.
9. Shipping: Finished products are ready to be delivered to customers.

The current state of production system is visualized via FVSM technique and given in Fig. 5. The following inferences are made when the current state map is analyzed.

WIP levels (i.e. lead-times) just before processes are shown in days as TFNs due to the uncertainty. While the workforce level in the painting process is high, the process time is high in the subtractive manufacturing process. The raw material and WIP levels are quite high in the system. Because raw materials cannot be supplied on time, high quantities are ordered. This situation causes the raw material inventory to occupy more space in the factory. Due to incomplete information flow, there is a discrepancy between departments and semi-finished products are transferred to the other processes in batches.

Step 2: A methodology based on AD and lean techniques has been adapted to turn the existing system into a customer-oriented production system. By doing so, the WIP level and therefore lead time decrease by eliminating wastes from the system. FRs, DPs, and their explanations along with the relations are given as follows.

Definitions

- FR: Construct a customer-oriented production system
  DP: Lean production methodology
- FR1: Analyze current state and enabling continuous improvements
  FR2: Boost the effectiveness of production system
  DP1: Current state map
  DP2: Future state map

$$\begin{bmatrix} FR1 \\ FR2 \end{bmatrix} = \begin{bmatrix} x & \\ x & x \end{bmatrix} \begin{bmatrix} DP1 \\ DP2 \end{bmatrix}$$

- FR21: Employee appropriate production system
  FR22: Provide material and information flow
  DP21: Cellular/Seru, functional production systems
  DP22: One-piece and continuous flow

$$\begin{bmatrix} FR21 \\ FR22 \end{bmatrix} = \begin{bmatrix} x & \\ x & x \end{bmatrix} \begin{bmatrix} DP21 \\ DP22 \end{bmatrix}$$

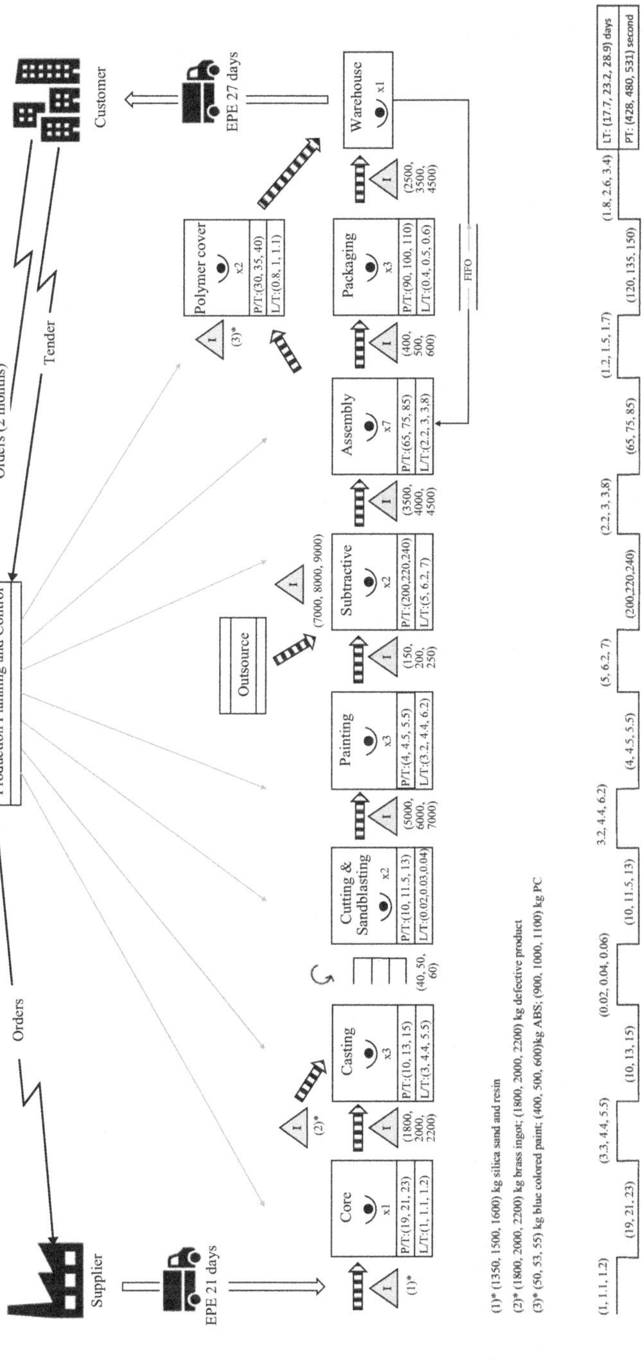

Fig. 5 Value stream map of current system

- FR211: Construct cellular/Seru production system
  FR212: Construct functional production system
  DP211: Part-machine grouping algorithms
  DP212: Layout for the machines which are not assigned to cells

$$\begin{bmatrix} \mathrm{FR211} \\ \mathrm{FR212} \end{bmatrix} = \begin{bmatrix} \mathrm{x} & \\ \mathrm{x} & \mathrm{x} \end{bmatrix} \begin{bmatrix} \mathrm{DP211} \\ \mathrm{DP212} \end{bmatrix}$$

- FR221: Provide pull production control system
  FR222: Construct movement system for raw material, semi-finished products, and products
  DP221: Pull production control system through Kanban
  DP222: Appropriate material handling system

$$\begin{bmatrix} \mathrm{FR221} \\ \mathrm{FR222} \end{bmatrix} = \begin{bmatrix} \mathrm{x} & \\ & \mathrm{x} \end{bmatrix} \begin{bmatrix} \mathrm{DP221} \\ \mathrm{DP222} \end{bmatrix}$$

- FR2111: Determine part-machine groups
  FR2112: Assign machines to cells
  FR2113: Prevent back-flows in cells
  DP2111: Part routes determination algorithms
  DP2112: Movable equipment/machines
  DP2113: Layout of machines in cells

$$\begin{bmatrix} FR2111 \\ FR2112 \\ FR2113 \end{bmatrix} = \begin{bmatrix} x & & \\ x & x & \\ x & x & x \end{bmatrix} \begin{bmatrix} DP2111 \\ DP2112 \\ DP2113 \end{bmatrix}$$

- FR2211: Eliminate Muda in system
  FR2212: Provide just-in-time and low-cost production
  DP2211: Mura and Muri activities
  DP2212: System implementation

$$\begin{bmatrix} \mathrm{FR2211} \\ \mathrm{FR2212} \end{bmatrix} = \begin{bmatrix} \mathrm{x} & \\ \mathrm{x} & \mathrm{x} \end{bmatrix} \begin{bmatrix} \mathrm{DP2211} \\ \mathrm{DP2212} \end{bmatrix}$$

- FR2221: Find the shortest path for part feedings
  FR2222: Determine appropriate material handling equipment
  DP2221: Route determination algorithms
  DP2222: Functional and low-cost handling equipment

$$\begin{bmatrix} \mathrm{FR2221} \\ \mathrm{FR2222} \end{bmatrix} = \begin{bmatrix} \mathrm{x} & \\ \mathrm{x} & \mathrm{x} \end{bmatrix} \begin{bmatrix} \mathrm{DP2221} \\ \mathrm{DP2222} \end{bmatrix}$$

- FR21121: Determine number of machines/equipment for cells
  FR21122: Provide machines/equipment flexibility
  DP21121: Sufficient number of machines/equipment
  DP21122: Movable and re-contractable machines/equipment

$$\begin{bmatrix} \mathrm{FR21121} \\ \mathrm{FR21122} \end{bmatrix} = \begin{bmatrix} \mathrm{x} & \\ \mathrm{x} & \mathrm{x} \end{bmatrix} \begin{bmatrix} \mathrm{DP21121} \\ \mathrm{DP21122} \end{bmatrix}$$

- FR22111: Reduce work-in-process (WIP)
  FR22112: Reduce setup times
  DP22111: Similar machine cycle times
  DP22112: Single-minute exchange of die (SMED)

$$\begin{bmatrix} \mathrm{FR22111} \\ \mathrm{FR22112} \end{bmatrix} = \begin{bmatrix} \mathrm{x} & \\ \mathrm{x} & \mathrm{x} \end{bmatrix} \begin{bmatrix} \mathrm{DP22111} \\ \mathrm{DP22112} \end{bmatrix}$$

- FR22121: Determine the pace-maker process
  FR22122: Provide continuous flow by controlling WIP
  FR22123: Provide visual management
  DP22121: Procedure for continuous flow
  DP22122: Supermarket system
  DP22123: Kanban application

$$\begin{bmatrix} FR22121 \\ FR22122 \\ FR22123 \end{bmatrix} = \begin{bmatrix} x & & \\ x & x & \\ & x & x \end{bmatrix} \begin{bmatrix} DP22121 \\ DP22122 \\ DP22123 \end{bmatrix}$$

All relations given in matrix format above are expressed in Fig. 6.

Step 3: The future state map showing the goal to be reached by the company after the implementation of the methodology is given in Fig. 7. The future state of the system is designed considering the following improvements:

- Necessary agreements are made with the supplier and the frequency of raw material procurement is adjusted twice a week to reduce the long-term raw material stock-keeping in the current state. Some arrangements are also made to ensure that the supplied raw materials are kept in the supermarkets for a maximum of 2.5 days.
- The products are grouped into three categories according to their diameters and length measurements and assigned to three manufacturing cells, respectively. By using cellular manufacturing, it is desired to meet different customer demands at the same time and reduce lead time.
- Core, casting, cut and sandblasting processes are combined into the same manufacturing cell. To complete these cells entirely, it is planned to provide one missing core and two cut and sandblasting equipment. Each cell is operated by only one worker. The positioning of the cells is determined considering the casting furnace in the factory.
- The products produced in the cells are transferred to the painting for dyeing. Supermarkets are placed separately for each cell between the manufacturing cells and the painting. The painting machine, which is operated every two days, is arranged to provide red and blue dyeing. Due to the operation of the machine, the dyeing of the meters is carried out in batches. Therefore, the signal Kanban is sent from the supermarket to the dyeing machine. There is one employee in the paint department.

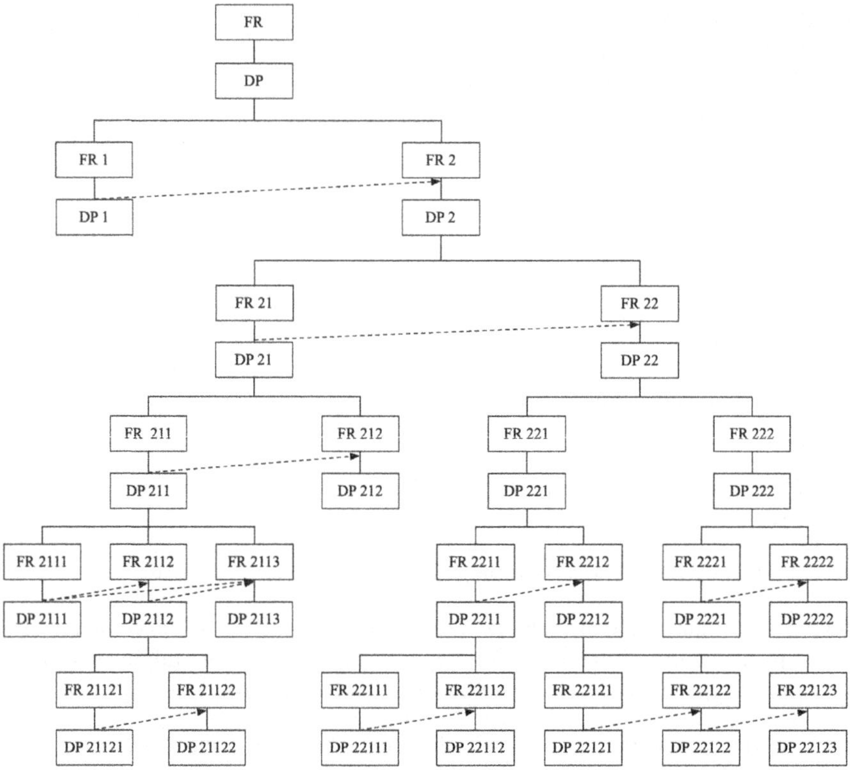

Fig. 6 Decomposition of lean production methodology

- It is decided to produce the upper clamp which is supplied as outsource within the enterprise due to the long lead time and forming inventory. To meet the demand, four machines are activated, which were not operated in the subtractive operation. A supermarket is inserted between the subtractive and the assembly processes and withdrawal from the assembly is applied. A total of nine workers are employed in this field.
- Sending the parts produced in the polymer cover to the shipping with FIFO causes extra transport costs and complexity. For this reason, a supermarket is located, and the products are withdrawal from the assembly. There are nine workers in this process.
- Seru production system is implemented in the assembly process where assembly and packaging processes are brought together. Since the cycle time of the cell could not meet the demand due to lack of capacity, three Serus are constructed. In addition, the employee flexibility of this system is considered, and three workers are assigned to each cell.

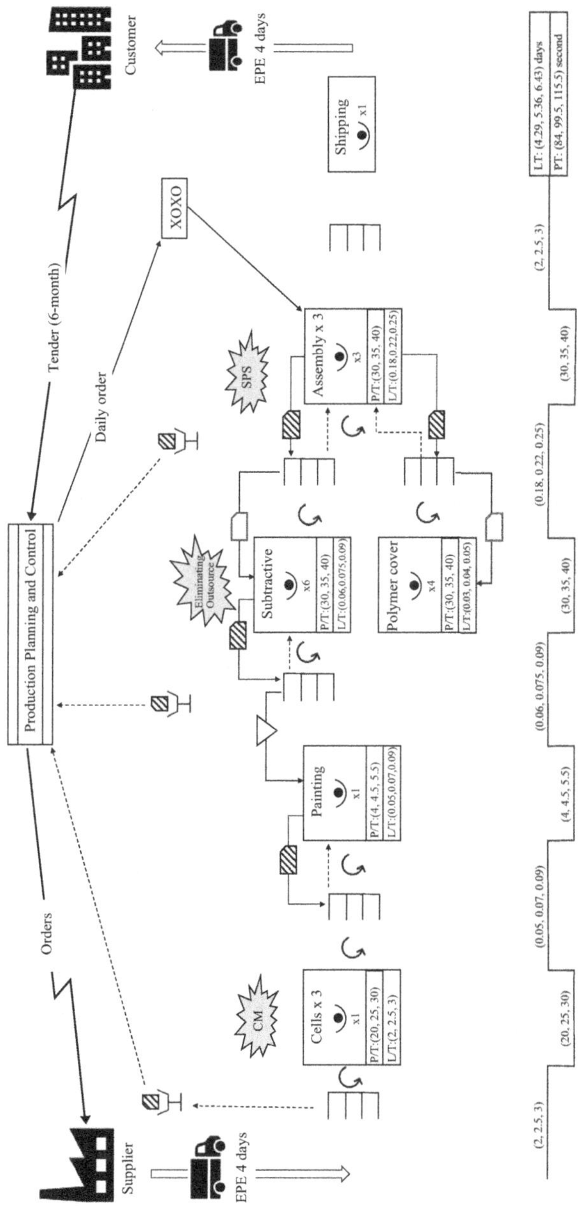

Fig. 7 Value stream map of future state

- The assembled products are taken to the supermarket and awaiting the order of withdrawal from the shipment. The products are delivered to the customer twice a week. Shipping is done by one worker.

As indicated earlier, the manufacturing lead-time is used for comparison purpose to measure the effectiveness of the proposed methodology. According to computational results, there are substantial differences between the lead times before and after the implementation of the methodology. The main reason for this situation is the elimination of waste resources and keeping raw material and WIP levels under control. The manufacturing lead-time is an indicator of the speed of responding to customer demands. As a result, a customer-oriented system is expected to have low lead times to react to the changes quickly. The proposed methodology provides a guideline to become a customer-oriented production system. In this regard, it aims to be a stepping stone for the researchers and practitioners in the industry.

5 Discussion

In this study, the proposed methodology based on the PDCA cycle is applied successfully to a production system so as to boost the performance of the system by considering a lean philosophy. The implemented Kaizens and the comparisons regarding current and future states are listed as follows.

- The raw material procurement, which is made every 21 days, has been rearranged to be made every four days.
- The core, casting and cut and sandblasting processes are combined into a single cell.
- Supermarkets are located in front of each cell to regulate inventory levels.
- While the total cycle times of the core, casting, cut and sandblasting processes are (39, 45.5, 51) s, it is reduced to (20, 25, 30) s by employing the principles of the cellular manufacturing system. The core process is the bottleneck that determines the cycle time of the cell.
- Ventilation is provided in the casting process.
- Improvements are made to make full use of the capacity of the painting machine and to paint different products according to the product type. Also, metal separators are placed between the products.
- Machines are classified according to product type in the subtractive process. The upper part, which was previously manufactured as an outsource, is also produced in the factory.
- The total process time of the subtractive process is reduced to (30, 35, 40) s by increasing the number of machines and eliminating the setup times thanks to SMED activities.
- The cycle time is reduced to (90, 105, 120) s by combining assembly and packaging processes and implementing the principles of the seru production system (SPS).

- In the current system, the production orders were sent to each process separately. With the implementation of the pull system, the production order is only sent to the final (assembly) process.
- It is made compulsory to use occupational health and safety equipment in all processes.
- The products coming out of the assembly cells wait (2, 2.5, 3) days to be shipped in the supermarket. The shipment to the customer is arranged as (every-part-every) EPE 4 days.
- The design of the supermarkets is based on the average size of the product. Supermarkets with a capacity of 180 products (6 shelves × 30 products) are located in the required areas.
- Kanban boxes are located.
- 5S activities (Sort-Set in order-Shine-Standardize-Sustain) are applied to all processes.
- Autonomous Maintenance (Jishu Hozen) is applied to reduce failures and time losses in the processes.
- While the lead time of current state is (17.7, 23.2, 28.9) days, the lead time of future state is planning to be reduced to (4.29, 5.36, 6.43) days.

6 Conclusion

This chapter introduces a lean-based methodology employing AD and FVSM to design a customer-oriented production system. The main purpose of this study is to fill the gap between theoretical studies and practical applications by employing the proposed holistic methodology by considering vagueness inherent of a real production system. The applicability of the methodology is demonstrated through a water-meter production system. In the case study, the processes and their relations are determined first. The WIP, process and lead times are defined using TFNs to cope with the lead time uncertainties. After these definitions, the current state of the system is visualized via FVSM. By doing so, the total process and lead times are calculated and the value adding and non-value adding processes are revealed. Considering these gains, the AD methodology is used to design a customer-oriented production system in terms of decreasing lead time. The hierarchical structure is constructed by following the questions of "what we want to achieve" and "how we satisfy the need" within AD principles to obtain a strong design that satisfy the customer needs. The main contributions of this study are stated within three folds. (i) A holistic methodology is provided with lean techniques combining with fuzzy concept, (ii) a real application is conducted and results are analyzed in detail to validate the methodology, and (iii) several managerial insights are provided for the practical applications. The future state of the system is drawn using FVSM after implementing the suggested improvements. Consequently, the improvements and total process and lead times are compared and discussed with respect to current and future states. The study can be extended to some interesting fields: (i) fuzzy axiomatic design approach can be employed within the proposed methodology, (ii) robust optimization model can be developed to cope with the uncertainty inherent of processes, (iii) the maintenance policies and employee flexibility issue can be added to the AD step.

References

1. Babic, B.: Axiomatic design of flexible manufacturing systems. Int. J. Prod. Res. **37**(5), 1159–1173 (1999)
2. Bahadir, M.C., Satoglu, S.I.: A novel robot arm selection methodology based on axiomatic design principles. Int. J. Adv. Manuf. Technol. **71**(9–12), 2043–2057 (2014)
3. Bi, Z.M., Lang, S.Y., Wang, L.: Design of reconfigurable manufacturing systems with strongly coupled nature. Chin. J. Mech. Eng. **20**(1), 91–95 (2007)
4. Braglia, M., Frosolini, M., Zammori, F.: Uncertainty in value stream mapping analysis. Int. J. Logist.: Res. Appl. **12**(6), 435–453 (2009)
5. Büyüközkan, G., Göçer, F.: Application of a new combined intuitionistic fuzzy MCDM approach based on axiomatic design methodology for the supplier selection problem. Appl. Soft Comput. **52**, 1222–1238 (2017)
6. Cevikcan, E., Durmusoglu, M.B.: Minimising utility work and utility worker transfers for a mixed-model assembly line. Int. J. Prod. Res. **49**(24), 7293–7314 (2011)
7. Cevikcan, E., Durmusoglu, M.B., Baskak, M.: Integrating parts design characteristics and scheduling on parallel machines. Expert Syst. Appl. **38**(10), 13232–13253 (2011)
8. Cochran, D.S., Eversheim, W., Kubin, G., Sesterhenn, M.L.: The application of axiomatic design and lean management principles in the scope of production system segmentation. Int. J. Prod. Res. **38**(6), 1377–1396 (2000)
9. Duda, J.W., Cochran, D.S., Castaneda-Vega, J., Baur, M., Anger, R., Taj, S.: Application of a lean cellular design decomposition to automotive component manufacturing system design (No. 1999-01-1620). SAE Technical Paper (1999)
10. Durmusoglu, M.B., Satoglu, S.I.: Axiomatic design of hybrid manufacturing systems in erratic demand conditions. Int. J. Prod. Res. **49**(17), 5231–5261 (2011)
11. Ertay, T., Satoğlu, S.I.: System parameter selection with information axiom for the new product introduction to the hybrid manufacturing systems under dual-resource constraint. Int. J. Prod. Res. **50**(7), 1825–1839 (2012)
12. Farid, A.M.: An axiomatic design approach to production path enumeration in reconfigurable manufacturing systems. In: 2013 IEEE International Conference on Systems, Man, and Cybernetics, pp. 3862–3869. IEEE (2013)
13. Farid, A.M., Ribeiro, L.: An axiomatic design of a multiagent reconfigurable mechatronic system architecture. IEEE Trans. Industr. Inf. **11**(5), 1142–1155 (2015)
14. Farid, A.M.: Measures of reconfigurability and its key characteristics in intelligent manufacturing systems. J. Intell. Manuf. **28**(2), 353–369 (2017)
15. Franco, G.N., Batocchio, A.: Towards an axiomatic framework to support the design of holonic systems. In: 12th International Workshop on Database and Expert Systems Applications, pp. 654–659 (2001)
16. Giambalvo, J., Vance, J., Hoffenson, S.: Toward a decision support tool for selecting engineering design methodologies. In: ASEE Annual Conference and Exposition, Columbus, Ohio, pp. 25–28 (2017)
17. Han, W.M., Zhao, J.L., Chen, Y.: A virtual cellular manufacturing system design model based on axiomatic design theory. In: Applied mechanics and materials, vol. 271, pp. 1478–1484. Trans Tech Publications (2013)
18. Hapke, M., Jaszkiewicz, A., Slowinski, R.: Fuzzy project scheduling system for software development. Fuzzy Sets Syst. **67**(1), 101–117 (1994)

19. Holzner, P., Rauch, E., Spena, P.R., Matt, D.T.: Systematic design of SME manufacturing and assembly systems based on axiomatic design. Procedia CIRP **34**, 81–86 (2015)
20. Houshmand, M., Jamshidnezhad, B.: An extended model of design process of lean production systems by means of process variables. Robot. Comput.-Integr. Manuf. **22**(1), 1–16 (2006)
21. Jadeja, S.B., Khandare, S.S., Batish, A.: B37 İmplementation of JIT methodology through axiomatic design approach (Advanced machining technology). In: Proceedings of International Conference on Leading Edge Manufacturing in 21st century: LEM21, vol. 2009, no. 5, pp. 705–709 (2009)
22. Jadeja, S.B., Khandare, S.S., Batish, A., Patel, D.: Integrating lean manufacturing and six sigma by means of axiomatic design principles. In: ASME 2008 International Mechanical Engineering Congress and Exposition, pp. 449–455 (2009b)
23. Kulak, O., Cebi, S., Kahraman, C.: Applications of axiomatic design principles: a literature review. Expert Syst. Appl. **37**(9), 6705–6717 (2010)
24. Luo, Z., Yu, X., Liu, J., Tang, H., Zhou, B., Chu, L.K.: Reconfigurability and reconfigurable design theory. J.-Tsinghua Univ. **44**(5), 577–580 (2004)
25. Matt, D.T.: Application of Axiomatic Design principles to control complexity dynamics in a mixed-model assembly system: a case analysis. Int. J. Prod. Res. **50**(7), 1850–1861 (2012)
26. Mokhtar, A., Houshmand, M.: Introducing a roadmap to implement the universal manufacturing platform using axiomatic design theory. Int. J. Manuf. Res. **5**(2), 252–269 (2010)
27. Molina, J., Sanchez, A.: Design of discrete-event controllers for flexible manufacturing system using informal product requirements. IFAC Proc. **45**(6), 218–223 (2012)
28. Puik, E., Telgen, D., van Moergestel, L., Ceglarek, D.: Assessment of reconfiguration schemes for reconfigurable manufacturing systems based on resources and lead time. Robot. Comput.-Integr. Manuf. **43**, 30–38 (2017)
29. Rauch, E., Dallasega, P., Matt, D.T.: Axiomatic design-based guidelines for the design of a lean product development process. Procedia CIRP **34**, 112–118 (2015)
30. Rother, M., Shook, J.: Learning to see: value stream mapping to add value and eliminate waste. Brazil Lean Enterprise Institute, Sao Paulo-SP (1998)
31. Salonitis, K.: Design for additive manufacturing based on the axiomatic design method. Int. J. Adv. Manuf. Technol. **87**(1–4), 989–996 (2016)
32. Seyedhosseini, S.M., Taleghani, A.E., Makui, A., Ghoreyshi, S.M.: Fuzzy value stream mapping in multiple production streams: a case study in a parts manufacturing company. Int. J. Manag. Sci. Eng. Manag. **8**(1), 56–66 (2013)
33. Shah, S.A., Smith, J.J., Cochran, D.S.: Guiding manufacturing enterprises to achieve long-term business sustainability using the collective system design approach. In: MATEC Web of Conferences, vol. 223, p. 01015 (2018)
34. Shirwaiker, R.A., Okudan, G.E.: Contributions of TRIZ and axiomatic design to leanness in design: an investigation. Procedia Eng. **9**, 730–735 (2011)
35. Suh, N.P.: The Principles of Design. Oxford University Press, New York (1990)
36. Suh, N.P.: Applications of axiomatic design. In: Integration of Process Knowledge into Design Support Systems, pp. 1–46. Springer, Dordrecht (1999)
37. Suh, N.P.: Axiomatic Design: Advances and Applications. Mit-Pappalardo Series in Mecha (2001)
38. Taha, Z., Soewardi, H., Dawal, S.Z.M.: Axiomatic design principles in analysing the ergonomics design parameter of a virtual environment. Int. J. Ind. Ergon. **44**(3), 368–373 (2014)
39. Uppala, A.K., Ranka, R., Thakkar, J.J., Kumar, M.V., Agrawal, S.: Selection of green suppliers based on GSCM practices: using fuzzy MCDM approach in an electronics company. In: Handbook of Research on Fuzzy and Rough Set Theory in Organizational Decision Making, pp. 355–375 (2017)
40. Vinodh, S.: Axiomatic modelling of agile production system design. Int. J. Prod. Res. **49**(11), 3251–3269 (2011)

41. Vinodh, S., Aravindraj, S.: Axiomatic modeling of lean manufacturing system. J. Eng. Des. Technol. **10**(2), 199–216 (2012)
42. Weber, J., Förster, D., Kößler, J., Paetzold, K.: Design of changeable production units within the automotive sector with axiomatic design. Procedia CIRP **34**, 93–97 (2015)
43. Yılmaz, Ö.F., Durmuşoğlu, M.B.: Evolutionary algorithms for multi-objective scheduling in a hybrid manufacturing system. In: Handbook of Research on Applied Optimization Methodologies in Manufacturing Systems, pp. 162–187. IGI Global (2018)

Website Design Using Pythagorean Fuzzy Axiomatic Design

Esra Ilbahar, Selcuk Cebi and Cengiz Kahraman

Abstract Competition has been significantly increasing in all the fields of business due to globalization. As the internet has become one of the preferred information and communication technologies, presence of companies on the internet has become increasingly important. In particular, web presence for banking industry is extremely significant. Although website usability for banking industry has been commonly examined, there are no studies comparing design performance of banking websites with respect to functional requirements which must be satisfied. For this purpose, axiomatic design method is used to analyze design characteristics of banking websites. Axiomatic design differs from the other multi-criteria decision making techniques in such a way that it takes both what designers want to achieve and what the system can offer into account to evaluate the probability of success of a design. The main advantage of utilizing Pythagorean fuzzy sets is that it enables decision makers to have more freedom for assigning membership and non-membership values. Therefore, Pythagorean fuzzy sets extension of axiomatic design is utilized in this study. In the application section, the website designs of Turkey's leading banks is analysed. This study shows that none of them, except for the second alternative, satisfy all functional requirements.

1 Introduction

Competition has significantly increased in all the fields of business due to globalization [1]. The rapid developments in information and communication technologies have enabled companies to take this competition to a new dimension. As the

E. Ilbahar (✉) · S. Cebi
Department of Industrial Engineering, Yildiz Technical University,
34349 Besiktas, Istanbul, Turkey
e-mail: esrailbahar@gmail.com; eilbahar@yildiz.edu.tr

E. Ilbahar · C. Kahraman
Department of Industrial Engineering, Istanbul Technical University,
34367 Macka, Istanbul, Turkey

C. Kahraman and S. Cebi (eds.), *Customer Oriented Product Design*, Studies in Systems, Decision and Control 279, https://doi.org/10.1007/978-3-030-42188-5_10

internet has become one of the most popular information and communication technologies, presence of companies on the internet has become increasingly important. In particular, web presence for banking industry is extremely crucial since e-banking systems provide several services to financial institutions and individuals such as access to accounts, business transactions or information about financial products and services [1].

Website usability is an important criterion to evaluate the web presence quality of a firm [2]. Various studies on website usability have been conducted in the literature. Al-Qeisi et al. [3] investigated website design and revealed that the general content, technical content and appearance are the most important dimensions of a website which are substantially related to usage behavior. In the study, website design was examined in the context of online banking to improve the unified theory of acceptance and use of technology. It was indicated that elements of website design quality such as appearance attractiveness, loading time, access, ease of navigation, organization, readability and content usefulness are significant attributes for a bank's website. Ageeva et al. [4] used exploratory factor analysis, confirmatory factor analysis and structural equation modeling to examine corporate website favorability, its reasons and consequences. Fang and Holsapple [5] investigated the impact of website navigation structures on website usability. It was indicated that navigation structure with a combined hierarchy or a usage-oriented hierarchy has a higher usability than subject-oriented hierarchies. Verkijika and De Wet [6] examined the usability of e-government websites in Sub-Saharan Africa. Liang et al. [7] assessed website quality of internet banking in Ghanaian banking industry by using Pythagorean fuzzy VIKOR approaches based on TODIM. Medyawati and Mabruri [1] analyzed service quality of e-banking websites by using the variables, adequacy of information, accessibility, lifestyle, interaction, usefulness of content and personality. In the study, discriminant analysis was employed to find a factor which differentiates the customers of two different banks. Chmielarz and Zborowski [8] used characteristic objects method to examine the banking websites in Poland. In another study of Chmielarza and Zborowski [9], TOPSIS method was employed to analyze the quality of e-banking websites. Agarwal and Venkatesh [2] proposed a heuristic evaluation procedure which employs usability guidelines developed by Microsoft to examine website usability. Lee and Chung [10] used structural equation modelling to reveal how information quality, system quality and interface design quality affect consumer satisfaction and trust in mobile banking. It was indicated that system and information quality have a significant impact on customer satisfaction and trust in mobile banking whereas interface design quality does not have any influence on it.

The literature on website design is not limited with the banking website usability studies. In the literature, there are various criteria considered for the analysis of website design performance except the ones mentioned above. For instance, Cyr and Head [11] investigated whether the gender differences affect website design preferences considering masculinity levels of countries. In other words, how cultural differences have an impact on preferences of different genders for website design was examined. Lin and Hsieh [12] examined the website interface design criteria with respect to different genders. It was indicated that there exist significant gender differences in terms of the importance of website interface design criteria. Minimal action, compatibility, and flexibility are found to be the most important design criteria for males whereas compatibility, learnability, and user guidance are selected as the most important design criteria for females. Hasan [13] investigated negative effects of perceived irritation due to website design characteristics on consumers' online shopping behavior. McDowell et al. [14] investigated whether there exists a link between website features and online conversion rates by using regression analysis. It was indicated that there is a considerable relationship between some website design features and online conversion rates. In addition, Cebi [15] examined both interactions among the website design parameters and importance degrees of these parameters.

Although website usability for banking industry has been commonly studied, there are no studies examining design performance of banking websites with respect to functional requirements which must be fulfilled. This is why axiomatic design method is adopted to assess banking websites. Axiomatic design outperforms the other multi-criteria decision making techniques in such a way that it considers both what designers want to attain and what the system can offer to assess the probability of success of a design. The main advantage of using Pythagorean fuzzy sets is that it enables decision makers to have a larger domain for assigning membership and non-membership values. Therefore, an integration of Pythagorean fuzzy sets and axiomatic design is employed in this study. The rest of this chapter is structured as follows: fundamentals of Pythagorean fuzzy sets, axiomatic design and Pythagorean fuzzy axiomatic design approach are provided in Sect. 2. Application and concluding remarks are given in Sects. 3 and 4, respectively.

2 Materials and Methods

In this section fundamentals of Pythagorean fuzzy sets, information axiom which is the second axiom of axiomatic design, and the steps of Pythagorean fuzzy axiomatic design are provided.

2.1 *Pythagorean Fuzzy Sets*

Pythagorean fuzzy sets were proposed as an extension to intuitionistic fuzzy sets not to force experts to assign membership and non-membership degrees whose sum is less than or equal to 1 [16, 17]. Namely, unlike the intuitionistic fuzzy sets, the sum of membership and non-membership degrees can be greater than 1 in Pythagorean fuzzy sets but the sum of their squares can not. It is expressed as follows:
Let X be a fixed set. A Pythagorean fuzzy set $\tilde{P}$ has the following form [18]:

$$\tilde{P} \cong \{\langle x, \mu_{\tilde{P}}(x), \nu_{\tilde{P}}(x)\rangle; x \epsilon X\} \quad (1)$$

where the functions $\mu_{\tilde{P}}(x) : X \mapsto [0, 1]$, $\nu_{\tilde{P}}(x) : X \mapsto [0, 1]$ represent the membership degree and the non-membership degree of the element $x \epsilon X$ to P, respectively. For every $x \epsilon X$, the following equation holds:

$$0 \leqslant \mu_{\tilde{A}}(x)^2 + \nu_{\tilde{A}}(x)^2 \leqslant 1 \quad (2)$$

2.2 Axiomatic Design

The axiomatic design method exploits two axioms, *information axiom* and *independence axiom* to improve the design of a product. Independence axiom brings forward that functional requirements are the minimum set of independent requirements describing the design goals and the independence of these functional requirements has to be fulfilled [19]. Information axiom asserts that the design which has the smallest information content is the best alternative among the designs fulfilling the independence axiom [20]. Information axiom is utilized to determine the best option when more than one design fulfilling the independence axiom exist. Namely, the best design can be determined by using both of these two axioms. Information content (I) represents the probability of fulfilling design goals [20]. The information content is calculated as follows:

$$I_i = -log_2\, p_i \quad (3)$$

where p_i shows the probability of achieving a functional requirement. In the case of a design, the probability of success is decided by comparing what designers want to attain in terms of range (i.e., design range) and what the system can offer (i.e., system range). Figure 1 demonstrates that the intersection of design range and system range is the common range involving the acceptable solution [21, 22]. Therefore, p_i might be shown as:

$$p_i = \frac{\text{Common range}}{\text{System range}} \quad (4)$$

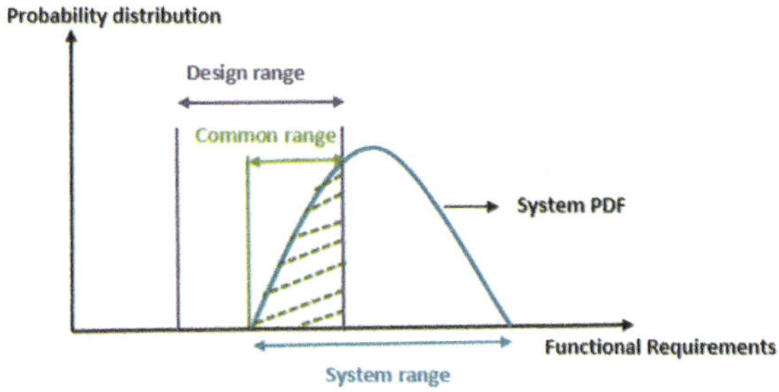

Fig. 1 Relationship between system range and design range

Consequently, information content might be calculated as follows:

$$I = log_2 \frac{\text{System range}}{\text{Common range}} \quad (5)$$

2.3 Pythagorean Fuzzy Axiomatic Design

The framework of the proposed approach [23] is given in Fig. 2. The steps of Pythagorean fuzzy axiomatic design (PFAD) are described as follows:

Step 1. Functional requirements are acquired from decision makers for all of the evaluation criteria by setting a limit to linguistic terms, e.g. "*At most M*" for cost criteria and "*At least H*" for benefit criteria.

Step 2. Assessments of the alternative designs with respect to identified criteria are collected in the form of linguistic terms and their corresponding μ and 1-ν functions are identified with respect to Tables 1 and 2.

Step 3. The corresponding μ and 1-ν functions of the linguistic terms are plotted by employing Tables 1 and 2. Illustration of a case, where linguistic evaluation of the alternative design is specified as "*High (H)*" and the functional requirement is determined by experts as "*At most M*", is given in Figs. 3–4.

Step 4. Intersection of membership functions of functional requirement and the alternative design (e.g. intersection in Fig. 3) is calculated through Eq. 6.

$$IA_\mu = \int_5^t \sqrt{1 - \left(\frac{x - 7.5}{2.5}\right)^2} dx + \int_t^{7.5} \sqrt{1 - \left(\frac{x - 5}{2.5}\right)^2} dx \quad (6)$$

Step 5. Intersection of 1-ν function of the alternative design and membership function of the functional requirement (e.g. intersection in Fig. 4) is calculated through Eq. 7.

$$IA_\nu = \int_5^k \frac{x - 5}{2.5} dx + \int_k^{7.5} \sqrt{1 - \left(\frac{x - 5}{2.5}\right)^2} dx \quad (7)$$

Step 6. The intersection areas calculated by Eqs. 6 and 7, IA_μ and IA_ν, are summed up.

Step 7. The areas under the function μ and the function 1-ν are calculated for each alternative design as given in Eqs. 8 and 9, respectively.

$$A_\mu = \int_5^{10} \sqrt{1 - \left(\frac{x - 7.5}{2.5}\right)^2} dx \quad (8)$$

$$A_\nu = \int_5^{7.5} \frac{x - 5}{2.5} dx + \int_{7.5}^{10} \frac{10 - x}{2.5} dx \quad (9)$$

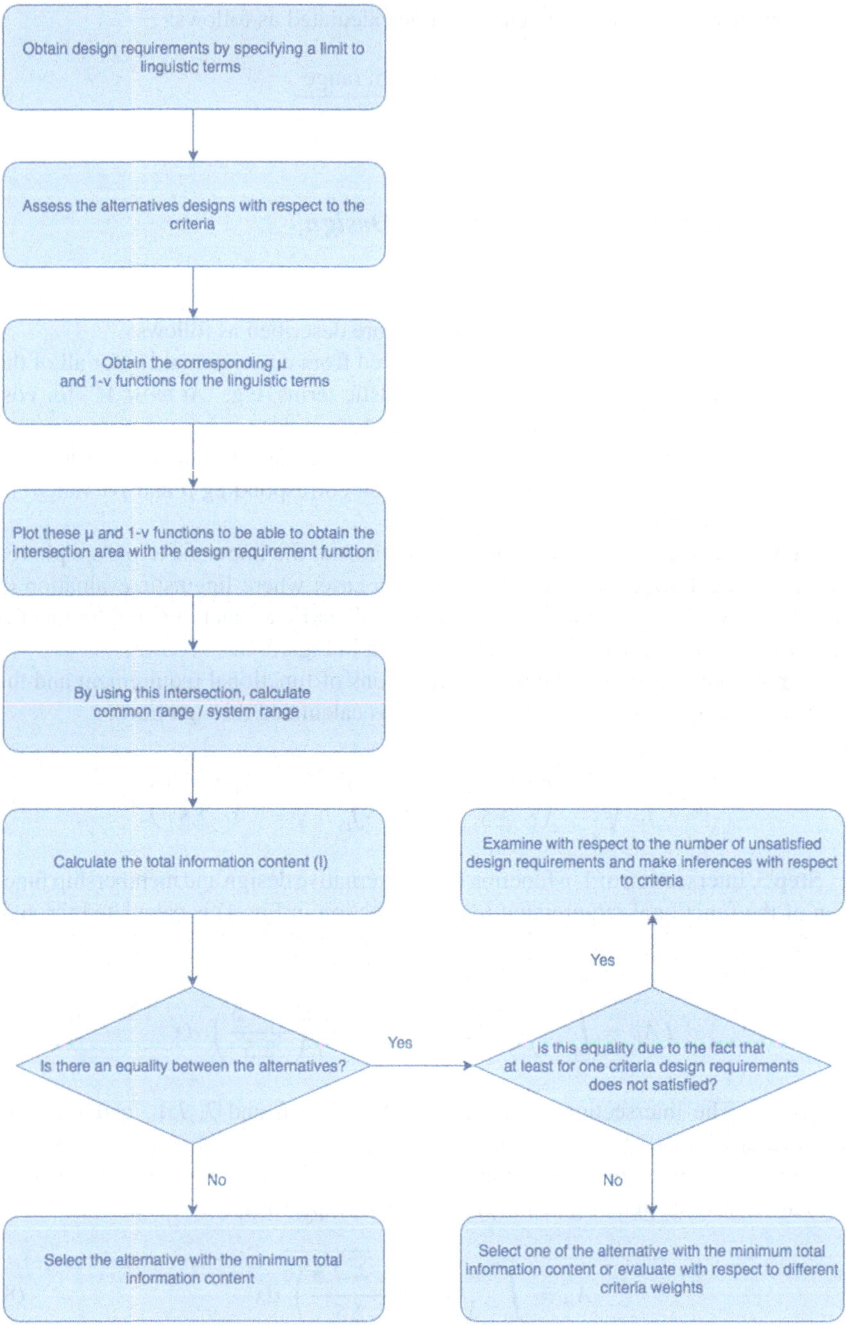

Fig. 2 Framework of the proposed assessment method

Table 1 Linguistic terms and corresponding membership (μ) functions

Linguistic terms	μ
Very low (VL)	$\sqrt{1-((x-0)/2.5)^2}$, $0 \le x \le 2.5$
Low (L)	$\sqrt{1-((x-2.5)/2.5)^2}$, $0 \le x \le 5$
Medium (M)	$\sqrt{1-((x-5)/2.5)^2}$, $2.5 \le x \le 7.5$
High (H)	$\sqrt{1-((x-7.5)/2.5)^2}$, $5 \le x \le 10$
Very High (VH)	$\sqrt{1-((x-10)/2.5)^2}$, $7.5 \le x \le 10$

Table 2 Linguistic terms and corresponding (1-ν) functions

Linguistic terms	1-ν
Very low (VL)	$(2.5-x)/2.5$, $0 \le x \le 2.5$
Low (L)	$(x-0)/2.5, 0 \le x \le 2.5$ $(5-x)/2.5, 2.5 \le x \le 5$
Medium (M)	$(x-2.5)/2.5, 2.5 \le x \le 5$ $(7.5-x)/2.5, 5 \le x \le 7.5$
High (H)	$(x-5)/2.5, 5 \le x \le 7.5$ $(10-x)/2.5, 7.5 \le x \le 10$
Very High (VH)	$(x-7.5)/2.5$, $7.5 \le x \le 10$

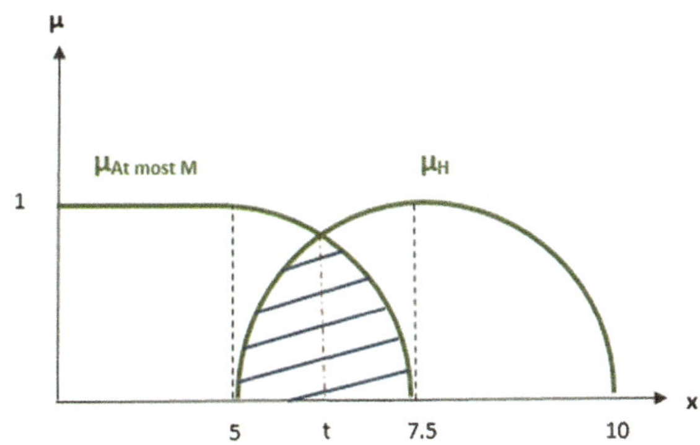

Fig. 3 Intersection of membership functions of system range and design range

Step 8. Information content value on the basis of each criterion is computed by using Eq. 10.

$$I_c = log_2 \frac{A_\mu + A_\nu}{IA_\mu + IA_\nu} \tag{10}$$

Step 9. Total information content (I) for each alternative is calculated by summing up all the information content values obtained from each criterion for the related alternative.

Step 10. Alternative which has the smallest total information content (I) is selected [23].

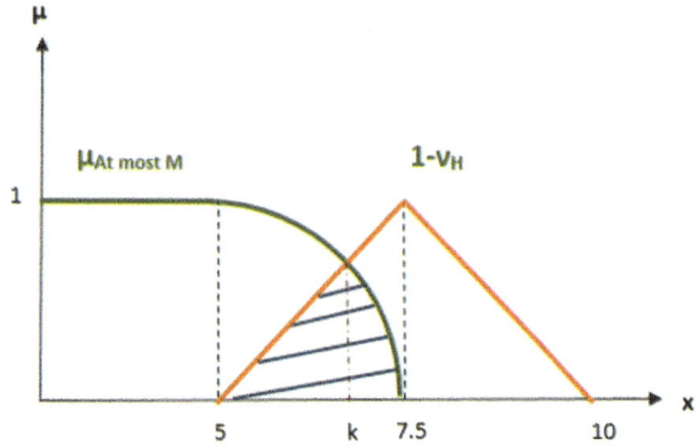

Fig. 4 Intersection of μ and $1 - \nu$ for system range and design range, respectively

3 Application

In this study, websites of the five leading banks in Turkey, two of which are state-funded, one of which is a state and private enterprise joint venture, and two of them belong to private enterprises, are examined. The evaluation criteria used in this study are identified as follows: ease of information access, ease of navigation, content usefulness, support services, accuracy, appearance attractiveness, readability, graphics and layout (See Fig. 5). Figure 6 shows the websites selected for the evaluation process.

Table 3 demonstrates the aggregated evaluation matrix of bank websites and the functional requirements in the form of linguistic terms. The linguistic terms are transformed to corresponding functions through Tables 1 and 2. These functions are plotted as described in Step 3 of Sect. 2.3 to compute the intersection areas. Table 4 shows IA_{μ}, IA_{ν}, A_{μ} and A_{ν} values calculated through Eqs. 6–9. Results of the analysis using the proposed PFAD approach are provided in Table 5. As a result of the calculations, *2nd alternative* is selected to be the best website whereas the other alternatives do not satisfy at least one functional requirement.

Although the alternatives with infinite information content are infeasible due to not satisfying at least one functional requirement, it is possible to examine them with respect to the number of functional requirements they can not satisfy and the total information content of the criteria having finite information content value. This kind of further analysis is plausible since we assume that each of these criteria has an equal effect on website design performance. Table 6 shows information content of the alternatives on the basis of each criterion. It can be seen that *1st, 3rd* and *4th alternatives* have two infinite information content whereas *5th alternative* has five infinite information content. When *1st, 3rd* and *4th alternatives* are compared with respect to information content of the criteria that do not have infinite information content, it is seen that *1st alternative* has the smallest information content, followed by *3rd and 4th alternatives*, respectively. In this way, the rank given in Table 5 can be extended as follows: *2nd alternative* is the best, followed by *1st, 3rd, 4th* and *5th alternative*, respectively.

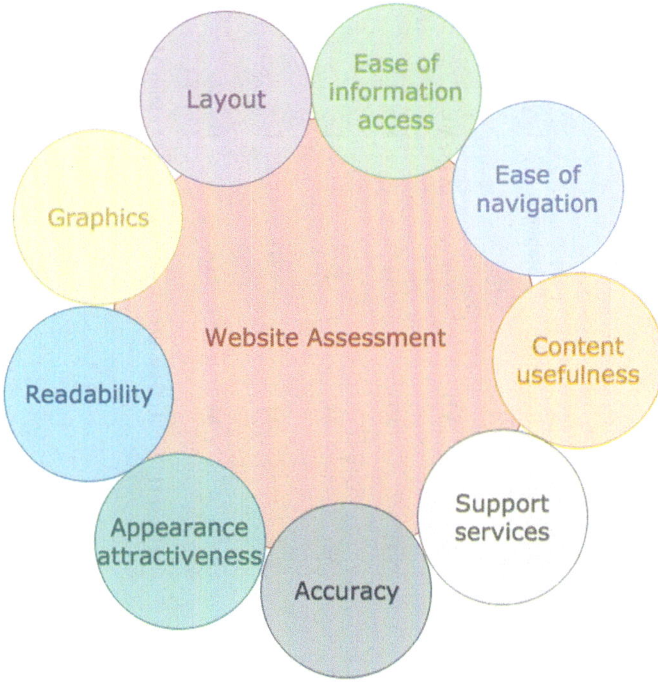

Fig. 5 Evaluation criteria for banking website design

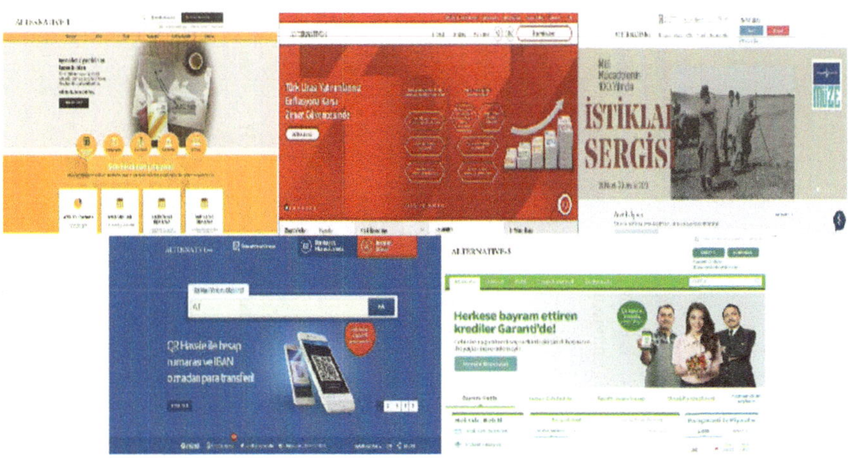

Fig. 6 Alternative websites

Table 3 Linguistic assessment of bank alternatives and functional requirements

Banks	Ease of information access	Ease of navigation	Content usefulness	Support services	Accuracy	Appearance attractiveness	Readability	Graphics	Layout
1st Alternative	VL	VH	VH	M	M	VH	VH	M	VL
2nd Alternative	VH	VH	VH	VH	VH	VH	VH	H	M
3rd Alternative	M	VL	VH	VH	VH	M	M	M	L
4th Alternative	M	VH	M	M	VH	L	M	VH	L
5th Alternative	L	M	M	L	VH	L	M	H	H
Functional requirement	At least H	At least VH	At least H	At least H	At least H	At least H	At least VH	At least M	At least M

Table 4 Calculations of the information content of the alternatives

	Alternatives	Ease of information access	Ease of navigation	Content usefulness	Support services	Accuracy	Appearance attractive-ness	Readability	Graphics	Layout
IA_{μ}	1st Alternative	0.00	1.96	1.96	1.54	1.54	1.96	1.96	3.93	0.00
	2nd Alternative	1.96	1.96	1.96	1.96	1.96	1.96	1.96	3.93	3.93
	3rd Alternative	1.54	0.00	1.96	1.96	1.96	1.54	0.00	3.93	1.54
	4th Alternative	1.54	1.96	1.54	1.54	1.96	0.00	0.00	1.96	1.54
	5th Alternative	0.00	0.00	1.54	0.00	1.96	0.00	0.00	3.93	3.93
IA_{ν}	1st Alternative	0.00	1.25	1.25	0.98	0.98	1.25	1.25	2.50	0.00
	2nd Alternative	1.25	1.25	1.25	1.25	1.25	1.25	1.25	2.50	2.50
	3rd Alternative	0.98	0.00	1.25	1.25	1.25	0.98	0.00	2.50	0.98
	4th Alternative	0.98	1.25	0.98	0.98	1.25	0.00	0.00	1.25	0.98
	5th Alternative	0.00	0.00	0.98	0.00	1.25	0.00	0.00	2.50	2.50

(continued)

Table 4 (continued)

	Alternatives	Ease of information access	Ease of navigation	Content usefulness	Support services	Accuracy	Appearance attractive-ness	Readability	Graphics	Layout
A_{μ}	1st Alternative	1.96	1.96	1.96	3.93	3.93	1.96	1.96	3.93	1.96
	2nd Alternative	1.96	1.96	1.96	1.96	1.96	1.96	1.96	3.93	3.93
	3rd Alternative	3.93	1.96	1.96	1.96	1.96	3.93	3.93	3.93	3.93
	4th Alternative	3.93	1.96	3.93	3.93	1.96	3.93	3.93	1.96	3.93
	5th Alternative	3.93	3.93	3.93	3.93	1.96	3.93	3.93	3.93	3.93
$I A_{\nu}$	1st Alternative	1.25	1.25	1.25	2.5	2.5	1.25	1.25	2.5	1.25
	2nd Alternative	1.25	1.25	1.25	1.25	1.25	1.25	1.25	2.5	2.5
	3rd Alternative	2.5	1.25	1.25	1.25	1.25	2.5	2.5	2.5	2.5
	4th Alternative	2.5	1.25	2.5	2.5	1.25	2.5	2.5	1.25	2.5
	5th Alternative	2.5	2.5	2.5	2.5	1.25	2.5	2.5	2.5	2.5

Table 5 Results of the analysis with the PFAD approach

Alternatives	Information content	Rank
1st Alternative	∞	2
2nd Alternative	0	1
3rd Alternative	∞	2
4th Alternative	∞	2
5th Alternative	∞	2

4 Conclusion

The developments in information and communication technologies have enabled companies to take the increasing competition to a new level. The internet has become one of the preferred information and communication technologies and consequently presence of companies on the internet has become more important. That is why website design for banking industry has been widely studied, yet there are no studies utilizing Pythagorean fuzzy sets, providing more freedom to decision makers for assigning membership and non-membership values, and axiomatic design, considering both what designers want to achieve and what the system can offer, in order to analyze website designs. Therefore, in this study, websites design of five leading banks in Turkey are analyzed by using Pythagorean fuzzy axiomatic design approach. It was concluded that only one design among the five alternatives can satisfy all of the functional requirements defined for banking websites, which is the *2nd alternative*. The design quality of the *1st alternative* should be improved under the ease of information access and layout criteria while ease of navigation and readability features should be improved for the *3rd alternative*. In addition to readability feature, appearance attractiveness of the *4th alternative* should be improved as well. The *5th alternative*, worst of all, should be redesigned for the all features except accuracy, graphics and layout criteria. In future studies, importance degrees of website design criteria can be taken into consideration.

Table 6 Information content of the alternatives on the basis of each criterion

Alternatives	Ease of information access	Ease of navigation	Content usefulness	Support services	Accuracy	Appearance attractiveness	Readability	Graphics	Layout
1st Alternative	∞	0.00	0.00	1.35	1.35	0.00	0.00	0.00	∞
2nd Alternative	0.00	0.00	0.00	0.00	0.00	0.00	0.00	0.00	0.00
3rd Alternative	1.35	∞	0.00	0.00	0.00	1.35	∞	0.00	1.35
4th Alternative	1.35	0.00	1.35	1.35	0.00	∞	∞	0.00	1.35
5th Alternative	∞	∞	1.35	∞	0.00	∞	∞	0.00	0.00

References

1. Medyawati, H., Mabruri, A.: Website quality: case study on local government bank and state own bank in Bekasi City. Procedia-Soc. Behav. Sci. **65**, 1086–1091 (2012)
2. Agarwal, R., Venkatesh, V.: Assessing a firm's web presence: a heuristic evaluation procedure for the measurement of usability. Inf. Syst. Res. **13**(2), 168–186 (2002)
3. Al-Qeisi, K., Dennis, C., Alamanos, E., Jayawardhena, C.: Website design quality and usage behavior: unified theory of acceptance and use of technology. J. Bus. Res. **67**(11), 2282–2290 (2014)
4. Ageeva, E., Melewar, T.C., Foroudi, P., Dennis, C.: Cues adopted by consumers in examining corporate website favorability: an empirical study of financial institutions in the UK and Russia. J. Bus. Res. **98**, 15–32 (2019)
5. Fang, X., Holsapple, C.W.: An empirical study of web site navigation structures' impacts on web site usability. Decis. Support Syst. **43**(2), 476–491 (2007)
6. Verkijika, S.F., De Wet, L.: A usability assessment of e-government websites in Sub-Saharan Africa. Int. J. Inf. Manag. **39**, 20–29 (2018)
7. Liang, D., Zhang, Y., Xu, Z., Jamaldeen, A.: Pythagorean fuzzy VIKOR approaches based on TODIM for evaluating internet banking website quality of Ghanaian banking industry. Appl. Soft Comput. **78**, 583–594 (2019)
8. Chmielarz, W., Zborowski, M.: On analysis of e-banking websites quality-comet application. Procedia Comput. Sci. **126**, 2137–2152 (2018)
9. Chmielarz, W., Zborowski, M.: Analysis of e-banking websites' quality with the application of the TOPSIS method-a practical study. Procedia Comput. Sci. **126**, 1964–1976 (2018)
10. Lee, K.C., Chung, N.: Understanding factors affecting trust in and satisfaction with mobile banking in Korea: a modified DeLone and McLean's model perspective. Interact. Comput. **21**(5–6), 385–392 (2009)
11. Cyr, D., Head, M.: Website design in an international context: the role of gender in masculine versus feminine oriented countries. Comput. Hum. Behav. **29**(4), 1358–1367 (2013)
12. Lin, C.J., Hsieh, T.L.: Exploring the design criteria of website interfaces for gender. Int. J. Ind. Ergon. **53**, 306–311 (2016)
13. Hasan, B.: Perceived irritation in online shopping: the impact of website design characteristics. Comput. Hum. Behav. **54**, 224–230 (2016)
14. McDowell, W.C., Wilson, R.C., Kile Jr., C.O.: An examination of retail website design and conversion rate. J. Bus. Res. **69**(11), 4837–4842 (2016)
15. Cebi, S.: Determining importance degrees of website design parameters based on interactions and types of websites. Decis. Support Syst. **54**(2), 1030–1043 (2013)
16. Atannasov, K.: Intuitionistic Fuzzy Sets: Theory and Applications. Physica-Verlag, New York (1999)
17. Yager, R.R.: Pythagorean fuzzy subsets. In: 2013 Joint IFSA World Congress and NAFIPS Annual Meeting (IFSA/NAFIPS), pp. 57–61. IEEE, Edmonton, AB (2013)
18. Yager, R.R.: Properties and applications of Pythagorean fuzzy sets. In: Imprecision and Uncertainty in Information Representation and Processing, pp. 119–136. Springer, Cham (2016)
19. Suh, H.P.: Axiomatic Design: Advances and Applications MIT-Pappalardo Series in Mechanical Engineering. Oxford University Press, New York (2001)
20. Suh, N.P.: The Principles of Design (Oxford Series on Advanced Manufacturing). Oxford University Press, New York (1990)
21. Kahraman, C., Kaya, I., Cebi, S.: A comparative analysis for multiattribute selection among renewable energy alternatives using fuzzy axiomatic design and fuzzy analytic hierarchy process. Energy **34**(10), 1603–1616 (2009)
22. Kahraman, C., Cebi, S.: A new multi-attribute decision making method: hierarchical fuzzy axiomatic design. Expert Syst. Appl. **36**(3), 4848–4861 (2009)
23. Ilbahar, E., Cebi, S., Kahraman, C.: Multi-criteria evaluation of biomass conversion technologies with a novel pythagorean fuzzy axiomatic design approach. Under review (2019)

Part IV
Customer Oriented Product Design Using Kano Model Method

The Kano Model in the Development of Customer Oriented Products

Maria Grazia Violante, Enrico Vezzetti and Francesca Nonis

Abstract In today's modern competitive market, the purpose of every business is to adopt strategies and technologies to manage their interactions with their current and potential customers. This is possible only when an organization listens to the customer's voice, pitching customer requirements with the best of their inputs. Kano's model offers an effective method for both industries and academic research in classifying different customer requirements into different categories based on their impact on Customer Satisfaction, obtaining competitive and accurate information about customer requirements. The aim of this chapter is to try to facilitate practitioners in future research about the further development of Kano's model and organizations in the future use of the Kano approach to correctly identify customer requirements and channelize their resources in the right direction.

1 Introduction

To survive in the present competitive market, the priority of a company is understanding, designing, and developing customer needs in a product [1]. The satisfaction of the customers is a factor to increase in sales: in fact, customers are expected to make repeat purchases and also to be a point of reference to others. So, since the organization's needs are dependent upon the satisfaction of the consumers, the consumer's satisfaction is reached by the organization providing a valuable product. With the intense competition in the market, customer satisfaction (CS) is the main objective of every enterprise which drives growth and success of the business. Now, companies can no longer depend on high volume and low-cost production for growth and survival [2], but they need to identify innovative and unique competitive plans which can attract and retain customers. As a decrease of 1% in customer satisfaction will lead to a reduction of 5% in the product profit [3], it is vital to consider customer satisfaction. CS is a key element that differentiates companies from others

M. G. Violante (✉) · E. Vezzetti · F. Nonis
Department of Management and Production Engineering, Politecnico Di Torino, Corso Duca Degli Abruzzi 24, 10129 Turin, Italy
e-mail: mariagrazia.violante@polito.it

C. Kahraman and S. Cebi (eds.), *Customer Oriented Product Design*, Studies in Systems, Decision and Control 279, https://doi.org/10.1007/978-3-030-42188-5_11

and helps in measuring their growth [4]. A practical method to classify the customer requirements and their satisfaction is the Kano model. The Kano model [5] describes a nonlinear relationship between customer satisfaction and product attributes and classifies customer requirements into different categories based on how they influence customers' satisfaction or dissatisfaction. By the use of a special questionnaire that consists of functional and dysfunctional questions, the Kano model classifies the customer requirements in five categories: (1) must-be requirements, necessary because customers will be extremely dissatisfied when these requirements are not met; (2) attractive requirements, which do not cause customers' dissatisfaction when they are not achieved but they arouse customers' great satisfaction if present; (3) one-dimensional requirements reveal the linear relation that the better implementation of these requirements results in higher satisfaction; (4) in-different elements are requirements that the customers simply don't care if they are present or absent, their satisfaction remains neutral under either circumstance (5) reverse requirements cause dissatisfaction when present and satisfaction when absent (they are scarce but do happen occasionally). The Kano model can be applied to new product development or new service creation. Till now, the model has been used to various fields such as product development projects, Web-community service quality, logistic service, supply chain performance, e-learning, reliability studies et so on [6–11]. To improve the original methodology and to understand customer needs in a more accurate manner, different approaches have been proposed to extend the traditional Kano model from qualitative descriptions to quantitative analysis [12–21]. Moreover, several studies have modified the questions and/or the answers using new terms [22–24]. Others have changed the subcategories of the original Kano categories [25] and others have integrated the Kano methodology with other methods such as QFD, SERVQUAL, FMEA, et so on [6, 17, 26–31]. This chapter describes how the Kano model works and reviews the existing literature related to research on the Kano methodology, the alternative approaches to classification of quality attributes and the relationship between the Kano methodology and other methods, such as QFD.

2 The Role of the Customer in the Product Development

New product development (NPD) is an essential issue for the company's overall growth and strategy [32]. One of the reasons for new product development failure is that the development process is driven by technology rather than customer [33]. A process that uses the user-centered design (UCD) principles can be an efficient method in developing new consumer product. In the UCD, the user is supported from the beginning of the design process defining which user needs should be taken part in any design decisions. In this way, since the typical users' needs are early taken into consideration in the design and then it is revised thanks to user testing, user-focused designers are able to realize better plans and, as a result, better products and better recognition from users [34, 35]. A product with high quality is designed to meet **customer needs** (where the customer is considered as a company user of the product or service), and without the "**voice of the customers**" (VoC), the quality of the product doesn't exist. The "voice of the customer" describes the stated and unstated customer needs or requirements, and then they are translated into design targets. If during the analysis of customer needs, poor understanding and inaccurate assumptions are made, the design and manufacturing of the products in terms of quality, lead time, and cost may adversely be influenced [6].

Direct discussion or interviews, surveys, focus groups, customer specifications, observation, warranty data, field reports, etc. are possible ways to capture the voice of the customers. According to Griffin and Hauser, at least 20–30 customers should be interviewed to obtain 90–95% of all potential customer needs [36].

So it is essential that in the development process of the product, customers play an active part in realizing a product with high **quality**. As the competition for new markets increased, **customer satisfaction** becomes the cause of business success and the major issue of the companies. Hanan and Karp [37] have stated that "Customer satisfaction is the ultimate objective of every business: not to supply, not to sell, not to service, but to satisfy the customer needs that drive companies to do business."

Customer satisfaction depends on the fulfillment of user needs. Since the **attributes** of a product/service do not play the same role in satisfying the customers' needs, identifying the critical factors that determine satisfaction is essential to the continued success of any organization [38, 39]. Every company tries to increase Customer Satisfaction by improving its products or services, but limited resources constrain it in the choices. The company needs to achieve the highest level of Customer Satisfaction, determining the most suitable way to allocate scarce resources. So in the phase of the design, it is necessary to demand that the requirements meet specific quality criteria. Otherwise, it will be more difficult to obtain quality in the following development phases [40]. Each user requirement impacts on user satisfaction in different ways, and for this reason, it is important to search which attributes of a product or service bring more satisfaction than others. Customer requirements are pulled out by the voice of customers, which means know what and how the customer wants by the product that he/she will use. Once the needs are identified, it is essential to categorize them. A valid method to do this classification, based on the satisfaction that the attributes bring to customers, is the **Kano model** [5]. Since its introduction in the 1980s, Kano's **two-dimensional model** has become one of the most popular quality models among marketing/management practitioners and researchers in a wide range of industries. The Kano model abandons a **strictly linear view** of the impact of product/service attribute-performance on customer satisfaction. The model identifies the attributes that have the potential to elicit customer satisfaction/delight and/or dissatisfaction/frustration.

3 The Kano Model

Traditionally, a **linear relationship** between the achievement of quality attributes and Customer Satisfaction is **assumed to be in research**. In particular, customers are satisfied if a specific quality attribute of a product or service is provided; otherwise, they are dissatisfied. With time, the hypothesis of linearity is challenged and proven to be inaccurate [5, 18, 24]. In fact, in a product/service quality, some attributes serve to fulfill minimum Customer Requirements, while others provide additional value. So not all characteristics are equally important to customers, and the achievement of quality attributes does not necessarily involve an improvement of overall Customer Satisfaction [19]. A non-linear relationship between the performance of quality attributes and overall Customer Satisfaction is present. The Kano model allows the identification of **exciting requirements**, usually associated with **innovations** [22] and brings a different perspective for the analysis of improvement opportunities in products and services. It considers the **asymmetry and non-linear relationship** between quality attribute performance and overall customer satisfaction [5, 6].

Kano model classifies quality attributes into five categories M-Must-be, O-One-dimensional, A-Attractive, I-Indifference, and R-Reverse concerning the level of impact of single attributes on overall Customer Satisfaction:

- *Attractive quality attributes (or Exciters or Delighters)* are neither explicitly expressed nor expected by the customer. They are the product criteria that have the most significant influence on how satisfied a customer will be with a given product. Fulfilling these features leads to more than proportional satisfaction. If they are not met, however, there is no feeling of dissatisfaction.
- *One-dimensional quality attributes* are usually explicitly demanded by the customer. Customer Satisfaction is proportional to the level of fulfillment of these features: the higher the level of accomplishment, the higher the customer's satisfaction, and vice versa.
- *Must-be quality attributes (or Basic Expectations)* are the essential criteria of a product/service, the prerequisites. The customer takes them for granted and therefore, does not explicitly demand them. On the other hand, their fulfillment will not increase his satisfaction. If they are not fulfilled, the customer will not be interested in the product, service, or process at all, and he will be extremely dissatisfied.
- *The Indifferent quality* and the *Reverse quality* attributes are, in the first case, features whose presence or absence does not cause any Customer Satisfaction or Dissatisfaction and, in the second case, features whose presence causes Customer Dissatisfaction, and whose absence results in Customer Satisfaction. Indifference and Reverse factors are quality attributes that have not much relevance to fulfill the improvement of overall satisfaction. Thus, the research tradition-ally excluded them and generally, focuses on the first three quality factors, Must-be, One-dimensional, Attractive.

Figure 1 displays three lines/curves illustrating the relationship between the extent to which a quality attribute is presented (x-axis) and customer satisfaction (y-axis) regarding the quality factors. Kano et al. theorized that the influence of the one-dimensional factor on overall satisfaction is linear with a constant slope while the

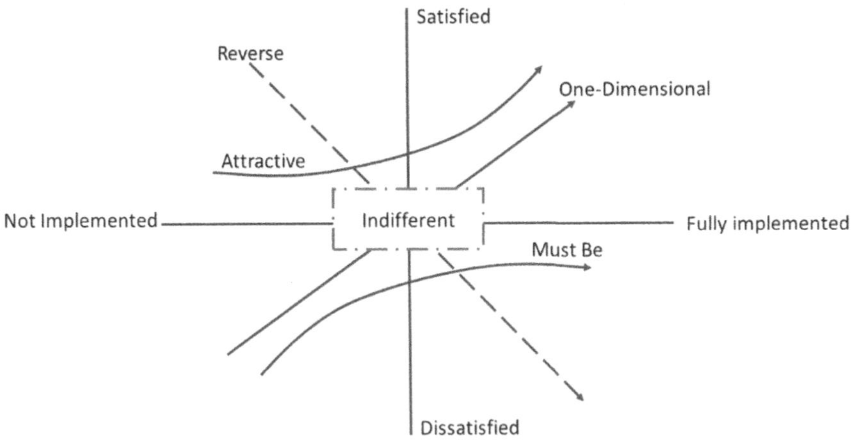

Fig. 1 The non-linear Kano model

impact of the Attractive factor and the Must-be factor are non- linear [5]. The influence of the Must-be factor is a downward relation curve in which the slope is steep when the attribute is absent and turns flat in the presence of the attribute, suggesting that a lack of this attribute results in customer dissatisfaction. Finally, the relation curve for the Attractive factor is upward with a steeper slope in the presence of the attribute, implying that a lack of this attribute has little impact on customer satisfaction, but the presence of the attribute provides excellent satisfaction. Before the development of Kano's model, all quality attributes were treated as One-dimensional factors with linear contributions to customer satisfaction.

From an operative point of view, to collect all the necessary information for the analysis, it has been necessary to employ a *questionnaire* (Table 1).

The Kano questionnaire design has formulated by a pair of questions for each product feature. In each couple of questions, the first one refers to a situation in which the attribute is met. This question is named functional question, and its formulation is similar to the: "if attribute X is present in the product, how do you feel?". The second one is called dysfunctional question, and it refers to the situation where the feature is not met. It is formulated as: "If attribute X is not present in the product, how do you feel?" For each question, the user can answer in one out of five different ways: (1) I like it that way; (2) I must be that way; (3) I am neutral; (4) I can live with it that way; (5) I dislike it that way. In Table 1, for example, the user responds: "*I can live with it that way*" to the functional question and "*I am neutral*" to the dysfunctional question.

The Table 2 shows that by combining the two answers in the Kano evaluation table, the features of the product/service can be classified into one of six quality categories: attractive quality (A), one-dimensional quality (O), must-be quality (M), indifferent quality (I), reverse quality (R), or questionable result (Q). In the Kano evaluation table (Table 2), the combination of the questions finds an "***I***", indicating that the attribute X is an *indifferent* quality item according to the user. In general, by

Table 1 Kano questionnaire-Step 1

STEP 1 KANO QUESTIONNAIRE	
Functional Question	
	□ I like it that way
If attribute X is present in the product, how do you feel?	□ It must be that way
	□ I am neutral
	□ I can live with it that way
	■ I dislike it that way
Dysfunctional Question	
	□ I like it that way
If attribute X is not present in the product, how do you feel?	□ It must be that way
	□ I am neutral
	■ I can live with it that way
	□ I dislike it that way

Table 2 Kano evaluation table-Step 2

Answer to functional question	Answer to dysfunctional question				
	Like	*Must be*	*Neutral*	*Live with*	*Dislike*
Like	Questionable **Q**	Attractive **A**	Attractive **A**	Attractive **A**	One-dimensional **O**
Must be	Reverse **R**	Indifferent **I**	Indifferent **I**	Indifferent **I**	Must be **M**
Neutral	Reverse **R**	Indifferent **I**	Indifferent I	Indifferent **I**	Must be **M**
Live with	Reverse **R**	Indifferent **I**	Indifferent **I**	Indifferent I	Must be **M**
Dislike	Reverse **R**	Reverse **R**	Reverse **R**	***Reverse R***	Questionable **Q**

combining the answers, we can obtain, in addition to the Attractive, One dimensional and Must be categories (described previously) also:

- category I, this means that the user is indifferent to this "feature x" and he/she does not care whether the attribute is present or not.
- category Q (questionable result), this signifies that the question was phrased incorrectly, or that the person interviewed misunderstood the question or crossed out a wrong answer by mistake.
- category R, this means that this "feature x" is not only not wanted by the user but he/she even expects the reverse.

Table 3 Kano results according to frequency—step 3

	Attractive A	Must be M	One-dimensional O	Reverse R	Questionable Q	Indifferent I	Total	Kano category
Attribute X				*1*			*1*	*A*
…								

By processing the results based on the **highest response frequency** obtained from the questionnaire, it is possible to classify each feature in the Kano quality category (Table 3). But if two or more Kano quality category had obtained **the same frequency value**, the authors have considered the following order M > O > A > I (that is the classification with the greatest impact on the product) to decide the quality category.

According to Kano, quality attributes are **dynamic** in the sense that customer favor and/or preferences change over time. In fact, a feature may change from satisfactory to unsatisfactory in the time if changes appear in customer expectations, familiarity with a product/service, product/service innovations, and so on. Nilsson-Witell and Fundin pointed out that, at different times, the category of the customer attributes changed accordingly and what now excites the customer (A) will be an expected requirement (O/M) in the future (A → O or A → M) [41, 42]. Moreover, what is now taken for granted (M) will possibly be a required attribute (O) or attractive attribute (A) as the customer begins to realize the importance of that particular attribute in the future (M → O or M → A), although this case is rather rare [42].

The Kano method offers the advantages to be applied not only to new product development (NPD) [6] but also to new service creation (NSC) [43]; it does not require that the customers have experience with the attributes and a limitation in the number of attributes to be analyzed. Unfortunately, the model does not distinguish attributes within the same category [44].

4 Proposed Improvements on Kano Model

The diverse relationships between customer requirement fulfillment and customer satisfaction, especially the **nonlinear relationships** are recognized by the Kano's model even if only qualitative descriptions of various relation-ship curves are focused by the classification method. The traditional Kano model discusses limited quantitative analysis or measurement of its relationship without precisely reflecting the extent to which the customers are satisfied and so lacking the quantitative measure of customer satisfaction.

Kano's classification enables to classify requirements into five different classes having different characteristics, and this doesn't provide a precise value of **customer satisfaction**. Different approaches, such as Berger's approach, Fuzzy Kano model, Analytical Kano model, Continuous Kano model and so on have been proposed to extend the Kano model from qualitative descriptions to quantitative analysis [31].

A convenient way to incorporate quantitative measures is to assign some scales in terms of Customer Satisfaction/Dissatisfaction level. Thanks to the integration of the Kano model and **Berger's CS-coefficient**, it is possible to identify the degree of satisfaction in the form of a percentage of satisfied or dissatisfied customers [45]. Berger et al. proposed customer satisfaction CS or satisfaction index SI and customer dissatisfaction CD or dissatisfaction index DI. These indexes are calculated for each requirement and determine the quantitative values of satisfaction and dissatisfaction coming from the fulfillment or non-fulfillment of a customer requirement. The two coefficients are calculated as follow:

$$CS = SI = \frac{A + O}{A + O + M + I} \tag{1}$$

$$CD = DI = \frac{O + M}{(A + O + M + I) \times (-1)} \tag{2}$$

where A, O, M and I are the attractive, one-dimensional, must-be and indifferent categories.

The dissatisfaction index (DI) ranges from −1 to 0; the closer the value is to −1, the higher the influence on customer dissatisfaction. The dissatisfaction index (DI) has a negative sign, emphasizing the negative influence that it has on customer satisfaction if the product attribute is not fulfilled.

The satisfaction index (SI) ranges from 0 to 1; if the value is closer to 1, it means that the influence on customer satisfaction is great.

The Kano model uses the functional and dysfunctional model to ask customers about product feelings, and one only single answer through the survey is permitted. Kano's model does not consider fuzzy descriptions but only crisp ones [13], because it is necessary to respond to the various Kano questions in a singular way that does not sufficiently reflect the complex thought of a person. In this way, it is not considered the fact that customer requirements are imprecise and uncertain due to their linguistic origins [12, 13, 46]. Customers often express their evaluations of customer requirements in linguistic terms, such as 'fair,' 'good,' 'between very poor and poor,' which are ambiguous and vague. Both human thought and languages contain vagueness and uncertainty [15, 47], and therefore, it is not appropriate to consider customers' evaluations of customer requirements as crisp values.

On the contrary, the **Fuzzy Kano model** uses a flexible way to answer questions even if it also uses the functional and dysfunctional model to ask customers about product feelings [47]. In this model, thanks to the use of the membership function, the users express the degree of their feelings based on their own choices and the answer presented will be closer to real human thinking [13, 31, 47, 48]. In this way, it is possible to get complete customers' information that better fits humanity thinking and assists the researcher in capturing the voice of the customer (Fig. 2).

The open-ended questions of the Fuzzy Kano questionnaire require more time and effort from the interviewers, and this can discourage the customers from responding to his questionnaire. Moreover, sometimes, it is hard for them to precisely express

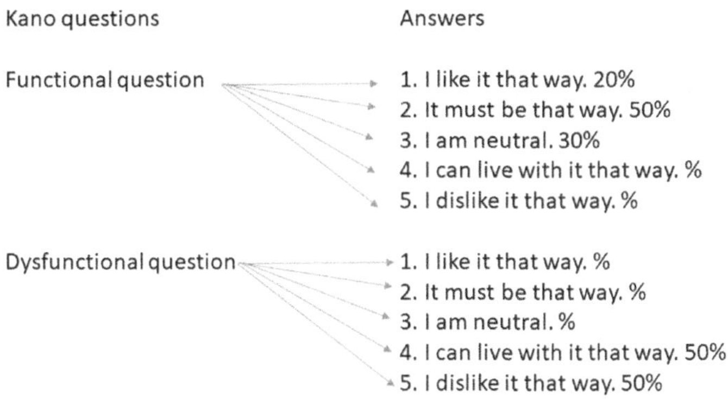

Fig. 2 Example of multiple answers with Fuzzy Kano model

their preferences in terms of percentage since their preferences are often ambiguous and unclear [20].

In the traditional Kano's model, designers make decisions based on the Kano categories in the following order: 'must-be,' 'one-dimensional,' 'attractive,' 'indifferent.' One strategy is to integrate all the 'must-be,' 'one-dimensional,' and 'attractive' attributes into the product. However, in the Kano model as also in the Fuzzy Kano model, it is difficult for designers to prioritize the attributes within the same category. So the accurate classification of the customer requirements in the presence of only a small statistical difference between two or more categories could be hard: it may suffer from a discontinuity problem [5, 20]. This means that with limited economic resources, it is not simple for designers to decide what specific attributes of the same category should be included in the product to meet the customer requirements.

The discontinuity problem and the imprecision and uncertainty in customer requirements can be simultaneously resolved, thanks to the **Continuous Fuzzy Kano model**. It integrates fuzzy techniques into Kano's model and introduces a quantitative assessment to help designers analyze customer requirements more accurately. In the Continuous Fuzzy Kano model, a modified fuzzy Kano's questionnaire, easier and quicker to complete than the Fuzzy one, deals with the imprecision and uncertainty in customer requirements and introduces an evaluation index to prioritize customer requirements. The format of the questionnaire encourages the respondents to proceed quickly through its pages since a single answer (not in terms of a percentage) to each question is requested [49].

Like the Fuzzy Kano model, the **modified fuzzy Kano questionnaire** also uses memberships functions to represent customers' preference to the standard answers ('enjoy', 'expect', 'neutral', 'live with', and 'dislike') (Fig. 3).

In conclusion, it is stated that the main differences between the fuzzy and the continuous Kano model are that:

- modified fuzzy Kano questionnaire consists of closed-ended questions, whereas the Fuzzy Kano questionnaire consists of open-ended ones.

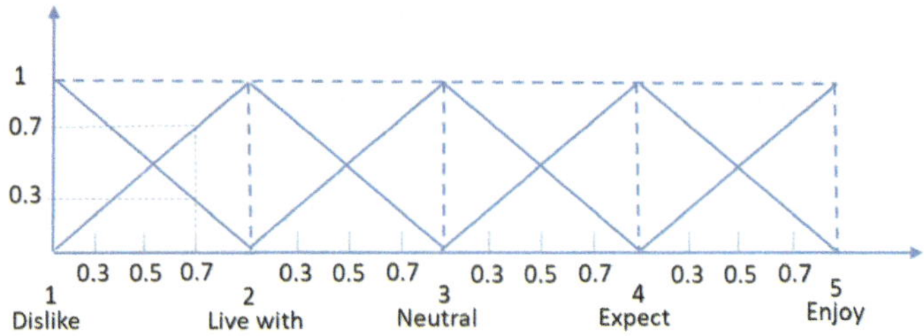

Fig. 3 Memberships functions

- modified Fuzzy Kano questionnaire is appropriate for use with self-administered surveys while the Fuzzy Kano model is more appropriate for use with interviewer-administered surveys.
- modified fuzzy Kano questionnaire demands less time and effort to complete from the respondents in self-administered surveys, and it saves costs without decreasing data quality in customer surveys for large-scale surveys.

5 Integration of Kano Model in QFD

Quality Function Deployment (QFD) is a method, developed in Japan by Akao, to help transform the voice of the customer into the product's specifications, translating customer requirements from initial planning stages through production control [50, 51]. QFD approach, widely used to plan new products and design product requirements, has diverse techniques and tools; it's of utmost importance to know which ones are best for a specific application. The QFD methodology is composed of 4 phases, or matrices, that represent a particular aspect of the product's requirements: product planning matrices, design planning matrix, process planning matrix and control planning matrix (Fig. 4).

Usually, the first matrix used in the QFD process is the House of Quality or product planning. Before starting with the designing process in the QFD, the first step is to capture the voice of the customer (VoC) [22]. The VoC should contain the customer needs, and surveys, personal interviews, or focus groups can be useful for helping capture new information about that. Getting good data from the customers is vital for the success of the entire process because the identified needs determine the product requirements.

The QFD process assumes that the customers can assess the importance and their satisfaction with the product requirements. In order to do this, customers must have previous experience with the product or service developed; for all those products that have innovative requirements, customers may not be able to express a competitive evaluation. The relationship between satisfaction and importance is linear and independent.

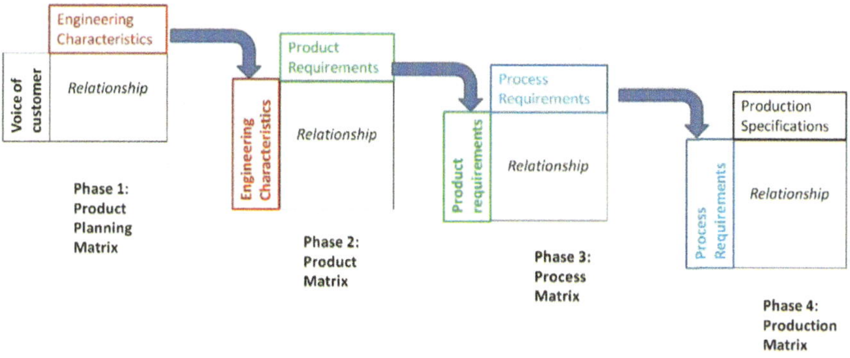

Fig. 4 QFD

In order to determine which requirements of a product bring more satisfactions to customers, the Kano model is used and integrated into the QFD [22]. The Kano model of customer satisfaction is an essential tool for understanding and classifying five types of customer requirements, or potential features, for new products and services. This model shows how each of these categories can influence satisfaction and dissatisfaction; in particular it can identify which are the requirements that the customers expect, which conditions of a product or service bring more than proportional satisfaction to customers, or which ones do not bring satisfaction when present, but bring dissatisfaction when they are not met. By integrating the Kano model into QFD, it's possible to improve the understanding of customer needs, leading to superior product design for three reasons.

First of all, it is able to provide a more authentic voice of the customer (VOC) as input to the QFD analysis, because customer needs are clearly defined, and each requirement is assigned to the reference category [52].

Second, in the QFD the section of the first matrix (House of Quality) that contains the design features or engineering characteristics or technical requirements ("Hows") can be used to determine the demand from the wants and needs of the customer, listed in the "Whats" section.

Third, thanks to the incorporation of Kano's model into the planning matrix of QFD to help deeply understand the nature of the VOC, it is possible to extend this new approach in service fields and use it for the development of service excellent [53]. The QFD aims to design new products increasing customer satisfaction, identified with the Kano model, and improving the competitive position on the market, with the compromise of the provider's capability and profitability.

The first part of the QFD, as anticipated, is the product planning matrix or House of Quality (HoQ), a useful tool used to define the relationship between customer desires and product or service design characteristics. Utilizing a relationship matrix, the House of Quality establishes customers' wants or "Whats" and translates them into appropriate technical and design requirements or "Hows." Before this, it is important to listen to the voice of the customer (VOC) to be able to collect what customers want (quantitative VOC) and how they prioritize their wants (qualitative VOC) [26].

One typical product planning matrix includes the *What analysis* with user needs, raw importance, competitive analysis, user satisfaction target, sales point, improvement ratio, absolute importance and absolute weight of whats and *How analysis* with engineering characteristics, relationship and absolute weight of hows (Fig. 5).

The integration of the *What* analysis between the QFD and the Kano model enables the total satisfaction of the client, which is the primary objective, to be effectively achieved. In this way, the quantitative VOC can be then reprioritized in the Kano model, providing a clear understanding of what requirements are needed to satisfy customers' wishes. As found in the literature, according to Tan and Shen's approach [26], shown in Fig. 6, Tontini [22] and Chaudha et al.'s approach [54], shown in Fig. 7, the results obtained from Kano model are integrated into the product planning matrix.

The terms of What analysis are formulated and regrouped in Table 4 and the detailed descriptions of "What analysis" and "How analysis" are presented in the following sections.

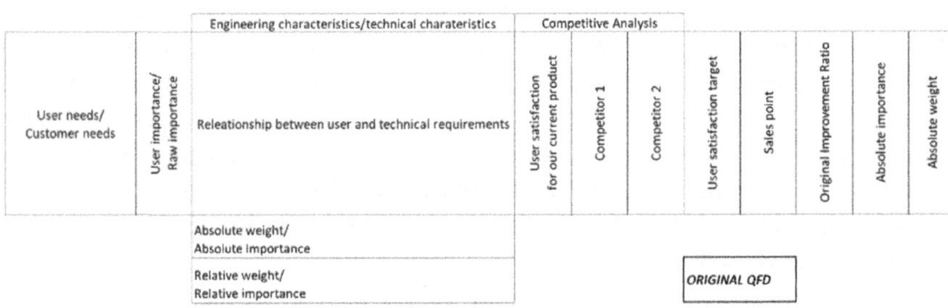

Fig. 5 Original HoQ matrix

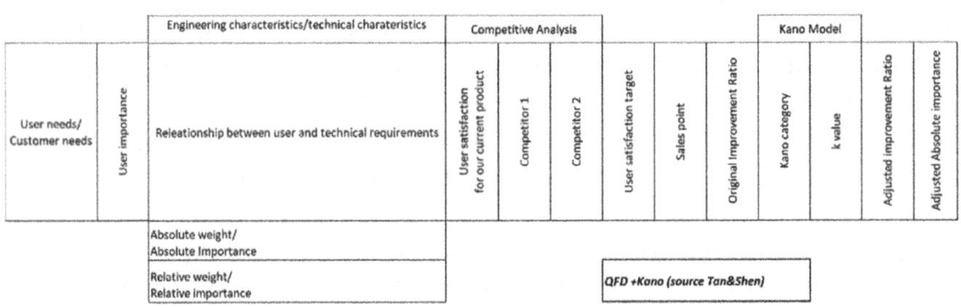

Fig. 6 QFD and Kano model according to Tan and Shen's approach

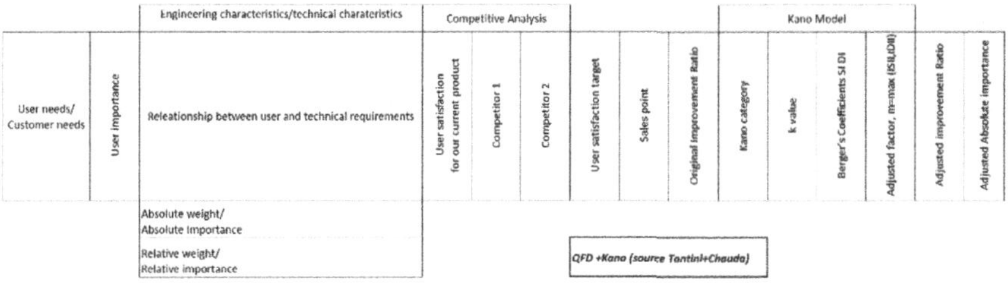

Fig. 7 QFD and Kano model according to Tontini and Chaudha's approach

5.1 WHAT Analysis

The first step in *What* analysis is the definition of the *What List or user requirement* that includes all the desirable qualities expected from the product or service. Two of the most commonly used methods for gathering user requirements are to interview customers and collect their complaints.

5.1.1 Importance of the Whats or Raw Importance

The first thing to do is to order all the user requirements by their importance; the importance of the Whats is also named "User importance", "Importance to customer", "Raw importance", "Self-stated importance" or, in our case, only A. In order to do this, the potential customers rate the importance of each aspect within a scale from 1 to 5, with 5 representing the highest level of importance, using a questionnaire survey or other means of data collection.

5.1.2 Competitive Analysis

Competitive analysis is one standard method to critically profile and compare the performance of the product or service in meeting customers' needs against the competition. Usually two key competitors are selected, and, within a competitive matrix, customers are asked to rate their satisfaction level for each customer requirements, for both our product or service and the competitors' product (here for our product we intend the product or service analyzed in the company/sector in which the analysis is carried out). In this way, QFD practitioners should be able to see the company's competitive landscape and the possible opportunities to differentiate the company's products from the competition. Moreover, they can benchmark their customer satisfaction and strategically set targets.

Table 4 Terms of what analysis

Feature	Methodology	Similar terms/equation
User importance	QFD	User importance; Importance to customer; Raw importance; Self-stated importance; importance rating; A
Competitive analysis	QFD	Current customer satisfaction, S_0; User satisfaction; Our current product
		Competitor 1
		Competitor 2
Target	QFD	Customer satisfaction target, S_1; User satisfaction target; Objective; Target
Sales	QFD	Sales point F; Sales argument
Original improvement ratio	QFD	Original improvement ratio, $IR_0 = S_1/S_0$
	Kano traditional	Kano category
	Tan and Shen; Chaudha's et al.	k value per Kano category
	Berger	Berger's coefficients SI DI
	Tontini	Adjusted factor, m = max (ISII, IDII)
Adjusted improvement ratio	Chaudha et al.	Adjusted improvement ratio, $IR_{adj\text{-}chaudha} = (1 + m)^k \times IR_0$
	Tan and Shen	Adjusted improvement ratio, $IR_{adj\text{-}tan} = (IR_0)^{1/k}$
Absolute weight/absolute importance	QFD	Absolute importance = $IR_0 \times A$
	QFD	Absolute weight = $A \times IR_0 \times F$
Adjusted absolute importance	Tan and Shen	Adjusted importance = $IR_{adj\text{-}tan} \times A = (IR_0)^{1/k} \times A$
	Garibay = Tan and Shen	Adjusted importance = $IR_{adj\text{-}tan} \times A = (IR_0)^{1/k} \times A$
	Chaudha et al.	Adjusted importance = $IR_{adj\text{-}chaudha} \times A$

(continued)

Table 4 (continued)

Feature	Methodology	Similar terms/equation
	Tontini	Adjusted importance = adjusted factor
Quality improvement index	Matzler	Quality improvement index, QI = Rx (evaluation of own product-evaluation of competitor's product)

5.1.3 Current Customer Satisfaction, S_0

This value is a measure of how products and services supplied by a company meet each user's requirement.

5.1.4 Customer Satisfaction Target, S_1

It is the target/planned expectation (performance) for each requirement, defined by the users themselves from the user importance survey.

5.1.5 Sales Point

In the column of the marketing "sales point," you will get a value between 1 and 1.5 if the mentioned WHAT a good sales pitch (1 = bad, 1.5 = good pitch) is. With this, we have included in our design not only the customer preferences (raw importance) but also our preferences as a company. So the sales point is sometimes used to intensify the weight of requirements that reinforce the company's sales strategy.

5.1.6 Original Improvement Ratio, IR_0

The original improvement ratio, interpreted as customer satisfaction improvement ratio, is calculated by the equation:

$$IR_0 = \frac{S_1}{S_0} \tag{3}$$

where S_1 is the customer satisfaction target (planned level) and S_0 is the current customer satisfaction level.

IR_0 values larger than 1 are good indicators of bad processes that highlight customer requirements that should receive priority attention in order to improve the process. Customer requirements with the highest value should be considered first and the lowest last.

5.1.7 Absolute Importance

The absolute importance permits to categorize the customer needs according the formula:

$$Absolute\,importance = IR_0 \times A \quad (4)$$

where A is the importance raw and IR_0 the improvement ratio

5.1.8 Absolute Weight

The absolute weight is the multiplication between the importance raw A, the improvement ratio IR_0 and the sales point F.

$$Absolute\,weight = IR_0 \times A \times F \quad (5)$$

5.1.9 Kano Category and K Value

In the case of integration of QFD and Kano, the analysis has been carried out considering the Tontini model [22], Chaudha et al.'s model [54] and the Tan and Shen's model [26].

The first step of the integration is to categorize the user requirements according to the Kano model. The questionnaire (Table 1), the evaluation table (Table 2) and the analysis of the highest response frequency (Table 3) permit to categorize customers' needs into five categories. The results of this categorization will be reported in the *column Kano category*, and for each class, the Kano adjustment parameter k can be fixed.

According to Tan-Shen [26], k is an important parameter that reflects different Kano categories and the choice of an appropriate numerical value for k is a critical issue, and it is dependent on QFD practitioners' experience and understanding of the relationships. It should be noted that QFD practitioners might choose different numerical values for k as long as they think the chosen values can appropriately reflect the real relationship. According to Tan-Shen [26], a possible set of k values could be 0.5, 1, and 2 for must-be (M), one-dimensional (O) and attractive attributes (A), respectively.

According to Chaudha et al. [54], the **k value** is defined as 0, 0.5, 1 and 1.5 for Indifferent (I), Must-be (M), One-dimensional (O) and Attractive (A) respectively.

5.1.10 Berger Coefficients

The Kano model is limited because it prevents a quantitative assessment of customer satisfaction concerning if requirements are met or unmet. Berger et al. suggested a solution in identifying relative values of achieving user satisfaction or not, by developing the customer satisfaction coefficient (CS) for a given requirement. The CS indicates how strongly a product feature may influence user satisfaction (satisfaction index SI), or how strongly the non-fulfillment may impact on user dissatisfaction (dissatisfaction index DS), preserving four Kano categories (A, O, M, I). The satisfaction index SI and the dissatisfaction index DI are defined by previous Eqs. (1) and (2).

Based on the above equations, it was easier to identify whether qualities offered will fulfill user satisfaction or prevent the user from dissatisfaction. According to Wang and Ji [15], calculating CS and CD values can reflect the average impacts of each quality provided to customers' feelings of satisfaction.

5.1.11 Adjusted Improvement Ratio_Tan and Shen, $\mathbf{IR_{adj\text{-}Tan\ and\ Shen}}$

Tan and Shen [26] proposed to re-prioritize every customer requirement from the QFD analysis by an improvement ratio using the equation:

$$IR_{adj\text{-}Tan\,and\,Shen} = (IR_0)^{1/k} \tag{6}$$

where IR_{adj} is the adjusted improvement factor, IR_0 is the original improvement ratio, and k is the Kano parameter for the three different categories. After classifying customer attributes into appropriate Kano categories, the corresponding value of k can be chosen. As said, Tan and Shen [26] leave the definition of the values of k to the design team but suggest that one possible set of values would be k 0.5, 1 and 2 for must-be, one-dimensional and attractive requirements respectively. Although adjusted improvement ratio larger than one indicate customer requirements that should be improved, no information regarding which customer requirement should take priority is provided.

Tan and Shen's proposal [26] consists of increasing the weight of the basic requirements, with $k = 0.5$, and decreasing the weight of the attractive ones, with $k = 2$. In this way, the design team will focus more on improving the basic requirements, giving less importance to the excitation ones. Customers tend to rate excitement requirements as less important due to lack of knowledge and previous experience with them. With the choice of $k = 2$ for excitation requirements, it is possible that the final product is excellent compared to the basic requirements, but at the same time will not bring superior satisfaction to the customers, due to lack of excitement requirements. Furthermore, the approach proposed by Tan and Shen [26] does not reduce the problems that arise with the traditional importance rating; on the contrary, it could increase them in some cases. For example, for basic requirements, when the performance is equal to competition, this method tends to put even more emphasis on the decision to improve them.

5.1.12 Adjustment Factor_Tontini, M

The adjustment factor suggested by Tontini [22] for the integration of the Kano model in the QFD, compared to the approach proposed by Tan and Shen [26] just analyzed, has less impact because it does not use the self-stated importance to calculate finals weights, and the Kano model questionnaire asks customers to state their satisfaction or dissatisfaction with a hypothetical situation. Tontini approach [22] defines the adjustment factor as:

$$Adjustment\ Factor = m = Max\,|SI, DI| \tag{7}$$

where SI and DI are the satisfaction and dissatisfaction indexes [45].

The adjustment factor is the higher absolute value of SI or DI, putting more weight on the requirements that bring more satisfaction when present or that bring more dissatisfaction when absent. In this case, excitement, performance, and basic needs will be taken into consideration depending on the degree of satisfaction or resentment that they could bring to customers.

5.1.13 Adjusted Improvement Ratio_Chaudha, $\mathbf{IR_{adj\text{-}Chaudha}}$

The important parameters deriving from the Kano's theory can both be used within the same function to adjust the traditional improvement ratio (target/our own customer satisfaction level).

The function proposed by Chaudha et al. [54] is:

$$IR_{adj-Chaudha} = (1+m)^k \times IR_0 \tag{8}$$

where IR_{adj} is the adjusted improvement ratio according to Chaudha and IR_0 is traditional improvement ratio of QFD, m = max ISII; IDII is the adjusted factor of Tontini [18], value of k can be taken as 0, 0.5, 1, 1.5 for indifferent, must-be, one dimensional and attractive requirements, respectively.

In this function, Chaudha et al. [54] chose to give the utmost importance to attractive attributes, while the lowest importance to indifferent ones. Not present in previous literature, the Kano parameter for the indifferent requirement is set as 0. These attributes may be innovative, and initially, when they are introduced, they do not result in either customer satisfaction or customer dissatisfaction. Kano, however, explained that indifferent requirements could not be neglected at all, because, in a lifecycle, they can turn into attractive conditions. They should be given more weight due to the modern customer's demand for innovative ideas and must-be requirements alone cannot make a competitively better product.

5.1.14 Adjusted Importance

Once the "What" analysis has been completed, it is possible to evaluate which are the most important aspects to be improved, looking at the weight of each one, and discard those that are less important. The adjusted importance provides a more unobstructed view of what customer's requirement should be tackled first to improve the quality of the product.

According the previous authors, it can be calculated with the following formulas:

- *Adjusted importance_Tan and Shen*

$$Adjusted\,importance = IR_{Tan\,and\,Shen} \times A \tag{9}$$

where IR_{adj} is the adjusted improvement ratio according to Tan and Shen [26] and A is the user importance of whats or self-stated importance or raw importance

- *Adjusted absolute importance_Tontini*

$$Adjusted\,absolute\,importance = Adjustment\ Factor = m = Max\,|SI, DI| \tag{10}$$

- *Adjusted absolute importance_Chaudha*

$$Adjusted\,importance = IR_{adj\text{-}Chaudha} \times A \tag{11}$$

where IR_{adj} is the adjusted improvement ratio according to Chaudha et al. [54] and A is the user importance of whats or self-stated importance

5.2 *HOW Analysis*

After considering the main aspects of the product or service under analysis, the next step is to define the technical requirements necessary to obtain it. To do this, a new list is created, called "How List", with the features that must have our product.

5.2.1 Relationship Between WHATs and HOWs

Basically, "Hows" are ways to get the "Whats." Within a matrix, the engineering characteristics (Hows) are correlated to each desirable qualities (Whats), to determine the strength of relationship and impact on the need. To do this, and fill the relationship matrix, a scoring system based on 9, 3, 1 and 0 is often used; these numbers indicate, respectively, a "strong", "moderate", "weak" or "none" relationship between user needs and technical characteristics.

The choice of numerical values 9, 3, 1, 0 creates a contrast between the "strong" relationship and the others so that strong correlations are valued heavily while associations that are weak or nonexistent carry little value. QFD used a 5, 3, 1, 0 or a 4, 2, 1, 0 scale, but this new system is believed to identify the drivers to the system better.

For this purpose, to make substantial impacts dominate the matrix, the value 9 was adopted. The ratio between "strong" (9) and "moderate" (3) is high so that it is less likely that the technical characteristics with moderate ratings have greater functional importance than technical ones with at least one strong rating [48].

The quality house, in particular applications, may include negative relationships that occur when a technical characteristic has a positive impact on one customer requirement, but a negative impact on another. Negative impacts require a scale containing −9, −3 and −1 for a "strong negative", "moderate negative" and "weak negative" impact, respectively, and they complicate the discussion and the analysis of the QFD. As a result, identifying technical characteristics that result in consistent positive ratings on all customer requirements is preferred.

5.2.2 Importance of the Hows

The next step is to determine which technical aspects of the product are most important to customers, and then what features will need to be implemented. To do this, it is necessary to take into consideration the relevance of each technical feature by calculating the importance of How. The absolute weight, based on its impact to fulfill the Whats, is calculated as the sum of the value of the relationship among Whats and Hows (9, 3 or 1) multiplied by the relative weight of the customer's need (importance of the Whats). The relative weight is the ratio of calculated importance divided by the total and expressed in percentage: importance weight divided by the total weight multiplied by 100.

The formulas are:

$$w_j = \sum_{i=1}^{n} d_i r_{ij} \quad w_j^* = w_j / \sum_{j=1}^{n} w_j \tag{12}$$

where w_j is the absolute technical importance rating of technical characteristics and w^*_j is the relative technical importance, correlation $r_{i,j}$ between the user needs and the technical characteristics and d_i is the relative importance of the user needs, n is the total number of user needs and m is the total number of technical characteristics.

5.3 Numerical Example

In this chapter, a case example is presented to illustrate how the Kano's model can be integrated into QFD by adjusting the raw importance of each customer attribute. The integrated approach is applied for the evaluation of 360° videos to support the user engagement. 360° videos are videos that place users in the director's chair of a cinematic experience and let users look in any direction they choose. Unlike traditional fixed-frame videos, which limit user's view to wherever the camera is pointing, 360° videos offer a full 360° × 180° panoramic field of view—placing user in the center of the action, completely surrounded by sights and sounds. These videos let users look up, down, and all around so users can explore a scene from any perspective they choose.

There are a number of ways to watch 360° Video that offer a broad range of experiences. From VR headsets that offer the most jaw-dropping experiences to 360 web players on popular sites like Facebook and Youtube, users can enjoy 360° videos whether or not they own any VR equipment.

The best, most immersive way to watch 360° videos is with a VR headset. Although headsets like these can be quite pricey, they offer a totally captivating experience that truly puts users in the middle of the action. With a VR headset, users can adjust users view simply by turning their head—as if they were really there.

If users don't have access to any type of VR headset, they can still enjoy 360° videos with their mobile phone. There are two ways to this. The first is to use their phone as a "magic window" of sorts. In this mode, users point their phone in whatever direction they wish to look, just as if they were recording their own video. The second is to simply tap and drag on the video to control change perspective.

The final way to watch 360° videos is with a web player like those available on both Facebook and YouTube. To change user perspective, user simply clicks and drags on the video. Although this isn't a true 3D experience, it's a quick and easy way to explore 360° videos and get a feel for what they can offer.

This chapter shows an example of an integrated approach between the Kano model and the QFD technique for the evaluation of immersive videos, also known as 360-degree videos. The purpose of this work is to improve the virtual reality experience of users by adjusting the importance assigned to each consumer attribute. 360-degree videos are made by a system formed by several radially arranged cameras that take reality at the same time in any direction, offering a panoramic 360° × 180° field of view. The viewer is immersed in the virtual space, surrounded by images and sounds, unlike traditional videos based on sequences of predetermined shots that create an experience of passive vision.

During the execution of the video it is possible to move and choose what to see without following the point of view of the director, but deciding how to use the multimedia content, all made possible by 360° technology. The best and most common way to watch this type of video is to wear a VR headset, which allows users to move around in the video by merely turning the head as if they were physically present in that place. Alternatively, if such devices are not available, consumers can

Table 5 Whats or customer requirements

Code	Customer requirements
C01	Easy to use
C02	Always available (24 h/24 h)
C03	Innovative
C04	Not required ICT competences for the use
C05	Playback via personal computers
C06	Playback via mobile devices such as smartphones
C07	Playback via dedicated head-mounted displays
C08	Playback via web player (Facebook and YouTube)
C09	Falicitate navigation and exploration
C10	Realism
C11	Improve interest, attention, concentration and retention

enjoy an immersive experience less intense with a computer or a mobile phone, pointing the device in the desired direction, or by touching and dragging the video to change perspective. This second mode is not a real virtual reality example, but a fast and straightforward way to try a new and engaging experience. YouTube and Facebook have introduced this technology, and other brands have contributed to the spread of this communication frontier.

The first step is the definition of the Whats or customer requirements (Table 5).

The second step is the definition of the Hows or technical features of a 360° video (Table 6).

Customers are asked to prioritize their needs. The raw importance's had been identified by using a 1–5 scale (Table 7).

Furthermore, two kinds of 360° video have been chosen to make a competitive analysis. Customers are asked to rate their satisfaction degree for our video and another using a 1–5 scale for each customer attribute. Through this way, customer attributes, raw importance's and competitive analysis are collected (Table 7).

By using the Kano model, the customers' needs are categorized into Kano's categories according to the **highest response frequency** obtained in the Kano questionnaire. In this way only a rough sketch of the customer's satisfaction is provided and this only allows qualitative assessment of product attributes [21, 55] and does not distinguish attributes within the same category [56].

Table 6 Hows or technical feature

Hows technical characteristics	Interactivity
	Mobility
	Photography
	Connectivity
	Audio

Table 7 Raw importance and competitive analysis

		Competitive analysis				Competitive analysis	
		Customer perception				Customer perception	
Customer needs	Self-stated importance	Our product	Competitor 1	Customer needs	Self-stated importance	Our product	Competitor 1
Whats	Raw importance	S_0		Whats	Raw importance	S_0	
C01	5.00	4.00	3.00	**C07**	3.00	4.00	5.00
C02	5.00	4.00	5.00	**C08**	5.00	4.00	5.00
C03	3.00	3.00	4.00	**C09**	4.00	3.00	5.00
C04	4.00	5.00	3.00	**C10**	3.00	2.00	4.00
C05	4.00	5.00	5.00	**C11**	4.00	3.00	4.00
C06	5.00	5.00	5.00				

A convenient way to incorporate quantitative measures is to assign some scales to kano categories. The use of Tan and Shen's approach [26] or Chaudha et al.'s approach [54] can help.

The following step is calculated the IR_0 Improvement ratio (3) according to traditional QFD and adjusted improvement ratio IR_{adj} according to Tan and Shen's approach (formula 6) and Chaudha et al.'s approach (formula 8) (Table 8).

The comparison with traditional QFD method shows that the integration of Kano in QFD with Tan and Shen's approach or Chaudha et al.'s approach is capable of providing more discrimination power to find which of the requirements are most

Table 8 Improvement ratio and adjusted improvement ratio

Customer Needs Whats	Self-stated importance raw importance	Quality Planning		CS-coefficient		Kano category	Adjusted improvement ratio Tan and Shen	Adjusted improvement ratio Chaudha et al.
				SI	DI			
		Target S_1	Improvement ratio $IR_0 = S_1/S_0$	$\frac{A+O}{A+O+M+I}$	$\frac{O+M}{(A+O+M+I)} \times (-1)$			
C01	5.00	5.00	1.25	0.57	−0.90	O	1.25	2.38
C02	5.00	5.00	1.25	0.67	−0.90	M	1.56	1.72
C03	3.00	4.00	1.33	0.97	−0.40	A	1.15	3.68
C04	4.00	5.00	1.00	0.90	−0.50	O	1.00	1.90
C05	4.00	5.00	1.00	0.43	−0.63	M	1.00	1.28
C06	5.00	5.00	1.00	0.43	−0.63	M	1.00	1.28
C07	3.00	5.00	1.25	0.43	−0.63	A	1.12	2.61
C08	5.00	5.00	1.25	0.43	−0.63	M	1.56	1.60
C09	4.00	5.00	1.67	0.43	−0.67	O	1.67	2.78
C10	3.00	4.00	2.00	0.77	−0.70	A	1.41	4.70
C11	4.00	4.00	1.33	0.63	−0.53	O	1.33	2.18

Table 9 Relative importance

Customer needs	Kano category	Relative importance (%)	Relative importance (%)	Relative importance (%)
Whats		Tan and Shen	Chaudha	QFD
C01	O	10.79	11.83	10.87
C02	M	13.48	8.58	10.87
C03	A	5.98	10.99	6.96
C04	O	6.90	7.57	6.96
C05	M	6.90	5.09	6.96
C06	M	8.63	6.37	8.70
C07	A	5.79	7.80	6.52
C08	M	13.48	7.96	10.87
C09	O	11.51	11.07	11.59
C10	A	7.32	14.04	10.43
C11	O	9.21	8.68	9.28

important for the customer. Values of IR_0 and IR_{adj} larger than 1 highlight customer requirements that should receive priority attention in order to improve the process. Customer requirements with the highest value of IR_0 and IR_{adj} should be considered first and the lowest last. However, to provide a more clear view of what customer requirement should be tackled first, it is necessary to calculate the adjusted importance (formula 4) and the adjusted importance with the Tan and Shen's approach (formula 9) or Chaudha et al.'s approach (formula 11) (Table 9).

In this chapter, the customer requirements with highest percentage in the traditional QFD are C09 "Falicitate navigation and exploration" (11.59%), C01 "Easy to use" (10.87%), C02 "Always available (24 h/24 h)" (10.87%) and C08 "Playback via web player (Facebook and YouTube)" (10.87%): they should be fulfilled first. However, C01 "Easy to use", C02 "Always available (24 h/24 h)" and C08 "Playback via web player (Facebook and YouTube)" have the same value of 10.87%. In this case, the integration of Kano in QFD permits to refine better these results. So the relative importance according to the Tan and Shan's approach, for example, highlights that C02 "Always available (24 h/24 h)" (13.48%) and C08 "Playback via web player (Facebook and YouTube) (13.48%) should be taken into consideration being Must be requirements.

After the classification of Whats, customer requirements, the following step is to evaluate the classification of the Hows, technical features of the 360° video. So first it is defined the correlation between hows and whats, with the value of 9 if there is high correlation, 3 medium correlation and 1 low correlation.

Then, it is possible to calculate the w_j, absolute technical weight of hows and w^*_j relative technical weight of hows (formula 12) (Table 10). Relative weight of Interactivity (30.79%) shows that this feature, if implemented in the 360° video, have

Table 10 Classification of the Hows

Customer needs	Self-stated importance	Hows				
		Technical characteristics				
Whats	Raw importance	Interactivity	Mobility	Photography	Connectivity	Audio
C01	5.00	9.00	3.00	1.00	1.00	1.00
C02	5.00	1.00	3.00	1.00	9.00	1.00
C03	3.00	9.00	9.00	9.00	3.00	9.00
C04	4.00	9.00	1.00	1.00	3.00	1.00
C05	4.00	9.00	1.00	1.00	9.00	1.00
C06	5.00	9.00	9.00	1.00	9.00	1.00
C07	3.00	9.00	9.00	1.00	9.00	1.00
C08	5.00	1.00	1.00	1.00	9.00	1.00
C09	4.00	9.00	9.00	1.00	1.00	1.00
C10	3.00	3.00	3.00	9.00	1.00	3.00
C11	4.00	9.00	9.00	9.00	1.00	9.00
Absolute technical weight		307.00	223.00	125.00	235.00	107.00
Relative technical weight (%)		30.79	22.37	12.54	23.57	10.73

a huge impact on the customer satisfaction, so designers and developers have to take into consideration during the creation of a video.

6 Conclusions

In any strategic development, the success of the organization depends principally on the understanding of what the customer wants and carefully listening to what a customer can save both time and money. Eighty percent of new products fail in the market place because no one paid attention to the customer. For this reason, collecting the VoC is the first thing an organization should embark on when entering the market with a new product or service, or in the redesign of an existing product or service. This chapter describes the different methodologies needed to analyze customer input. In this context, the Kano model provides a practical approach to prioritize customer requirements and to help understand the nature of them. The combination of QFD and the Kano model is a suitable methodology for listening and understanding customer requirements better. A critical step when applying only QFD is to define customer requirements and to assign priorities to them. Priorities should reflect what customers want the most. Therefore, it is necessary to have real information about what satisfies customers and to provide sufficient analysis to establish priorities properly. The integration of Kano and QFD is the right solution that permits to prioritize better customer needs. Moreover, the refinements of the Kano model such as the use of Berger's coefficients, the Fuzzy Kano model and the Continuous Kano model permit to quantify the customer needs and to reflect the complex thought of a person sufficiently. A numerical example has been described in the chapter in order to show all the steps necessary to make an integration of the Kano model with QFD with Tan and Shen's approach and Chaudha et al.'s approach.

References

1. Prananto, A., McKay, J., Marshall, P.: Frameworks to Support eBusiness Growth Strategy. In: ECIS 2001 Proceedings, p. 85 (2001)
2. Meng, Q., Jiang, X., Bian, L.: A Decision-making method for improving logistics services quality by integrating fuzzy Kano model with importance-performance analysis. J. Serv. Sci. Manage. **08**(03), 10 (2015)
3. Anderson, E.W., Mittal, V.: Strengthening the satisfaction-profit chain. J. Serv. Res. **3**(2), 107–120 (2000)
4. Gitman, L.J., McDaniel, C.: The Future of Business. South-Western (2005)
5. Kano, N., et al.: Attractive quality and must-be quality. J Japan. Soc. Qual. Control **14**(2), 39–48 (1984)
6. Matzler, K., Hinterhuber, H.H.: How to make product development projects more successful by integrating Kano's model of customer satisfaction into quality function deployment. Technovation **18**(1), 25–38 (1998)
7. Kuo, Y.-F.: Integrating Kano's Model into Web- community Service Quality. Total Qual. Manage. Bus. Excellence **15**(7), 925–939 (2004)
8. Hines, P., Zokaei, K.: Achieving consumer focus in supply chains. Int. J. Phys. Distrib. Logistics Manage. **37**(3), 223–247 (2007)
9. Sahin Basfirinci, C., et al.: An application of integrating SERVQUAL and Kano's model into QFD for logistics services: A case study from Turkey. Asia Pac. J. Mark. Logistics **21**(1), 106–126 (2009)
10. Chen, L.-H., Kuo, Y.-F.: Understanding e-learning service quality of a commercial bank by using Kano's model. Total Qual. Manage. Bus. Excellence **22**(1), 99–116 (2011)
11. Violante, M.G., Vezzetti, E.: Virtual interactive e-learning application: an evaluation of the student satisfaction. Comput. Appl. Eng. Educ. **23**(1), 72–91 (2015)
12. Chen, L.-F.: A novel approach to regression analysis for the classification of quality attributes in the Kano model: an empirical test in the food and beverage industry. Omega **40**(5), 651–659 (2012)
13. Chen, L.-H., Ko, W.-C.: Fuzzy approaches to quality function deployment for new product design. Fuzzy Sets Syst. **160**(18), 2620–2639 (2009)
14. Chen, L.-H., Ko, W.-C.: A fuzzy nonlinear model for quality function deployment considering Kano's concept. Math. Comput. Model. **48**(3), 581–593 (2008)
15. Wang, T., Ji, P.: Understanding customer needs through quantitative analysis of Kano's model. Int. J. Qual. Reliab. Manage. **27**(2), 173–184 (2010)
16. Ji, P., et al.: Quantification and integration of Kano's model into QFD for optimising product design. Int. J. Prod. Res. **52**(21), 6335–6348 (2014)
17. Lee, Y.-C., Huang, S.-Y.: A new fuzzy concept approach for Kano's model. Expert Syst. Appl. **36**(3, Part 1), 4479–4484 (2009)
18. Lin, L.-Z., Yeh, H.-R., Wang, M.-C.: Integration of Kano's model into FQFD for Taiwanese Ban-Doh banquet culture. Tour. Manag. **46**, 245–262 (2015)

19. Lin, S.-P., et al.: Refining Kano's 'quality attributes–satisfaction' model: a moderated regression approach. Int. J. Prod. Econ. **126**(2), 255–263 (2010)
20. Wu, M., Wang, L.: A continuous fuzzy Kano's model for customer requirements analysis in product development. Proc. Inst. Mech. Eng. B J. Eng. Manuf. **226**(3), 535–546 (2011)
21. Xu, Q., et al.: An analytical Kano model for customer need analysis. Des. Stud. **30**(1), 87–110 (2009)
22. Tontini, G.: Integrating the Kano model and QFD for designing new products. Total Qual. Manage. Bus. Excellence **18**(6), 599–612 (2007)
23. Chen, J.-K., Lee, Y.-C.: A new method to identify the category of the quality attribute. Total Qual. Manage. Bus. Excellence **20**(10), 1139–1152 (2009)
24. Schvaneveldt, S.J., Enkawa, T., Miyakawa, M.: Consumer evaluation perspectives of service quality: evaluation factors and two-way model of quality. Total Qual. Manage. **2**(2), 149–162 (1991)
25. Shahin, A., Zairi, M.: Kano model: A dynamic approach for classifying and prioritising requirements of airline travellers with three case studies on international airlines. Total Qual. Manage. Bus. Excellence **20**(9), 1003–1028 (2009)
26. Tan, K.C., Shen, X.X.: Integrating Kano's model in the planning matrix of quality function deployment. Total Qual. Manage. **11**(8), 1141–1151 (2000)
27. Pawitra, T.A., Tan, K.C.: Integrating SERVQUAL and Kano's model into QFD for service excellence development. Manag. Serv. Qual. Int. J. **11**(6), 418–430 (2001)
28. Pawitra, T.A., Tan, K.C.: Tourist satisfaction in Singapore—a perspective from Indonesian tourists. Manag. Serv. Qual. Int. J. **13**(5), 399–411 (2003)
29. Tan, K.C., XIE, M., SHEN, X.-X.: Development of innovative products using Kano's model and quality function deployment. Int. J. Innov. Manag. **03**(03), 271–286 (1999)
30. Roslan, N.A.A., Wahab, E., Abdullah, N.H.: Service quality: a case study of logistics sector in Iskandar Malaysia using SERVQUAL Model. Procedia Social Behav. Sci. **172**, 457–462 (2015)
31. Violante, M.G., Vezzetti, E.: Kano qualitative vs quantitative approaches: an assessment framework for products attributes analysis. Comput. Ind. **86**, 15–25 (2017)
32. Ernst, H.: Success factors of new product development: a review of the empirical literature. Int. J. Manage. Rev. **4**(1), 1–40 (2002)
33. Veryzer, R.W., Borja de Mozota, B.: The impact of user-oriented design on new product development: an examination of fundamental relationships. J. Prod. Innov. Manag. **22**(2), 128–143 (2005)
34. Gould, J.D., Boies, S.J., Ukelson, J.: How to design usable systems. Handbook Hum. Comput. Interact. **2**, 231–254 (1988)
35. Gould, J.D., Boies, S.J., Lewis, C.: Making usable, useful, productivity-enhancing computer applications. Commun. ACM **34**(1), 74–85 (1991)
36. Griffin, A., Hauser, J.R.: The voice of the customer. Market. Sci. **12**(1), 1–27 (1993)
37. Hanan, M., Karp, P.: Customer satisfaction: how to maximize, measure, and market your Company's ultimate Product. American Management Association (1989)
38. Violante, M.G., Vezzetti, E.: Implementing a new approach for the design of an e-learning platform in engineering education. Comput. Appl. Eng. Educ. **22**(4), 708–727 (2014)
39. Violante, M.G., Vezzetti, E.: A methodology for supporting requirement management tools (RMt) design in the PLM scenario: an user-based strategy. Comput. Ind. **65**(7), 1065–1075 (2014)
40. Génova, G., et al.: A framework to measure and improve the quality of textual requirements. Requirements Eng. **18**(1), 25–41 (2013)
41. Nilsson-Witell, L., Fundin, A.: Dynamics of service attributes: a test of Kano's theory of attractive quality. Int. J. Serv. Ind. Manag. **16**(2), 152–168 (2005)
42. Raharjo, H., Brombacher, A.C., Xie, M.: Dealing with subjectivity in early product design phase: a systematic approach to exploit Quality Function Deployment potentials. Comput. Ind. Eng. **55**(1), 253–278 (2008)

43. Bhattacharyya, S., Rahman, Z.: Capturing the customer's voice, the centerpiece of strategy making: a case study in banking. Eur. Bus. Rev. **16**(2), 128–138 (2004)
44. Lee, M., Newcomb, J.: Applying the Kano methodology to meet customer Requirements: NASA's microgravity science program. Qual. Manage. J. **4**, 95–106 (1997)
45. Berger, C., et al.: Kano's methods for understanding customer-defined quality. Center Qual. Manage. J. **2**(4), 3–36 (1993)
46. Chen, L.-H., Kuo, Y.-F.: Understanding e-learning service quality of a commercial bank by using Kano's model. Total Qual. Manage. **22**(1), 99–116 (2011)
47. Lee, Y.-C., Huang, S.-Y.: A new fuzzy concept approach for Kano's model. Expert Syst. Appl. **36**(3), 4479–4484 (2009)
48. Yadav, H., et al.: Prioritization of aesthetic attributes of car profile. Int. J. Ind. Ergon. **43**(4), 296–303 (2013)
49. Amritesh, S., Misra, C., Chatterjee, J.: Quality framework for credence-based informational services: applying Kano's method. Total Qual. Manage. Bus. Excellence. 1–32 (2016)
50. Akao, Y.: QFD: Past, present, and future. In: International Symposium on QFD (1997)
51. Cohen, L., Quality Function Deployment: How to Make QFD Work for You. Addison-Wesley (1995)
52. Sireli, Y., Kauffmann, P., Ozan, E.: Integration of Kano's model into QFD for multiple product design. Eng. Manage. IEEE Trans. **54**(2), 380–390 (2007)
53. Tan, K.C., Pawitra, T.A.: Integrating SERVQUAL and Kano's model into QFD for service excellence development. Manag. Serv. Qual. **11**(6), 418–430 (2001)
54. Chaudha, A., et al.: Integration of Kano's Model into quality function deployment (QFD). Int. J. Adv. Manuf. Technol. **53**(5–8), 689–698 (2011)
55. Riviere, P., et al.: Adaptive preference target: contribution of Kano's model of satisfaction for an optimized preference analysis using a sequential consumer test. Food Qual. Prefer. **17**(7), 572–581 (2006)
56. Xu, Q., et al.: Customer requirement analysis based on an analytical Kano model. In: 2007 IEEE International Conference on Industrial Engineering and Engineering Management. IEEE (2007)

Mobile Application Design by Using Modified Fuzzy Kano Model

Gulberk Demir and Selcuk Cebi

Abstract With the developing technology, access to the internet via mobile devices has become widespread day by day since mobile phones present a more easy and simple way to reach the internet at any time. This has led to the development and widespread of mobile applications. Therefore, firms aim to develop their mobile applications to include various features in order to keep their products in the global market. Although many features have been developed to attract customer attention, customer expectations may differ from each other. The main reason for this is that customer expectations and perceptions influence the features offered by mobile applications. Consequently, the effects of mobile application features on customers should be measured. A questionnaire-based Kano Model Method was proposed to the literature for this purpose. However, the Kano Model questionnaire includes linguistic evaluation and the traditional Kano Model Method is weak at handling uncertainty in linguistic evaluation. Therefore, in the scope of this chapter, a modified Fuzzy Kano Model Method has been proposed to overcome this difficulty. Furthermore, the proposed method has been applied to classify features of mobile device applications. The obtained results have been compared with the traditional Kano Model.

1 Introduction

Since the introduction of smartphones and the widespread use of the internet, mobile applications that make life easier have become an indispensable element of everyday life. For this, a rapid digital transformation has begun in this area. A new mobile application development process has already started offering solutions to people's needs. This process has created new competitive investment areas for telecommunication companies. Those that have been able to analyze customer needs and desires accurately have managed to turn this into profit.

G. Demir (✉) · S. Cebi
Department of Industrial Engineering, Yildiz Technical University, Besiktas, Istanbul, Turkey
e-mail: d.gulberk@hotmail.com

C. Kahraman and S. Cebi (eds.), *Customer Oriented Product Design*, Studies in Systems, Decision and Control 279, https://doi.org/10.1007/978-3-030-42188-5_12

In Turkey, telecommunication companies that intend to increase their market size have established their digital teams by following the trends of this new digital technology. Just like the banks, they have combined the online transactions that millions of their customers are using in one application. The goal of combining people's frequent online transactions into a single application is to satisfy their customers by facilitating their work and to attract new customers to their market in a competitive environment. Telecommunication companies need to know which features should be in the applications and how much the new developments will satisfy the customers' expectations. Sometimes, while a new product has been developed, companies do not perform market research correctly, focusing on features that are not so important to the customer, and exerting unnecessary effort. Therefore, focusing on product features that do not create added value for the customer is a waste of time and money for companies.

According to Google Play July 2019 data, mobile applications belonging to companies operating in Turkey have been downloaded over 10 million times by users [1–3]. Therefore, customer-oriented features that will be added to these applications, which are used by millions of people, have a significant impact on increasing the satisfaction and loyalty of the users. To satisfy the users and to enable them to use the online operator application continuously, features that will meet the needs of the customers must be identified and should be added to the application. In order to analyze customer requirements and demands, and to measure the satisfaction level of the features, the customers' expectations have been the subject of academic studies. Various methods have been developed in this area and are still being developed. One of these methods is the Kano Model developed by Dr. Kano [4]. The model evaluates the customers' answers to the functional and dysfunctional questions according to a 5×5 table. The main purpose of this assessment is to measure the customer's interest of product features. In this way, the method aims to develop customer-oriented services and products by using the customer feedback obtained through the survey. Therefore, customer satisfaction also increases in direct proportion.

In the Kano Model, quality elements are grouped into five categories. The model has been applied in many case studies to figure out the customers' perceptions of product features. For this reason, the Kano Model has been the subject of much research. Although Kano Model has been used in many academic studies aimed at determining product needs and customer needs in recent years, the Kano Model and Fuzzy Set Theory have been used together in a limited number of studies in order to analyze the uncertainty of the answers given by the customers for the functional and dysfunctional questions [5].

In order to better understand customer expectations, the studies that come to the forefront of the Kano Model applications using Fuzzy Set Theory are as follows: Lee and Huang [5] proposed a Fuzzy questionnaire since the Kano Model does not consider customers' mental and emotional thoughts adequately. Therefore, instead of Kano's known two-dimensional questionnaire, Fuzzy surveys where customer responses are expressed as percentages, were applied for the first time. Matzler and Hinterhuber [7] applied the Fuzzy Kano Model, which included customer perceptions, in deciding the main features that should be in the design of the new product.

While analyzing the results, they modified the satisfaction and dissatisfaction formulas proposed by Berger in 1993 [6–9]. In another study, Ilbahar and Cebi [10] utilized the fuzzy set theory for the Kano Model in the classification of product properties. By the proposed method, they make it possible to evaluate each product feature in more than one class based on the membership degree of the feature.

The main objective of the presented chapter is to analyze various features of mobile online transaction applications including Top-Up Purchase (Mobile Recharge), Monitoring Usage, Online Live Support, Winning Weekly Gifts, Paying Bills, etc., which were published by three major telecommunications companies that still operate in Turkey. Thus, this chapter aims to define the product features that are the most important for the customers, and which should be taken into consideration while developing the telecommunication applications of the enterprises. For this, first of all, the detailed market information on product characteristics has been researched, and later a web-based Kano survey has been developed. Then, the obtained data was analyzed by using both the traditional Kano Model and the modified Fuzzy Kano Model which will be proposed for the first time in this chapter. Finally, the product features are classified according to the customers' expectations and obtained results from these methods have been compared with the each other.

The rest of this chapter is organized as follows: The second part offers a detailed literature review of the studies on the Kano Model. Section 3 describes the Traditional Kano Model and the proposed Fuzzy Kano Model. Section 4 gives the application of the proposed approach. Section 5 provides the findings and discussion. Finally, Sect. 6 summarizes the results of this chapter.

2 Literature Review

In this section, we will talk about the history of the Fuzzy Kano Model and how it is used. Several studies have been conducted in the past regarding the Fuzzy Kano Model. In the scope of this studies, various solution methods using this model, along with other have been developed.

Lee and Huang [5], who first criticized Kano's two-dimensional survey model, presented a Fuzzy survey approach in 2009. In the questionnaires designed according to the Traditional Kano Model, the uncertainty of human thought was ignored because people were asked to choose a definite answer. Lee and Huang [5], in their study, aimed to express the feelings and thoughts of the participants with the right degrees [5]. They used Fuzzy Set Theory using Zadeh's [12] concept of α-cut standard to make the results of their work more satisfactory and descriptive.

In 2012, Florez-Lopez and Ramon-Jeronimo [11] proposed a new framework including the Kano Model, fuzzy distances, and the linguistic two-tuples model in order to manage customer service logistics more effectively. Florez-Lopez and

Ramon-Jeronimo [11] have found new strategies that can effectively manage logistics customer service and improve customer satisfaction by providing an integrative fuzzy-linguistic structure to better measure the relationship between product characteristics and customer satisfaction [11].

Some studies combine the Kano Model and multi-criteria decision making. Ghorbani et al. [13] used a Kano Model and fuzzy multi-criteria decision-making methods for supplier selection. In the first stage, they proposed a three-stage approach in which the importance of the criteria was calculated using the Fuzzy Kano Survey and Fuzzy AHP. Then, talented suppliers were eliminated with the fuzzy TOPSIS technique [13].

Wang [14] presented a mixed framework of customer satisfaction and product configuration by conducting a case study on the configuration of smart pad types. In the study, Fuzzy Kano Model was used to reveal customer perception in product features and to achieve customer satisfaction. Wang [14] later found the importance of product features using information entropy. In the case of similarity to the ideal solution, competitor design alternatives were prioritized with the order preference technique [14].

Vinodh et al. [15] used a multi-criteria and uncertainty-based logic with the Fuzzy Set Theory. In their study, they used Fuzzy Kano Model to assess sustainability in an automotive parts manufacturing organization in India. The approach was proposed in order to assess sustainability. Firstly, the necessary criteria were taken into consideration by evaluating the economic, environmental, and social sustainability of providers. The requirements were then analyzed with the Kano model to make it more objective. As a result of the evaluations, it determined how important the economic, environmental, and social sustainability criteria are for the customer [15].

Wang and Wang [16] proposed a mixed framework combining Fuzzy AHP, Zero Integer Programming, and a Fuzzy Kano Model. Fuzzy AHP is used to reveal customer preferences for core features, while Fuzzy Kano Model is used to disclose customer perceptions of optional features. The zero-one integer programming developed in the study optimizes the product range by considering production costs and the importance of the feature for the customers [16].

Shafia and Abdollahzadeh [17] identified and classified the customer needs of Iran Code, a system of commodity and services standardization, with Fuzzy TOPSIS. With the Fuzzy Kano Model, they listed and classified the functional needs of the standardization system of commodities and services. This ranking helps to determine investment priority in different groups of tools [17].

Meng et al. [18] conducted a Fuzzy Kano Model study in China to prioritize the quality of logistics services and to anticipate customer satisfaction. In the survey study, a detection method was found by integrating the Fuzzy Kano Model with the IPA (interpretative phenomenological analysis) method [18].

In another study, Meng et al. [19] proposed a new method to provide classification of service quality elements by integrating Fuzzy set theory into the Kano Model. The method was found more successful in processing the psychology of uncertainties than the traditional Kano Model [19].

Lin et al. combined Fuzzy QFD and Fuzzy Kano Model to gain competitive advantage in their study to improve the service process of the Taiwan banquet culture, and to show the relationship between service characteristics and customer responses. This study provided an opportunity to focus on the quality of service features that satisfy customers and to determine how much they could prioritize the development of service design to meet customer expectations in Ban-Doh banquet services [20].

Wang and Fong [21] presented a new framework for the Fuzzy Kano Model called Statistical Regression, and IPA (interpretative phenomenological analysis) method to achieve the goals. Initially, Fuzzy Kano Model is used to detect customer perceptions of service features and convert them into quantitative degrees of customer satisfaction. Multiple regression and logistic regression, which were found by Matzler et al. [22, 23], are used to find the weights of the service characteristics and then determine the basic service characteristics. Finally, IPA was conducted to present internal managerial opinions [21].

In a study conducted in 2017, Cebi and Ilbahar [10] determined the parameters affecting website usage with a Fuzzy Kano Model and found the effects on customer satisfaction. In the study, the survey type of the Classical Kano Model was used and based on the results obtained from the survey outputs. They then used the Fuzzy Set Theory when defining the feature classes. In the study, for the first time, the presence and absence scores of four different websites were determined by the satisfaction of the customers.

Shahin et al. [24] used Fuzzy Set Theory and evaluated the uncertainty hidden in human judgment. In the scope of the study, they developed a systematic approach to classify and weight the necessary factors affecting innovation in the management organization of a steel manufacturing company in India. The study firstly found a comprehensive set of factors from the literature, and the factors obtained were specified for the company through a case study. In the second step, the quality characteristics of the Kano Model were applied, and the alternatives were graded with the AHP method based on extent analysis method [25]. Finally, a hypothesis test on expert opinions was performed to confirm the validity of the results [24].

Jain and Singh [26] conducted the fuzzification of the Kano Model in order to classify the supplier selection criteria. Stepwise Weight Assessment Ratio Analysis (SWARA) and Range of Values (ROV) method were used to determine the weight of the criteria and the order of alternatives. In brief, an integrated method, including SWARA, Fuzzy Kano Model, and ROV methods, has been proposed to solve supplier selection [26].

He et al. [27] proposed a Fuzzy Kano Model to analyze customer needs under uncertain environment. They used two-tuple linguistic variables in a multi-granularity linguistic environment to define the uncertainty and ambiguity of customers' assessments. So, they developed a two-tuple linguistic Fuzzy Kano survey. It is a method that integrates subjective decisions made by decision-makers, actual weights based on maximizing the deviation method, and customer satisfaction contribution [27].

Yeh et al. [28], developed a Kano Model with fuzzy linguistic concepts to demonstrate consumers' expectations about the culture of temple streets. In the first stage of this two-stage study, the Fuzzy Kano Model was used to list the priorities of different cultural elements in the design process. Consumer satisfaction related to traditional cultural aspects of the Fuzzy Kano Model was then examined. In the second stage, according to the Satisfaction Increment Index (SII) and Dissatisfaction Decrement Index (DDI), the Kano Model was applied to calculate each weight of traditional cultural elements [28].

Zhang et al. [29], obtained different attitudes of users towards product aspects by performing sensitivity analysis with an Aspect Sentiment Collaborative Filtering (ASCF) algorithm. Then, based on the Fuzzy Kano Model, they introduced a new similarity measurement method that analyzes the user's desire and importance for each feature [29].

Li et al. [30] presented a two-stage method for partitioning a service module based on the Fuzzy Kano Model. First, an entropy weight method was added to the Fuzzy Kano Model to determine the importance of service requirements. Fuzzy Clustering was used in the cluster modules based on the correlation analysis results of the service process. Second, an edge transformation model was presented to the Kano Model to transfer fuzzy indifference requirements. Thus, a nonlinear programming method was used to reclassify service modules to take into account service time constraints and satisfaction.

Pai et al. [31] investigated the perceived importance and satisfaction of consumers' service quality in the chain restaurant industry by combining it with the IPA and the Fuzzy Kano Model on the foodservice scale (DINESERV). They then found out the requirements to be prioritized in service quality.

Jain and Singh [32, 33] found a systematic method of separating sustainability criteria into different useful categories in 2019. In the first stage of the process, the economic and social sustainability criteria were collected and specified for India's iron and steel industry. In the second stage, the criteria were classified according to the Fuzzy Kano Model. In the third stage, the "Must-Be" and "Attractive" criteria derived from the second stage were divided into three categories using the Type IV Kano Model [32, 33].

The common point of this literature review is that the Fuzzy Theory and the Kano Model approach have been combined and applied to different case studies. Focusing on the wrong points on a subject that requires improvement in the sector causes cost increases and loss of time. In other words, focusing on criteria that will not directly affect the customer is a great effort and material loss for companies and employers. In addition to the Fuzzy Kano Model, the researchers included techniques that have been the subject of literature, like multi-criteria decision-making methods, statistics, entropy, integer programming, importance-performance analysis, and so on. Significant criteria and variables were identified in case studies and then weighted. As can be seen, the Fuzzy Kano Model includes customer satisfaction, production, industry, informatics, software, healthcare, service systems, logistics, among other things, may be applied in many fields. Ultimately, to increase customer satisfaction, the interaction of service features and service design should be developed in parallel with customer wishes.

3 Methodology

3.1 Traditional Kano Model

The Kano Model [4], which was proposed in 1984, uses a two-dimensional method to simulate the concept of unsymmetrical customer perceptions according to functional presence and dysfunctional absence as given in Fig. 1. The method works purely customer-oriented, and the customer's views on the relevant feature are collected through a Kano-specific survey. Participants in the Kano survey are forced to choose an answer for both functional and dysfunctional situations from the options given in Table 1, such as "I like it that way", "It must be that way", "I am neutral", "I can live with it", "I dislike it that way" [4]. After the answers are given, the category of the related feature is defined according to the combination of answers given to functional and dysfunctional questions using Table 2.

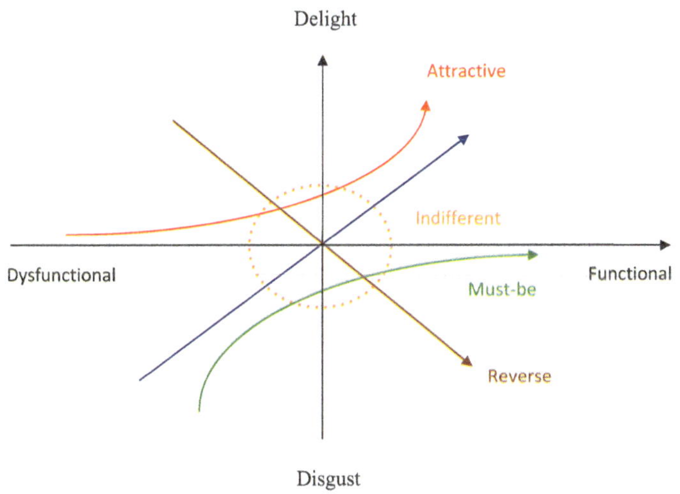

Fig. 1 Kano model diagram

Table 1 A model of traditional Kano questionnaire

How do you feel about this attribute?						
		I like it that way	It must be that way	I am neutral	I can live with it	I dislike it that way
Attribute 1	Functional		√ (100%)			
	Dysfunctional				√ (100%)	
Attribute *n*	Functional		√ (100%)			
	Dysfunctional					√ (100%)

Table 2 Kano evaluation table

		Dysfunctional absence				
		Like	Must-be	Neutral	Live-with	Dislike
Functional presence	Like	Q	A	A	A	O
	Must-be	R	I	I	I	M
	Neutral	R	I	I	I	M
	Live-with	R	I	I	I	M
	Dislike	R	R	R	R	Q

As given in Table 2, related features are categorized according to six characteristics, such as Attractive (A), One dimensional (O), Must-be (M), Indifferent (I), Reverse (R), and Questionable (Q) [7]. Kano categories may be defined briefly as follows:

- Must-be Attributes: These are the qualities that should be available to customers and which are expected from the product. The absence of these qualities leads to dissatisfaction in the customer, while the presence of these qualities does not increase customer satisfaction because customers accept these characteristics as the default.
- One-dimensional Attributes: Having these features increases customer satisfaction in direct proportion, while the absence of these features decrease customer satisfaction in proportion.
- Attractive Attributes: The presence of these features creates excitement for customers. The lack of such features does not lead to dissatisfaction with the customer, while the presence of such features makes the customer very satisfied.
- Reverse Attributes: The presence of these features cause dissatisfaction to the customer, while the absence of these features in the product satisfy the customer. Therefore, it is necessary to remove these features that have a negative impact on the customer.
- Indifferent Attributes: The presence or absence of this feature has no positive or negative impact on customer satisfaction. These are the features about which the customer does not care.
- Questionable Attributes: This feature indicates that there was a mistake in the survey question or that the customer gave an unreasonable answer to the question.

After completing the Kano questionnaire, results indicate not only a description of the feature itself, but also the degree of satisfaction of the customer. In Eq. 1, the relationship between product performance and customer satisfaction is given [34].

$$\Delta S/S = k\Delta(P/P) \tag{1}$$

The s and p given in the equation shows *the customer satisfaction degree* and *the product performance*, respectively. In the statement given in Eq. 1, it is assumed that there is a linear relationship between $\Delta S/S$ and $\Delta(P/P)$ and the degree of the relationship is measured by the coefficient k. For the feature examined, $k > 1$ defines the property in the "attractive attribute" category; $k = 1$ falls into the "one-dimensional attribute" category; $0 < k < 1$ shows the property in the "must-be attribute" category [14].

In the creation of the visual diagram given in Fig. 1, satisfaction (delight; positive effect $\left(d_j^+\right)$) and dissatisfaction (disgust; negative effect $\left(d_j^-\right)$) can be calculated by classifying the desired property class [6–9]. Equations 2 and 3 present satisfaction value and dissatisfaction value, respectively.

$$d_j^+ = \frac{A_j + O_j - R_j}{A_j + O_j + M_j + R_j + I_j} \tag{2}$$

$$d_j^- = -\frac{O_j + M_j - R_j}{A_j + O_j + M_j + R_j + I_j} \tag{3}$$

where A_j, O_j, M_j, R_j and I_j represent values corresponding to the Kano categories.

The answers obtained from the Traditional Kano Model are evaluated by using Matzler and Hinterhuber's model [7] to find the feature's Kano category class. The frequency of answers for each Kano category class is found. If $(O + A + M) > (I + R + Q)$, the maximum value in O, A, M is selected as the category. Otherwise, the Kano with the maximum value from I, R, Q should be chosen as the category class [35]. If there are classes with the same frequency value, the Kano category class is determined according to the priority order $(M > O > A > I)$ [36]. To measure the degree of satisfaction and dissatisfaction of the participants, Matzler and Hinterhuber use Eqs. (4) and (5), which determine the customer satisfaction coefficient [7].

$$\textit{Enhanced Satisfaction Coefficients} = \frac{A + O}{A + O + M + I} \tag{4}$$

$$\textit{Reduced Dissatisfaction Coefficients} = \frac{O + M}{A + O + M + I} \tag{5}$$

3.2 *Fuzzy Kano Model*

Since the responses given in the Kano Model are based on perception, and the expression of perceptions involves uncertainty, in recent years, the studies combining the Fuzzy Set Theory and the Kano Model are increasing day by day. Fuzzy Set Theory is an appropriate model to manage uncertainty in people's decisions [12]. Lee and Huang [5] have included a fuzzy approach to the traditional Kano Model. The common point of the Traditional Kano Model and the Fuzzy Kano Model is to ask dysfunctional and functional questions. In the traditional Kano Model, customers can answer a single question while participants can respond more than one preference

Table 3 A model of Fuzzy Kano questionnaire

How do you feel about this attribute?						
		I like it that way	It must be that way	I am neutral	I can live with it	I dislike it that way
Attribute 1	Functional	85%	15%			
	Dysfunctional				30%	70%
Attribute *n*	Functional	30%	60%	10%		
	Dysfunctional				65%	35%

by defining a ratio for their assessment in the Fuzzy Kano Model developed by Lee and Huang [5]. Thus, it is claimed in the study that survey participants can better express the emotional dimensions of their wishes by using membership value or numerical value. Table 3 shows the sample data representation with fuzzy responses to the Fuzzy Kano Questionnaire model [5].

In a hybrid study, Wang and Wang [16] applied a Fuzzy Kano survey model to measure customer perceptions of the features of smart cameras, integrating fuzzy AHP, integer programming, and Fuzzy Kano. The answers for each question do not contain exact values but are designed to respond with probabilities to better reflect the thinking of the customer in each question [5, 16]. In 2017, Cebi and Ilbahar [10] found the design parameters that affect the choices of website users by asking the Traditional Kano survey model questions. Using the proprietary formulas that they proposed, fuzzy outputs of customers' input values were obtained [10]. In the scope of the presented study, features of the online mobile transaction applications which belong to three telecommunications companies, serving an active role in the Turkey market, and including millions of users, will be analyzed with a survey to find the importance degrees of mobile application features for users. When the studies in the literature are examined, it is seen that the questionnaire studies are designed to answer each question with probabilities that are not very easy to apply to users. It is a challenging task for the participants to respond to both functional and dysfunctional questions. Possibility values are used to figure out how much they are wanted by customers. This process is both time consuming and confusing for the user. In this chapter, a new approach has been presented which simplifies the completion of the questionnaire and eliminates the uncertainty in the survey evaluation in order to overcome the deficiency in the literature mentioned above. Thus, the collected data is better evaluated, and an improvement is made on current Fuzzy Kano Models. The steps of the proposed method are as follows:

Step 1. Identification of features: Search for features in mobile online transaction applications belonging to telecommunication companies and identify important features for the application.

Step 2. Kano Questionnaire Design and Implementation: At first, the Kano Questionnaire is created by preparing functional and dysfunctional questions for the features defined in the first step. Then, it should be decided how the survey will be implemented, and it should be defined the potential customer type of the product to be developed.

Step 3. Grouping of Participant Responses: Taking into account the total number of participants (n), the frequency of the five answer groups (j) that can be given to the relevant feature is calculated according to functional $n(i)_j^+$ and dysfunctional $n(i)_j^-$ questions. The answer groups are divided into five types as "*I like it that way*", "*It must be that way*", "*I am neutral*", "*I can live with it*", "*I dislike it that way*".

The frequencies calculated for all answer groups that can be assigned to the relevant feature are divided by the total number of participants, and calculated according to both functional and dysfunctional questions; thus, obtaining $P(i)_j^+$ and $P(i)_j^-$ values (Eqs. 6 and 8). The sum of the $P(i)_j^+$ and $P(i)_j^-$ values of the functional and dysfunctional questions for each feature must be equal to 1 (Eqs. 7 and 9) [7].

$$p(i)_j^+ = \frac{n(i)_j^+}{n} \quad i = 1, 2, 3, \ldots, m, \; j = 1, 2, 3, \ldots., k \tag{6}$$

$$\sum_{j=1}^{k} P(i)_j^+ = 1 \quad i = 1, 2, 3, \ldots, m \tag{7}$$

where $n(i)_j^+$ represents the number of participants who answer jth to the functional question of feature ith while $p(i)_j^+$ represents the ratio of those who answer jth to the functional question of property ith. Furthermore, m, n, and j represents the number of evaluated attributes, the total number of participants, and the answer group, respectively.

$$p(i)_j^- = \frac{n(i)_j^-}{n} \quad i = 1, 2, 3, \ldots, m, \; j = 1, 2, 3, \ldots., k \tag{8}$$

$$\sum_{j=1}^{k} P(i)_j^- = 1 \quad i = 1, 2, 3, \ldots, m \tag{9}$$

where $n(i)_j^-$ is the number of participants who answer jth to the dysfunctional question of feature ith while $p(i)_j^-$ represents the ratio of those who answer jth to the dysfunctional question of property ith.

Step 4. Creating the relationship (*R*) matrix of the product properties: In the Fuzzy Kano Model, the membership function values for the ith feature are defined as $fun = (p(i)_{j=1}^+, p(i)_{j=2}^+, p(i)_{j=3}^+, p(i)_{j=4}^+, p(i)_{j=5}^+)$ ve $dys = (p(i)_{j=1}^-, p(i)_{j=2}^-, p(i)_{j=3}^-, p(i)_{j=4}^-, p(i)_{j=5}^-)$. By multiplying the matrix product, the transpose of the *fun* function and the *dys* function are obtained by multiplying the R matrix [5, 12]. In this research, membership degrees of two-dimensional modified Fuzzy Kano matrix are obtained by using Eq. 10.

$$R = fun^{t} \times dys \tag{10}$$

Step 5. Calculation of membership degree to classes: After obtaining the R matrix, two-dimensional property classification will be obtained based on the Matzler and Hinterhuber [7] model. Referring to Table 2 or Eq. 11, new membership degrees of the M, A, O, R, Q, I Kano category classes are obtained.

$$Kano = \begin{bmatrix} Q\,A\,A\,A\,O \\ R\,I\,I\,I\,M \\ R\,I\,I\,I\,M \\ R\,I\,I\,I\,M \\ R\,R\,R\,R\,Q \end{bmatrix} \tag{11}$$

Step 6. Calculation of the severity of the characteristics: The satisfaction (delight $\left(d_j^+\right)$) and dissatisfaction (disgust $\left(d_j^-\right)$) of the customer for the related feature are calculated by using Eqs. 2 and 3. In this chapter, instead of immediately determining Kano categories by alpha-cut $\langle \alpha - cut \rangle$ method on membership values, it was aimed to protect membership values throughout the calculation [16]. Therefore, the modified positive delight $\left(d_j^+\right)$ and negative disgust $\left(d_j^-\right)$ formulas will be used for satisfaction and dissatisfaction situations, respectively [6–9]. And then, the degrees of quantitative delight and disgust will be found.

4 Case Study

Telecommunication companies aim to reach online customers and increase customer satisfaction through online mobile transaction applications. They devote significant resources to develop these practices. In recent years, competition among telecommunication companies has become increasingly popular among online mobile transaction applications, which have millions of users. Therefore, mobile online transaction applications have been chosen for the method developed within the scope of the chapter. For this purpose, this chapter aims to classify the features of mobile online transaction applications according to the expectations and perceptions of users. The implementation steps of the chapter are as follows:

Step 1. Identification of features: Existing studies of Turkey's three largest telecommunications companies which examine the features of online mobile transaction applications are listed in Table 4. Although the nomenclature of application features shows differences among the enterprises, the service offered to users by the product

Table 4 Feature pool

Feature no	Feature name in application
F01	Showing remaining usage information
F02	Weekly random gift winning
F03	Package/Plan change and purchase
F04	Device and accessories sales
F05	Showing owned package/plan and service information
F06	Showing current and past billing information and top up mobile detail
F07	Updating personal information
F08	Managing incoming/outgoing call operations
F09	Making application language selection and service language selection
F10	Finding the desired answer in matters that need help
F11	Widget
F12	Top up mobile and paying bills
F13	Viewing all accounts individually/corporate cellular phone line
F14	Viewing 4.5 G subscription information and starting new subscription
F15	Calling abroad/seeing and managing properties related to use abroad
F16	Service purchasing
F17	Searching within the application
F18	Managing incoming messages
F19	List of orders in application
F20	Accessing live support and customer service

features and those features have been categorized and renamed within the scope of the chapter.

Step 2. Kano Questionnaire Design and Implementation: In this step, functional and dysfunctional questions were prepared and shared with the users in accordance with the Traditional Kano Model [4] shown in Table 1 for each feature identified in Table 4. As an example, a functional and dysfunctional question structure related to "Showing Remaining Usage Information," which is one of the online mobile transaction application features are given below.

Functional Question: How do you feel if an online mobile transaction application has the feature of "*Showing Remaining Usage Information*"?

Dysfunctional Question: How do you feel if an online mobile transaction application does not have the feature of "*Showing Remaining Usage Information*"?

In this chapter, 108 participants, 45% of whom were women and 55% of whom were men, participated in the survey developed for online mobile transaction applications, and 86% of the participants stated that they are actively using the online mobile transaction application, which belongs to the operator. 5% of the participants were under age 20, 59% were between the ages of 20 and 30, 27% were between the ages of 30 and 40, 6% were between the ages of 40 and 50, and 3% were over the age of 50. 65% of the participants hold undergraduate degrees, 9% are high school graduates, 5% are associate degree graduates, 19% have completed postgraduate studies, 2% are doctorate graduates. 2% of the participants reported using the internet for less than 1 h/day, 26% of the participants used the internet for between 1 and 3 h/day, 39% of the participants used the internet for between 3 and 6 h/day, 14% of the participants used the internet for between 6 and 9 h/day, and 19% of the participants used the internet for 9 h/day or more.

Step 3. Grouping of Participant Responses: Participants' responses to the questions about the product characteristics asked in the questionnaire are summarized in Table 5 using Eqs. 6 and 8.
Step 4. Creating the relationship (*R*) matrix of the product properties: For each feature, the obtained values are given in Table 5 by using Eq. 10, and the matrix R of the related feature is obtained. The matrix R is a value that reflects the emotion and thought of all participants for the feature. To illustrate the calculation made in the chapter, "F01. Showing the Remaining Usage Information" feature R relationship matrix is obtained as follows [5];

$$R^{F01} = \{0.444\,0.528\,0.0\,0.028\,0.0\} \times \{0.102\,0.102\,0.009\,0.120\,0.067\}$$

$$R^{F01} = \begin{bmatrix} 0.045 & 0.045 & 0.004 & 0.053 & 0.296 \\ 0.054 & 0.054 & 0.005 & 0.064 & 0.352 \\ 0 & 0 & 0 & 0 & 0 \\ 0.003 & 0.003 & 0 & 0.003 & 0.019 \\ 0 & 0 & 0 & 0 & 0 \end{bmatrix}$$

Step 5. Calculation of membership degree to classes: The numerical values in the matrix R show the degree of belonging to the classes defined by the Kano Model of the feature.

$$R = \begin{bmatrix} Q & A & A & A & O \\ R & I & I & I & M \\ R & I & I & I & M \\ R & I & I & I & M \\ R & R & R & R & Q \end{bmatrix}$$

$$R = \begin{bmatrix} 0.045 & 0.045 & 0.004 & 0.053 & 0.296 \\ 0.054 & 0.054 & 0.005 & 0.064 & 0.352 \\ 0 & 0 & 0 & 0 & 0 \\ 0.003 & 0.003 & 0 & 0.003 & 0.019 \\ 0 & 0 & 0 & 0 & 0 \end{bmatrix}$$

Table 5 Feature Kano category frequency possibilities

Feature no	Question type	I like it that way	It must be that way	I am ne utral	I can live with it	I dislike it that way
F01	Func.	0.444	0.528	0	0.028	0
	Dys	0.102	0.102	0.009	0.12	0.667
F02	Func.	0.731	0.157	0.065	0.037	0.009
	Dys	0.093	0.037	0.093	0.435	0.343
F03	Func.	0.463	0.426	0.083	0.019	0.009
	Dys	0.083	0.046	0.102	0.12	0.648
F04	Func.	0.204	0.222	0.389	0.102	0.083
	Dys	0.074	0.083	0.315	0.361	0.167
F05	Func.	0.463	0.426	0.074	0.028	0.009
	Dys	0.056	0.083	0.074	0.139	0.648
F06	Func.	0.463	0.472	0.028	0.028	0.009
	Dys	0.046	0.046	0.074	0.139	0.694
F07	Func.	0.324	0.509	0.111	0.037	0.019
	Dys	0.056	0.019	0.204	0.157	0.565
F08	Func.	0.398	0.361	0.167	0.074	0
	Dys	0.074	0.056	0.157	0.25	0.463
F09	Func.	0.37	0.463	0.102	0.065	0
	Dys	0.046	0.056	0.083	0.343	0.472
F10	Func.	0.444	0.444	0.065	0.028	0.019
	Dys	0.046	0.056	0.111	0.148	0.639
F11	Func.	0.352	0.306	0.231	0.083	0.028
	Dys	0.028	0.037	0.25	0.361	0.324
F12	Func.	0.519	0.389	0.056	0.028	0.009
	Dys	0.083	0.009	0.056	0.12	0.731
F13	Func.	0.38	0.389	0.139	0.065	0.028
	Dys	0.074	0.046	0.185	0.222	0.472
F14	Func.	0.417	0.407	0.13	0.046	0
	Dys	0.065	0.046	0.139	0.296	0.454
F15	Func.	0.454	0.426	0.083	0.037	0
	Dys	0.046	0.046	0.065	0.176	0.667
F16	Func.	0.454	0.352	0.12	0.056	0.019
	Dys	0.065	0.046	0.176	0.25	0.463
F17	Func.	0.389	0.398	0.157	0.037	0.019
	Dys	0.037	0.028	0.139	0.343	0.454
F18	Func.	0.435	0.343	0.157	0.037	0.028

(continued)

Table 5 (continued)

Feature no	Question type	I like it that way	It must be that way	I am ne utral	I can live with it	I dislike it that way
	Dys	0.065	0.056	0.176	0.269	0.435
F19	Func.	0.417	0.269	0.241	0.046	0.028
	Dys	0.065	0.065	0.259	0.306	0.306
F20	Func.	0.5	0.426	0.037	0.037	0
	Dys	0.065	0.037	0.046	0.148	0.704

By using Table 2, the likelihood of membership in classes for R^{F01} is obtained as P^{F01} [5, 7, 12].

$$P^{F01} = \left\{ \frac{0.103}{A}, \frac{0.371}{M}, \frac{0.296}{O}, \frac{0,129}{I}, \frac{0,057}{R}, \frac{0,045}{Q} \right\}$$

The obtained vector shows the membership values of the related feature to the Kano classes. In other words, the "*F01. Showing Remaining Usage Information*" feature has membership value to "*Attractive Attributes*" class as 0.103, has membership value to "*Must-Be Attributes*" class of 0.371, has membership value to "*One-dimensional Attributes*" class of 0.296, has membership value to "*Reverse Attributes*" class of 0.057 and shows membership value to "*Indifferent Attributes*" class of 0.12. As can be seen from this result, it is not correct to include any feature directly into any category.

Table 6 presents the degree of membership of the Kano Model classes showing the feelings and thoughts of all participants as related to each feature examined in the chapter.

Step 6. Calculation of the severity of the characteristics: In the previous step, after calculating the membership degrees of each feature for the Kano Model classes, the satisfaction level of each feature should be calculated. Equations 2 and 3 were used to obtain the satisfaction $\left(d_j^+\right)$ and dissatisfaction $\left(d_j^-\right)$ values of the users. Satisfaction and dissatisfaction values of each trait and the order of priority obtained according to these values are given in Table 7 [6–9].

According to the Fuzzy Kano Model, the priority order of the products can be reached for all users after the calculation of positive delight $\left(d_j^+\right)$ and negative disgust $\left(d_j^-\right)$ formulas are obtained. According to the obtained results it is seen that the most important feature for all users is the ability to "*Top Up Mobile and Paying Bills*" while the most insignificant feature is "*Device and Accessory Sales*" made through the application.

Table 6 Possibilities of Fuzzy Kano categories of all characteristics

Feature no	A	M	O	R	Q	I
F01	0.103	0.371	0.296	0.057	0.045	0.129
F02	0.413	0.089	0.25	0.03	0.071	0.146
F03	0.124	0.342	0.3	0.047	0.044	0.142
F04	0.155	0.119	0.034	0.122	0.029	0.541
F05	0.137	0.342	0.3	0.033	0.032	0.156
F06	0.12	0.367	0.322	0.027	0.027	0.137
F07	0.123	0.371	0.183	0.045	0.029	0.249
F08	0.184	0.279	0.184	0.045	0.029	0.279
F09	0.178	0.298	0.175	0.029	0.017	0.303
F10	0.14	0.343	0.284	0.032	0.033	0.169
F11	0.228	0.201	0.114	0.036	0.019	0.402
F12	0.096	0.346	0.38	0.042	0.05	0.088
F13	0.172	0.28	0.179	0.059	0.041	0.269
F14	0.201	0.265	0.189	0.038	0.027	0.281
F15	0.13	0.364	0.303	0.025	0.021	0.157
F16	0.214	0.244	0.21	0.044	0.038	0.249
F17	0.198	0.269	0.176	0.033	0.023	0.301
F18	0.218	0.234	0.189	0.051	0.04	0.269
F19	0.263	0.17	0.127	0.055	0.036	0.35
F20	0.116	0.352	0.352	0.032	0.032	0.116

5 Findings and Discussions

Within the scope of the chapter, both the Traditional Kano Model and the Fuzzy Kano Model proposed in the study have been utilized to evaluate the data collected via the survey which was prepared according to the Traditional Kano Model. Table 8 presents the results obtained from the application of the Traditional Kano Model.

In order to measure the degree of satisfaction and dissatisfaction among the participants, the equations (Eqs. 4 and 5) proposed by Matzler and Hinterhuber [7] were used. The order of features according to the Traditional Kano Model is shown in Table 9. Based on these results, it is seen that the most significant feature for users is "*Top Up Mobile and Paying Bills.*" The most insignificant feature for users was "*Device and Accessories Sales.*" According to the Traditional Kano Model, "*Weekly Random Gift Winning*" was found to be the only attractive feature.

Table 10 shows the prioritized rankings obtained according to the Traditional Kano Model and Fuzzy Kano Model. If the rankings in the table are taken into consideration, there is no significant difference between the results of either method. However, there exist significant differences in the Kano classes to which the features belong. These results allow the business to conduct a cost-benefit analysis

Table 7 Priority order of the properties examined

Feature No	Delight	Disgust	Range	Rank
F12	0.456	−0.718	1.174	1
F20	0.45	−0.694	1.145	2
F6	0.427	−0.68	1.107	3
F15	0.417	−0.656	1.073	4
F5	0.417	−0.629	1.046	5
F10	0.405	−0.615	1.020	6
F3	0.395	−0.623	1.018	7
F2	0.682	−0.333	1.015	8
F1	0.358	−0.638	0.996	9
F16	0.395	−0.427	0.822	10
F7	0.269	−0.524	0.793	11
F14	0.361	−0.427	0.789	12
F9	0.33	−0.452	0.781	13
F17	0.349	−0.422	0.771	14
F8	0.333	−0.43	0.763	15
F18	0.37	−0.387	0.758	16
F13	0.304	−0.417	0.722	17
F19	0.347	−0.251	0.598	18
F11	0.312	−0.284	0.596	19
F4	0.069	−0.032	0.101	20

for investment in the product feature. If the classification results obtained using Fuzzy Kano Model and Traditional Kano Model are compared, "*Showing Remaining Usage Information*" is defined as a "*Must-Be Attribute*" according to the Traditional Kano model. Furthermore, according to the proposed Fuzzy Kano Model, this feature has membership value to the "*Must-Be Attribute*" class as 37.10%, to the "*One Dimensional Attribute*" class as 29.60%, to the "*Attractive Attribute*" class as 10.30%, to the "*Indifferent Attribute*" class as 12.90%, and shows membership value to "*Reverse Attribute*" class as 5.70%. Additionally, it has a membership value to the "*Questionable Attribute*" class as 4.50%. As can be seen, we cannot include features in a category by using exact expressions when evaluating with the Fuzzy Kano Model. Thus, the proposed method provides an advantage in which the feelings and expectations of the participants may be expressed with greater realism.

If we examine the order of the product properties given in Table 10, "*Top Up Mobile and Paying Bills*," "*Accessing Live Support and Customer Service*," and "*Showing Current and Past Billing Information and Top Up Mobile Detail*" are the top three features in both the Fuzzy Kano Model and the Traditional Kano Model, respectively. These three features are the most important features for both models,

Table 8 Determination of Kano categories based on traditional Kano model

Feature No	A	M	O	R	Q	I	Total	Category
F01	5	38	34	2	9	20	108	M
F02	39	6	30	0	11	22	108	A
F03	7	33	37	3	6	22	108	O
F04	6	7	11	9	5	70	108	I
F05	5	31	39	1	6	26	108	O
F06	7	36	39	2	4	20	108	O
F07	5	33	26	2	6	36	108	M
F08	13	25	25	3	5	37	108	M
F09	11	26	25	1	4	41	108	O
F10	9	33	35	2	5	24	108	O
F11	17	15	19	3	3	51	108	I
F12	8	37	42	3	6	12	108	O
F13	13	28	22	2	7	36	108	M
F14	14	24	25	1	6	38	108	O
F15	4	31	41	1	4	27	108	O
F16	15	20	30	3	4	36	108	O
F17	13	24	25	2	4	40	108	O
F18	15	18	29	5	3	38	108	O
F19	16	9	24	4	5	50	108	I
F20	5	32	44	2	5	20	108	O

and telecommunication companies can make the necessary investments and put forth effort primarily on these features.

Likewise, features such as the "*List of Orders in Application*," "*Widget*," and "*Device and Accessories Sales*" are least important according to both the Fuzzy Kano Model and the Traditional Kano Model, respectively. These three features are the most insignificant for both models. According to the Traditional Kano Model, these features are directly included in the "*Indifferent Attributes Kano Category*," whereas in the Fuzzy Kano Model, these features have membership values to "*Indifferent Attribute Category*" as 35%, 40.20%, and 54.10%, respectively. Telecommunication companies should not give priority to these features when making the necessary investments or developments.

The significant differences between the results with the Fuzzy Kano Model and the Traditional Kano Model are as follows: "*Showing Remaining Usage Information*" ranks 5th in the Traditional Kano Model and 9th in the proposed Fuzzy Kano Model, four places behind the other. The feature "*Making Application Language Selection and Service Language Selection*" ranks 17th in the Traditional Kano Model and 13th in the proposed Fuzzy Kano Model. In other words, it is a more important feature in the Fuzzy Kano Model, in which customers' perceptions and expressions are

Table 9 Satisfaction coefficients and priority ranks of features according to the traditional Kano

Feature No	Category	ESC	RDC	Range	Rank
F12	O	0.505	−0.798	1.303	1
F20	O	0.485	−0.752	1.238	2
F06	O	0.451	−0.735	1.186	3
F3	O	0.444	−0.707	1.152	4
F01	M	0.402	−0.742	1.144	5
F15	O	0.437	−0.699	1.136	6
F05	O	0.436	−0.693	1.129	7
F10	O	0.436	−0.673	1.109	8
F02	A	0.711	−0.371	1.082	9
F16	O	0.446	−0.495	0.941	10
F18	O	0.44	−0.47	0.91	11
F07	M	0.31	−0.59	0.9	12
F08	M	0.38	−0.5	0.88	13
F14	O	0.386	−0.485	0.871	14
F13	M	0.354	−0.505	0.859	15
F17	O	0.373	−0.48	0.853	16
F09	O	0.35	−0.495	0.845	17
F19	I	0.404	−0.333	0.737	18
F11	I	0.353	−0.333	0.686	19
F04	I	0.181	−0.191	0.372	20

better reflected. The "*Managing Incoming Messages*" feature ranks 11th in the Traditional Kano Model and 16th in the recommended Fuzzy Kano Model. According to the Fuzzy Kano Model, the feature is five places behind in customers' ranking of importance, as compared to the other.

6 Conclusion

This chapter proposes a new approach to identify and classify product features that can meet the expectations of consumers using online mobile transaction applications. The proposed approach in this chapter is an improved extension of the current method, based on the current Fuzzy Kano Model, which overcomes the difficulty of collecting survey responses. In the implementation phase of the proposed method, researchers examined the characteristics of mobile online transaction applications belonging to three of the largest telecommunications companies in Turkey in order to identify the design characteristics of mobile online transaction applications and the Kano Model questionnaire has been developed. Then, the Fuzzy Kano Model,

Table 10 Comparisons of the results

Feature no	Fuzzy rank	Traditional rank	Difference
F01	9	5	−4
F02	8	9	1
F03	7	4	−3
F04	20	20	0
F05	5	7	2
F06	3	3	0
F07	11	12	1
F08	15	13	−2
F09	13	17	4
F10	6	8	2
F11	19	19	0
F12	1	1	0
F13	17	15	−2
F14	12	14	2
F15	4	6	2
F16	10	10	0
F17	14	16	2
F18	16	11	−5
F19	18	18	0
F20	2	2	0

proposed by Lee and Huang [5] and Wang and Wang [16], was extended to provide more flexible analysis of the participants' responses to the Kano Model questionnaires. In the current Fuzzy Model approach, participants are requested to reflect their feelings on the product characteristics as a proportion. However, this is a more complicated task than simply choosing desirable or undesirable situations. To overcome this challenge, the responses from the participants were collected as applied in the Traditional Kano Model, and the participants' responses were expressed with possibilities to their class membership. Subsequently, the priorities of the results related to the relevant characteristics were evaluated with the formulas of satisfaction and dissatisfaction (Eqs. 2 and 3). The results obtained from the proposed Fuzzy Model in the study were compared with the results obtained from the Traditional Kano Model [35, 36]. In conclusion, in both methods, the features "*Top Up Mobile and Paying Bills*," "*Accessing Live Support and Customer Service*," and "*Showing Current and Past Billing Information and Top Up Mobile Detail*" has been founded as the most important features.

For further studies, improvements may be added to the model to take into account the interdependencies between design features and consistency of responses from participants.

References

1. https://play.google.com/store/apps/details?id=com.ttech.android.onlineislem&hl=en_US
2. https://play.google.com/store/apps/details?id=com.vodafone.selfservis&hl=en_US
3. https://play.google.com/store/apps/details?id=com.tmob.AveaOIM&hl=en_US
4. Kano, N.: Attractive quality and must-be quality. Hinshitsu (Quality. J. Jpn. Soc. Qual. Control). **14** 39–48 (1984)
5. Lee, Y.C., Huang, S.Y.: A new fuzzy concept approach for Kano's model. Expert Syst. Appl. **36**(3), 4479–4484 (2009)
6. Berger, C.: Kano's methods for understanding customer-defined quality. Cent. Qual. Manag. J. **2**(4), 3–36 (1993)
7. Matzler, K., Hinterhuber, H.H.: How to make product development projects more successful by integrating Kano's model of customer satisfaction into quality function deployment. Technovation. **18**(1), 25–38 (1998)
8. Clegg, B., Wang, T., Ji, P.: Understanding customer needs through quantitative analysis of Kano's model. Int. J. Qual. Reliab. Manag. (2010)
9. Delice, E.K., Güngör, Z.: A mixed integer goal programming model for discrete values of design requirements in QFD. Int. J. Prod. Res. **49**(10), 2941–2957 (2011)
10. Ilbahar, E., Cebi, S.: Classification of design parameters for E-commerce websites: a novel Fuzzy Kano approach. Telematics Inform. **34**, 1814–1825 (2017)
11. Florez-Lopez, R., Ramon-Jeronimo, J.M.: Managing logistics customer service under uncertainty: an integrative Fuzzy Kano framework. Inf. Sci. **202**, 41–57 (2012)
12. Zadeh, L.A.: Fuzzy sets. Inf. Control **8**(3), 338–353 (1965)
13. Ghorbani, M., Mohammad Arabzad, S., Shahin, A.: A novel approach for supplier selection based on the Kano model and fuzzy MCDM. Int. J. Prod. Res. **51**(18), 5469–5484 (2013)
14. Wang, C.H.: Incorporating customer satisfaction into the decision-making process of product configuration: a Fuzzy Kano perspective. Int. J. Prod. Res. **51**(22), 6651–6662 (2013)
15. Vinodh, S., Jayakrishna, K., Girubha, R.J.: Sustainability assessment of an automotive organisation using fuzzy Kano's model. Int. J. Sustain. Eng. **6**(1), 1–9 (2013)
16. Wang, C.H., Wang, J.: Combining fuzzy AHP and fuzzy Kano to optimize product varieties for smart cameras: a zero-one integer programming perspective. Appl. Soft Comput. **22**, 410–416 (2014)
17. Shafia, M.A., Abdollahzadeh, S.: Integrating fuzzy kano and fuzzy TOPSIS for classification of functional requirements in national standardization system. Arab. J. Sci. Eng. **39**(8), 6555–6565 (2014)
18. Meng, Q., Jiang, X., Bian, L.: A decision-making method for improving logistics services quality by integrating fuzzy Kano model with importance-performance analysis. J. Serv. Sci. Manag. **8**(03), 322 (2015)
19. Meng, Q., Jiang, X., He, L., Guo, X.: Integrating fuzzy theory into Kano model for classification of service quality elements: a case study of machinery industry in China. J. Ind. Eng. Manag. (JIEM) **8**(5), 1661–1675 (2015)
20. Lin, L.Z., Yeh, H.R., Wang, M.C.: Integration of Kano's model into FQFD for Taiwanese Ban-Doh banquet culture. Tour. Manag. **46**, 245–262 (2015)
21. Wang, C.H., Fong, H.Y.: Integrating Fuzzy Kano model with importance-performance analysis to identify the key determinants of customer retention for airline services. J. Ind. Prod. Eng. **33**(7), 450–458 (2016)
22. Matzler, K., Bailom, F., Hinterhuber, H.H., Renzl, B., Pichler, J.: The asymmetric relationship between attribute-level performance and overall customer satisfaction: a reconsideration of the importance–performance analysis. Ind. Mark. Manag. **33**(4), 271–277 (2004)
23. Matzler, K., Sauerwein, E., Heischmidt, K.: Importance-performance analysis revisited: the role of the factor structure of customer satisfaction. Serv. Ind. J. **23**(2), 112–129 (2003)
24. Shahin, A., Barati, A., Geramian, A.: Determining the critical factors of radical innovation using an integrated model of Fuzzy analytic hierarchy process-Fuzzy Kano with a case study in Mobarakeh steel company. Eng. Manag. J. **29**(2), 74–86 (2017)

25. Chang, D.Y.: Applications of the extent analysis method on fuzzy AHP. Eur. J. Oper. Res. **95**(3), 649–655 (1996)
26. Jain, N., Singh, A.R.: Fuzzy Kano integrated MCDM approach for supplier selection based on must be criteria. Int. J. Supply Chain. Manag. **6**(2), 49–59 (2017)
27. He, L., Ming, X., Li, M., Zheng, M., Xu, Z.: Understanding customer requirements through quantitative analysis of an improved Fuzzy Kano's model. Proc. Inst. Mech. Eng.. Part B: J. Eng. Manuf. **231**(4), 699–712 (2017)
28. Yeh, H.R., Lin, L.Z., Lu, C.F.: Classification of traditional cultural elements in temple street festivals using the fuzzy Kano model. Curr. Issues Tour. 1–26 (2017)
29. Zhang, J., Chen, D., Lu, M.: Combining sentiment analysis with a Fuzzy Kano model for product aspect preference recommendation. IEEE Access. **6**, 59163–59172 (2018)
30. Li, L., Lin, Y., Wang, X., Guo, T., Zhang, J., Lin., H., Nan, F.: A clustering-classification two-phase model on service module partition oriented to customer satisfaction. Eng. Lett. **26**(1)
31. Pai, F.Y., Yeh, T.M., Tang, C.Y.: Classifying restaurant service quality attributes by using Kano model and IPA approach. Total. Qual. Manag. Bus. Excel. **29**(3–4), 301–328 (2018)
32. Jain, N., Singh, A.R.: Sustainable supplier selection criteria classification for Indian iron and steel industry: a Fuzzy modified Kano model approach. Int. J. Sustain. Eng. 1–16 (2019)
33. Shahin, A., Pourhamidi, M., Antony, J., Hyun Park, S.: Typology of Kano models: a critical review of literature and proposition of a revised model. Int. J. Qual. Reliab. Manag. **30**(3), 341–358 (2013)
34. Tan, K.C., Shen, X.X.: Integrating Kano's model in the planning matrix of quality function deployment. Total Qual. Manag. **11**(8), 1141–1151 (2000)
35. Huang, C.C., Guan, S.S.: Application of Kano Model in study of satisfaction with quality of website browsing environment. Bull. Jpn. Soc. Sci. Des. **59**(1), 1_49–1_58 (2012)
36. Qiting, P., Uno, N. Kubota, Y.: Kano Model analysis of customer needs and satisfaction at the Shanghai Disneyland (2011)

Classification of Laptop Design Attributes Using Fuzzy Kano Model

Esra Ilbahar, Selcuk Cebi and Cengiz Kahraman

Abstract The laptop computers have become one of the preferred products in the consumer electronics market. To be successful in a market, it is not enough to have knowledge about competitors and their product performances, but also it is crucial to know the impressions that products create. A large part of the laptop consumer market is comprised of college and graduate students. That is why, in this study, design preferences of college students are analyzed by using fuzzy Kano Model, and the results are compared to the results of traditional Kano Model. It is shown that traditional Kano Model and fuzzy Kano Model yield different results for most of the design attributes.

1 Introduction

To be successful in a market it is not enough to have knowledge about competitors and their product performances, but also to know the impressions that products evoke [1]. Popular concepts that a designer can utilize to provide a better product design are Kansei engineering, design for manufacturing, quality function deployment, design of experiments, value engineering, failure mode and effects analysis, simulation techniques, computer aided design, reliability engineering plans and solid modelling [2].

Laptops are one of the most popular products in consumer electronics market. As the popularity and sales volumes of laptops have increased, the studies on the design of laptops have also increased. Sharma et al. [1] studied the link between customer emotions and laptop computer properties such as screen size, keyboard design, fan

E. Ilbahar (✉) · S. Cebi
Department of Industrial Engineering, Yildiz Technical University, Besiktas, 34349 Istanbul, Turkey
e-mail: esrailbahar@gmail.com; eilbahar@yildiz.edu.tr

E. Ilbahar · C. Kahraman
Department of Industrial Engineering, Istanbul Technical University, Macka, 34367 Istanbul, Turkey

C. Kahraman and S. Cebi (eds.), *Customer Oriented Product Design*, Studies in Systems, Decision and Control 279, https://doi.org/10.1007/978-3-030-42188-5_13

position and colour. Jiang et al. [3] introduced an innovative design process and proposed a notebook with a detachable keyboard and a mobile host to improve some design problems such as difficulty to detach, repair and clean by utilizing the conceptual design method. Randall et al. [4] investigated user design concept by taking design parameters such as memory, hard drive, weight, display, resolution, processor and their causal relations into consideration. In the study of Shiau et al. [5], a demand model obtained from a choice-based conjoint survey was used to propose a new laptop computer design by analyzing the preferences of respondents. In that study, different demand model specifications, linear-in-parameters and discrete part-worth specifications for aggregate multinomial logit and mixed logit models, were compared. Then, optimization was performed to maximize profit by combining an engineering model with each demand model. Design variables used in that study [5] were body thickness, body depth, LCD size, body width, battery volume ratio whereas maximum total weight, minimum battery life, LCD aspect ratio,battery weight-to-volume ratio and body weight-to-volume ratio were considered as design parameters. The results indicated that longer battery life, larger LCD screen, lower price and lighter weight are commonly preferred whereas the respondents are less sensitive to increase in thickness than increase in weight [5].

Laptop computers have been designed in a variety of shapes, sizes and specifications to fulfill different consumer preferences and priorities [5]. In each design, a particular set of parameters is enhanced to be able to provide better design trade-offs for potential consumers. Both laptop computer designers and consumers benefit from a better comprehension of these trade-offs and consumer priorities [5]. That is why it is important to examine their priorities and preferences while choosing laptops. In this study, design preferences of college students are analyzed by using fuzzy Kano Model. The rest of the chapter is structured as follows: Sect. 2 provides the fundamental knowledge on Kano Model and fuzzy sets. The steps of fuzzy Kano Model and its application to laptop design are given in Sects. 3 and 4, respectively. Finally, concluding remarks are presented in Sect. 5.

2 Materials and Methods

In this section, after the fundamentals of Kano Model and fuzzy sets are provided, the steps of fuzzy Kano Model are given.

2.1 Kano Model

Kano model [6] is essentially applied by classifying quality attributes via a questionnaire containing pairs of questions. These question pairs consist of functional and dysfunctional questions. Functional questions aim at analyzing the consumer's feelings in the case of fulfillment of an attribute. However, dysfunctional question examines consumer's feelings in the case of non-fulfillment of an attribute [7]. In such a questionnaire, participants are expected to answer both functional and dysfunctional questions by selecting one of the following linguistic terms: "dislike", "live-with", "neutral", "must-be" and "like" [9]. The analysis on the collected data is performed by using an evaluation table. The evaluation table given in Table 1 enables researchers to categorize attributes for each respondent. In Table 1, "Q, R, I, A, O, M" indicate Questionable, Reverse, Indifferent, Attractive, One-Dimensional and Must-be attributes, respectively. Finally, majority of categorizations made by respondents are used to determine the final classification of attributes [7]. According to Kano Model, attributes of a product can be categorized as follows:

- Non-fulfillment of *Must-be attributes* creates extreme customer dissatisfaction whereas fulfillment of Must-be attributes does not generate any increase in satisfaction level of customers.
- Customer satisfaction is proportional to the level of fulfillment of *One-Dimensional attributes*. In other words, a higher customer satisfaction can be attained by a higher level of fulfillment.
- Fulfillment of the *Attractive attributes* yields more than proportional customer satisfaction whereas non-fulfillment of these attributes does not lead to any dissatisfaction [8].
- An attribute is qualified as *Indifferent*, if existence or absence of an attribute does not have an influence on satisfaction level of customers.
- An attribute is called *Reverse attribute*, if existence of an attribute has a negative impact on customer satisfaction.
- *Questionable attributes* indicates that either the respondent provides an unreasonable answer or the question is described incorrectly [9].

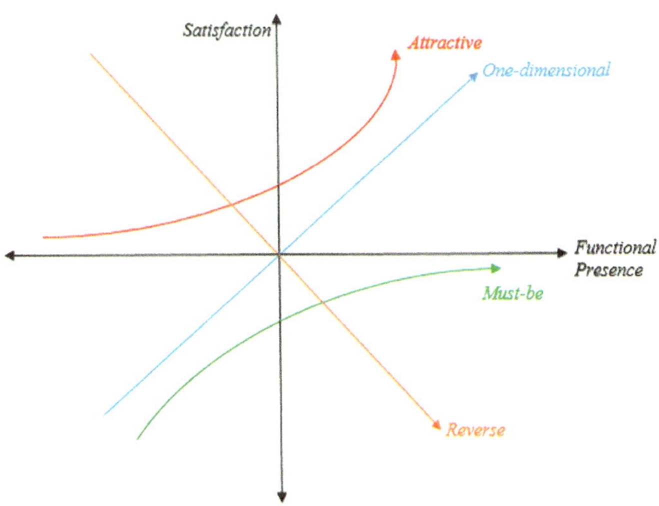

Fig. 1 Relationship between satisfaction level and fulfillment for different kinds of attributes

Table 1 Frequency of answers and attributes' classes according to Kano model

Customer response		Dysfunctional question				
		Like	Must be	Neutral	Live-with	Dislike
Functional question	Like	Q	A	A	A	O
	Must be	R	I	I	I	M
	Neutral	R	I	I	I	M
	Live-with	R	I	I	I	M
	Dislike	R	R	R	R	Q

The relationships between satisfaction level and fulfillment for different kinds of attributes are given in Fig. 1.

2.2 *Fuzzy Sets*

Fuzzy sets were proposed by Zadeh [10] to represent uncertainty and vagueness in human thinking. Fuzzy sets have been commonly employed as integrated to different methods. Kano Model is one of these methods combined with fuzzy approach in the literature. Lee et al. [11] adopted quality function deployment based on both Kano Model and fuzzy approach for product life-cycle assessment. It was shown that fuzzy logic facilitates Kano Model to produce a more objective weighting. Chen and Ko [12] proposed a fuzzy nonlinear model to identify the performance of design requirements for maximizing customer satisfaction. Li [13] utilized Kano Model along with trapezoidal fuzzy number similarity measures to classify quality attributes of mobile classrooms. In the study of Beheshtinia and Azad [14], a hybrid model involving the fuzzy house of quality, SERVQUAL and Kano Model is introduced to improve the quality of hotels under budget constraint. Florez-Lopez and Ramon-Jeronimo [15] introduced an approach involving Kano Model, fuzzy distances, and 2-tuple fuzzy-linguistic model for effective management of customer-service logistics.

The disadvantage of Kano Model is that it takes the majority of responses into account to identify the class of an attribute. Suppose that a survey of 170 participants is examined and it is shown that *A, O, M, I, R*, and *Q* values for a particular attribute are 49, 44, 48, 28, and 1, respectively. This particular attribute is classified as *A*, because the value of *A* is the highest. On the other hand, 48 people categorize this parameter as *M*. In this case, only one participant determines the class of the related attribute as *A*. Hence, this section of Kano Model requires fuzzification. Each attribute has to be defined by the values representing the membership degree of the attribute to each class. For this purpose, an integrated method consisting of fuzzy sets theory and Kano Model has been introduced [16].

3 Fuzzy Kano Model

The utilized approach mainly consists of Kano Model and fuzzy sets. The main steps of the approach are as follows:

Step 1. Conduct a thorough research to determine main design attributes of the product.

Step 2. Prepare and conduct a questionnaire for Kano Model by using the attributes determined in Step 1.

Step 3. Determine the weights of the attributes according to formulas given in Eqs. (1)–(3) since A, O, M, I, R, Q values are obtained by using the results of the questionnaire conducted in Step 2. Use Eq. (1) when $(A + O + M) \geq (I + R + Q)$ holds, otherwise utilize Eqs. (2)–(3).

$$W_i = \frac{A + O}{A + 2 \times O + M} \quad \text{for} \quad i \epsilon \{A, O, M\} \tag{1}$$

$$W_i = 0 \quad \text{for} \quad i \epsilon \{I\} \tag{2}$$

$$W_i = \frac{-R}{I + R + Q}, \quad \text{for} \quad i \epsilon \{R\} \tag{3}$$

Step 4. Obtain normalized values for a class by dividing the responses belong to this specific class by the total number of responses.

Step 5. For each attribute, place the normalized values belonging to either the group "A-O-M" or "I-R-Q" on the graph according to the group with the majority (see Fig. 2). Then, perform defuzzification by using the center of gravity formula given in Eq. (4) [17] to identify membership degrees of an attribute to these classes.

$$COG(F) = \frac{\int_X xF(x)dx}{\int_X F(x)dx} \tag{4}$$

After a crisp number is acquired as a result of Eq. (4), membership degrees of an attribute to classes are computed based on distances of the obtained crisp number to numbers corresponding to these classes. Eqs. (5)–(7) are utilized when $(A + O + M) \geq (I + R + Q)$ holds to obtain the membership degrees to classes *M, O, A*; otherwise Eqs. (8)–(9) are employed to determine the membership degrees to classes *R, I*.

$$\mu_{ij} = \max((1 - |COG - 1|), 0) \quad \text{for} \quad j \epsilon \{M\} \tag{5}$$

$$\mu_{ij} = \max((1 - |COG - 2|), 0) \quad \text{for} \quad j \epsilon \{O\} \tag{6}$$

$$\mu_{ij} = \max((1 - |3 - COG|), 0) \quad \text{for} \quad j \epsilon \{A\} \tag{7}$$

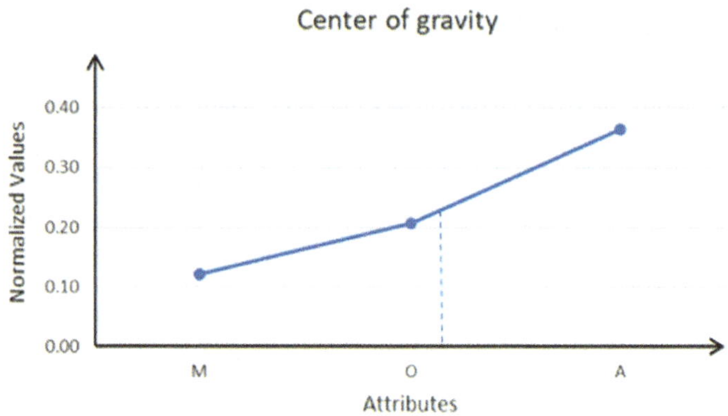

Fig. 2 Sample graph for center of gravity process given in Step 5

$$\mu_{ij} = \max((1 - |COG - 1|), 0) \quad \text{for} \quad j \in \{R\} \tag{8}$$

$$\mu_{ij} = \max((1 - |COG - 2|), 0) \quad \text{for} \quad j \in \{I\} \tag{9}$$

where μ_{ij} represents membership degree of ith design attribute to class j.

4 Application

The laptop computers have become one of the preferred products in the consumer electronics market, and a large part of the laptop consumer market is comprised of college and graduate students [5]. College and graduate students can be classified as a separate customer segment because of their needs for portability and financial constraints [5]. That is why fuzzy Kano Model is applied to identify the priorities of college students for laptop computers. The framework of the study is presented in Fig. 3.

A questionnaire on laptop attributes was prepared and conducted with 82 respondents. 30 and 70% of the participants are female and male, respectively. 98 and 2% of the participants are between the ages of 15 & 25 and 26 & 35, respectively. In addition to being a student, 10% of the participants work at different companies. 98% of the participants use laptops. Figure 4 shows that 47%, 40% and 13% of the participants use it for less than 2 h, 2–4 h and 5–10 h, respectively. According to Fig. 5, 7%, 28%, 27%, 28%, and 10% of the participants use it with the purpose of access to social media, doing research on the internet or reading news, doing homework, watching movies and playing games, respectively. Figure 6 indicates that 1%, 4%, 23% and 72% of the participants give the responses, less than 1 year, 2 years, 3 years and more than 4 years, respectively, to the question "How often do you change your laptop?". Figure 7 demonstrates that the market value of the current laptop of

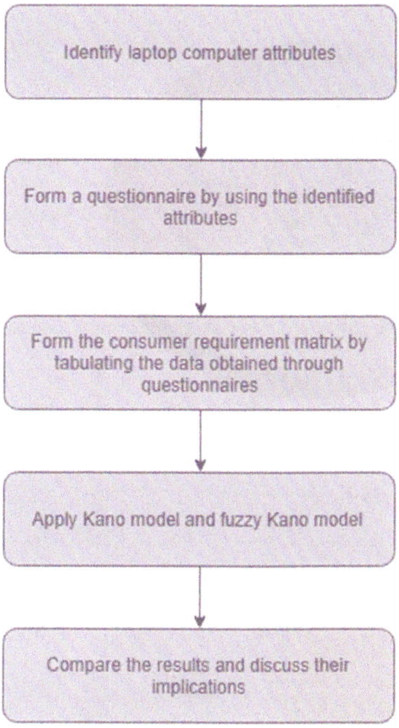

Fig. 3 Framework of the study

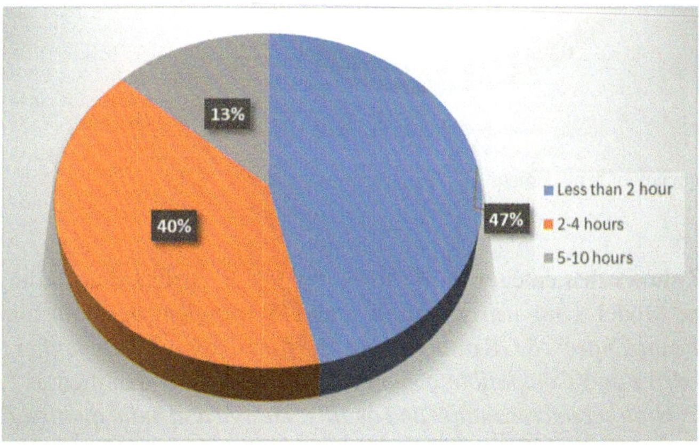

Fig. 4 Daily spent time using laptops

76%, 22% and 2% of the participants are less than 5000 TL, 5000–10,000 TL and more than 20,000 TL, respectively.

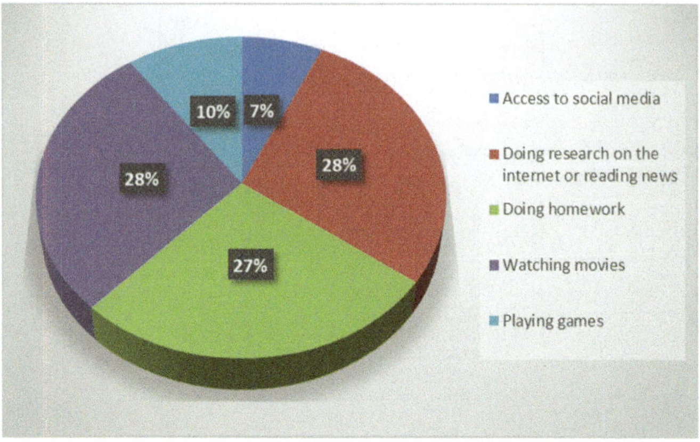

Fig. 5 Purpose of usage

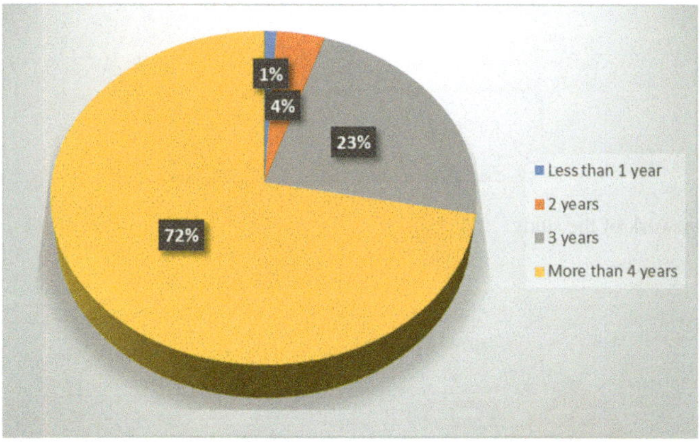

Fig. 6 Frequency of replacement

Table 2 shows that categories of the laptop design attributes according to traditional Kano Model. *Long warranty period* and *i5 or higher processor* are classified as "*Must-be attributes*". *8 GB or higher RAM*, *Hard drive capacity higher than 1 TB*, *SSD hard drive* and *Compatible computer accessories* are qualified as "*Attractive attributes*". *High screen resolution and quality*, *Design and build quality*, *Battery life longer than 3 h* are classified as "*One dimensional attributes*" whereas *Heavier than 2 kg* is identified as a "*Reverse attribute*". The rest of these attributes is qualified as "*Indifferent attributes*".

Table 3 demonstrates the comparison of the results obtained from Kano Model and fuzzy Kano Model. The attributes, *Long warranty period, i5 or higher processor, 8 GB or higher RAM, Hard drive capacity higher than 1 TB, SSD hard drive, High screen resolution and quality, Compatible computer accessories, Market availabil-*

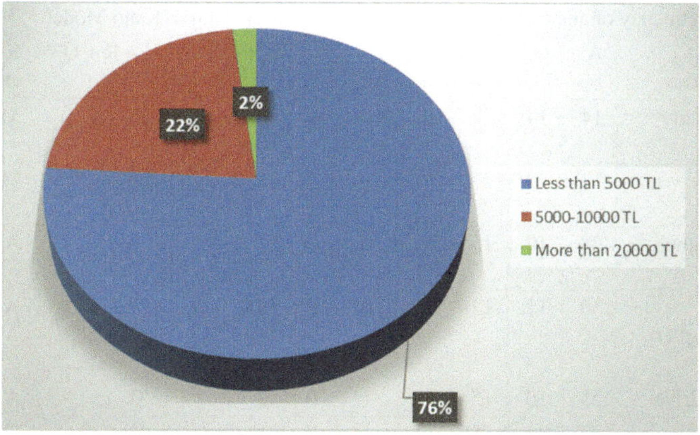

Fig. 7 Market values of the laptops

ity, Design and build quality, Battery life longer than 3 h and *Screen larger than* 15", are mostly qualified as "*One dimensional attributes*" since their membership degrees to one dimensional class is quite high. The attributes, *Touch screen feature, Having fingerprint reader*, approximately equally belong to both "*One dimensional*" and "*Attractive attributes*". The attribute, *Heavier than* 2 kg, approximately equally belongs to both "*Indifferent*" and "*Reverse attributes*".

According to the classical Kano evaluation, the long warranty period is a Must-be attribute, while it is a one-dimensional attribute according to the fuzzy Kano Model. Since customer satisfaction and warranty period are expected to increase at the same time, it seems more appropriate to belong one-dimensional class. Similarly, the attribute 8 GB or higher RAM is classified as attractive attribute according to the classical Kano Model, while the membership degree to the one-dimensional class is high in the fuzzy Kano Model. Since at least 8 GB RAM already exists on today's computers, the higher the RAM attribute is, the better customer satisfaction is expected to improve. In other words, the fact that the the attribute 8 GB or higher RAM belongs to one-dimensional class seems to be more appropriate to customer expectations. As a result, it is possible to say that the fuzzy Kano Model gave more consistent results than the classical Kano Model in reflecting the voice of customers.

5 Conclusions

The most important factor that keeps a product in the market is to analyze customer needs successfully. In this study, the effects of computer design parameters on users are measured and the important computer attributes have been tried to be revealed. Students currently studying at the university and newly graduated students were chosen as the focus group of this study. This group does not need highly advanced

Table 2 Frequency of answers and classes of attributes according to Kano Model

Assessed characteristics	A	O	M	I	R	Q	A + O + M	I + R + Q	Kano classes
Long warranty period	14	14	28	26	0	0	56	26	M
i5 or higher processor	13	22	27	16	1	3	62	20	M
8 GB or higher RAM	28	19	15	20	0	0	62	20	A
Hard drive capacity higher than 1 TB	30	12	13	27	0	0	55	27	A
SSD hard drive	25	10	15	31	1	0	50	32	A
High screen resolution and quality	22	24	15	20	1	0	61	21	O
Brand awareness	24	7	6	45	0	0	37	45	I
Heavier than 2 kg	1	0	0	30	51	0	1	81	R
Touch screen feature	20	2	0	50	10	0	22	60	I
Compatible computer accessories	23	14	10	35	0	0	47	35	A
Market availability	13	11	5	51	2	0	29	53	I
Design and build quality	24	24	13	21	0	0	61	21	O
Battery life longer than 3 h	16	39	20	7	0	0	75	7	O
Having fingerprint reader	23	6	1	52	0	0	30	52	I
Screen larger than 15”	20	7	9	41	5	0	36	46	I

computers since they generally use computers for research, homework and movies. In order to determine the characteristics of laptop computers that address this group, Kano Model was preferred. In this study, both crisp Kano Model and fuzzy Kano Model were employed and it was shown that fuzzy Kano Model produces more consistent results when both models were compared. This is because the classical Kano Model provides information about the class of an attribute just by looking at the majority, while the fuzzy Kano Model also considers the values related to other classes to produce a result.

Nowadays, laptop computers can be used for various purposes such as game, coding, animation, product development. In accordance with its purpose, the user’s expectation from the product is changing. Therefore, the analysis can be repeated for different customer segments in future studies.

Table 3 Comparison of attributes' classes by traditional Kano model and the fuzzy Kano model

Design attributes	Class in Kano model	Fuzzy Kano model				
		$\mu(M)$	$\mu(O)$	$\mu(A)$	$\mu(R)$	$\mu(I)$
Long warranty period	M	0.13	0.87	0.00	0.00	0.00
i5 or higher processor	M	0.11	0.89	0.00	0.00	0.00
8 GB or higher RAM	A	0.00	0.89	0.11	0.00	0.00
Hard drive capacity higher than 1 TB	A	0.00	0.83	0.17	0.00	0.00
SSD hard drive	A	0.00	0.89	0.11	0.00	0.00
High screen resolution and quality	O	0.00	0.95	0.05	0.00	0.00
Brand awareness	I	0.00	0.73	0.27	0.00	0.00
Heavier than 2 kg	R	0.00	0.00	0.00	0.46	0.54
Touch screen feature	I	0.00	0.44	0.56	0.00	0.00
Compatible computer accessories	A	0.00	0.86	0.14	0.00	0.00
Market availability	I	0.00	0.87	0.13	0.00	0.00
Design and build quality	O	0.00	0.91	0.09	0.00	0.00
Battery life longer than 3 h	O	0.02	0.98	0.00	0.00	0.00
Having fingerprint reader	I	0.00	0.59	0.41	0.00	0.00
Screen larger than 15"	I	0.00	0.83	0.17	0.00	0.00

References

1. Sharma, V., Prakash, N.R., Kalra, P.: User sensory oriented product form design using kansei engineering and its methodology for laptop design. In International Journal of Scientific Research in Science, Engineering and Technology (IJSRSET) (2016). ISSN 2395-1990
2. Lutters, E., van Houten, F.J., Bernard, A., Mermoz, E., Schutte, C.S.: Tools and techniques for product design. CIRP Ann. **63**(2), 607–630 (2014)
3. Jiang, S., Liang, W., Li, J., Lu, C.: Innovative design of a new laptop. Procedia Eng. **29**, 2932–2937 (2012)
4. Terwiesch, C., Ulrich, K., Randall, T.: User design of customized products. Marketi. Sci. **26**(2), 268–280 (2007)
5. Shiau, C.S., Tseng, I.H., Heutchy, A.W., Michalek, J.: Design optimization of a laptop computer using aggregate and mixed logit demand models with consumer survey data. In: ASME 2007 International Design Engineering Technical Conferences and Computers and Information in Engineering Conference, pp. 175–185 (2007). American Society of Mechanical Engineers
6. Kano, N.: Attractive quality and must-be quality. Hinshitsu (Qual. J. Japan. Soc. Qual. Control) **14**, 39–48 (1984)
7. Mikulic, J., Prebezac, D.: A critical review of techniques for classifying quality attributes in the Kano model. Manag. Serv. Qual. Int. J. **21**(1), 46–66 (2011)
8. Matzler, K., Hinterhuber, H.H.: How to make product development projects more successful by integrating Kano's model of customer satisfaction into quality function deployment. Technovation **18**(1), 25–38 (1998)
9. Wang, C.H., Wang, J.: Combining fuzzy AHP and fuzzy Kano to optimize product varieties for smart cameras: A zero-one integer programming perspective. Appl. Soft Comput. **22**, 410–416 (2014)
10. Zadeh, L.A.: Fuzzy sets. Inf. Control **8**(3), 338–353 (1965)

11. Lee, Y.C., Sheu, L.C., Tsou, Y.G.: Quality function deployment implementation based on Fuzzy Kano model: an application in PLM system. Comput. Indus. Eng. **55**(1), 48–63 (2008)
12. Chen, L.H., Ko, W.C.: A fuzzy nonlinear model for quality function deployment considering Kano's concept. Math. Comput. Model. **48**(3–4), 581–593 (2008)
13. Li, X.: KANO quality attribute classification method based on trapezoidal fuzzy number similarity measures. J. Intell. Fuzzy Syst. **33**(5), 2869–2876 (2017)
14. Beheshtinia, M.A., Farzaneh Azad, M.: A fuzzy QFD approach using SERVQUAL and Kano models under budget constraint for hotel services. Total Qual. Manage. Bus. Excellence **30**(7–8), 808–830 (2019)
15. Florez-Lopez, R., Ramon-Jeronimo, J.M.: Managing logistics customer service under uncertainty: an integrative fuzzy Kano framework. Inf. Sci. **202**, 41–57 (2012)
16. Ilbahar, E., Cebi, S.: Classification of design parameters for E-commerce websites: a novel fuzzy Kano approach. Telematics Inform. **34**(8), 1814–1825 (2017)
17. Wang, X., Ruan, D., Kerre, E.E.: Mathematics of Fuzziness-Basic Issues, vol. 245 (2009). Springer

Part V
Customer Oriented Product Design Using MCDM and Heuristics Algorithm

Application of MADM Techniques to Develop Customer Oriented Product

Sanjaykumar Gangurde

Abstract This chapter provides an approach to apply a multi-attribute decision making (MADM) method to evaluate the product design alternatives in respect to the customer and designer preferences. Customers employ a variety of heuristics when evaluating product alternatives. Different customer may have different choices for the particular product out of set of multi attribute alternatives of different brands or within the same brand also. Product design selection can be considered as a complex multi-attribute decision problem since the expectations differ from customer's as well as designer's point of view. Therefore, a proper mechanism is to be applied to decide on the most suitable product for the customer. MADM methods provide more scientific way to evaluate and select the best product design alternative. In this chapter, an example of personal data assistant (PDA) selection is considered as complex MADM problem. The alternatives of personal data assistant are evaluated using different MADM methods. The alternative are ranked on the basis of composite score 'Product Design Selection Index', PDSI (Pi). The ranking provided by these various methods is then compared. A novel quantitative approach is proposed for the comparison of alternatives in case of conflict in ranking.

1 Introduction

The economic success of a manufacturing firm depends on their ability to identify and satisfy customer needs. Products that meet the needs should be designed and produced at low cost. Product development initially starts with an idea, which is then transformed into artifact after going through various stages like, concept development and selection, embodiment and detailed design. Product is a set of attributes such as quality, features and design. In order to improve sales or profits, product attributes or features must be added, modified or withdrawn, and design (or styling) might be changed. Globalization has forced companies to offer wide variety of products to

S. Gangurde (✉)
Department of Production Engineering, K. K. Wagh Institute of Engineering Education and Research, Nasik, Maharashtra, India
e-mail: gangurdesanjay@rediffmail.com

C. Kahraman and S. Cebi (eds.), *Customer Oriented Product Design*, Studies in Systems, Decision and Control 279, https://doi.org/10.1007/978-3-030-42188-5_14

satisfy spectrum of customer demands. Most of the survey has identified '*new product development*' as a leading driver for revenue, profit, and market share growth. One of the important reasons for this is that majority of the new product cost is decided by the decisions made during the product conceptual stage.

In consumer products like mobile phones, family cars, laptops, washing machines etc., customer voices are diverse which needs to be captured as customer needs to develop a product closer to their expectations (Example, what features of the product are preferred, what is the worth of each feature from customer's point of view). Product configuration has been recognized as an effective means in meeting a customer's needs. It attempts to deliver products at lowest cost while maximizing customer satisfaction. Product configuration is obtained from a predefined set of design units, like building blocks, parts, modules, and assemblies, by relating them with each other. Addition/deletion of features to create a new product configuration have a great impact on cost and lead time of product development. Therefore, careful product configuration planning is essential.

2 Generic Approach to Product Development

A product development process is a sequence of steps or activities which an enterprise employs to conceive, design, and commercialize a product. The six phases of the generic product development process are shown in Fig. 1. These are briefly summarized below:

Planning: This phase begins with corporate strategy and includes assessment of technology developments and market objectives. The output of the planning phase is the project mission statement, which specifies the target market of product, business goals, key assumptions, and constraints.

Concept development: In this phase, the needs of the target market are identified, alternative concepts are generated and evaluated, and one or more concepts are selected for further development and testing. A concept is a description of the form, function and features of a product. The concept development process includes the following activities:

- Identifying customer needs: The goal of this activity is to understand customer needs.
- Establishing target specifications: Specifications are the translation of the customer needs into technical terms.
- Concept generation: The goal of concept generation is to generate product concepts that may satisfy the customer needs. A product concept is an approximate description of technology, working principles, and form of the product. The degree to which a product satisfies customers and can be successfully commercialized depends to a large measure on the quality of the concept. Concept generation is relatively inexpensive and can be done relatively quickly in comparison to the rest of the development process. The concept generation process begins with a

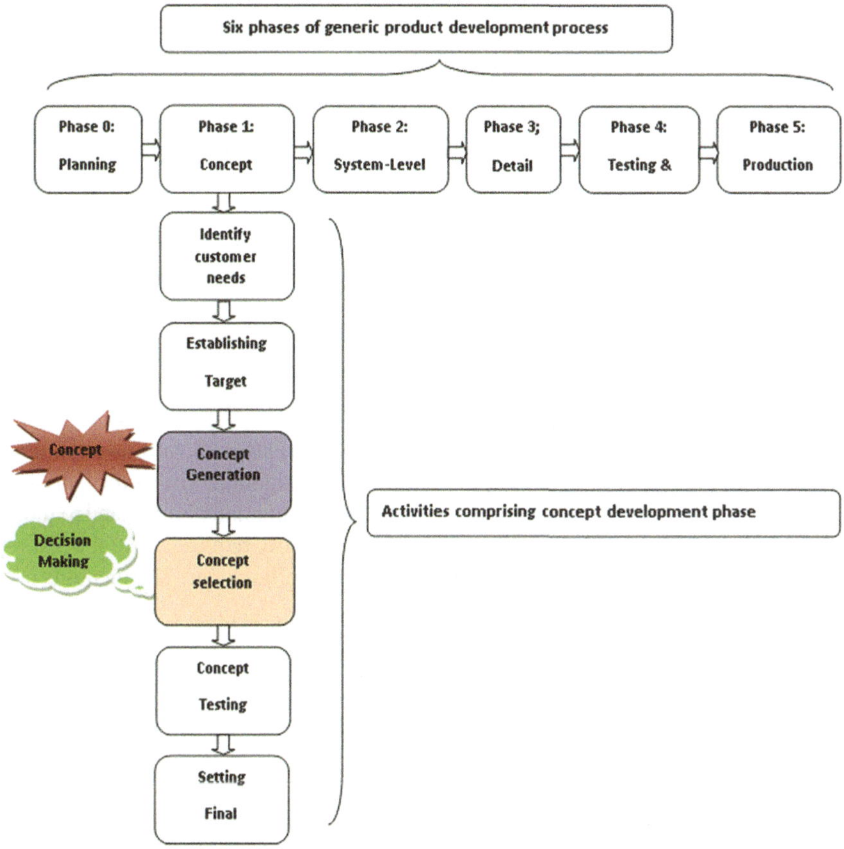

Fig. 1 Six phases of generic product development process and activities comprising concept development phase [1]

set of customer needs and results in a set of product concepts from which the development team will make a final selection. In most cases, team will generate hundreds of concepts of which 5–20 will merit serious consideration during concept selection activity. The concept generation method consists of five steps:

- Clarify the problem: Many design challenges are too complex to solve as a single problem and can be usually divided into several simpler sub problems (i.e. problem decomposition).
- Search externally: External search aimed at finding existing solutions to both the overall problem and the sub problems identified during the problem clarification step. The external search for solutions is essentially an information gathering process. Information is to be gathered from published literature (journals, conference proceedings, trade magazines, market/consumer/product information),

internets, competitors, patents, lead users, experts, and related benchmarked products.
  - Search internally: Use individual and group methods to retrieve and adapt the knowledge of the team.
  - Explore systematically: As a result of the external and internal search activities, the team may generate hundreds of concept fragments. These fragments are to be organized and synthesized by considering all of the combinations of the fragments associated with each sub problems. If more the number of combinations, then it will be a difficult task. However, some of the combinations do not make sense.
  - Reflect on the solutions and the process: Identify opportunities for improvement in subsequent iterations or future projects.

- Concept selection: Concept selection is the process of evaluating concepts with respect to customer needs and comparing these concepts with various criteria and then selecting one or more concepts for further investigation. This is the activity in which various product concepts are analyzed and sequentially eliminated to indentify the most promising concept(s). The process usually requires several iterations and may initiate additional concept generation and refinement. The methods used for concept selections are:

  - External decision: Concepts are turned over to the customer, client, or some other external entity for selection
  - Product champion: An influential member of product development team chooses a concept based on personal preference.
  - Intuition: The concept is chosen by its feel.
  - Multivoting: Each member of the team votes for several concepts. The concept with the most votes is selected.
  - Pros and cons: The team lists the strengths and weakness of each concept and makes a choice based upon group opinion.
  - Prototype and test: The organization builds and tests prototypes of each concept, making a selection based upon test data.
  - Decision matrices: The team rates each concept against specific selection criteria, which may be weighted.

- Concept testing: One or more concepts are then tested to verify that the customer needs have been met, assess the market potential of the product, and identify any shortcomings which must be remedied during further development.
- Setting final specifications: The target specifications set earlier in the process are revisited. At this stage, the product development team must commit to specific values of the metrics reflecting the constraints inherent in the product concept, limitations identified through technical modeling and trade-offs between cost and performance.

System-level design: In this phase, the product is decomposed into subsystems and components. The final assembly scheme for production system is defined during this phase.

Detail design: This phase includes the complete specification of the geometry, materials and tolerances of all of the unique parts in the product. The output of this phase is control documentation (i.e. drawings, computer files, tooling required, standard or purchased part list of the product, process plan for fabrication and assembly etc.).

Testing and refinement: It involves the construction and evaluation of multiple preproduction versions of the product (prototype). The prototype is tested to determine whether the product will work as designed and also tested by customers to check whether the product satisfies the key customer needs.

Production ramp-up: In the production ramp-up phase, the product is made using intended production system and then supplied to preferred customers to identify the remaining flaws. During the transition from production ramp-up to ongoing production, the product is launched.

3 Multi-Attribute Decision Making (MADM)

To be competitive in the market and to face the global challenges, manufacturing industries have to select appropriate manufacturing strategies, product designs, manufacturing processes, work piece and tool materials, machinery and equipment, etc. The selection decisions are complex as it involves more than one dimension and hence decision making is more challenging today. In most of the situations, decision maker has to assess a wide range of alternatives based on a set of conflicting criteria to choose the best alternative by including dimensions like technological, economic, ethical, political, legal, and social etc. Hence, MADM methodology/tools are the most scientific way to solve such problems. It involves a preparation of decision table (also called decision matrix), which has four main parts, namely: (a) alternatives (b) attributes/criteria (c) weight or relative importance of each attribute, and (d) measures of performance of alternatives with respect to the attributes. The decision matrix is shown in Table 1. It shows alternatives, A_i (for i = 1, 2, …, n), attributes, B_j (for j = 1, 2, …, m), weights of attributes, w_j (for j = 1, 2, …, m) and the measures of performance of alternatives, y_{ij} (for i = 1, 2, …, n; j = 1, 2, …, m). It may be

Table 1 Decision matrix in MADM methods [2]

Attributes (Bi)		B_1	B_2	B_3	…	B_m
Weights (wi)		(w_1)	(w_2)	(w_3)	…	(w_m)
Alternatives (A_i)	A_1	y_{11}	y_{12}	y_{13}	…	y_{1m}
	A_2	y_{21}	y_{22}	y_{23}	…	y_{2m}
	A_3	y_{31}	y_{32}	y_{33}	…	y_{3m}
	:	:	:	:	:::	:
	A_n	y_{n1}	y_{n2}	y_{n3}	…	y_{nm}

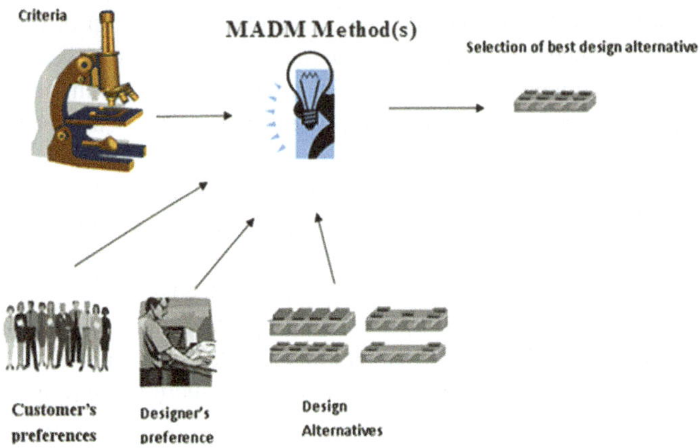

Fig. 2 Overall product design selection framework [10]

added here that all the elements in the decision table must be normalized to the same units so that all the possible attributes in the decision problem can be considered.

Important MADM methods includes Simple Additive Weighing, SAW [3]; Weighted Product Method, WPM [4]; Technique for Order Preference by Similarity to Ideal Solution, TOPSIS [5]; Modified TOPSIS [6], Grey Relational Analysis, GRA [7]; Preference Ranking Organization Method for Enrichment Evaluations, PROMETHEE [8] and Compromise Solution Method, also known as the VIKOR (VIšekriterijumsko KOmpromisno Rangiranje) method [9]. These are briefly discussed in Sect. 4. The overall frame work of MADM approach for selection of best product design alternative is shown in Fig. 2.

Several market-based decision support methodologies have been reported in the literature to aid product selection. The overall frame work of MADM approach for selection of best product design alternative is shown in Fig. 2.

Isiklar and Buyukozkan [11] used analytic hierarchy process (AHP) to determine the relative weights of evaluation product related criteria (basic requirements, physical characteristics and technical features) and user related criteria (functionality, brand choice and customer excitement) and TOPSIS method is applied to rank the mobile phone alternatives. Lin et al. [12] proposed a framework that integrated AHP and TOPSIS to evaluate and select the best product design alternative. Five design alternatives are compared on the basis of twelve customer criteria for personal data assistant (PDA). Besharati et al. [10] considered ten design alternatives of a power electronic device for the selection and three criteria manufacturing cost, junction temperature, and thermal cycles to failure for evaluation of the alternatives. The proposed methodology is based on customers purchase behavior. Hsiao [13] proposed a fuzzy decision-making method for selecting an optimum design of a juicer by considering different criteria namely; technology (part number, assembly efficiency, assembly time, manufacturability, esthetic sense (style, color, humanization, aggregative esthetic), real function (performance, noise, easy to operate, reliability, efficiency, safety) and economics (low cost, material saving, save manufacturing time). The tabulated judgment method [14] is used to determine relative importance (weightage) of these criteria.

Vanegas and Labib [15] presented a systematic method for deriving optimum targets of engineering characteristics (ECs) in QFD. Fuzzy numbers are used to represent the imprecise nature of the judgments for determining the importance levels of customer attributes (CAs). Also, instead of addressing the relationships between ECs and CAs in a descriptive and subjective manner (strong positive, medium negative, etc.), these are defined more appropriately through fuzzy sets, indicating the levels of satisfaction that a certain set of values of an EC gives to a certain CA. Constraints such as cost, technical difficulty and market position are also integrated in the process of determination of targets. An example of a car door is presented to show the application of the method.

The Analytic Hierarchy Process (AHP) is a multi-criteria decision-making approach and was introduced by Saaty [16]. The AHP is a decision support tool which can be used to solve complex decision problems. It uses a multi-level hierarchical structure of objectives, criteria, sub-criteria, and alternatives. The pertinent data are derived by using a set of pair-wise comparisons. These comparisons are used to obtain the weights of importance of the decision criteria, and the relative performance measures of the alternatives in terms of each individual decision criterion. If the comparisons are not perfectly consistent, then it provides a mechanism for improving consistency. The AHP has attracted the interest of many researchers in product design. Saaty [16] and Lu et al. [17] recommended that designers apply analytic hierarchy process (AHP) to determine the importance degrees of customer requirements. Kalargeros and Gao [18] further proposed a fuzzy analytic hierarchy process (FAHP) to evaluate the importance weightings of customer attributes. Kuo et al. [19] presented green fuzzy design analysis method (GFDA) to evaluate product design alternative based on environmental consideration using fuzzy logic. The hierarchical structure was constructed using AHP which consists of five criteria: energy, recycling, toxicity, cost and material. The most desirable design alternative is selected based on fuzzy-multi-attribute decision technique (FMADM). The model is validated using desktop computers as an example.

Fung et al. [20] proposed approach for analyzing customer attributes and projecting them into the relevant design, engineering and product attributes, in order to facilitate decision-making and to guide downstream manufacturing planning and control activities. The proposed hybrid system incorporates the principles of QFD, AHP and fuzzy set theory to tackle the complex and often imprecise problem domain encountered in customer requirement management, mainly VoC. The problem domain is to determine the target value of the product attribute on the 'Engine Power' required to satisfy specified customer attributes on the 'Top Speed' and the 'Seating Capacity' of a given model of motor car.

The review of literatures reveals that limited research has been made to generate product design alternatives based on customer preferences and evaluating and ranking them using multi-attribute decision making (MADM) methods to meet the customer requirements.

4 MADM Methods for Selection of Product Design Alternative

This section provides an approach to apply a multi-attribute decision making methods (MADM) to evaluate the product design alternatives in respect to the customer preferences and designer preferences. Customers employ a variety of heuristics when evaluating product alternatives in the marketplace. Many products (for example, computers, cell phone, laptops, washing machines, vacuum cleaners, insurance policies etc.) are made up of different features and are mainly differentiated by brand, packaging, and price. Different customer may have different choices for the particular product out of set of multi attribute alternatives of different brands or within the same brand also. As such, a quantitative basis for comparison and selection of the best product design solution among a host of alternatives could greatly impact on the eventual success or failure of a product in the market. Product design selection can be considered as a complex multi-attribute decision problem since the expectations differ from customer's point of view as well as designer's point of view. Therefore, a proper mechanism is to be applied to decide on the most suitable product for the customer. MADM method is the more scientific way to evaluate and select the best product design alternative.

In this section, the examples of two products, vacuum cleaner and personal data assistant (PDA) selection are considered as complex multi-attribute decision making problem. The alternatives are evaluated using SAW [3], WPM [4], TOPSIS [5], Modified TOPSIS [6], GRA [7], PROMETHEE [8] and VIKOR [9] method. The composite score of the alternative is termed as the 'Product Design Selection Index', PDSI (P_i) in the present report. The ranking of the alternative is given on the basis of the values of P_i. Some of the MADM methods used in this work are discussed below. The ranking provided by these various methods is then compared. A novel quantitative approach is proposed for the comparison of alternatives in case of conflict in ranking. This quantitative approach provides better guidelines to the decision maker, than that provided by qualitative approach applied by earlier researchers. Different Multi-Attribute Decision Making (MADM) methods are discussed as follows:

4.1 Simple Additive Weighing (SAW) Method

This method is also called Weighted Sum Method developed by Fishburn [3] in 1967. All the elements of the decision table are normalized, and then SAW can be used for any type and any number of attributes. The overall or composite score P_i of the alternative A_i is determined by Eq. (1).

$$P_i = \sum_{j=1}^{m} w_j \left(y_{ij}\right)_{normal} \tag{1}$$

where $(y_{ij})_{normal}$ represents the normalized value of y_{ij}. The composite score of the product design alternative is termed as the 'Product Design Selection Index (PDSI)' in the present report. The alternative with the highest composite score (PDSI), P_i is considered as the best alternative.

4.2 *Weighted Product Method (WPM)*

This method is developed by Miller and Starr [4] in 1969 is similar to SAW. However, the main difference is that instead of addition in the model there is multiplication. The overall or composite performance score (PDSI), P_i of an alternative A_i is determined by Eq. (2).

$$P_i = \prod_{j=1}^{m} [(y_{ij})_{normal}]^{\overrightarrow{\leftarrow} w_j} \quad (2)$$

4.3 *Technique for Order Preference by Similarity to Ideal Solution (TOPSIS) Method*

TOPSIS method, which is based on choosing the best alternative having the shortest distance to the ideal solution and the farthest distance from the negative-ideal solution, was first proposed by Hwang and Yoon [5]. The ideal solution is the solution that maximizes the benefit; on the contrary; the negative-ideal solution is the solution that minimizes the benefit. For the first step of this methodology, the decision matrix, representing the performance values of each alternative with respect to each criterion, is computed. Next, these performance values are multiplied with the criteria weights calculated. The ideal solution consists of the best values of alternatives and the negative-ideal solution consists of the worst values of alternatives. Subsequently, the alternatives are ranked with respect to their relative closeness to the ideal solution.

The TOPSIS procedure consists of the following steps:

Step 1: The first step is to determine the objective and to identify the pertinent evaluation attributes.

Step 2: This step represents a matrix based on all the information available on attributes. This matrix is nothing but the decision table as shown in Table 1.

Step 3: The first step of the procedure involves the calculation of the normalized decision matrix. The normalized value (R_{ij}) is calculated using Eq. (3a).

$$R_{i \underset{\leftarrow}{\rightarrow} j} = \frac{y_{i \underset{\leftarrow}{\rightarrow} j}}{\sqrt{\sum_{j=1}^{m} y^2_{i \underset{\leftarrow}{\rightarrow} \underset{\leftarrow}{\rightarrow} j}}} \tag{3a}$$

Step 4: Decide the relative importance (i.e., weights) of different attributes with respect to the objective. A set of weights w_j (for j = 1, 2, …, m) such that,

$$\sum_{j=1}^{m} w_j = 1 \tag{3b}$$

Step 5: In the next step, the weighted normalized decision matrix is calculated. The weighted normalized value (V_{ij}) is calculated using Eq. (4).

$$V_{ij} = w_j \times R_{ij} \tag{4}$$

Step 6: In this step; the ideal (best), V_j^+ and negative – ideal (worst), V_j^- solutions are determined using Eq. (5) and Eq. (6) respectively.

$$V^+ = \left\{ \left(\max_i V_{ij} | j \in J \right), \left(\min_i V_{ij} | j \in J' \right) \right\} \tag{5}$$

$$V^+ = \left\{ V_1^+, V_2^+, V_3^+, \ldots, V_M^+ \right\} \tag{5a}$$

$$V^- = \left\{ \left(\min_i V_{ij} | j \in J \right), \left(\max_i V_{ij} | j \in J' \right) \right\} \tag{6}$$

$$V^- = \left\{ V_1^-, V_2^-, V_3^-, \ldots, V_M^- \right\} \tag{6a}$$

where J is associated with beneficial attributes, and J′ is associated with non-beneficial attributes.

Step 7: The separation measures are calculated using the n-dimensional Euclidean distance. The separation of each alternative from the ideal solution is given by Eqs. (7) and (8).

$$S_i^+ = \sqrt{\sum_{j=1}^{m} \left(V_{i \underset{\leftarrow}{\rightarrow} j} - V_j^+ \right)^2} \tag{7}$$

$$S_i^- = \sqrt{\sum_{j=1}^{m} \left(V_{i \underset{\leftarrow}{\rightarrow} j} - V_j^- \right)^2} \tag{8}$$

Step 8: The relative closeness (PDSI) of a particular alternative to the ideal solution, P_i can be expressed in this step using Eq. (9).

$$P_i = \frac{S_i^-}{\left(S_i^+ + S_i^- \right)} \tag{9}$$

Step 9: At the final step, the preference order is ranked. In this step, a set of alternatives is made in the descending order according to the value of P_i indicating the most preferred and least preferred feasible solutions. P_i is an overall or composite performance score (PDSI) of alternative A_i.

4.4 Modified TOPSIS Method

Deng et al. [6] presented the weighted Euclidean distances rather than creating a weighted decision matrix as in case of TOPSIS method. In this method, the positive ideal solution (R^+) and the negative ideal solution (R^-) are not dependent on the weighted decision matrix and can be represented by Eq. (10) and Eq. (11) respectively.

$$R^+ = \left\{ \left(\max_i R_{\underset{\leftarrow}{\rightarrow} i \underset{\leftarrow}{\rightarrow} j} | j \in J \right), \left(\min_i R_{i \underset{\leftarrow}{\rightarrow} j} | j \in J' \right) \right\} \tag{10}$$

$$R^+ = \left\{ R_1^+, R_2^+, R_3^+, \ldots, R_m^+ \right\} \tag{10a}$$

$$R^- = \left\{ \left(\min_i R_{i \underset{\leftarrow}{\rightarrow} \underset{\leftarrow}{\rightarrow} j} | j \in J \right), \left(\max_i R_{i \underset{\leftarrow}{\rightarrow} j} | j \in J' \right) \right\} \tag{11}$$

$$R^- = \left\{ R_1^-, R_2^-, R_3^-, \ldots, R_m^- \right\} \tag{11b}$$

The weighted Euclidean distances are calculated using Eqs. (12) and (13)

$$D_i^+ = \sqrt{\sum_{j=1}^{m} w_j \left(R_{i \underset{\leftarrow}{\rightarrow} \underset{\leftarrow}{\rightarrow} j} - R_j^+ \right)^2} \tag{12}$$

$$D_i^- = \sqrt{\sum_{j=1}^{m} w_j \left(R_{i \underset{\leftarrow}{\rightarrow} j} - R_j^- \right)^2} \tag{13}$$

The relative closeness (PDSI) of particular alternative to the ideal solution P_i can be determined by Eq. (14).

$$P_i = \frac{D_i^-}{\left(D_i^+ + D_i^-\right)} \tag{14}$$

The ranking of the alternatives is given in descending order of the values of $P_{i.}$

4.5 Grey Relational Analysis (GRA)

The grey system theory is proposed by Deng [7]. GRA solves MADM problems by combining the entire range of performance attribute values being considered for every alternative into one, single value. This reduces the original problem to a single attribute decision making problem. Therefore, alternatives with multiple attributes can be compared easily after the GRA process. The procedure of grey relational analysis is given below.

Step 1: Grey relational generating: When the units in which performance is measured are different for different attributes, the influence of some attributes may be neglected. Therefore, processing all performance values for every alternative into comparability sequence, in a process analogous to normalization, is necessary. This processing is called grey relational generating in GRA.

For MADM problem, the *i*th alternative can be expressed as $A_i = (y_{i1}, y_{i2}, y_{i3}, \ldots, y_{ij}, \ldots y_{im})$ where y_{ij} is the performance value of attribute *j* of alternative *i*. The term y_i can be translated into the comparability sequence $X_i = (x_{i1}, x_{i2}, x_{i3}, \ldots, x_{ij}, \ldots, x_{im.})$ by Eqs. (15) and (16).

$$x_{i \rightleftarrows j} = \frac{y_{i \rightleftarrows j} - \min\left\{y_{i \rightleftarrows j}, i = 1, 2, \ldots, n\right\}}{\max\left\{y_{i \rightleftarrows j}, i - 1, 2, \ldots, n\right\} - \min\left\{y_{i \rightleftarrows j}, i = 1, 2, \ldots, n\right\}} \tag{15}$$

$$x_{i \rightleftarrows j} = \frac{\max\left\{y_{i \rightleftarrows j}, i = 1, 2, \ldots, n\right\} - y_{i \rightleftarrows j}}{\max\left\{y_{i \rightleftarrows j}, i = 1, 2, \ldots, n\right\} - \min\left\{y_{i \rightleftarrows j}, i = 1, 2, \ldots, n\right\}} \tag{16}$$

Equation (15) is used for larger-the-better attributes and Eq. (16) for the smaller-the better attributes.

Step 2: Reference sequence definition: After the grey relational generating procedure, all the performance values are scaled into [0, 1] An alternative will be the best choice if all of its performance values are closest to or equal to 1, however, such type of alternative may not exist. The reference sequence X_o is to be defined as $(x_{o1}, x_{o2}, x_{o3},$

…, x_{oj}, …, x_{om}.) = (1, 1, …, 1, …, 1), and then aims to find the alternative whose comparability sequence is the closest to reference sequence.

Step3: Calculate grey relational coefficient: Grey relational coefficient is used for determining how close x_{ij} and x_{oj}. The larger the grey relational coefficient, the closer x_{ij} and x_{oj} are. The grey relational coefficients can be calculated by Eq. (17).

$$\gamma \overset{\rightarrow}{\underset{\leftarrow}{\rightleftarrows}} \left(x_{0 \rightleftarrows \rightleftarrows j}, x_{ij} \right) = \frac{\Delta_{\min} + \zeta \Delta_{\max}}{\Delta_{ij} + \zeta \Delta_{\max}} \tag{17}$$

where γ (x_{oj}, x_{ij}) is the grey relational coefficient between x_{oj} and x_{ij}, and

$\Delta_{ij} = |x_{oj} - x_{ij}|$, $\Delta_{\min} = \min\{\Delta_{ij}, i = 1, 2, \ldots, n; j = 1, 2, \ldots, m\}$, $\Delta_{\max} = \max\{\Delta_{ij}, i = 1, 2, \ldots, n; j = 1, 2, \ldots, m\}$, ζ is the distinguishing coefficient, $\zeta \in [0, 1]$.

The distinguishing coefficient can be taken by the decision maker exercising judgment. The rank order of alternative remains always same though the different coefficient is adopted. After grey relational generating, $\Delta_{\max}$ will be equal to 1 and $\Delta_{\min}$ will be equal to 0. In this work, the distinguishing coefficient is set as 0.5.

Step 4: Determine grey relational grade: After calculating the entire grey relational coefficient γ (x_{oj}, x_{ij}), grey relational grade can be determined using Eq. (18).

$$\Gamma(X_O, X_i) = \sum_{j=1}^{m} w_j \gamma \overset{\rightarrow}{\underset{\leftarrow}{\rightleftarrows}} \left(x_{o \rightleftarrows j}, x_{i \rightleftarrows j} \right) \; for\; i = 1, 2, \ldots, n \tag{18}$$

$\Gamma(X_o, X_i)$ is the grey relational grade (i.e. PDSI) between X_o and X_i. It represents the level of correlation between the reference sequence and the comparability sequence. The grey relational grade indicates the degree of similarity between the reference sequence and the comparability sequence. If the comparability sequence for an alternative gets the highest grey relational grade with the reference sequence, it means that the comparability sequence is most similar to the reference sequence, and that alternative would be the best choice. The ranking of the alternatives is given in descending order of the values of grey relational grade.

4.6 Preference Ranking Organization Method for Enrichment Evaluations (PROMETHEE)

The PROMETHEE method was developed by Brans et al. [8]. It proceeds to pair-wise comparison of alternatives in each single criterion in order to determine partial binary relations denoting the strength of preference of an alternative A_1 over alternative A_2. The steps in PROMETHEE method are described below:

Step 1: Identify the selection criteria for the considered decision making problem and short-list the alternatives on the basis of the identified criteria satisfying the requirements.
Step 2: Prepare a decision table including the measures or values of all criteria for the short-listed alternatives and the relative importance of the criteria is assigned.
Step 3: The next step is to have information on the decision maker preference function, which he/she uses when comparing the contribution of the alternatives in terms of each separate criterion. The preference function (F_j) translates the difference between the evaluations obtained by two alternatives (A_1 and A_2) in terms of particular criterion, into preference degree ranging from 0 to 1. Let $F_{j,A1A2}$ preference function associated to the criterion B_j.

$$F_{j,A_1A_2} = G_j\left[B_j(A_1) - B_j(A_2)\right] 0 \le F_{j,A_1A_2} \le 1$$

where, G_j is a non decreasing function of the observed deviation between two alternatives A_1 and A_2 over the criterion B_j.

In order to facilitate the selection of a specific preference function for a criterion, six basic types were proposed [8]. These include 'usual function', 'U-shape function', 'V-shape function', 'level function', 'linear function', and 'Gaussian function'. Preference 'usual function' is equal to the simple difference between the values of the criterion B_j for alternatives A_1 and A_2. The 'usual function' is an easy to use preference function and is generally used with quantitative criteria. In this work the preference 'usual function' for all attribute is used. If two alternatives have a difference $d \neq 0$ in criterion B_j, then a preference value ranging between 0 and 1 is assigned to the 'better' alternative whereas the 'worse' alternative receives a value 0. If $d = 0$, then they are indifferent which results in an assignment of 0 to both alternatives. The pair wise comparison of all the alternatives of each criterion is to be carried out.

Let the decision maker have specified a preference function F_j and weight w_j for each criterion B_j ($j = 1, 2, \ldots, m$) of the problem. The multiple attribute preference index Π_{A1A2} is then defined as the weighted average of the preference functions F_j:

$$\Pi_{A_1A_2} = \sum_{j=1}^{m} w_j F_{j,A_1A_2}$$

Π_{A1A2} represents the intensity of preference of the decision maker of alternative A_1 over alternative A_2, when considering simultaneously all the attribute. Its value ranges from 0 to 1. This preference index determines a valued outranking relation on the set of actions.

For PROMETHEE outranking relations, the leaving flow, entering flow and the net flow for an alternative a belonging to a set of alternatives A are defined by the following equations.

$$\phi^{+}(a) = \sum_{x \in A} \prod_{xa} \tag{19}$$

$$\phi^{-}(a) = \sum_{x \in A} \prod_{xa} \tag{20}$$

$$\phi(a) = \phi^{+}(a) - \phi^{-}(a) \tag{21}$$

φ^{+}(a) is called the leaving flow, φ^{-}(a) is called the entering flow and φ(a) is called the net flow. φ^{+}(a) is the measure of the outranking character of a (i.e. dominance of alternative a over all other alternatives) and φ^{-}(a) gives the outranked character of a (i.e. degree to which alternative a is dominated by all other alternatives). The net flow, φ(a), represents a value function, whereby a higher value reflects a higher attractiveness of alternative 'a'. The net flow values are used to indicate the outranking relationship between the alternatives. The ranking of the alternatives is given in the descending order of the net flow values.

4.7 Compromise Ranking Method (VIKOR)

The foundation for compromise solution was established by Yu [9]. The compromise solution is a feasible solution that is the closest to the ideal solution, and a compromise means an agreement established by mutual concession. VIKOR is a helpful tool in MADM, particularly in a situation where the decision maker is not able, or does not know how to express preference at the beginning of system design. The obtained compromise solution could be accepted by the decision makers because it provides a maximum 'group utility' (represented by $E_{\text{i-min}}$) of the 'majority' and a minimum of individual regret (represented by $K_{\text{i-min}}$) of the 'opponent'. The main procedure of the VIKOR method is described below:

Step 1: The first step is to determine the objective, and to identify the pertinent evaluation attributes. Also determine the best, i.e., $(y_{ij})_{max}$, and the worst, i.e., $(y_{ij})_{min}$, values of all attributes.
Step 2: Calculate the values of E_i and K_i using Eq. (22) and Eq. (23) respectively

$$E_i = \sum_{j=1}^{m} w_j \frac{[(y_{ij})(y_{ij})_{\max}[]]}{\left[(y_{ij})(y_{ij})_{\min_{\max}}\right]} \tag{22}$$

$$K_i = \max of\ w_j \frac{[(y_{ij})(y_{ij})_{\max}[]]}{\left[(y_{ij})(y_{ij})_{\min_{\max}}\right]} \tag{23}$$

Step 3: Calculate the values of Pi using Eq. (24)

$$P_i = \frac{v(E_i - E_{i-\min}())}{(Ei - \min_{i-\max}) + \frac{(1-v)(K_i - K_{i-\min}())}{(Ki - \min_{i-\max})}} \quad (24)$$

where, $E_{\text{i-max}}$ is the maximum value and $E_{\text{i-min}}$ the minimum value of E_{i} respectively; whereas

$K_{\text{i-max}}$ is the maximum value and $K_{\text{i-min}}$ is the minimum value of K_{i} respectively. v is introduced as weight of the strategy of 'the majority of attributes'. Usually, the value of v is taken as 0.5. However, v can take any value from 0 to 1.

Step 4: Arrange the alternatives in the ascending order, according to the values of P_{i} (Product Design Selection Index; *PDSI*). The best alternative, ranked by P_{i}, is the one with the minimum value of P_{i}.

5 Case Study: Design of Personal Data Assistant

The example presented in this work is the selection of personal data assistance (PDA) design, same as considered by Lin et al. [12] and Gangurde and Akarte [21]. According to customer requirements, five personal data assistant design alternatives A_1, A_2, A_3, A_4 and A_5 selected for evaluation are shown in Fig. 3.

The weightage for various criteria are decided in this work using analytical hierarchy process as used by Lin et al. [12] are shown in Table 2. The measures of performance of alternatives, y_{ij} values of the decision matrix are shown in Table 3.

5.1 *Simple Additive Weighing (SAW) Method*

The normalized matrix for SAW method is shown in Table 4. The overall or composite score (PDSI), P_i of the PDA design alternatives is calculated using Eq. (1). The ranking is arranged in descending order of the values of P_i as shown in Table 11.

Fig. 3 Personal data assistant design alternatives [12, 21]

Table 2 List of customer requirements and their relative weightage [12]

Customer requirements (Criteria)	Weightage (w_j)
R_1: form aspects	
R_{11}: A variety of forms	0.285
R_{12}: Well fit the whole shape on function keys	0.123
R_{13}: Good color combination	0.065
R_{14}: Good quality of color screen	0.125
R_2: ergonomic aspects	
R_{21}: Comfortable feeling on touch	0.040
R_{22}: Easy to hold and push buttons	0.059
R_{23}: Easy to take and store	0.038
R_{24}: Easy to set up and use	0.092
R_{25}: Avoidable of carelessly turn off buttons	0.053
R_3: other extra functions	
R_{31}: Can combine with other add-on functions	0.500

5.2 Weighted Product Method (WPM)

The normalized matrix for WPM method is shown in Table 4. The overall or composite score (PDSI), P_i of the alternatives is determined using Eq. (2) and the ranking is given in descending order of P_i values as shown in Table 11.

5.3 TOPSIS Method

Step 1: The objective is to evaluate the personal data assistant design alternatives.
Step 2: Prepare decision matrix to represent all the information available on customer requirements and design alternatives as shown in Table 3.
Step 3: The performance measure values of the decision matrix are to be normalized using Eq. (3a, 3b). The normalized matrix (R_{ij}) is shown in Table 5.
Step 4: The relative importance weights are assigned to all the attributes as shown in Table 2.
Step 5: The weighted normalized decision matrix (V_{ij}) is calculated using Eq. (4).
Step 6: Obtain the ideal (best), V_j^+ and negative ideal (worst), V_1^- solutions using Eqs. (5) and (6). These values are given as:

Table 3 Decision matrix for customer requirements and design alternatives

	R_{11}	R_{12}	R_{13}	R_{14}	R_{21}	R_{22}	R_{23}	R_{24}	R_{25}	R_{31}
A_i	0.285	0.123	0.065	0.125	0.040	0.059	0.038	0.09	0.053	0.115
A_1	0.556	0.611	0.667	0.611	0.667	0.556	0.667	0.611	0.611	0.500
A_2	0.556	0.444	0.500	0.667	0.556	0.722	0.667	0.611	0.667	0.500
A_3	0.778	0.333	0.611	0.667	0.444	0.556	0.389	0.500	0.667	0.722
A_4	0.667	0.278	0.722	0.667	0.722	0.556	0.722	0.833	0.722	0.611
A_5	0.500	0.667	0.444	0.667	0.556	0.778	0.556	0.611	0.389	0.500

Table 4 Normalized matrix for SAW and WPM methods

	R_{11}	R_{12}	R_{13}	R_{14}	R_{21}	R_{22}	R_{23}	R_{24}	R_{25}	R_{31}
A_i	0.285	0.123	0.065	0.125	0.041	0.060	0.038	0.092	0.053	0.116
A_1	0.715	0.916	0.924	0.916	0.924	0.715	0.924	0.733	0.846	0.693
A_2	0.715	0.666	0.693	1.000	0.770	0.928	0.924	0.733	0.924	0.693
A_3	1.000	0.499	0.846	1.000	0.615	0.715	0.539	0.600	0.924	1.000
A_4	0.857	0.417	1.000	1.000	1.000	0.715	1.000	1.000	1.000	0.846
A_5	0.643	1.000	0.615	1.000	0.770	1.000	0.770	0.733	0.539	0.693

Table 5 Normalized matrix for TOPSIS and modified TOPSIS method

	R_{11}	R_{12}	R_{13}	R_{14}	R_{21}	R_{22}	R_{23}	R_{24}	R_{25}	R_{31}
A_i	0.285	0.123	0.065	0.125	0.041	0.060	0.038	0.092	0.053	0.116
A_1	0.401	0.557	0.499	0.416	0.500	0.388	0.488	0.425	0.439	0.390
A_2	0.401	0.405	0.374	0.455	0.417	0.504	0.488	0.425	0.479	0.390
A_3	0.562	0.304	0.457	0.455	0.333	0.388	0.284	0.348	0.479	0.563
A_4	0.482	0.253	0.540	0.455	0.541	0.388	0.528	0.580	0.519	0.476
A_5	0.361	0.608	0.332	0.455	0.417	0.543	0.406	0.425	0.280	0.390

$$
\begin{aligned}
V_1^+ &= 0.1602 \quad V_1^- = 0.1029\\
V_2^+ &= 0.0750 \quad V_2^- = 0.0312\\
V_3^+ &= 0.0352 \quad V_3^- = 0.0217\\
V_4^+ &= 0.0569 \quad V_4^- = 0.0521\\
V_5^+ &= 0.0220 \quad V_5^- = 0.0135\\
V_6^+ &= 0.0324 \quad V_6^- = 0.0232\\
V_7^+ &= 0.0203 \quad V_7^- = 0.0109\\
V_8^+ &= 0.0533 \quad V_8^- = 0.0320\\
V_9^+ &= 0.0276 \quad V_9^- = 0.0148\\
V_{10}^+ &= 0.0651 \quad V_{10}^- = 0.0451
\end{aligned}
$$

Step 7: Determine the separation measures using Eqs. (7) and (8), and these separation measures are:

$$
\begin{aligned}
S_1^+ &= 0.0536 \quad S_1^- = 0.0434\\
S_2^+ &= 0.0589 \quad S_2^- = 0.0282\\
S_3^+ &= 0.0463 \quad S_3^- = 0.0626\\
S_4^+ &= 0.0512 \quad S_4^- = 0.0476\\
S_5^+ &= 0.0654 \quad S_5^- = 0.0459
\end{aligned}
$$

Step 8: The relative closeness of a particular alternative (PDSI) to the ideal solution is calculated using Eq. (9). The relative closeness of a particular alternative (PDSI) to the ideal solution, P_i is shown in Table 11.
Step 9: Finally, the preference order is ranked in descending order of the P_i values as shown in Table 11.

5.4 *Modified TOPSIS Method*

The Modified TOPSIS method is then applied. The weighted Euclidean distances calculated using Eqs. (12) and (13) are shown below:

$$
\begin{aligned}
D_1^+ &= 0.1245 \quad D_1^- = 0.1352\\
D_2^+ &= 0.1440 \quad D_2^- = 0.0945\\
D_3^+ &= 0.1495 \quad D_3^- = 0.1364\\
D_4^+ &= 0.1401 \quad D_4^- = 0.1416\\
D_5^+ &= 0.1555 \quad D_5^- = 0.1361
\end{aligned}
$$

The relative closeness of particular alternative (PDSI) to the ideal solution, P_i is shown in Table 11 and the ranking is given in descending order of the P_i values.

5.5 Grey Relational Analysis (GRA)

Step 1: Grey relational generating

The values of grey relational generating using Eqs. (15) and (16) are shown in Table 6.

Step 2: Reference sequence definition

All the performance values are scaled into [0, 1] as discussed in Sect. 5.

Step 3: Calculate grey relational coefficient

The grey relational coefficients are calculated using Eq. (17), are given in Table 7.

Step 4: Determine grey relational grade

The grey relational grade, Γ (PDSI) is determined using Eq. (18). The ranking is given in the descending order of the values of grey relational grade as shown in Table 11.

6 PROMETHEE

Step 1: The problem considering ten criteria and five design alternative of personal data assistant is shown in Table 3.
Step 2: A decision matrix including the performance measures of all criteria and of five design alternatives along with their weightage is prepared as shown in Table 3.
Step 3: The decision maker compares the contribution of the alternatives in terms of each separate criterion. For example, the pair wise comparison for the criteria 'A variety of forms (R_{11})' is shown in Table 8.

The leaving flow; $\varphi^{+}(a)$, entering flow; $\varphi^{-}(a)$ and the net flow values; $\varphi(a)$ for different alternatives are calculated using Eqs. (19)–(21) as shown in Table 9. The ranking of the alternatives is arranged in the descending order of the net flow values (i.e. PDSI) as given in Table 11.

6.1 Compromise Ranking Method (VIKOR)

Step 1: The best, i.e., $(y_{ij})_{max}$, and the worst, $(y_{ij})_{min}$, values of all attributes are determined.
Step 2: The values of E_i and K_i are determined using Eqs. (22) and (23) are shown in Table 10.
Step 3: The values of P_i (PDSI) are calculated using Eq. (24) are shown Table 11.
Step 4: Ranking of the alternatives is carried out in the ascending order of the values of Pi.

Table 6 The grey relational generating

	R_{11}	R_{12}	R_{13}	R_{14}	R_{21}	R_{22}	R_{23}	R_{24}	R_{25}	R_{31}
A_i	0.285	0.123	0.065	0.125	0.041	0.060	0.038	0.092	0.053	0.116
A_1	0.201	0.856	0.802	0.000	0.802	0.000	0.835	0.333	0.667	0.000
A_2	0.201	0.427	0.201	1.000	0.403	0.748	0.835	0.333	0.835	0.000
A_3	1.000	0.141	0.601	1.000	0.000	0.000	0.000	0.000	0.835	1.000
A_4	0.601	0.000	1.000	1.000	1.000	0.000	1.000	1.000	1.000	0.500
A_5	0.000	1.000	0.000	1.000	0.403	1.000	0.502	0.333	0.000	0.000

Table 7 The grey relational coefficient

	R_{11}	R_{12}	R_{13}	R_{14}	R_{21}	R_{22}	R_{23}	R_{24}	R_{25}	R_{31}
A_i	0.285	0.123	0.065	0.125	0.041	0.060	0.038	0.092	0.053	0.116
A_1	0.385	0.776	0.716	0.333	0.716	0.333	0.752	0.429	0.600	0.333
A_2	0.385	0.466	0.385	1.000	0.456	0.665	0.752	0.429	0.752	0.333
A_3	1.000	0.368	0.556	1.000	0.333	0.333	0.333	0.333	0.752	1.000
A_4	0.556	0.333	1.000	1.000	1.000	0.333	1.000	1.000	1.000	0.500
A_5	0.333	1.000	0.333	1.000	0.456	1.000	0.501	0.429	0.333	0.333

Table 8 Pair-wise comparison for the criteria 'A variety of forms'

	A_1	A_2	A_3	A_4	A_5
A_1	–	–	0	0	1
A_2	–	–	0	0	1
A_3	1	1	–	1	1
A_4	1	1	0	–	1
A_5	0	0	0	0	–

Table 9 The leaving flow, entering flow and the net flow values

	A_1	A_2	A_3	A_4	A_5	$\varphi^+(a)$	$\varphi^-(a)$	$\varphi(a)$
A_1	–	0.229	0.359	0.123	0.417	1.128	1.940	−0.812
A_2	0.238	–	0.354	0.183	0.442	1.217	1.568	−0.351
A_3	0.579	0.466	–	0.524	0.519	2.088	1.356	0.732
A_4	0.815	0.690	0.289	–	0.690	2.484	1.013	1.471
A_5	0.308	0.183	0.354	0.183	–	1.028	2.068	−1.040

Table 10 The values of E_i and K_i

E_i		K_i	
E_1	0.6522	K_1	0.2277
E_2	0.5819	K_2	0.2277
E_3	0.3714	K_3	0.1059
E_4	0.3547	K_4	0.1233
E_5	0.6239	K_5	0.2851

Table 11 Ranking of personal data assistant

A_i	SAW	WPM	TOPSIS	Modified TOPSIS	GRA	PROMETHEE	VIKOR
A_1	0.800	0.795	0.447	0.521	0.482	−0.812	0.840
A_2	0.775	0.769	0.323	0.396	0.522	−0.351	0.722
A_3	0.835	0.810	0.575	0.477	0.724	0.732	0.028
A_4	0.851	0.826	0.482	0.503	0.692	1.471	0.049
A_5	0.768	0.754	0.412	0.467	0.558	−1.040	0.952
Rank	4-3-1-2-5	4-3-1-2-5	[a]3-4-1-5-2	1-4-3-5-2	3-4-5-2-1	4-3-2-1-5	3-4-2-1-5

[a]As suggested by Lin et al. [12]

The ranking of personal data assistant design alternatives using MADM approaches is shown in Table 11.

7 Results and Discussion

In case of PDA design selection problem; Table 11 shows that product design alternative, A_4 is the most preferred choice among the five alternatives which is suggested by SAW, WPM, and PROMETHEE. However, TOPSIS, GRA and VIKOR method suggest the product design alternative A_3 as the best choice. It indicates three methods are in favor of alternative A_4 and three methods are favoring alternative A_3. As ranking suggested by Modified TOPSIS method provides design alternative, A_1 as best choice which is not the case for all remaining six methods, is ignored. It is also reported in the literature [22, 23] that in case of conflict between the alternatives, the ranking suggested by more number of methods is an appropriate choice. In this particular example, a tie situation occurs among the MADM methods themselves. Considering this as conflict between alternative A_4 and alternative A_3; the qualitative approach is applied to resolve this conflict.

In qualitative approach; the conflicting alternatives are compared individually (i.e. one to one) on the basis of same criteria as already considered for the study and the corresponding performance measures. When applying the qualitative approach to PDA problem; the criteria considered in Table 3 for comparison of the alternatives A_3 and A_4, shows that A_3 is much better than A_4 in the case of three criteria: R_{11}, R_{12} and R_{31}, and A_4 is better than A_3 in the case of other five criteria namely: R_{13}, R_{21}, R_{23}, R_{24} and R_{25}. This type of comparison indicates that A_4 is mostly important than A_3. However, the above comparison is qualitative only and hence may not necessarily provide the correct choice. Hence, quantitative approach for comparison of the alternatives is developed and applied.

In the quantitative approach, the two alternatives are to be compared on the basis of superiority of the performance value of one alternative over the other. The positive value indicates the superiority of an alternative over other whereas negative value indicates inferiority. The overall superiority index (OSI) is then calculated by considering algebraic sum of all the superiority values. If the overall superiority index is positive then the alternative is said to be superior over other. This procedure of comparing the alternatives quantitatively will certainly help decision maker to choose correct alternative, than comparing the alternatives qualitatively. The quantitative approach is then applied to the product design selection problem of PDA. The superiority of A_4 over A_3 is shown in Table 12. The overall superiority index (OSI) of A_4 is positive (i.e. +0.833), and therefore, the alternative A_4 is said to be superior over alternative A_3. The superiority of alternative A_4 over A_3 is shown in Table 12.

From Table 12, it is understood that the overall superiority index (OSI) of A_4 is positive (i.e. +0.833), therefore, alternative A_4 is said to be superior over alternative A_3. From this point of view, it is clear that the ranking suggested by Lin et al. [12] using TOPSIS method as well as qualitative approach is not appropriate. This clearly shows the usefulness of the quantitative approach over the qualitative approach. Using this approach the ranking A_4-A_3-A_1-A_2-A_5 is more logical. Thus, from the customer's requirements point of view, personal data assistant A_4 is the most preferred choice.

Table 12 Superiority of alternative A_4 over alternative A_3

A_i	R_{11}	R_{12}	R_{13}	R_{14}	R_{21}	R_{22}	R_{23}	R_{24}	R_{25}	R_{31}	
A_3	0.778	0.333	0.611	0.667	0.444	0.556	0.389	0.500	0.667	0.722	OSI
A_4	0.667	0.278	0.722	0.667	0.722	0.556	0.722	0.833	0.722	0.611	
Sup.	−0.111	−0.055	0.111	0	0.278	0	0.333	0.333	0.055	−0.111	+0.833

(+) Sign indicates superiority, (−) Sign indicates inferiority, *Sup.* Superiority of A_4 over A_3

8 Conclusion

Organizations develop products to serve the customers with different needs. Product configuration (combinations of features) plays an important role in accepting the product by the customers. This chapter has addressed selection of product design based on the customer preferences to increase the customer satisfaction. Following are the main contributions of this chapter:

- **Multi-Attribute model to select the best product design**: This chapter has demonstrated a systematic multi-attribute decision making methodology for selecting the most suitable product design from the different design alternatives (product configurations).
- **Resolving the conflict in the raking of the alternatives**: A novel quantitative approach has been demonstrated to resolve the conflict in the ranking of the alternatives. This approach is also used to select the best alternative when there is a tie in the ranking obtained as a result of different MADM methods.

References

1. Ulrich, K.T., Eppinger, S.D.: Product Design and Development. Tata McGraw-Hill Publishing, New Delhi (2003)
2. Rao, R.V.: Decision Making in the Manufacturing Environment: Using Graph Theory and Fuzzy Multiple Attribute Decision Making Methods. Springer, London (2007)
3. Fishburn, P.C.: Additive Utilities with Incomplete Product Set: Applications to Priorities and Assignments. Operations Research Society of America, Baltimore (1967)
4. Miller, D.W., Starr, M.K.: Executive Decisions with Operations Research. Prentice Hall, Englewood Cliffs, New Jersey (1969)
5. Hwang, C.L., Yoon, K.: Multiple Attribute Decision Making-Methods and Applications. Springer, Heidelberg (1981)
6. Deng, H., Yeh, C.H., Willis, R.J.: Inter-company comparison using modified TOPSIS with objective weights. Comput. Oper. Res. **27**, 963–973 (2000)
7. Deng, J.: Control problems of grey systems. Syst. Control Lett. **1**, 288–294 (1982)
8. Brans, J.P., Mareschal, B., Vincke, P.: PROMETHEE: a new family of outranking methods in multi-criteria analysis. Proc. Oper. Res. 477–490. Amsterdam, North Holland (1984)
9. Yu, P.L.: A class of solutions for group decision problems. Manag. Sci. **19**, 936–946 (1973)
10. Besharati, B., Azarm, S., Kannan, P.K.: A decision support system for product design selection: a generalized purchase modeling approach. Decis. Support Syst. **42**(1), 333–350 (2006)
11. Isiklar, G., Buyukozkan, G.: Using a multi-criteria decision making approach to evaluate mobile phone alternatives. Comput. Stand. Interfaces **29**(2), 265–274 (2007)
12. Lin, M., Wang, C., Chen, M., Chang, C.A.: Using AHP and TOPSIS approaches in customer-driven product design process. Comput. Ind. **59**, 17–31 (2008)
13. Hsiao, S.W.: Fuzzy logic based decision model for product design. Int. J. Ind. Ergon. **21**, 103–116 (1998)
14. Chien, C.C., Feng, M., Chu, C.M.: Methodology in Industrial Design. Beijing University Press, Beijing (in Chinese) (1993)
15. Vanegas, L.V., Labib, A.W.: A fuzzy quality function deployment (FQFD) model for deriving optimum targets. Int. J. Prod. Res. **39**(1), 99–120 (2001)

16. Saaty, T.L.: The Analytic Hierarchy Process. McGraw-Hill, New York (1980)
17. Lu, M., Madu, C.N., Kuei, C., Winokur, D.: Integrating QFD, AHP, and benchmarking in strategic marketing. J. Bus. Ind. Mark. **9**(1), 41–50 (1994)
18. Kalargeros, N., Gao, J.X.: QFD: focusing on its simplification and easy computerization using fuzzy logic principles. Int. J. Veh. Des. **19**, 315–325 (1998)
19. Kuo, T.C., Chang, S.H., Huang, S.H.: Environmentally conscious design by using fuzzy multi-attribute decision-making. Int. J. Adv. Manuf. Technol. **29**, 209–215 (2006)
20. Fung, R.Y.K., Popplewell, K., Xie, J.: An intelligent hybrid system for customer requirements analysis and product attribute targets determination. Int. J. Prod. Res. **36**(1), 13–34 (1998)
21. Gangurde, S.R., Akarte, M.M.: Ranking of product design alternatives using multi-criteria decision making. In: Tenth International Conference on Operations and Quantitative Management, SIOM, INFOMS, AIMS, pp. 101–111, 28–30 June 2011
22. Kuo, Y., Yang, T., Huang, G.W.: The use of grey relational analysis in solving multiple attribute decision-making problems. Comput. Ind. Eng. **55**, 80–93 (2008)
23. Wu, H.H.: A comparative study of using grey relational analysis in multiple attribute decision making problems. Qual. Eng. **15**, 209–217 (2002)

Advanced Fuzzy Sets and Multicriteria Decision Making on Product Development

Cengiz Kahraman, Fatma Kutlu Gündoğdu, Ali Karaşan and Eda Boltürk

Abstract Product design is typically a critical component in the early stages of a product development process. Product development has received much attention over the last decade. Classical multi criteria decision making methods are used in product design and development in the literature. Fuzzy sets approaches have a significant place among these works. Intuitionistic fuzzy, hesitant fuzzy, and type-2 fuzzy Quality Function Deployment (QFD) have been already developed in the literature. In this chapter, we employ the very recent extension of ordinary fuzzy sets, spherical fuzzy sets, in the house of quality development. For the comparison of competitive firms, spherical fuzzy TOPSIS (SF-TOPSIS) has been used. An illustrative application has also been given.

1 Introduction

In recent years, market dynamics and technological changes have forced firms to be successful in developing new products. This situation is the key to growth for small and medium-sized companies in a competitive environment, while in large-scale companies it has become the most important way of advancing the current presence in the market as well as protecting it.

The steps of product development were described many forms by many researchers, but their basic stages are resembled. Crawford and Di Benedetto [14] introduced the steps of product development as follows:

- Determination of opportunities in the market
- Concept development
- Concept/Project evaluation

C. Kahraman (✉) · F. K. Gündoğdu · A. Karaşan · E. Boltürk
Industrial Engineering Department, Istanbul Technical University, 34367 Besiktas, Istanbul, Turkey
e-mail: kahramanc@itu.edu.tr

A. Karaşan
Graduate School of Science and Engineering, Yildiz Technical University, 34220 Davutpasa, Istanbul, Turkey

C. Kahraman and S. Cebi (eds.), *Customer Oriented Product Design*, Studies in Systems, Decision and Control 279, https://doi.org/10.1007/978-3-030-42188-5_15

- Product development
- Product presentation.

Haines [20] introduced the new product development stages as follows:

- New product planning
  - Concept
  - Compliance
  - Identification.
- Startup of the new product
  - Development
  - Market presentation.

Since the development process is composed of multi stage and in every stage there are many criteria should be taken into consideration by the experts, this process can be analyzed in the context of multi criteria decision making (MCDM) problem. The first step in the MCDM problems is to determine the set of criteria and the set of alternatives based on the available data and the problem environment. As a next step, selection of the appropriate method for the application is another important issue. Since the aim of this step is to assess the data accurately, it becomes the most important step of the evaluation process. But, in the most of the MCDM problems conducting both qualitative and quantitative data is not fully possible. Therefore, some extensions are required for the data to be handled appropriately.

One of the best efficient tools for MCDM problems, fuzzy logic, is a way of handling uncertainty by representing it with the degree of the membership function was introduced by Zadeh [50]. It is applied to obtain efficient solutions by the researchers for the areas of engineering, mathematics, computer sciences, medical sciences, business and economics, social sciences, human behaviors, urban planning. In order to increase this efficiency, the fuzzy sets are extended in many types in light of fuzzy logic: Type-2 fuzzy sets [51]; intuitionistic fuzzy sets [4], neutrosophic fuzzy sets [37], hesitant fuzzy sets [41], Pythagorean fuzzy sets [48], and orthopair fuzzy sets [13], spherical fuzzy sets [29].

The number of published papers MCDM methods with neutrosophic fuzzy sets, Pythagorean fuzzy sets and spherical fuzzy sets are increasing in recent years. In this chapter, the main contribution is to introduce the used multi criteria decision making methods with the extension of fuzzy sets and to propose spherical fuzzy QFD method as a new MCDM method.

The rest of this chapter is organized as follows. In Sect. 2, advanced fuzzy sets are detailed and intuitionistic fuzzy set, Pythagorean fuzzy sets, neutrosophic set and spherical fuzzy sets are introduced. In Sect. 3, the literature review of classical MCDM approaches on Product development is given. In Sect. 4, the MCDM approaches using advanced fuzzy sets on Product development is presented. An application on Product development using SFS is given in Sect. 5. The conclusion about this study is given in Sect. 6 with future suggested works.

2 Advanced Fuzzy Sets

Fuzzy sets have been very popular in almost all branches of science since it has emerged in 1965 [50]. Researchers have introduced many extensions of ordinary fuzzy sets in the literature. It starts from ordinary fuzzy sets and extends to recently developed types of fuzzy sets as shown in Fig. 1.

2.1 *Intuitionistic Fuzzy Sets*

Intuitionistic fuzzy set (IFS) theory is one of the significant extension of the ordinary fuzzy set (FS) theory to deal with the vagueness in the data, which utilizes a membership degree and a non-membership degree, whose sum is less than or equal to 1. Let a set U be a universe of discourse. An IFS $\tilde{A}$ is an object having the form, $\tilde{A} = \{\langle u, (\mu_{\tilde{A}}(u), v_{\tilde{A}}(u)) | u \in U\}$ where the function $\mu_{\tilde{A}}(u) : U \to [0, 1], v_{\tilde{A}}(u) : U \to [0, 1]$ and $0 \leq \mu_{\tilde{A}}(u) + v_{\tilde{A}}(u) \leq 1$ are the degree of membership, non-membership of u to $\tilde{A}$, respectively. For any IFS $\tilde{A}$ and $u \in U$, $\pi_{\tilde{A}}(u) = 1 - \mu_{\tilde{A}}(u) - v_{\tilde{A}}(u)$ is called degree of indeterminacy of u to $\tilde{A}$.

In order to present a larger preference area to decision makers (DMs), Yager [48] proposed a novel concept called Pythagorean Fuzzy Sets (PFS).

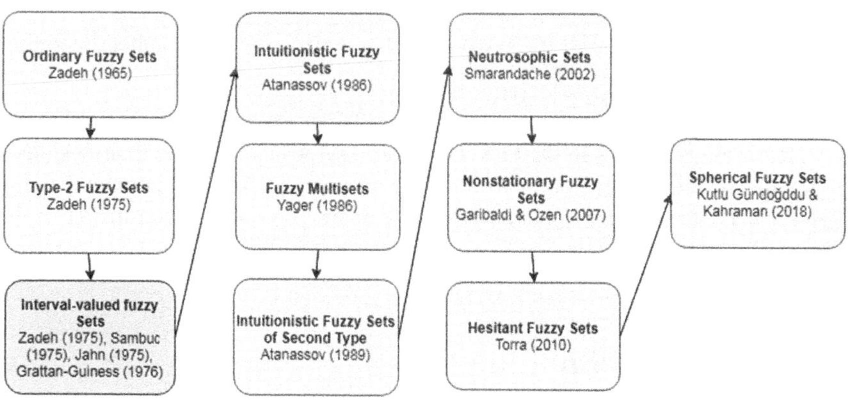

Fig. 1 Extension of fuzzy sets

2.2 Pythagorean Fuzzy Sets

In the real life, the decision makers might express their preferences about membership degrees and non-membership degrees of an alternative with respect to a criterion dissatisfies the condition that the sum of the membership and non-membership degrees should be less than or equal to 1. For instance, an expert may provide their preference about the degree of an alternative on a criterion in terms of membership is 0.5, non-membership degree may be 0.6. It is clearly understood that sum of 0.5 and 0.6 is not less than 1; hence, IFS could not describe this situation. Instead of requiring the decision makers to change their preference information, Pythagorean fuzzy sets (PFS) developed by Yager [48], which had been called as Intuitionistic type-2 fuzzy sets (IFS2) by Atanassov previously [3], are characterized by a membership degree and a non-membership degree satisfying the condition that their squared sum is at most equal to one, which is a generalization of Intuitionistic Fuzzy Sets (IFS) [48]. This concept provides a larger preference area for decision makers. In other words, all the intuitionistic fuzzy degrees are a part of the Pythagorean fuzzy Sets, which shows that the PFS is more powerful to handle the uncertain problems.

Definition of PFS is given as follows:

Let a set U be a universe of discourse. A PFS $\tilde{P}$ is an object having the form,

$\tilde{P} = \{\langle u, (\mu_{\tilde{P}}(u), \nu_{\tilde{P}}(u))\big| u \in U\}$ where the function $\mu_{\tilde{P}}(u) : U \to [0, 1]$, $\nu_{\tilde{P}}(u) : U \to [0, 1]$ and

$0 \leq \mu^2_{\tilde{P}}(u) + \nu^2_{\tilde{P}}(u) \leq 1$ are the degree of membership, non-membership of u to P, respectively.

For any PFS $\tilde{A}$ and $u \in U$, $\pi_{\tilde{P}}(u) = \sqrt{1 - \mu^2_{\tilde{P}}(u) - \nu^2_{\tilde{P}}(u)}$ is called degree of hesitancy of u to $\tilde{P}$.

As seen as from the above formula, hesitancy degree are based on membership and non-membership degrees in the PFS. However, DM wants to define their hesitancy degree like a third dimension as irrevelant from membership and nonmembership degrees. It is not allowed to define hesitancy degree by the decision-maker in the PFS.

2.3 Neutrosophic Sets

Similar to IFS and PFS, neutrosophic sets (NS) are represented by the three dimensions: a truthiness degree, an indeterminacy degree, and a falsity degree [38]. NS do not only deal with the hesitancy of the system but also decrease indecisiveness of inconsistent information. Thus, the truthiness, falsity and indeterminacy values can be independently assigned [38].

Let U be a universe of discourse. Neutrosophic set $\tilde{A}$ in U is an object having the form, $\tilde{A} = \{\langle u, (T_{\tilde{A}}(u), I_{\tilde{A}}(u), F_{\tilde{A}}(u))\big| u \in U\}$ where, $T_{\tilde{A}}$ is the truth-membership function, $I_{\tilde{A}}$ is the indeterminacy-membership function and $F_{\tilde{A}}$ is the falsity-membership function. There is no restriction on their sum and so $0 \leq T_{\tilde{A}}(u) + I_{\tilde{A}}(u) + F_{\tilde{A}}(u) \leq 3$.

In the following, we introduce a last extension of fuzzy sets that are Spherical Fuzzy Sets (SFS), which provides a larger preference domain for decision makers. DMs can also define their hesitancy information about of an alternative with respect to a criterion independently.

2.4 *Spherical Fuzzy Sets*

Intuitionistic and Pythagorean fuzzy membership functions are composed of membership, non-membership and hesitancy parameters, which can be calculated by $\pi_{\tilde{I}} = 1 - \mu - \vartheta$ or $\pi_{\tilde{p}} = (1 - \mu_{\tilde{p}}^2(u) - \nu_{\tilde{p}}^2(u))^{1/2}$, respectively. Neutrosophic membership functions are also defined by three parameters *truthiness, falsity* and *indeterminacy,* whose sum can be between 0 and 3, and the value of each is between 0 and 1 independently. In spherical fuzzy sets, while the squared sum of *membership*, non-membership and hesitancy parameters can be between 0 and 1, each of them can be defined between 0 and 1 independently to satisfy that their squared sum is at most equal to 1. Figure 2 illustrates the differences among IFS, PFS, NS and SFS. For example, an expert may provide their preference about the degree of an alternative with respect to a criterion in terms of membership is 0.5, non-membership degree may be 0.6. It is clearly understood that sum of 0.5 and 0.6 is not less than 1; hence, IFS could not describe this situation while PFS (Pythagorean Fuzzy Sets) can describe it. Third dimension of PFSs is hesitancy degree as 0.6245, which can be calculated by using $\pi_{\tilde{p}}(u) = \sqrt{1 - \mu_{\tilde{p}}^2(u) - \nu_{\tilde{p}}^2(u)}$. In SFS, Decision Makers should define hesitancy just like other dimensions that are membership and non-membership. For instance, a decision maker may express their precedence about the degree of an alternative with respect to a criterion in terms of membership is 0.5, non-membership degree may be 0.6 and hesitancy degree may be 0.2. SFSs are characterized by a membership degree, a non-membership degree, a hesitancy degree satisfying the condition that the square sum of its membership degree and non-membership degree is equal to

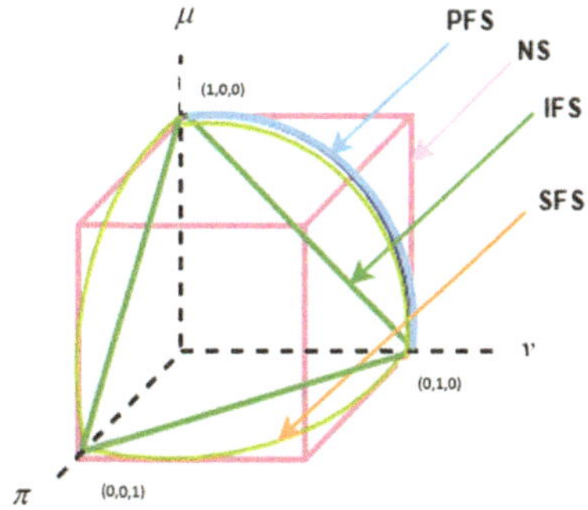

Fig. 2 Geometric representations of IFS, PFS, NS and SFS

or less than one, which is a generalization of Pythagorean Fuzzy Sets (PFS) Kutlu Gündoğdu and Kahraman [29].

A Spherical Fuzzy Sets (SFS) $\tilde{A}_S$ of the universe of discourse U is given by;

$$\tilde{A}_S = \left\{\langle u, (\mu_{\tilde{A}_S}(u), \nu_{\tilde{A}_S}(u), \pi_{\tilde{A}_S}(u)) \middle| u \in U\right\} \tag{1}$$

$$\tilde{A}_S = \left\{\langle u, (\mu_{\tilde{A}_S}(u), \nu_{\tilde{A}_S}(u), \pi_{\tilde{A}_S}(u)) \middle| u \in U\right\}$$

where

$$\mu_{\tilde{A}_S}(u) : U \to [0, 1], \quad \nu_{\tilde{A}_S}(u) : U \to [0, 1], \quad \pi_{\tilde{A}_S}(u) : U \to [0, 1]$$

and

$$0 \leq \mu^2_{\tilde{A}_S}(u) + \nu^2_{\tilde{A}_S}(u) + \pi^2_{\tilde{A}_S}(u) \leq 1 \quad \forall u \in U \tag{2}$$

For each u, the numbers $\mu_{\tilde{A}_S}(u)$, $\nu_{\tilde{A}_S}(u)$ and $\pi_{\tilde{A}_S}(u)$ are the degree of *membership, non-membership* and *hesitancy* of u to $\tilde{A}_S$, respectively [29].

3 Classical MCDM Approaches on Product Development

Product development papers are issued with MCDM methods. Ayag [5] used analytic hierarchy process (AHP) method in conceptual design for a new product development. Honkala et al. [22] used MCDM methods in concept selection for new products. Chatterjee et al. [11] used Vlse Kriterijumska Optimizacija Kompromisno Resenje (VIKOR) and ELimination and Et Choice Translating REality (ELECTRE) methods in products development and design. Kuan et al. [28] applied DEMATEL approach to find the relevance of the quality factors of new product development process. Chang [10] integrated the analytic network process (ANP) and a technique for order preference by similarity to the ideal solution (TOPSIS) to help Taiwanese managers in century-old food industry companies make better decisions for new product development project selection. Rosli et al. [34] used a combined an AHP-TRIZ method for automotive door panel design. Vinodh et al. [43] used VIKOR method for sustainable concept selection. Jou and Yuan [26] used of Decision making trial and evaluation laboratory (DEMATEL) and ANP in order to analyse dependence and feedback among market, product, and technology and find the criteria weights. Buyukozkan and Görener [8] assessed partners in product development with AHP and VIKOR methods. [25] used comprising modified Delphi method, DEMATEL method, and an ANP for project portfolio-selection problem. Bouzon et al. [7] used Interpretive Structural Modeling in reverse logistics. Favi e al. [16] applied MCDM method in order to evaluate of the best design option product development. Dursun and Arslan [15] used 2-tuple fuzzy linguistic MCDM model, linguistic hierarchies and QFD in order to decide significance of selection criteria in material selection. Ayağ [6] applied the TOPSIS and the ANP methods in new product development. Sanyang et al. [36] used MCDM for alleviating material selection problems. Favi et al. [17] presented a model with assemble, manufacturing, cost and material selection, MCDM with multi-objective criteria. Yousefi et al. [49] studied critical success factors in product development Iranian pharmaceutical industry in order to prioritize them using the methodology with AHP method. Wang et al. [44] used a strategic innovation included design structure matrix, ANP and the multi-dimensional game theory are introduced with an attempt to find the most suitable customized new style in product development. Hellebrandt et al. [21] introduced a methodology with the ANP approach in order to identify the best knowledge management solutions for complaint knowledge transfer to product development. Lo and Liou [31] introduced a model that uses multiple-criteria decision-making in combination with grey theory for FMEA in order to a solution for product development. Wang et al. [45] used an integrated MCDM method based on regret theory and TODIM (an acronym in Portuguese for interactive MCDM) is developed to prioritize failure modes taking experts' psychological behaviors into account in Failure mode and effects analysis (FMEA) with an application in product development.

4 MCDM Approaches Using Advanced Fuzzy Sets on Product Development

In order to reach the appropriate number of samples, we have emphasized the two main features of the production development articles available in the literature; the number of citations and been published in the recent years. To identify the articles, the most comprehensive academic search engines, SCOPUS and Science Direct, are used. The pursuit of the related articles on our study is based on keywords-driven search. We selected the articles mainly cited prominently by using keywords-driven search. The used search pattern consists the below keywords:

Product Development AND *MCDM*

After search results, we also used some proxies of MCDM which are:

TOPSIS OR *AHP* OR *multi criteria*

After the analysis, we introduce brief information about the studies which is given in Table 1 as follows:

Özdemir et al. [33] applied an integrated fuzzy goal programming and TOPSIS methodology for the product development in bakery production. Chyu and Fang[12]

Table 1 Literature review of the studies based on product development analysis using fuzzy approaches

Author and Year	Methodology	Subject
[32]	AHP and Fuzzy TOPSIS	Green Product Development for Small and Medium Enterprises
[40]	Neutrosophic TOPSIS-based QUALIFLEX	Green Product Development
[1]	Fuzzy Shannon's entropy and Modified TOPSIS	Lean product development for Automotive Company
[42]	Fuzzy TOPSIS	Sustainable product development
[9]	Intuitionistic AHP and Intuitionistic TOPSIS	Product development partner selection
[2]	Fuzzy reasoning approach and Fuzzy-AHP and Fuzzy-TOPSIS	Lean product development
[46]	Fuzzy Extent Analysis and Fuzzy TOPSIS	Green Product Development
[39]	Fuzzy MCDM methods	Sustainable product development
[12]	Fuzzy Kano method and fuzzy DEMATEL and fuzzy ANP	Product development
[33]	Fuzzy Goal Programming and Fuzzy TOPSIS	Product development in Bakery
[30]	Case-Based Reasoning and fuzzy AHP	Product development in Chemical industry
[27]	Evaluation Grid Method and fuzzy QFD and fuzzy Kano model and fuzzy AHP	Product development for minicars
[23]	QFD and heuristics-AHP and ANN and fuzzy logic	Product development for miniature circuit breaker

developed a hybrid fuzzy ANP approach for the new product development selection problem. Soota [39] applied an integrated fuzzy MCDM method for the sustainable product development. Wang et al. [46] applied an integrated Fuzzy Extent Analysis and Fuzzy TOPSIS for the assessment of environmental performance with respect to different product designs. Aikhuele and Turan [2] studied a hybrid fuzzy model consist of Fuzzy Reasoning Approach, and the extension of Fuzzy-AHP and Fuzzy-TOPSIS to evaluate lean product development performance measurement. Büyüközkan and Güleryüz [9] applied an integrated intuitionistic fuzzy AHP and TOPSIS methodology for the product development partner selection. Turan et al. [42] studied criteria assessment model by using fuzzy TOPSIS for the sustainable product development. Aikhuele [1] introduced an integrated approach consist of Fuzzy Shannon's entropy and Modified TOPSIS for the development of lean products for an automotive related company. Hundal and Kant [23] applied an integrated QFD approach with heuristics-AHP, ANN and fuzzy logic for product design development of miniature circuit breaker. Tian et al. [40] studied Green Product Development by using an integrated neutrosophic TOPSIS based QUALIFLEX. Oliveira et al. [32] developed an integrated AHP and fuzzy TOPSIS methodology to evaluate new product development focused on small and medium enterprises. Kang et al. [27] applied an integrated Evaluation Grid Method, fuzzy QFD together with fuzzy Kano model combined with fuzzy AHP for the development of new products. Lee [30] proposed a knowledge based system by using Case-Based Reasoning with fuzzy AHP to support chemical product development.

5 Application: Product Development Using SFS

The proposed spherical fuzzy QFD method is composed of several steps as given as follows:

Step 1. Define linguistic customer requirements (CRs) and constitute pairwise comparison matrices using spherical fuzzy judgment matrices based on the linguistic terms given in Table 2.

Equations (3) and (4) are used to obtain the score indices in Table 2.

$$SI = \sqrt{\left|100 * \left[\left(\mu_{\tilde{A}_s} - \pi_{\tilde{A}_s}\right)^2 - \left(v_{\tilde{A}_s} - \pi_{\tilde{A}_s}\right)^2\right]\right|} \tag{3}$$

for *AMI, VHI, HI, SMI*, and *EI*.

Table 2 Linguistic and corresponding spherical scale

Linguistic terms	(μ, v, π)	Score Index (SI)
Absolutely More Importance (AMI), Absolutely More Relation (AMR), Absolutely More Satisfactory (AMR)	(0.9, 0.1, 0.1)	9
Very High Importance (VHI), Very High Relation (VHR), Very High Satisfactory (VHR)	(0.8, 0.2, 0.2)	7
High Importance (HI), High Relation (HR), High Satisfactory (HS)	(0.7, 0.3, 0.3)	5
Slightly More Importance (SMI), Slightly More Relation (SMR), Slightly More Satisfactory (SMS)	(0.6, 0.4, 0.4)	3
Equally Importance (EI), Medium Relation (MR), Medium Satisfactory (MS)	(0.5, 0.5, 0.5)	1
Slightly Low Importance (SLI), Slightly Low Relation (SLR), Slightly Low Satisfactory (SLS)	(0.4, 0.6, 0.4)	1/3
Low Importance (LI), Low Relation (LR), Low Satisfactory (LS)	(0.3, 0.7, 0.3)	1/5
Very Low Importance (VLI), Very Low Relation (VLR), Very Low Satisfactory (VLS)	(0.2, 0.8, 0.2)	1/7
Absolutely Low Importance (ALI), Absolutely Low Relation (ALR), Absolutely Low Satisfactory (ALS)	(0.1, 0.9, 0.1)	1/9

$$\frac{1}{SI} = \frac{1}{\sqrt{\left|100 * \left[\left(\mu_{\tilde{A}_s} - \pi_{\tilde{A}_s}\right)^2 - \left(v_{\tilde{A}_s} - \pi_{\tilde{A}_s}\right)^2\right]\right|}} \tag{4}$$

for *EI, SLI, LI, VLI*, and *ALI*.

Step 2. Check for the consistency of each pairwise comparison matrix. To do that, convert the linguistic terms in the pairwise comparison matrix to their corresponding score indices. Then, apply the classical consistency check. The threshold of the CR is 10%. If the crisp pairwise comparison matrix is consistent, we continue with its corresponding spherical fuzzy pairwise comparison matrix. For instance, the pairwise comparison matrix $J = \begin{array}{c|ccc|} C_1 & EI & SLI & SMI \\ C_2 & SMI & EI & HI \\ C_3 & SLI & LI & EI \end{array}$ is converted to $J = \begin{array}{c|ccc|} C_1 & 1 & 1/3 & 3 \\ C_2 & 3 & 1 & 5 \\ C_3 & 1/3 & 1/5 & 1 \end{array}$ and the consistency ratio is calculated by using the classical way and found to be 0.03457, which indicates that the pairwise comparison matrix is consistent.

Step 3. Calculate the spherical fuzzy global weight of each customer requirement and normalize them.

Determine the importance evaluation (*IE*) of each customer requirement using SWAM operator given in Eq. (5).

$$\begin{aligned} SWAM_w(\tilde{A}_{S1}, \ldots, \tilde{A}_{Sn}) &= w_1\tilde{A}_{S1} + w_2\tilde{A}_{S2} + \cdots + w_n\tilde{A}_{Sn} \\ &= \Bigg\langle [1 - \prod_{i=1}^{n}(1 - \mu^2_{\tilde{A}_{Si}})^{w_i}]^{1/2}, \prod_{i=1}^{n} v^{w_i}_{\tilde{A}_{Si}}, [\prod_{i=1}^{n}(1 - \mu^2_{\tilde{A}_{Si}})^{w_i} \\ &\quad - \prod_{i=1}^{n}(1 - \mu^2_{\tilde{A}_{Si}} - \pi^2_{\tilde{A}_{Si}})^{w_i}]^{1/2} \Bigg\rangle \end{aligned} \tag{5}$$

where $w = 1/n$.

CRs are rated by three decision makers as in Table 3 by using the scale in Table 2.

Step 4. Define the design requirements (DRs), and determine the direction of improvement of DRs (Fig. 3). Then, fill in the correlation matrix among DRs as given in Fig. 1 by using linguistic scale as in Table 2. Positive correlation (bluc color) and negative correlation (red color) are represented with *pozc* and *negc*, respectively. If there is no correlation, the cell is left empty which means change in one DR does not affect the other DR. In Fig. 1, each cell shows three evaluations from three experts. We aggregate these linguistic terms for getting aggregated positive and negative correlations of each design requirement based on SWAM operator as given in Table 4.

Step 5. Aggregate all positive and negative correlations separately by using Eq. (5). Defuzzify the aggregate of positive and negative correlations separately based on Eq. (6). Finally, CIF_j can be obtained from Eq. (7) as given in Table 4. In our example, blue and pink colors denote the aggregated positive and negative correlations of DRs, respectively.

Table 3 Customer requirements' importance ratings

DMs	Customer Requirements (CR)	CR1	CR2	CR3	CR4	Importance evaluations
DM1	CR1	EI	SMI	VLI	SLI	(0.70, 0.18, 0.36)
DM2		EI	SLI	ALI	ALI	
DM3		EI	HI	AMI	SMI	
DM1	CR2	SLI	EI	ALI	LI	(0.63, 0.22, 0.40)
DM2		SMI	EI	SLI	LI	
DM3		LI	EI	SMI	LI	
DM1	CR3	VHI	AMI	EI	HI	(0.93, 0.03, 0.19)
DM2		AMI	SMI	EI	SMI	
DM3		ALI	SLI	EI	VLI	
DM1	CR4	SMI	HI	LI	EI	(0.88, 0.06, 0.25)
DM2		AMI	HI	SLI	EI	
DM3		SLI	HI	VHI	EI	

$$defuzz\left(\tilde{A}_s\right) = \left(2\mu_{\tilde{A}_s}\right)^2 - v_{\tilde{A}_s}^2 - \pi_{\tilde{A}_s}^2/2 \tag{6}$$

$$CIF_{DR_j} = \beta + \left(poz\bar{c}_{DR_j} - neg\bar{c}_{DR_j}\right) * \left(n_j/(m-1)\right) \quad j = 1, 2, \ldots, m \tag{7}$$

where n_j is the number of correlations in one DR and β value can be used as the smallest integer to make CIF_{DR_j} value positive. In our chapter, β value is 1.0.

Step 6. Construct the relationship matrix (R) between DRs and CRs as given in Fig. 4. Aggregate the linguistic terms of each relationship between DRs and CRs using SWAM operator given in Eq. (5).

Step 7. To obtain the absolute importance $\left(A\tilde{I}_j\right)$ for each DR, multiply the aggregated linguistic importance evaluations $\left(I\tilde{E}\right)$ of CRs by the aggregated linguistic terms in the relationship matrix (R) by using Eq. (8). Then multiply the result by the aggregated correlation impact factor $\left(CIF_j\right)$ based on Eq. (9) as in Fig. 5.

$$\tilde{A}_S \otimes \tilde{B}_S = \left\{ \begin{array}{c} \mu_{\tilde{A}_S}\mu_{\tilde{B}_S}, \left(v_{\tilde{A}_S}^2 + v_{\tilde{B}_S}^2 - v_{\tilde{A}_S}^2 v_{\tilde{B}_S}^2\right)^{1/2}, \\ \left(\left(1 - v_{\tilde{B}_S}^2\right)\pi_{\tilde{A}_S}^2 + \left(1 - v_{\tilde{A}_S}^2\right)\pi_{\tilde{B}_S}^2 - \pi_{\tilde{A}_S}^2\pi_{\tilde{B}_S}^2\right)^{1/2} \end{array} \right\} \tag{8}$$

$$\lambda \cdot \tilde{A}_S = \left\{ \left(1 - \left(1 - \mu_{\tilde{A}_S}^2\right)^{\lambda}\right)^{1/2}, v_{\tilde{A}_S}^{\lambda}, \left(\left(1 - \mu_{\tilde{A}_S}^2\right)^{\lambda} - \left(1 - \mu_{\tilde{A}_S}^2 - \pi_{\tilde{A}_S}^2\right)^{\lambda}\right)^{1/2} \right\} \tag{9}$$

$$A\tilde{I}_j = \left\{ \left(\oplus_{i=1}^{n} I\tilde{E}_i \otimes \tilde{R}_{ij}\right) \otimes CIF_j \right\} \tag{10}$$

where $i = 1, 2, \ldots, n$ and $j = 1, 2, \ldots, m$

Table 4 Correlation impact factor calculations

Design requirements (DR_j)	Aggregated correlations of each DR with other DRs (μ, v, π)	Aggregated of positive and negative correlations separately (μ, v, π)	Defuzzified aggregate of positive and negative correlations	$poz\overline{c}_j - neg\overline{c}_j$	CIF_j
DR1	(0.82,0.18,0.19)				
	(0.47, 0.53, 0.47)				
	(0.82, 0.18, 0.19)	(0.75,0.26,0.28)	2.148	2.148	**2.289**
DR2	(0.57, 0.43, 0.43)				
	(0.47, 0.53, 0.47)	(0.52,0.48, 0.45)	0.768		
	(0.71, 0.29, 0.30)	(0.71, 0.29, 0.30)	1.913	1.145	**1.687**
DR3	(0.82, 0.18, 0.19)				
	(0.57, 0.43, 0.43)	(0.57, 0.43, 0.43)	1.021		
	(0.74, 0.26, 0.27)				
	(0.38, 0.63, 0.40)	(0.71, 0.31, 0.28)	1.857	0.835	**1.669**
DR4	(0.47, 0.53, 0.47)				
	(0.74, 0.26, 0.27)				
	(0.38, 0.63, 0.40)	(0.57, 0.44, 0.38)	1.049		
	(0.38, 0.63, 0.40)	(0.38, 0.63, 0.40)	0.117	0.933	**1.746**
DR5	(0.82, 0.18, 0.19)				
	(0.47, 0.53, 0.47)				
	(0.38, 0.63, 0.40)				
	(0.38, 0.63, 0.40)	(0.62, 0.41, 0.32)	1.300		
	(0.74, 0.26, 0.27)	(0.64, 0.37, 0.37)	1.415	–0.115	**0.885**
DR6	(0.71, 0.29, 0.30)	(0.71, 0.29, 0.30)	1.913		
	(0.38, 0.63, 0.40)				
	(0.74, 0.26, 0.27)	(0.61, 0.40, 0.33)	1.293	0.619	**1.372**

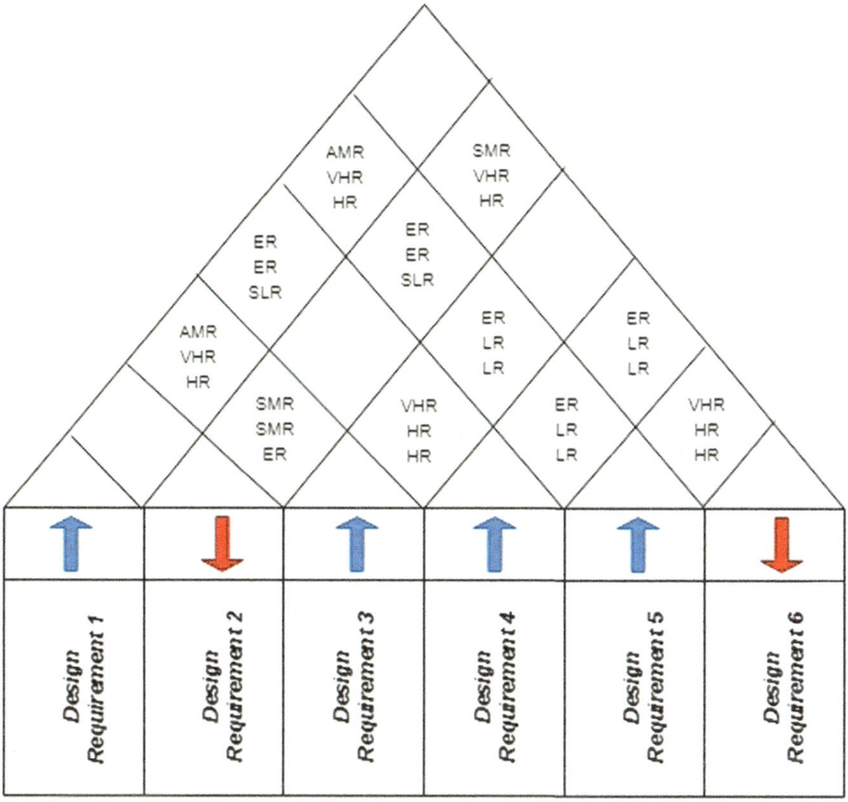

Fig. 3 DRs and their directions of improvement

Customer Requirements (WHATS) / HOWS	Design Requirement 1	Design Requirement 2	Design Requirement 3	Design Requirement 4	Design Requirement 5	Design Requirement 6
Customer Requirement 1	AMR.VHR. HR	HR, HR, HR	MR, VHR, AMR			
Customer Requirement 2				VHR, AMR, MR	VHR, HR, HR	MR, MR, SLR
Customer Requirement 3				VLR, SMR, SMR	SMR, HR, VHR	HR, SMR, VLR
Customer Requirement 4			AHR, VHR, HR			

Fig. 4 The relationship matrix between DRs and CRs

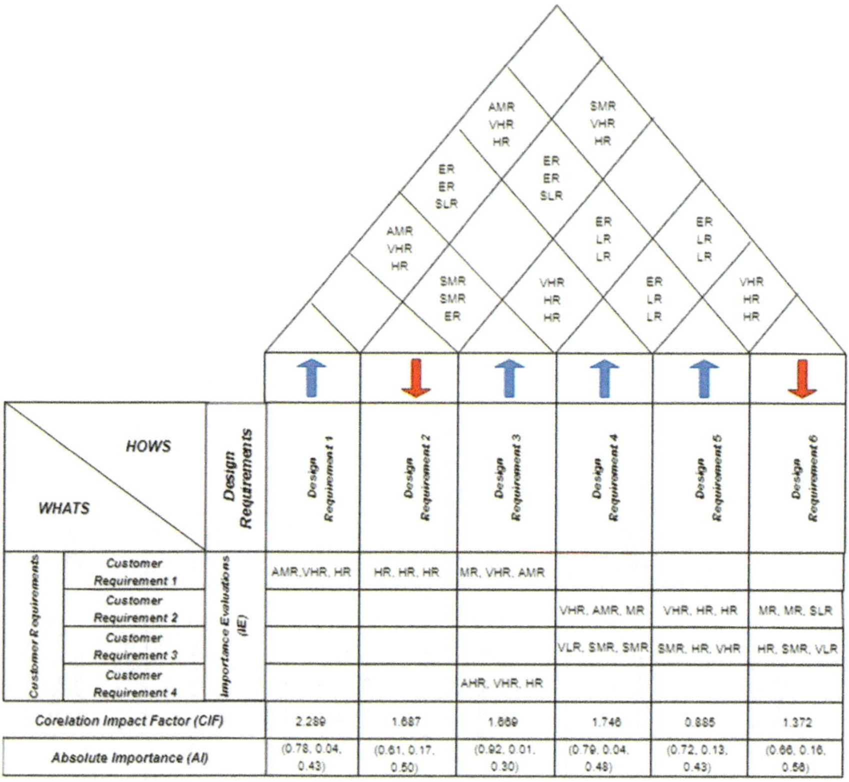

Fig. 5 Absolute importance spherical fuzzy values

Step 8. Determine the linguistic customer ratings of the alternatives by using Spherical fuzzy scale given in Table 2. Table 5 indicates that the linguistic customer ratings are assigned by three decision makers.

Step 9. Aggregate the judgments of each decision maker (DM) using Spherical Weighted Arithmetic Mean (SWAM) that are given in Eq. (5).

Step 10. Construct aggregated spherical fuzzy decision matrix based on the opinions of decision makers.

Denote the evaluation values of each alternative $X_i (i = 1, 2 \ldots ..m)$ with respect to design requirement $A\tilde{I}_j$ by $AI_j(\tilde{X}_i) = (\mu_{ij}, v_{ij}, \pi_{ij})$ and $\tilde{D} = \left(AI_j(\tilde{X}_i)\right)_{m \times n}$ is a spherical fuzzy decision matrix. For a MCDM problem with SFS, decision matrix $\tilde{D} = \left(AI_j(\tilde{X}_i)\right)_{m \times n}$ should be constructed as in Eq. (12).

Table 5 Linguistic customer ratings of the alternatives

Alternatives	DR1	DR2	DR3
Our Company	HS, AMS, VHS	HS, HS, HS	AMS, MS, VHS
Company A	AMS, SLS, SLS	MS, VHS, AMS	HS, VHS, HS
Company B	HS, LS, SLS	SMS, VLS, SMS	VHS, SMS, HS
Company C	LS, HS, VLS	MS, VHS, SLS	LS, VHS, VHS
Alternatives	DR4	DR5	DR6
Our Company	AMS, SMS, SMS	HS, AMS, VHS	HS, HS, HS
Company A	SLS, MS, MS	AMS, SLS, SLS	ES, VHS, AMS
Company B	VLS, HS, SMS	HS, LS, SLS	SMS, VLS, SMS
Company C	HS, SMS, LS	LS, HS, VLS	MS, VHS, SLS

$$\tilde{D} = \left(AI_j(\tilde{X}_i)\right)_{m\times n} = \begin{pmatrix} (\mu_{11}, v_{11}, \pi_{11}) & (\mu_{12}, v_{12}, \pi_{12}) & \ldots & (\mu_{1n}, v_{1n}, \pi_{1n}) \\ (\mu_{21}, v_{21}, \pi_{21}) & (\mu_{22}, v_{22}, \pi_{22}) & \ldots & (\mu_{2n}, v_{2n}, \pi_{2n}) \\ . & . & & . \\ . & . & & . \\ (\mu_{m1}, v_{m1}, \pi_{m1}) & (\mu_{m2}, v_{m2}, \pi_{m2}) & \ldots & (\mu_{mn}, v_{mn}, \pi_{mn}) \end{pmatrix} \tag{12}$$

Step 11. Construct the weighted aggregated spherical fuzzy decision matrix by multiplying absolute importance matrix $\left(A\tilde{I}_j\right)$ and aggregated spherical fuzzy decision matrix $\left(\tilde{D} = \left(AI_j(\tilde{X}_i)\right)_{m\times n}\right)$ as given in Table 6. Then defuzzify the values of weighted aggregated decision matrix by using Eq. (7).

Table 6 Weighted decision matrix

Alternatives	DR1	DR2	DR3	DR4	DR5	DR6
Our Company	(0.64, 0.19, 0.45)	(0.43, 0.34, 0.54)	(0.73, 0.22, 0.38)	(0.60, 0.25, 0.53)	(0.59, 0.22, 0.46)	(0.46, 0.33, 0.59)
Company A	(0.54, 0.33, 0.47)	(0.49, 0.27, 0.53)	(0.68, 0.26, 0.39)	(0.37, 0.53, 0.58)	(0.50, 0.35, 0.47)	(0.52, 0.26, 0.58)
Company B	(0.40, 0.50, 0.48)	(0.32, 0.52, 0.53)	(0.66, 0.29, 0.41)	(0.45, 0.46, 0.52)	(0.37, 0.51, 0.48)	(0.34, 0.52, 0.57)
Company C	(0.38, 0.55, 0.44)	(0.38, 0.42, 0.55)	(0.65, 0.30, 0.36)	(0.45, 0.44, 0.53)	(0.35, 0.56, 0.44)	(0.41, 0.42, 0.59)

$$\tilde{D} = \left(Al_j(\tilde{X}_{iw})\right)_{m \times n} = \begin{pmatrix} (\mu_{11w}, v_{11w}, \pi_{11w}) & (\mu_{12w}, v_{12w}, \pi_{12w}) & \ldots & (\mu_{1nw}, v_{1nw}, \pi_{1nw}) \\ (\mu_{21w}, v_{21w}, \pi_{21w}) & (\mu_{22w}, v_{22w}, \pi_{22w}) & \ldots & (\mu_{2nw}, v_{2nw}, \pi_{2nw}) \\ . & . & & . \\ . & . & & . \\ (\mu_{m1w}, v_{m1w}, \pi_{m1w}) & (\mu_{m2w}, v_{m2w}, \pi_{m2w}) & \ldots & (\mu_{mnw}, v_{mnw}, \pi_{mnw}) \end{pmatrix} \quad (13)$$

Step 12. Determine the Spherical Fuzzy Positive Ideal Solution (SF-PIS) and the Spherical Fuzzy Negative Ideal Solution (SF-NIS), based on the defuzzified values obtained in Step 11.

For the SF-PIS, Eq. (14) is used to find the maximum scores in the decision matrix. Based on the crisp maximum scores, the corresponding SF numbers are determined as in Eq. (15).

$$X_j^* = \left\{ Al_j, \max_i < Score(Al_j(X_i)) > \middle| j = 1, 2 \ldots m \right\} \quad (14)$$

$$\tilde{X}_j^* = \left\{ \langle Al_1, (\mu_1^*, v_1^*, \pi_1^*) \rangle, \langle Al_2, (\mu_2^*, v_2^*, \pi_2^*) \rangle \cdots \langle Al_m, (\mu_m^*, v_m^*, \pi_m^*) \rangle \right\} \quad (15)$$

For the SF-NIS, Eq. (16) is used to find the minimum scores in the decision matrix. Based on the crisp minimum scores, the corresponding SF numbers are determined as in Eq. (17).

$$X_j^- = \left\{ Al_j, \min_i < Score(Al_j(X_i)) > \middle| j = 1, 2 \ldots m \right\} \quad (16)$$

$$\tilde{X}_j^- = \left\{ \langle Al_1, (\mu_1^-, v_1^-, \pi_1^-) \rangle, \langle Al_2, (\mu_2^-, v_2^-, \pi_2^-) \rangle \cdots \langle Al_m, (\mu_m^-, v_m^-, \pi_m^-) \rangle \right\} \quad (17)$$

According to our example, best and worst scores are given in Table 7.

Step 13. Calculate the distances between alternative X_i and the SF-PIS as well as SF-NIS, respectively. Xu and Zhang's distance can be used for this step.

For the SF-NIS:

$$D\left(\tilde{X}_{ij}, \tilde{X}_j^*\right) = \frac{1}{2m} \sum_{j=1}^{m} \left(\left| \mu_{\tilde{X}_{ij}}^2 - \mu_{\tilde{X}_j^*}^2 \right| + \left| v_{\tilde{X}_{ij}}^2 - v_{\tilde{X}_j^*}^2 \right| + \left| \pi_{\tilde{X}_{ij}}^2 - \pi_{\tilde{X}_j^*}^2 \right| \right) \quad (18)$$

For the SF-PIS:

$$D\left(\tilde{X}_{ij}, \tilde{X}_j^-\right) = \frac{1}{2m} \sum_{j=1}^{m} \left(\left| \mu_{\tilde{X}_{ij}}^2 - \mu_{\tilde{X}_j^-}^2 \right| + \left| v_{\tilde{X}_{ij}}^2 - v_{\tilde{X}_j^-}^2 \right| + \left| \pi_{\tilde{X}_{ij}}^2 - \pi_{\tilde{X}_j^-}^2 \right| \right) \quad (19)$$

Step 14. Determine the closeness ratio and the optimal ranking order of the alternatives and identify the optimal alternative based on Eq. (20).

Table 7 SF-PIS and SF-NIS

	DR1	DR2	DR3	DR4	DR5	DR6
$\tilde{X}_j^*$	(0.64, 0.19, 0.45)	(0.49, 0.27, 0.53)	(0.73, 0.22, 0.38)	(0.60, 0.25, 0.53)	(0.59, 0.22, 0.46)	(0.52, 0.26, 0.58)
$\tilde{X}_j^-$	(0.38, 0.55, 0.44)	(0.32, 0.52, 0.53)	(0.66, 0.29, 0.41)	(0.37, 0.53, 0.58)	(0.35, 0.56, 0.44)	(0.34, 0.52, 0.57)

Table 8 Distances to SF-PIS, SF-NIS, and ranking

Alternatives	PIS	NIS	Closeness	Ranking
Our Company	0.0173	0.1862	0.9150	1
Company A	0.0818	0.1255	0.6054	2
Company B	0.1764	0.0350	0.1656	4
Company C	0.1610	0.0508	0.2399	3

$$\xi(X_i) = \frac{D\left(\tilde{X}_{ij}, \tilde{X}_j^-\right)}{D\left(\tilde{X}_{ij}, \tilde{X}_j^-\right) + D\left(\tilde{X}_{ij}, \tilde{X}_j^*\right)} \tag{20}$$

The closeness ratios indicate that the best alternative is Company A and overall ranking is *Our Company* > *Company A* > *Company B* > *Company C* as seen in Table 7 (Table 8).

6 Conclusion

Product development is a tremendous process based on several design requirements, customer requirements and their relationships. This process requires linguistic terms to be used in the house of quality. Linguistic terms can be incorporated into numerical analysis through the fuzzy set theory. IFS, hesitant fuzzy sets, Pythagorean fuzzy sets, spherical fuzzy sets are the extensions fuzzy sets. In this chapter, we employed the latest extension of ordinary fuzzy sets, namely spherical fuzzy sets, in the design phase. Aggregation operators for spherical fuzzy sets took an important place in this process. A spherical fuzzy multi criteria decision analysis has been used to compare our company with the competitors. Spherical fuzzy TOPSIS was successfully employed for this aim.

For further research, we suggest interval-valued spherical fuzzy QFD analysis to be developed. Comparison of competitors for CRs can be also incorporated into the model.

References

1. Aikhuele, D.: Systematic model for lean product development implementation in an automotive related company. Manag. Sci. Lett. **7**(7), 337–350 (2017)
2. Aikhuele, D.O., Turan, F.M.: A hybrid fuzzy model for lean product development peformance measurement. In: IOP Conference Series: Materials Science and Engineering, vol. 114, no. 1, p. 012048. IOP Publishing (2016)
3. Atanassov, K.: Geometrical interpretation of the elements of the intuitionistic fuzzy objects, Preprint IM-MFAIS (1989) 1–89, Sofia. Reprinted: Int. J. Bioautomation **20**(S1):27–42
4. Atanassov, K.T.: Intuitionistic fuzzy sets. Fuzzy Sets Syst. **20**(1), 87–96 (1986)
5. Ayag, Z.: An integrated approach to evaluating conceptual design alternatives in a new product development environment. Int. J. Prod. Res. **43**(4), 687–713 (2005)
6. Ayağ, Z.: An integrated approach to concept evaluation in a new product development. J. Intell. Manuf. **27**(5), 991–1005 (2016)
7. Bouzon, M., Govindan, K., Rodriguez, C.M.T.: Reducing the extraction of minerals: reverse logistics in the machinery manufacturing industry sector in brazil using ISM approach. Resour. Policy **46**, 27–36 (2015)
8. Büyüközkan, G., Görener, A.: Evaluation of product development partners using an integrated AHP-VIKOR model. Kybernetes **44**(2), 220–237 (2015)
9. Büyüközkan, G., Güleryüz, S.: A new integrated intuitionistic fuzzy group decision making approach for product development partner selection. Comput. Ind. Eng. **102**, 383–395 (2016)
10. Chang, K.: Combined MCDM approaches for century-old taiwanese food firm new product development project selection. Br. Food J. **115**(8), 1197–1210 (2013)
11. Chatterjee, P., Athawale, V.M., Chakraborty, S.: Selection of materials using compromise ranking and outranking methods. Mater. Des. **30**(10), 4043–4053 (2009)
12. Chyu, C.C., Fang, Y.C.: A hybrid fuzzy analytic network process approach to the new product development selection problem. Math. Probl. Eng. (2014)
13. Ciucci, D.: Orthopairs and granular computing. Granul. Comput. **1**(3), 159–170 (2016)
14. Crawford, M., Di Benedetto, A.: New Products Management, 9th edn. McGraw-Hill International Edition, New York (2008)
15. Dursun, M, Arslan, Ö.: A fuzzy MCDM approach to determine critical evaluation criteria in washing liquid material selection procedure. In: Paper Presented at the Lecture Notes in Engineering and Computer Science, vol. 2223, pp. 73–77
16. Favi, C., Germani, M., Mandolini, M.: A multi-objective design approach to include material, manufacturing and assembly costs in the early design phase. Pap. Present. Procedia CIRP. **52**, 251–256 (2016)
17. Favi, C., Germani, M., Mandolini, M.: Multi-objective conceptual design: an approach to make cost-efficient the design for manufacturing and assembly in the development of complex products (2017). https://doi.org/10.1007/978-3-319-45781-9_7
18. Garibaldi, J.M., Ozen, T.: Uncertain fuzzy reasoning: A case study in modelling expert decision making. IEEE Trans. Fuzzy Syst. **15**(1), 16–30 (2007)
19. Grattan-Guinness, I.: Fuzzy membership mapped onto intervals and many-valued quantities. Zeitsehr Math. Logik und Grundlagen Math. **22**, 149–160 (1976)
20. Haines, S.: The Product Manager's Desk Reference. McGraw-Hill International Edition, New York (2009)
21. Hellebrandt, T., Heine, I., Schmitt, R.H.: Knowledge management framework for complaint knowledge transfer to product development. Pap. Present. Procedia Manuf. **21**, 173–180 (2018)
22. Honkala, S., Hämäläinen, M., Salonen, M.: Comparison of four existing concept selection methods. In: Paper presented at the Proceedings of ICED 2007. The 16th International Conference on Engineering Design, DS 42 (2007)
23. Hundal, G.P.S., Kant, S.: Product design development by integrating QFD approach with heuristics-AHP, ANN and fuzzy logics-a case study in miniature circuit breaker. Int. J. Prod. Qual. Manag. **20**(1), 1–28 (2017)

24. Jahn, K.U.: Intervall-wertige Mengen. Math. Nach. **68**, 115–13 (1975)
25. Jeng, D.J.F., Huang, K.: Strategic project portfolio selection for national research institutes. J. Bus. Res. **68**(11), 2305–2311 (2015)
26. Jou, G.T., Yuan, B.J.C.: The application of DEMATEL and ANP to technology roadmapping in an RandD institute: a case study of anti-static fiber. Taiwan Text. Res. J. **23**(1), 1–19 (2013)
27. Kang, X., Yang, M., Wu, Y., Ni, B.: Integrating evaluation grid method and fuzzy quality function deployment to new product development. Math. Probl. Eng. (2018)
28. Kuan, M., Hsiang, C., Tzeng, G.: Probing the innovative quality system structure model for NPD process based on combining DANP with MCDM model. Int. J. Innov. Comput. Inf. Control. **8**(8), 5745–5762 (2012)
29. Kutlu Gündoğdu, F., Kahraman, C.: Spherical fuzzy sets and spherical fuzzy TOPSIS method. J. Intell. Fuzzy Syst. **36**(1), 337–352 (2019)
30. Lee, C.K.H.: A knowledge-based product development system in the chemical industry. J. Intell. Manuf. **30**(3), 1371–1386 (2019)
31. Lo, H., Liou, J.J.H.: A novel multiple-criteria decision-making-based FMEA model for risk assessment. Appl. Soft Comput. J. **73**, 684–696 (2018)
32. Oliveira, G.A., Tan, K.H., Guedes, B.T.: Lean and green approach: An evaluation tool for new product development focused on small and medium enterprises. Int. J. Prod. Econ. **205**, 62–73 (2018)
33. Özdemir, R.G., Çiçek, B., Özenç, D.: Multi objective new product development in bakery production under fuzzy demand parameters. J. Intell. Fuzzy Syst. **21**(5), 303–316 (2010)
34. Rosli, M.U., Ariffin, M.K.A., Sapuan, S.M., Sulaiman, S.: Integrated AHP-TRIZ innovation method for automotive door panel design. Int. J. Eng. Technol. **5**(3), 3158–3167 (2013)
35. Sambuc, R.: Fonctions ϕ-floues. Application l'aide au diagnostic en pathologie thyroidienne. Univ, Marseille, France (1975)
36. Sanyang, M.L., Mansor, M.R., Sapuan, S.M., Ahmed Ali, B.A.: Conceptual design of biocomposites for automotive components. Green Biocomposites 101–126 (2017)
37. Smarandache, F.: Neutrosophic Logic and Set (1995)
38. Smarandache, F.: Neutrosophy: neutrosophic probability, set, and logic: analytic synthesis and synthetic analysis (1998)
39. Soota, T.: Integrated approach to multi-criteria decision making for sustainable product development. Int. J. Qual. Research. **8**(4), 543–555 (2014)
40. Tian, Z.P., Wang, J., Wang, J.Q., Zhang, H.Y.: Simplified neutrosophic linguistic multi-criteria group decision-making approach to green product development. Group Decis. Negot. **26**(3), 597–627 (2017)
41. Torra, V.: Hesitant fuzzy sets. Int. J. Intell. Syst. **25**(6), 529–539 (2010)
42. Turan, F.M., Johan, K., Nor, N.H.M.: Criteria assessment model for sustainable product development. IOP Conf. Ser.: Mater. Sci. Eng. **160**(1), 012004 (2016)
43. Vinodh, S., Kamala, V., Shama, M.S.: Compromise ranking approach for sustainable concept selection in an indian modular switches manufacturing organization. Int. J. Adv. Manuf. Technol. **64**(9–12), 1709–1714 (2013)
44. Wang, C.C., Yang, C.H., Wang, C.S.: A modular and innovative concept design process in decision making for a new bike development. In: Paper presented at the Proceedings of International Conference on Computers and Industrial Engineering, CIE (2018)
45. Wang, L., Hu, Y., Liu, H., Shi, H.: A linguistic risk prioritization approach for failure mode and effects analysis: a case study of medical product development. Qual. Reliab. Eng. Int. (2019). https://doi.org/10.1002/qre.2472
46. Wang, X., Chan, H.K., Li, D.: A case study of an integrated fuzzy methodology for green product development. Eur. J. Oper. Res. **241**(1), 212–223 (2015)
47. Yager, R.R.: On the theory of bags. Int. J. Gen. System. **13**(1), 23–37 (1986)
48. Yager, R.R.: Pythagorean fuzzy subsets. In: Joint IFSA World Congress and NAFIPS Annual Meeting (IFSA/NAFIPS), Edmonton, AB, Canada, 24–28 June, pp. 57–61 (2013)

49. Yousefi, N., Mehralian, G., Rasekh, H.R., Yousefi, M.: New product development in the pharmaceutical industry: evidence from a generic market. Iran. J. Pharm. Res. **16**(2), 831–843 (2017)
50. Zadeh, L.A.: Fuzzy sets. Inf. Control. **8**(3), 338–353 (1965)
51. Zadeh, L.A.: The concept of a linguistic variable and its application to approximate reasoning. Inf. Sci. **8**, 199–249 (1975)

A Marketing Decision Support System for Product Design Based on an Outranking Approach

Juan Carlos Leyva López, Juan Francisco Figueroa Pérez, Edgar Omar Pérez Contreras and Pedro Sánchez Sánchez

Abstract In marketing decision support systems (MDSSs), basic information is obtained through market surveys and, to some extent, by using appropriate marketing models. Marketing models with MDSSs may represent the responses of the market to the various actions of the decision-makers. In this way, decision-makers and marketers can predict the reactions of the customers before trying to push a product in the market. In this chapter, an MDSS is presented. This MDSS has a user-friendly and easy-to-learn, menu-driven interface. The purpose of this MDSS is to assist a marketer in designing a new product when the customer preferences are not compensatory. The MDSS uses the share of preference frequency criterion. In this criterion, the new product is determined to maximize the relative preference frequency at which customers choose the new product. The MDSS examines different scenarios using the "what if analysis". Additionally, the MDSS finds near-optimal solutions for realistically sized problems using evolutionary algorithms. It is not necessary for the user to be familiar with the underlying models. A case study that applies the MDSS to the design of a new product for a Mexican company is presented.

J. C. Leyva López (✉) · E. O. Pérez Contreras
Department of Economic and Management Sciences, Universidad Autónoma de Occidente, 80020 Culiacán, Sinaloa, México
e-mail: juan.leyva@udo.mx

E. O. Pérez Contreras
e-mail: edgar.perez@uas.edu.mx

J. F. Figueroa Pérez
Faculty of Engineering—Los Mochis, Universidad Autónoma de Sinaloa, 80013 Los Mochis, Sinaloa, México
e-mail: juanfco.figueroa@uas.edu.mx

P. Sánchez Sánchez
Department of Computer Languages and Systems, Universidad de Jaen, 23071 Jaén, España
e-mail: pedroj@ujaen.es

C. Kahraman and S. Cebi (eds.), *Customer Oriented Product Design*, Studies in Systems, Decision and Control 279, https://doi.org/10.1007/978-3-030-42188-5_16

1 Introduction

Currently, the economic environment in which companies operate has become more competitive than ever. The globalization of markets, the shorter product life cycles, and the rapid technological development place high pressure on companies' profitability. To survive under such circumstances, a company must continuously launch new products or redesign current products. However, such procedures entail risk. A new product is costly and difficult to change, and if the product results in commercial failure, the company's viability may be endangered. Poor design constitutes one of the most frequent reasons for the failure of a new product [1]. For this reason, one of the most important decisions in marketing is the decision of new products [2]. Then, it becomes obvious that a company should put special care in the eventual design, production, and introduction of a new product in the market. This necessary attention is because new products play a key role in the growth of sales. The question is: what can a company do to ensure the success of its new products?

An important part of the new product development problem is the optimal product design problem [3]. In a way, many researchers and marketing managers have addressed the optimal product design. However, even the correct choices regarding the attributes of a product cannot guarantee the success of the new product [4, 5].

Normally, each product consists of a number of attributes at specific levels, and the consumer preferences regarding the various attribute levels are considered as known. Using these consumer preferences as inputs, optimization algorithms are used to design optimal product profiles. Different objectives can be employed, such as maximization of the products' market share or maximization of the company's profit. In real-world applications, as the number of attributes and levels increases, the number of candidate product profiles can grow to be uncontrollably large, making the managerial task of selecting the appropriate combination of attribute levels practically infeasible. According to Kohli and Krishnamurti [6], the share of choices for the single product design problem is NP-hard, which means that the complete enumeration of the solution space is practically infeasible in tractable amounts of time.

Substantial developments in computer technology, mathematical modeling and information handling using computer-based techniques have allowed for the development of decision support systems that can play a crucial role in the success of a company [7]. Therefore, there is a clear need for tools that can improve marketing decision and a growing interest in the use of marketing decision support systems (MDSSs) that are designed for use in complex marketing decision-making problems [8].

To help marketing managers address critical problems in real-time, more powerful web- and desktop-based tools are required to help analyze and address such a complex decision-making problem. To this end, decision support systems (DSSs) offer such a tool, which can be easy to employ and interact with for marketing managers.

An MDSS is defined as "a coordinated collection of data, systems, analytic tools and techniques with supporting software and hardware by which an organization gathers and interprets relevant information from businesses and the environment and turns this information into a basis for marketing decisions" [9].

In this chapter, we present an MDSS to support the design of a new product. The proposed MDSS uses evolutionary algorithms (EAs) to find near-optimal solutions in a reasonable time. The aim of the MDSS is to help a marketing manager, who does not need to be familiar with the underlying models, to address questions such as "what will happen if...?" or "what should happen...?". Then, the proposed MDSS can predict the effect of marketing actions using multicriteria, multiobjective and behavioral models to compute the outcome of different actions. Additionally, the MDSS aims to find the best marketing strategy in each situation. In short, the MDSS can facilitate the decision-making process and improve the quality of the decision.

From a strategic marketing point of view, the proposed system is the first to introduce a multicriteria methodology application in the product design process based on the outranking approach. The MDSS automates the use of a new product design multicriteria methodology based on customer preferences that were proposed by [10–12]. The prototype is composed of elements of marketing decision support systems, multiobjective evolutionary algorithms, and multicriteria decision analysis methods. Here, customer satisfaction is modeled using flexible multicriteria outranking approaches through a new aggregation/disaggregation preferences model, a market segmentation model, and a brand choice model. In this way, unlike other proposals in the literature, when a product is evaluated, the principle of noncompensatoriness does not lie and the importance of each criterion in the evaluation is considered. It is believed that this approach will produce an appropriate representation of customer preferences in the software.

In the remainder of the chapter, it is reviewed the related work, the product design problem, the marketing decision support system, and the conclusion.

2 Related Work

In recent years, researches proposed different MDSS to support new product development. Among these are those that does not support consumer satisfaction (CS) for decision support (DS) or distributed decision-making support (DDMS) as the presented by Lei and Moon [13] which is a decision support system for market-driven product positioning and design, based on market data and design parameters which determines market segments for new products using Principal Component Analysis (PCA), K-means, and AdaBoost classification. The system combines a database, the unparalleled analytical capability of Matlab suite and a Graphical User Interface to allow users to explore and evaluate alternative scenarios during product development. A case study using USA automotive market data was conducted; others support only CS for DS, but they not taking into account DDMS, such is the case of that presented by Li et al. [14], which is a system that can extract and consolidate the reviews expressed via social media and derive insights (product feature specification and feature importance) to help enterprises make decisions on developing next-generation products by analyzing the reviewers' knowledge and authority and their opinion sentiment toward the target products. The experimental results obtained show that the proposed mechanism is good for trend prediction and customer acceptance. Other support only DDMS, but they not taking into account CS for DS in their models, such as the case of that presented by Yu et al. [15], which is a multi-agent DSS for new product development composed of eight modules: data collection, data management, statistical analysis, breakeven analysis, risk prediction analysis, resource optimization scheduling, expert system, and information exchange. Only a few of the reviewed systems take into account both CS for DS and DDMS, but they have the disadvantage that methods implemented to model CS not taking into account the parameter of importance that is assigned by a customer to each criterion to evaluate a product, ignoring with it valuable information to model accurately consumer preferences. Such is the case of that presented by Guo et al. [16], who developed a DSS based on the methodology of Kansei Engineering composed of a friendly human-computer interface, a database management module and a model management section. We consider this an important finding to highlight and area of opportunity for new developments.

3 The Product Design Problem

A product is anything that can be offered to a market to satisfy a want or need. When the customers' preferences are approximately homogeneous, it is sufficient to design a single product. The product design problem is a critical issue faced by organizations seeking to introduce a new product or refine existing products in the market. Heuristic solutions to the product design problem have been proposed by several researchers in the past (see Albritton and McMullen [17] and Green and Krieger [18] for a review of issues in the domain).

Before making a decision among competing products, the aggregation model of preferences is constructed, it is represented by a fuzzy outranking relation that is exploited with a multi-objective evolutionary algorithm to derive a recommendation in the form of a ranking of products by decreasing level of preference [19]. On the basis of the above representation scheme, the preference frequency of a product is the sum of the preference frequency of the corresponding frequency levels. The higher the product's frequency, the higher the possibility that it is chosen. Suppose that two customers should choose between PC1 (14″, core-i7 3.3 GHz Turbo, 256 solid-state drive, 8 GB RAM, cable mouse, no camera) and PC2 (24″, core-i7 2.7 GHz, 1 TB HD, 6 GB RAM, wireless mouse, HD camera). Customer 1 will probably choose PC1 (Borda count = 38) over PC2 (Borda count = 25), whereas Customer 2 will probably choose PC2 (Borda count = 34) over PC1 (Borda count = 18). The Borda counts are converted to choice frequencies for each product through choice models and are then aggregated across the whole customer base to provide hypothetical market shares. If we know the performance matrix (the assessment of the products by the criteria) for a population of customers, we can simulate the introduction of different product configurations (combinations of criteria values) to the market and estimate conditional market shares. With the use of optimization algorithms, we can find the product that maximizes a company's market share, given the customer preferences and the configuration of competitive products in the market (for a more detailed explanation of the entire procedure, the reader can review [11] and [10].

An example could be a company that produces cooking oil that is interested in introducing a new corn oil that will provide the highest-possible volume sales. The customer preferences are usually revealed through market surveys, the results of which are entered into preference disaggregation methods based on the outranking approach of multicriteria decision analysis, which estimate preference parameters for each customer. In the optimal product design problem, the performance matrix for each customer, as well as the competitive product profiles, are considered to be known, and the aim is to find the product configuration that maximizes the company's preference frequency criteria.

According to the outranking approach for the new product design, the customers' preferences are described in terms of criteria, assessment of the products by the criteria, weights, and thresholds. The assessment of the products denotes the values assumed by the criteria over a precise scale. In the outranking-based methodology, the aggregation model of preferences for a set of products in the market is represented by a fuzzy outranking relation. The rank order of the candidate product can be computed based on the relative preference frequency. The aim of the outranking-based methodology is to identify the "best" market share for a new product. It is assumed that customers prefer products that maximize their relative preference frequency.

Recently, an evolutionary algorithm (EA) based solution method was proposed for the product design problem [11]. The evolutionary algorithm uses the share of preference frequency criterion. The EAs have several advantages in the way they perform the search process. The EA works with "complete" product profiles. In the EAs for each criteria value, there is a possibility to change. Moreover, EAs work with a large set of candidate solutions. The computational results showed that the EAs can find near-optimal solutions in reasonable computational time. The EA-based solution method is included in the model base of the proposed MDSS.

4 The Marketing Decision Support System

A user-friendly MDSS based on customer preferences was developed for the design of a new product using the share of preference frequency criterion (http://midss.net). The share of preference frequency criterion is an emerging criterion used in product design. Brand preference is a compulsory ingredient for customer satisfaction. Interestingly, the term "customer satisfaction" is most important for any marketing manager who wishes to improve the market share by improving brand preference through customer satisfaction. The market share is essential for the survival of an enterprise, and it has a significant influence on the profit [20]. The MDSS uses a preference choice rule based on outranking relations, where the product associated with the greater (maximum) preference frequency is nondeterministically chosen [10]. Outranking relations are models that capture the customers' preferences in pairwise evaluations. In the share of preference frequency criterion, the new product is determined to maximize the number of new customers. The MDSS does not refer to any specific product. The implementation was performed using C, Java and PHP languages for evolutionary algorithms, multiagent architecture and the dialog component, respectively. The MS Sql Server database was used to store data.

The architecture of the MDSS is presented in Fig. 1. To present the MDSS, the

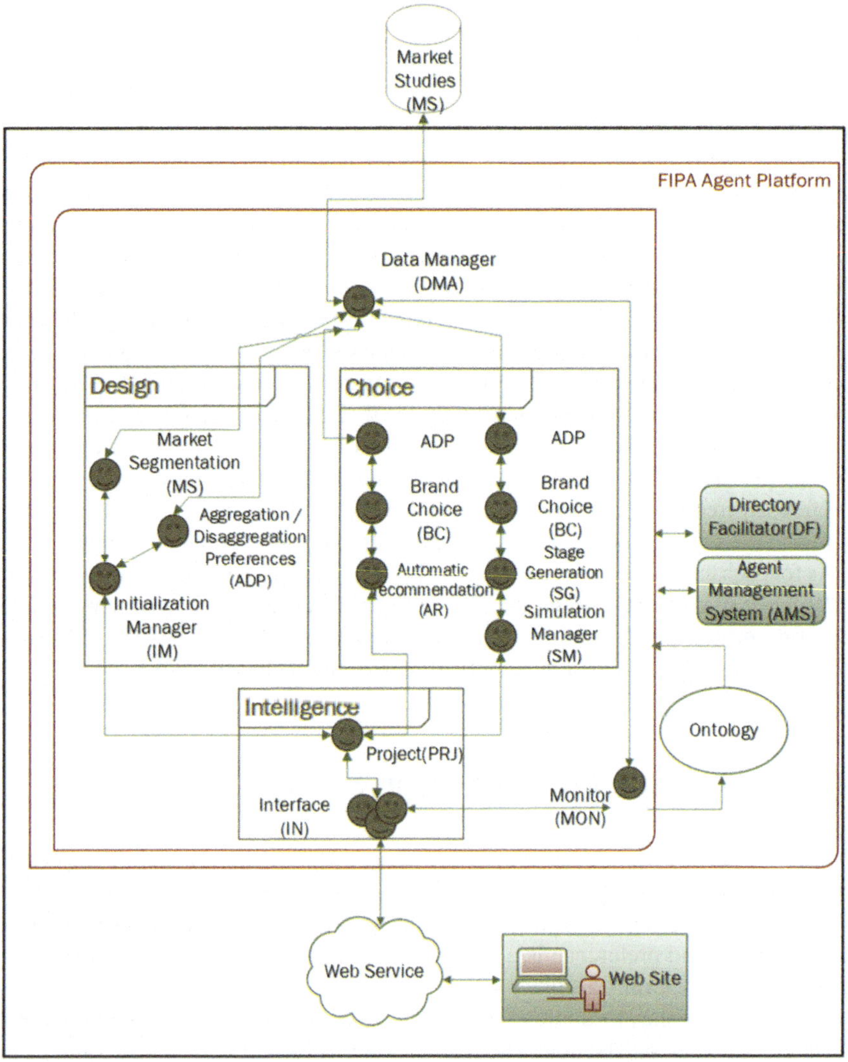

Fig. 1 The architecture of the MDSS

design of new corn oil is used as an example. The example is not based on a real application. Before the user interacts with the MDSS, we assume that:

- The good or service to be designed has been identified.
- A representative set of products in the market has been identified.
- A representative family of criteria and the associated scales that fully describe the good or service have been identified.
- An experimental design has been constructed that is appropriate for those criteria and scales, either from a design catalog or via a software program.
- A survey has been constructed and administered to a sample of respondents via a web survey.
- The corresponding data from the survey have been stored in the database of the MDSS.

4.1 Data

The MDSS permits the systematic storage, request, retrieval and maintenance of all necessary and available data. The data of a new product design problem are stored in a relational database. However, during operation of the system, the user must only enter the name of the design problem.

4.1.1 The Product Criteria and Criteria Values

Let $A = \{a_1, a_2, \ldots, a_m\}$ be a set of products available to customers, which can be analyzed by a family of criteria $G = \{g_1, g_2, \ldots, g_n\}$, where $g_j(a_i)$ represents the evaluation of the product a_i by criterion g_j. We can further extend this formulation to the value scale within each criterion and for each customer: $g^k_{jm}(a_i)$, which represents an evaluation of product a_i on criterion j with value scale (level) m for consumer k. These evaluations are based on the customers' judgments and should be directly collected from the customers; these evaluations become the customer's performance matrix. We assume that the performance matrix $\left[g^k_{jm}(a_i)\right]$ for a set of customers k and the market share of the alternatives A are both known.

We assume that the aggregation model of preferences for each customer $k_i, i = 1, 2, \ldots, K$ can be represented by a fuzzy outranking relation $S_A^{\sigma_{ki}}$ and that each customer k_i can rank the products in decreasing order of preference R_{ki}, producing a total order, total preorder, or a partial preorder depending on the characteristics of the products and preferences of the customers. We also assume that customers can be segmented by their profile $S = \{s_1, s_2, \ldots, s_t\}$. Finally, let us assume that we know in advance the market share of each one of the products in A.

Input data

To run the optimization phase, the model should contain the following input data:
$A = \{a_1, a_2, \ldots, a_m\}$, the set of products that comprise the market.
$G = \{g_1, g_2, \ldots, g_n\}$, the set of n criteria that comprise the product.
$\Phi_r = (1, 2, \ldots, J_r)$, the set of J_r values on the scale of criterion r.
$\Theta = (1, 2, \ldots, K)$, the set of K consumers.
w_{ktr}, the assessment that consumer $k \in \Theta$ assigns to value on the scale $t \in \Phi_r$ of criterion $g_r \in G$, where

$$x_{tr} = \begin{cases} 1 : \text{if the value on the scale of criterion } \mathrm{g_r} \text{ of the new product is t} \\ 0 \quad \text{otherwise} \end{cases}$$

$T^{(k)}(i, j), (1 \le i, j \le m)$, the number of times that a product a_j was found at a certain place in the ranking of the individual associated with the members of the final restricted Pareto for a consumer k.
$w_i^{(k)}$, the weights according to the importance of the products to be ranked for a consumer k. $(1 \le i \le m)$

Output
The output of the genetic algorithm (GA) consists of a set of values of the binary variables

$$x_{tr} = \begin{cases} 1 : \text{if the value on the scale of criterion } r \text{ of the new product is } t \\ 0 \quad \text{otherwise} \end{cases}$$

for the design of a new product.
In the example used in the chapter, it is considered that the corn oil model can be described in terms of 6 attributes. The determination of the appropriate criteria and the values on the scale for each criterion is a very important task in a multicriteria decision analysis study.

4.1.2 Customers' Relative Preferences of Products

As previously mentioned, the fuzzy outranking relation S_A^σ defined as $A \times A$, where A is the set of products under consideration in a specific product design problem, can be inferred using the preference disaggregation algorithm proposed by [21].

A fuzzy outranking relation S_A^σ can be exploited with the multiobjective evolutionary algorithm proposed by [11]. As an output, we obtain a restricted Pareto front $PF_{known}^{restricted}$ and the associated final set of solutions (ranking) returned by the MOEA at the termination $P_{known}^{restricted}$ represented in Table 1. $P_{known}^{restricted}$ is the region of interest (RoI) of the Pareto front.

The number $T(i, j)$, $(1 \le i, j \le m)$ of times (frequency) that a product a_j was found at a certain place in the ranking of the individual is associated with the members of the final restricted Pareto front, shown in Table 2. This table summarizes the results of all individual preferences over a set of products in an evolutionary algorithm.

Table 1 Restricted Pareto set $P_{known}^{restricted}$ of the solution space

Individuals: Final set of solutions returned by the MOEA	$\hat{p}_1$	$\hat{p}_2$	$\hat{p}_3$	...	$\hat{p}_{r-1}$	$\hat{p}_r$
	$a_{k_1}^{(1)}$	$a_{k_1}^{(2)}$	$a_{k_1}^{(3)}$	...	$a_{k_1}^{(r-1)}$	$a_{k_1}^{(r)}$
	$a_{k_2}^{(1)}$	$a_{k_2}^{(2)}$	$a_{k_2}^{(3)}$	...	$a_{k_2}^{(r-1)}$	$a_{k_2}^{(r)}$
	$a_{k_3}^{(1)}$	$a_{k_3}^{(2)}$	$a_{k_3}^{(3)}$	...	$a_{k_3}^{(r-1)}$	$a_{k_3}^{(r)}$
	$\vdots$	$\vdots$	$\vdots$	$\vdots$ ...	$\vdots$	$\vdots$
	$a_{k_m}^{(1)}$	$a_{k_m}^{(2)}$	$a_{k_m}^{(3)}$		$a_{k_m}^{(r-1)}$	$a_{k_m}^{(r)}$

where $a_{k_j}^{(i)} \in A, \forall i = 1, 2, \ldots, r, \quad j = 1, 2, \ldots, m, a_{k_j}^{(i)} \neq_{i \neq l} a_{k_l}^{(i)}$ and $[k_1^{(i)}, k_2^{(i)}, \ldots, k_m^{(i)}]$ is a permutation of $[1, 2, \ldots, m], i = 1, 2, \ldots, r$.

Table 2 The number of times that a product was found at a certain place in the ranking

$B_j = \sum_{i=1}^{m} w_i T(i, j)$	Weight w_i	Rank	a_1	a_2	...	a_m
	w_1	1	$T(1, 1)$	$T(1, 2)$	...	$T(1, m)$
	w_2	2	$T(2, 1)$	$T(2, 2)$	...	$T(2, m)$
	$\vdots$	$\vdots$	$\vdots$	$\vdots$	$\vdots$	$\vdots$
	w_{m-1}	$m-1$	$T(m-1, 1)$	$T(m-1, 2)$	...	$T(m-1, m)$
	w_m	m	$T(m, 1)$	$T(m, 2)$	...	$T(m, m)$
Borda count	$B_j = \sum_{i=1}^{m} w_i T(i, j)$		B_1	B_2	...	B_m

where $\sum_{i=1}^{m} T(i, j) = \left|P_{known}^{restricted}\right| \quad \forall j = 1, 2, \ldots, m.$

Given that the ranking of the products provides information regarding the individual preferences of customers for a given product, the number of times that a product is found at a certain place in the ranking is relevant to finding the relative preferences for that customer.

To capture the importance of products, we implement a Borda count (based on Table 2) using the following procedure. The frequency that a product is found at a certain place in the ranking is weighted according to the importance of the products to be ranked. It is reasonable to conclude that in certain cases; the rank of the products would not be of equal importance. Relative importance could be reflected in a weighting procedure where the values w_i are dependent on each customer.

The total satisfaction value of the product a_j, denoted as B_j, is defined as the weighted sum $B_j = \sum_{i=1}^{m} w_i T(i, j), \quad j = 1, 2, \ldots, m$, where the Borda count B_j is the relative preference of the product, given the individual customer obtained from the nondominated rankings from the outranking relations. We could consider this as the market preferences of individual customers.

However, there could be a differentiated customer response based on the consistency of a brand, the quality or other nontractable features. To account for this, for each customer, we estimate the products' preference relative weighted frequencies as a proxy to market sales:

$$F_{kj} = \frac{B_{kj}}{\sum\limits_{i \in C} B_{ki}} \tag{1}$$

where F_{kj} is the preference relative frequency that customer k chooses product j, $B_{kj} = \sum_{i=1}^{m} w_i T(i, j)$ is the total satisfaction value the customer assigns to the product a_j, and C is the set of considered products.

4.1.3 Competitive Environment and Current Company Products

The product that a customer prefers before the introduction of the new product is called the status quo product. Customers do not have a status quo product and do not deterministically choose the highest-value product. Instead, we assume that each of the m products has a certain preference frequency to be selected, which is calculated with the use of the preference frequency model based on the Borda count. The preference frequency based on the Borda count is denoted B_j, which is the relative preference of the product, given the individual consumer obtained from the nondominated rankings from the outranking relations. A customer prefers a new product to other products in the market only if it has a better rank order than the others.

The models of the proposed MDSS can answer the following questions:

1. "What product should the company design to maximize the share of preference frequency?"
2. "What will happen if the company designs a particular product?"

To answer the first question, the MDSS uses an optimization solution approach. For the optimization criteria used in the system, the model base contains evolutionary algorithms.

To answer the second question, the MDSS allows the user to examine different scenarios performing the "What if Analysis". That is, the user can insert a product profile, and the system computes the additional market share (as a percentage). Additionally, the MDSS computes the number of customers, the number of segments and the segments that are expected to become new customers of the company. Previously, the user had to estimate the optimal product design using the appropriate evolutionary algorithms. The optimal product design functions as a reference point for the user who is looking for a product design that is the most appropriate for the conditions of the company.

The MDSS presented in the current chapter uses a GA for solving the product design problem. In the EA used by the MDSS, each individual of the population corresponds to a product profile. In each generation of the GA, using the share of choice preference frequency criterion, the new population elements are produced using the selection method, the uniform crossover operator and the mutation operator. When the best candidate solution does not improve in 20 consecutive generations, the GA terminates. The population size is set equal to 180, that is, M = 180. These parameters influence the performance of the GAs. From our experiences with a wide range of problem sizes, the selected parameters perform well. For more details about the GAs used in the MDSS, see Alexouda [8].

4.2 User-Interface

The user-interface of a decision support system can influence its acceptance by the user. To achieve the acceptance of a decision support system, the user's skills, needs, and expectations must be seriously considered in the design and implementation of the user-interface [22]. The interaction style can influence the user's attitudes toward DSSs. The more common the interface of a DSS, the less effort and training is needed for a user to learn how to use it [23].

To improve the MDSS's chances for success, a menu-driven user-interface with common, easy-of-use features such as grid formats, navigators for grids and pop-up menus was chosen. Shortcuts are used to perform an action quickly. Tools provide an easy to understand, visible way to present options to the user. It is assumed that the end-users are familiar with using the Web environment. No special knowledge about the underlying models is required to use the system. The end-users must insert the necessary data, and the results of the models appear in an easy to understand form. There are many benefits to be gained by the selected user-interface, such as facilitating learning, reducing error rates, etc. To explain these benefits, some of the windows of the MDSS are presented in the following figures.

The first screen when entering the System is the Login window, through which the user must authenticate himself or herself (Fig. 2).

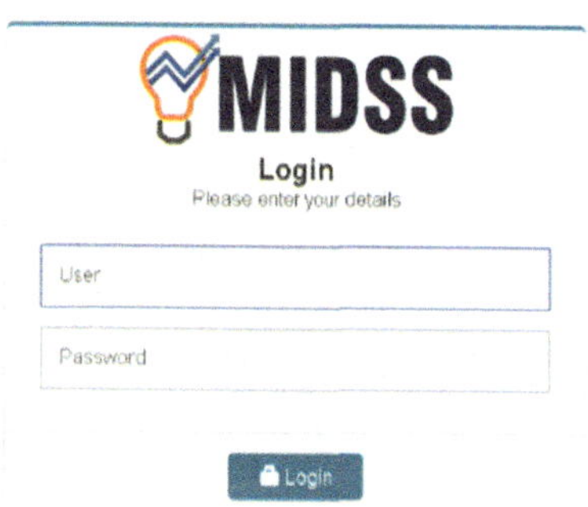

Fig. 2 The login window

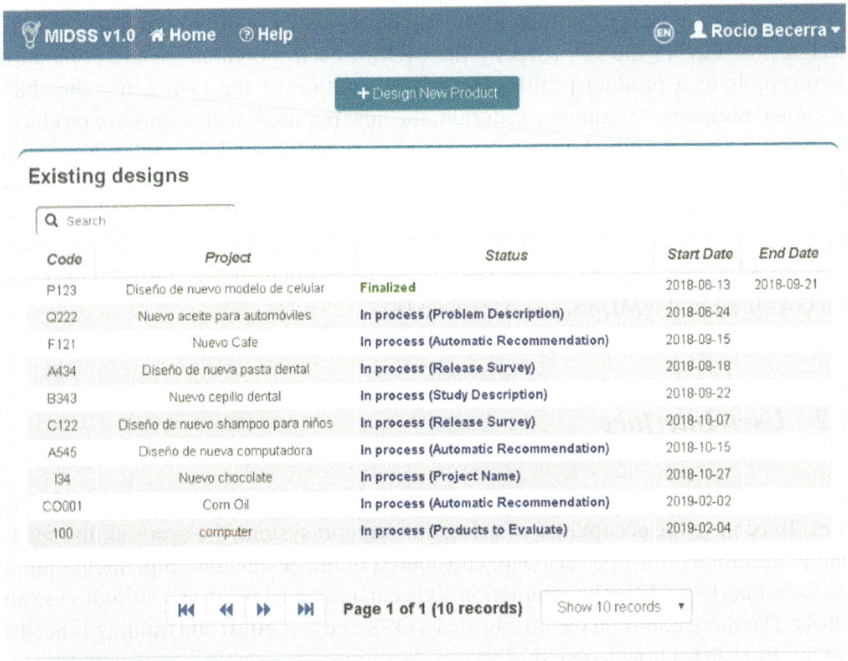

Fig. 3 The dashboard window

Once the user enters the System, a dashboard is presented in which the list of the user's projects is displayed, see Fig. 3. In this window, it is possible to create a new project by pressing the <Design New Product> button or manage an existing project by clicking on the name of the desired project. In this example, the process of designing a new corn oil is shown, so the content of the remaining example windows is related to the design of this product.

4.2.1 The Main Menu

When creating a new project or selecting an existing one, the main window of the MDSS, shown in Fig. 4, is presented. It contains the following sections: in the upper part is the *title bar*, which shows the system version, the home menu, the help system, the search system and the user's name; under the title bar is the *navigation bar*, in which the path of the currently selected option and the name of the current project are displayed; below the navigation bar is the *progress bar*, in which the status of the current project is displayed. In the left panel, the main menu appears, which includes the options *Projects*, *Marketing Study*, *New Product Design* and *Summary*. Finally, in the right panel, which occupies most of the window, it is the area destined to display the information of the selected menu option.

Fig. 4 The main window of the MDSS

4.2.2 The Projects Menu

This menu contains the following options: *Project Name*, *Problem Description*, *Product Attributes* and *Products to Evaluate*. In Fig. 5, the *Project Name* option is presented, in which code and a name are assigned to the project for the design of a new product. This option is used mainly to register a new project. From here, the user can enter the remaining options by clicking directly on the menu option or by navigating between the screens using the <Previous> and <Next> buttons.

The description of the problem and the market study are conducted through the *Problem Description* option, presented in Fig. 6.

In the next screen, see Fig. 7, the user must enter the set of attributes. The screen has options to add or remove an attribute with visible buttons. If the user wants to modify an existing attribute, it is necessary to select the attribute by selecting it in

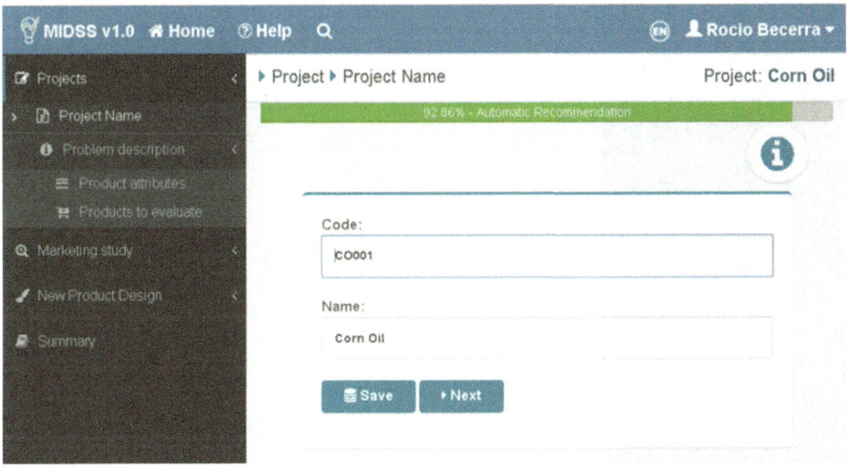

Fig. 5 The *Project Name* option of the Projects menu

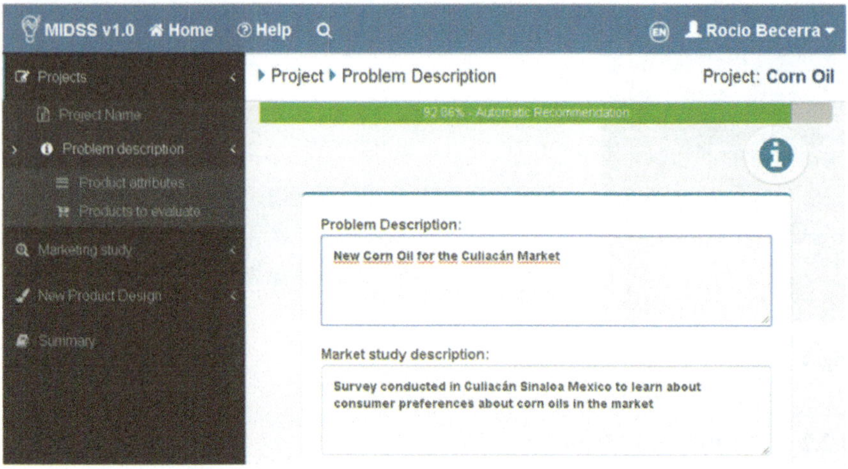

Fig. 6 The *Problem Description* option of the Projects menu

the grid, after which a pop-up window will appear with the data for the selected attribute, allowing the user to modify any of its fields (such as name, label, description, orientation (minimize, maximize), type (integer, decimal, date, hour, scale) and scale).

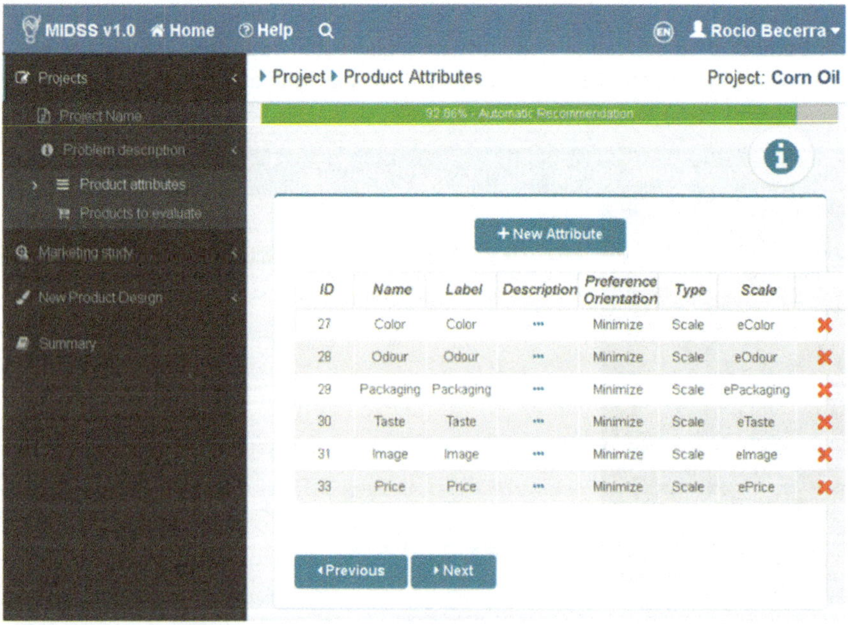

Fig. 7 The *Product Attributes* option of the Projects menu

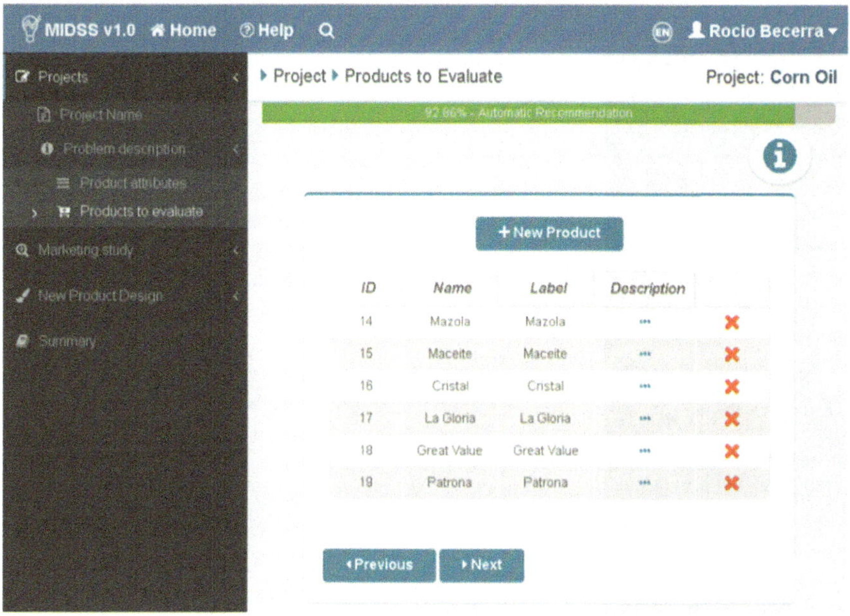

Fig. 8 The *Products to Evaluate* option of the Projects menu

Figure 8 shows the option that allows the user to register the set of products to be evaluated by consumers. This operation is similar to the screen in which the attributes are registered, Fig. 7.

4.2.3 The Marketing Study Menu

This menu contains a group of options that constitute the main data set for a new product design problem, as described in Sect. 3. The fields are named *Study Description*, *Design Survey*, *Sample Size*, *Release Survey,* and *Monitor Application*, with which the application of the surveys is administered to obtain the preferences of the consumers. Figure 9 shows the screen of the first option of this menu, in which the date and place of application of the survey are defined, in addition to the methodology that will be used.

The next screen, in Fig. 10, allows the user to design the survey through an external subsystem based on the Web that is accessible through the <Load Survey Subsystem> button. Additionally, it allows the user to indicate who will be responsible for applying the survey. For this, the pollster must be selected from the left panel and passed to the right panel using the buttons (> add one to the list, < remove one from the list, > > add all to the list, < < remove all from the list). It should be noted that the survey is applied through a mobile application for tablets.

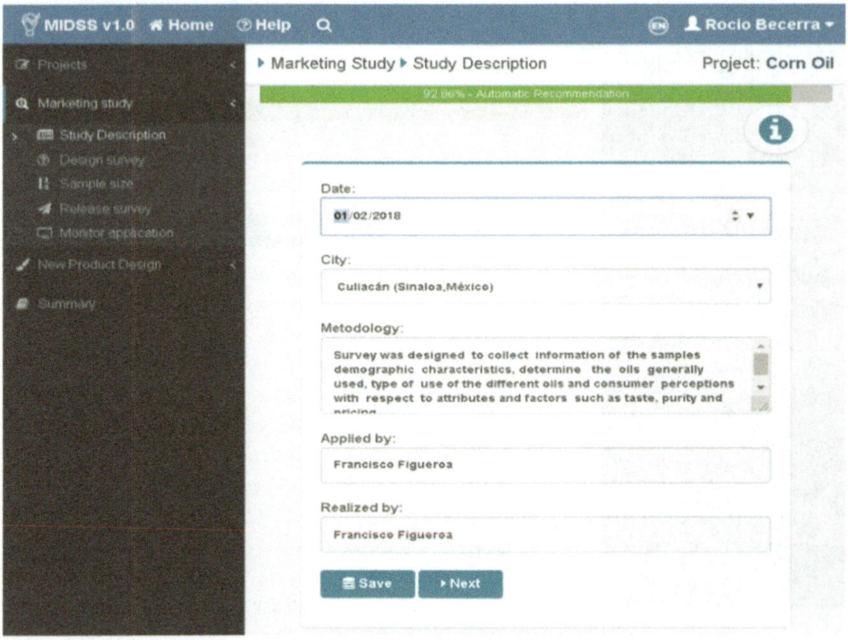

Fig. 9 The *Study Description* option of the marketing study menu

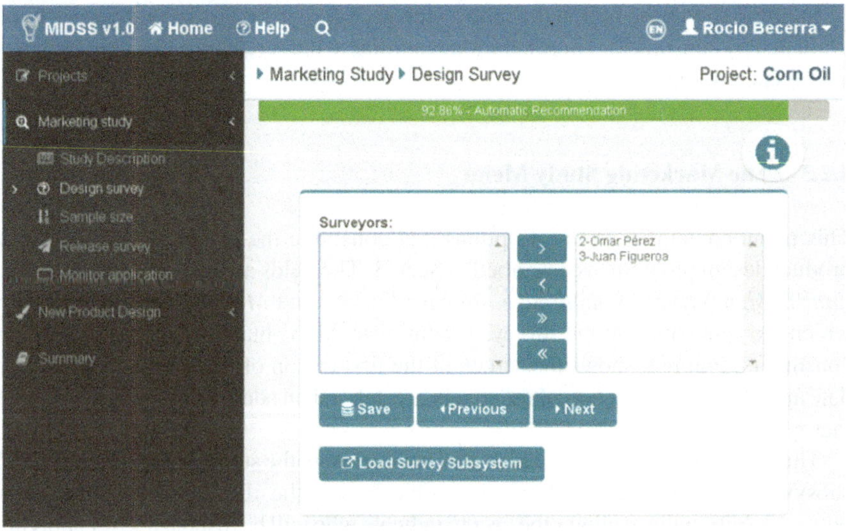

Fig. 10 The *Design Survey* option of the Marketing Study menu

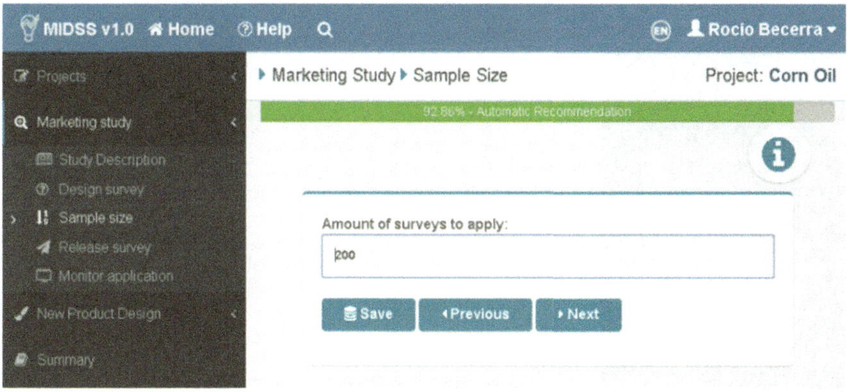

Fig. 11 The *Sample Size* option of the marketing study menu

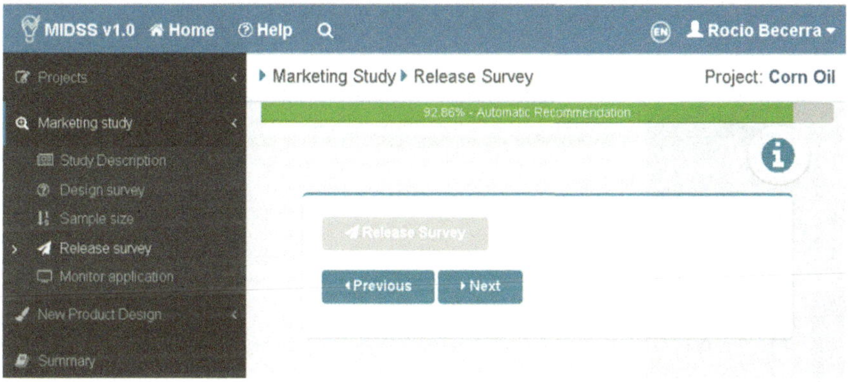

Fig. 12 The *Release Survey* option of the marketing study menu

An important step when applying the survey is to define the number of consumers that will be surveyed; this is defined in the screen shown in Fig. 11.

Once the survey is designed and approved by the decision-maker, the survey must be released for the application using the screen shown in Fig. 12, using the <Release Survey> button.

Finally, the progress of the application of the survey can be monitored through the screen shown in Fig. 13, where the number of surveys completed until the time of the consultation is shown.

4.2.4 The New Product Design Menu

This menu concentrates the core MDSS options that form the basis of the models, named Aggregate/Disaggregate, Segment Market, Market Share, and Automatic

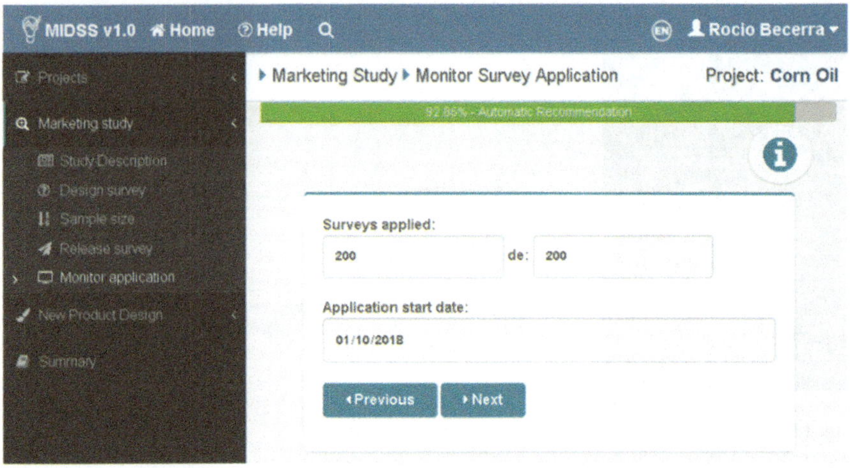

Fig. 13 The *Monitor Survey Application* option of the marketing study menu

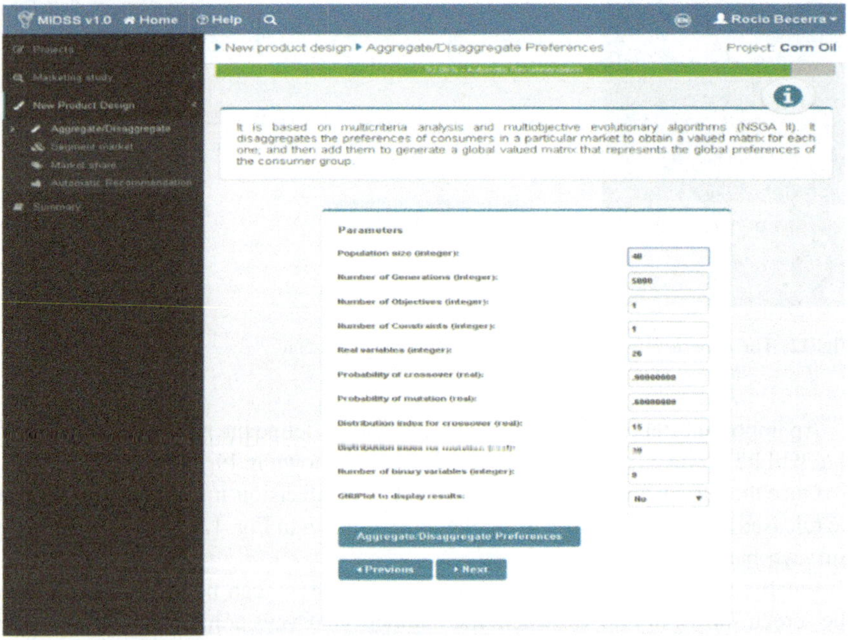

Fig. 14 The *Aggregate/Disaggregate* option of the new product design menu

Recommendation. Figure 14 shows the first option of this menu; in this screen, the user defines the parameters of the evolutionary algorithm for the aggregation and disaggregation of preferences for each consumer. The parameters are characteristic of a genetic algorithm. Currently, the system presents default values for each of the

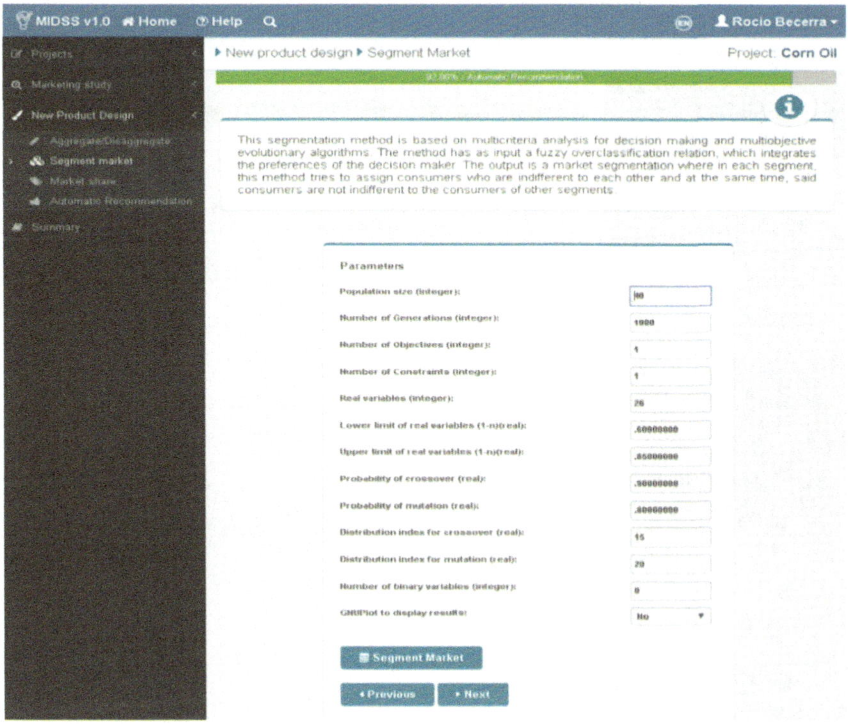

Fig. 15 The *Segment Market* option of the new product design menu

parameters based on our experience with a wide range of problems, for which it has been shown that these parameters perform well. However, the user is allowed to modify the values to perform the "What If Analysis". It should be noted that, although it has been shown that algorithms can find a good solution in a reasonable amount of time, it is possible that the delay does not allow for the user to advance to the next process until the current process ends.

The screen shown in Fig. 15, which corresponds to the evolutionary algorithm for market segmentation, is very similar to the screen presented in Fig. 14 because it requires the same parameters based on its genetic-type algorithm. Default values are also displayed here, which have already been confirmed to perform well, but the user is also allowed to modify them. Once the values are established, the algorithm is invoked using the <Segment Market> button, which can also take considerable time and will not allow the user to advance to the next step until the current step is completed.

The last model used is the Market Share, accessible from the screen shown in Fig. 16. Similar to that of Aggregation/Disaggregation and Segmentation, this model requires certain parameters that contain default values, and the algorithm can be invoked with the button <Market Share> .

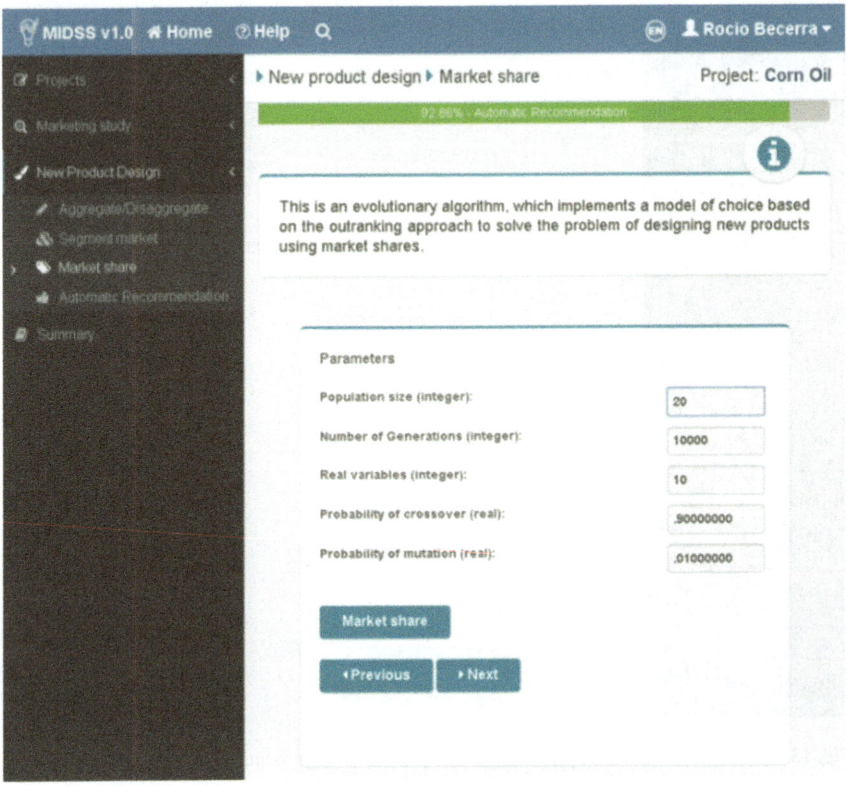

Fig. 16 The *Market Share* option of the new product design menu

Finally, the result of the simulation for the design of the new product is shown on the screen of Fig. 17, which is divided into three sections. In *Setting Grid*, the optimal values for each attribute of the new product are displayed in a grid format, where the first column is the identifier of the attribute, the second column is the name of the attribute, the third is the weight assigned to the attribute and the last column is the proposed value for the attribute. In *Market Share Grid*, the market share obtained for the new product is shown in a grid format based on the products that were evaluated by consumers, where the first column represents the position in the ranking of the new product relative to the competition, the second column shows the name of the product and the last column shows the market share for each product. In *Expected Market Share*, the same information is shown from the Market Share Grid section, but in pie chart format, in which the market share of the new product can be observed visually relative to all products evaluated.

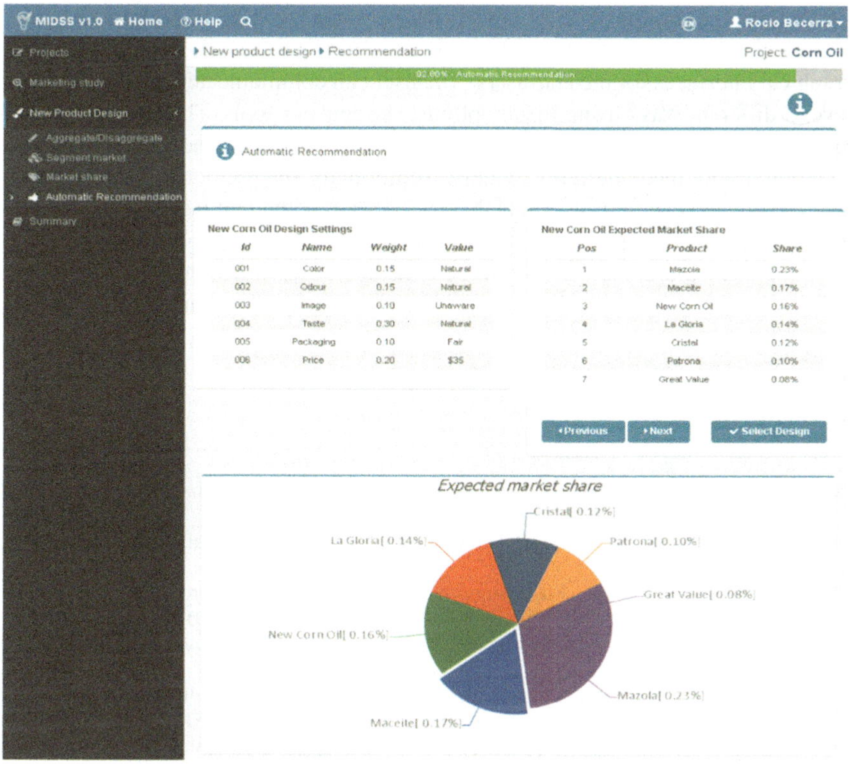

Fig. 17 The Recommendation option of the new product design menu

5 Conclusion

To summarize, this chapter presented a new decision support system to support new products design in a real-world context. It is described as a novel MDSS which implements a new methodology for designing new products based on consumer preferences. Their models are based in ELECTRE III multicriteria method and were built in the software through multi-objective evolutionary algorithms. Unlike other approaches in the literature, consumer preferences are modeled in the methodology using a more flexible approach which allows considering the importance assigned by a customer to each criterion to evaluate a product with what is intended to model more appropriately consumer preferences than in the current proposals.

This MDSS can be used as a marketing manager's consultant. The MDSS allows the decision-maker to examine different scenarios using "What if analysis". Moreover, the system uses EAs, which can find near-optimal solutions for realistically sized problems in reasonable amounts of time. The MDSS addresses the share of preference frequency problem.

The Web environment and Laravel, the open-source PHP web framework that is used for the development of the web system, offer friendly and easy-to-learn graphical interfaces for the end-users. The user can communicate with the MDSS in several different ways using menu options, shortcuts or tools. The system does not require that the user has special knowledge about computers, and the user need not be familiar with the underlying models. Additionally, the user can obtain help from the system. However, it is assumed that the user has some familiarity with using the Web environment.

In the near future, the MDSS will be extended to work with groups of decision-makers that are geographically distributed and meet asynchronously.

References

1. Kotler, P., Keller, K.: Marketing Management 12th Edition México. Pearson Education, London (2008)
2. Lilien, G.L., Kotler, P., Moorthy, K.S.: Marketing Models. Prentice Hall, Upper Saddle River (1995)
3. Tsafarakis, S., Delias, P., Matsatsinis, N.: A service-oriented approach for the optimal product/service design business process. Int. J. Inf. Syst. Serv. Sect. (IJISSS) **5**, 68–81 (2013)
4. Rackham, N.: From experience: why bad things happen to good new proucts. J. Prod. Innov. Manag.: Int. Publ. Prod. Dev. Manag. Assoc. **15**, 201–207 (1998)
5. Korgaonkar, P., O'Leary, B.: Product, marketing and web site attribute discriminating between successful and failed internet businesses. J. Internet Commer. **7**, 485–512 (2008)
6. Kohli, R., Krishnamurti, R.: Optimal product design using conjoint analysis: computational complexity and algorithms. Eur. J. Oper. Res. **40**, 186–195 (1989)
7. Power, D.J., Burstein, F., Sharda, R.: Reflections on the past and future of decision support systems: perspective of eleven pioneers. In: Proceedings of the Conference on Decision Support, pp. 25–48. Springer, New York (2011)
8. Alexouda, G.: A user-friendly marketing decision support system for the product line design using evolutionary algorithms. Decis. Support Syst. **38**, 495–509 (2005)
9. Wiid, J., Diggines, C.: Marketing Research. Juta and Company Ltd, Cape Town (2010)
10. Leyva Lopez, J.C., León Santiesteban, M., Ahumada Valenzuela, O., Romero Serrano, A.M.: A choice model for the product design problem based on the outranking approach. In: Liu, J., Lu, J., Xu, Y., Martinez, L., Kerre, E. (eds.) Data Science and Knowledge Engineering for Sensing Decision Support, vol. 11, pp. 1018–1025. World Scientific, Singapore (2018)
11. Leyva Lopez, J.C., León Santiesteban, M., Ahumada Valenzuela, O., Solano Noriega, J.J.O.: The new product design problem using a novel preference approach. In: Liu, J., Lu, J., Xu, Y., Martinez, L., Kerre, E. (eds.) Data Science and Knowledge Engineering for Sensing Decision Support, vol. 11, pp. 987–994. World Scientific, Singapore (2018)
12. Gastelum Chavira, D.A., Leyva Lopez, J.C., Larreta Ramírez, E.V.: A multi-criteria and multi-objective approach for the market segmentation problem. In: Liu, J., Lu, J., Xu, Y., Martinez, L., Kerre, E. (eds.) Data Science and Knowledge Engineering for Sensing Decision Support. vol. 11, pp. 1003–1009. World Scientific, Singapore (2018)
13. Lei, N., Moon, S. K.: A decision support system for market-driven product positioning and design. Decis. Support Syst. **69**, 82–91 (2015)
14. Li, Y.-M., Chen, H.-M., Liou, J.-H., Lin, L.-F.: Creating social intelligence for product portfolio design. Decis. Support Syst. **66**, 123–134 (2014)
15. Yu, T., Zhou, J., Xu, F., Gong, Y., Wang, W.: Decision support system of product development based on multi-agent. In: 2009 International Conference on Information Technology and Computer Science, vol. 2, 0–3. IEEE, Washington, DC (2009)

16. Guo, F., Ren, L., He, Z., Wang, H.: Decision support system for industrial designer based on Kansei engineering. In: Internationalization, Design and Global Development, pp. 47–54. Springer, Berlin (2011)
17. Albritton, M.D., McMullen, P.R.: Optimal product design using a colony of virtual ants. Eur. J. Oper. Res. **176**, 498–520 (2007)
18. Green, P.E., Krieger, A.M.: Recent contributions to optimal product positioning and buyer segmentation. Eur. J. Oper. Res. **41**, 127–141 (1989)
19. Leyva Lopez, J.C., Aguilera Contreras, M.A.: A multiobjective evolutionary algorithm for deriving final ranking from a fuzzy outranking relation. In: Coello Coello, C.A., Zitzler, E., Hernández Aguirre, A. (eds.) 3rd International Conference Evolutionary Multi-Criterion Optimization, EMO 2005. Lecture Notes in Computer Science, vol. 3410, pp. 235–249. Springer, Berlin (2005)
20. Genchev, E.: Effects of market share on the bank's profitability. Rev. Appl. Socio-Econ. Res. **3**, 87 (2012)
21. Alvarez Carrillo, P., Leyva López, J.C., Lopez Parra, P.: A new disaggregation preference method for new products design. In: Liu, J., Lu, J., Xu, Y., Martinez, L., Kerre, E. (eds.) Data Science and Knowledge Engineering for Sensing Decision Support, pp. 1010–1017. World Scientific, Singapore (2018)
22. Marakas, G.M.: Decision support systems in the 21st-century, vol. 134. Prentice Hall, Upper Saddle River, NJ (2003)
23. Wiedenbeck, S., Davis, S.: The influence of interaction style and experience on user perceptions of software packages. Int. J. Hum Comput Stud. **46**, 563–588 (1997)

Analysis of a New Product Development Strategy Based on a Heuristic Multi-criteria Methodology

Fethullah Göçer

Abstract In the recent years, there has been a significant attention among researchers and practitioners to new product development (NPD) strategies. The development of a new product also has long been categorized as the key function of companies in an increasingly competitive global market. However, the initiation of a new product is a process involving risk and uncertainty. That is why companies needs to adapt more accurate product development strategies and evaluate the launching of a new product carefully. One way to cope with this risk is to use novel product development strategies. Therefore, this control problem is formulated as a systematic decision process in order to select the more rational candidate to be launched as a new product. Basically in this chapter, the determination of a comparable new product alternatives and the selection of the best one is done through an integrated approach based on a heuristic multi-criteria decision methodology. The Pythagorean Fuzzy sets (PFSs) are used as an objective world environment since its definite advantages in handling vagueness and uncertainty. A significant focus of the chapter is the dependency of decision criteria and to reflect this situation, the Pythagorean Fuzzy based heuristic approach is proposed for the first time as a combination of AHP (Analytic Hierarchy Process) and ARAS (Additive Ratio Assessment). A production system is considered where manufacturing of a new product is performed in a Group Decision Making (GDM) setting. Literature reviewed in this chapter presents the current state of the art and discusses the potential future research trends. Finally, a practical case study is presented to demonstrate the potential of the methodology and validate the outcome.

F. Göçer (✉)
Industrial Engineering Department, Kahramanmaraş Sütçü İmam University, 46040 Kahramanmaraş, Turkey
e-mail: gocer.fethullah@gmail.com

C. Kahraman and S. Cebi (eds.), *Customer Oriented Product Design*, Studies in Systems, Decision and Control 279, https://doi.org/10.1007/978-3-030-42188-5_17

1 Introduction

The market dynamics, technology shifts, and competitive globalization have brought transformation into every sector and New Product Development (NPD) strategy has been made one of the most complex activities of these changes. The companies have been affected by the monumental changes that constantly impact themselves and enforced to innovate in a hustle with a more efficient manner and better qualitied outputs. In turn, this has made NPD to be among the one of the most complex, difficult but core functions of companies. Using a well-considered NPD strategy, time waste, money lost, and misplaced resources can be avoided so that NPD strategy helps to organize product plan, capture consumers' viewpoints and expectations, resource the NPD project accurately. Companies seeking to remain competitive, profitable, and efficacious must constantly conceive NPD that not only exist in the market, but also bring greater value to target consumers. That is, it is quite significant for companies to comprehend what their audience want, whether related output already exist or how their outputs delivers superiority to those developed by the competitors [31]. A novel type of NPD strategies is observed in newly emerging areas due to digital transformation in daily life. Smartphone is a type of tool at its core. That's why myriad of developers have created tons of tools or apps for them. Mobile apps or applications are software designated to run on mobile devices. NPD for apps is a cumbersome process due to diversity of utility apps, which can do quite a few things actually. Incorporating apps with daily life requires robust NPD strategies. Instead of hastening with app launch, app developers should search for thoughtful and patient strategies to achieve the best new products.

All the Decision-making tasks in a real life are based on identifying and selection of candidates in an uncertain and imprecise environments by the preferences and expectations of the decision maker(s) (DMs) [10, 11]. There are several steps in a decision-making process [31]: problem definition, requirements determination, goal establishment, candidate identification, criteria determination, solution methodology selection, candidate evaluation with respect to defined criteria, and conclusively solution validation. In a real-world, these steps are ordinarily automated and too complicated in nature. Namely, the evaluation of the real decision-making problems is not usually done by the conventional decision making techniques. The appropriate description of the real-world decision-making problems in the context of mathematical representation is one of the most crucial requirements of DMs. The conventional techniques represent the DMs judgments in ordinary crisp sets. However, for the most everyday instances, the given information is ill-structured and ambiguous, or the DMs are incapable of assigning crisp values for their judgments due to subjectivity and qualitative nature of evaluation criteria. Because of the difficulty of expressing the DMs' judgments by ordinary crisp sets, DMs mostly prefer linguistic evaluations in their preferences [14]. Besides, the conventional techniques seem to have less effect in dealing with uncertain nature of linguistic evaluations.

The real life decision making problems are too complex and involve vagueness and fuzziness. Therefore, Zadeh [56] introduced the concept of fuzzy sets and represent the data rather fuzzy, but not precise. A mathematical validity is rationalized by this theory to capture the uncertainty associated with the human cognitive. Bellman and Zadeh [6] has developed the term of the fuzzy decision making as meaning the process in which the objectives or the constraints are fuzzy in nature but not necessarily the system. Also, given that the data can be incomplete, resulting in an inadequate and imprecise information at hand, the ordinary crisp sets or type 1 fuzzy sets cannot truly describe real-world problems. The weaknesses of ordinary crisp sets and type 1 fuzzy sets is addressed by the utilization of the Intuitionistic Fuzzy (IF) theory. Atanassov [4] has presented an extension of the fuzzy logic and developed IF set theory by assigning degrees of membership, non-membership, and hesitancy for all of the elements in the set. The key significance of IF sets is the differentiation of the positive indications and the negative indications to each element's membership and non-membership in the set [10, 11]. Pythagorean Fuzzy Sets (PFSs) theory is the extended versions of IF sets theory, which is developed recently by Yager in order define elements by membership and non-membership degrees with the square sum of them is being maximally 1 [52, 54]. The PFSs have been widely applied to several types of Multi-Criteria Decision-Making (MCDM) problems due to their ability to successfully represent the fuzzy character of things by involving the uncertainty and vagueness associated with them. The benefits of the PFSs theory compared to other sets can be expressed as its elasticity in the assignment of the membership degree and non-membership degree.

Several new types of solution approaches have been developed up to now with an objective to cope with the complications of the NPD problems. One of the best promising approaches to the evaluation and analysis of such problems is based on the notion of heuristic algorithms. Although heuristic methods are designed to solve problems in a faster and more efficient manner, if they are not arranged adequately, they do not guarantee that the best will be found and usually find a solution close to the best one. Heuristic methods combining several MCDM methods can play an indispensable role in being a cure to this approximate solution problem. An accurate solution can be achieved by combining two MCDM techniques. In order to obtain a well-defined criteria set and preference structure, hierarchical analyses should be carried out and the Analytical Hierarchy Process (AHP) [46], is one the best tools to do that. The evaluation of alternative sets and their performance can be ranked using several MCDM techniques. The Additive Ratio Assessment (ARAS) [57] technique is also one of the best tools to rank alternatives based on DMs judgments. ARAS technique lean on the notion of simple relative evaluations can be used to solve the complex domains with conflicting criteria [12]. In addition, PF values are applied to deliver a prospect for a much more adequate modelling to solve complex real case problems.

The proposed heuristic MCDM methodology uses PF values to express the performance ratings of criteria and alternatives in a Group Decision Making (GDM) environment. GDM is mostly better equipped as far as information is concerned rather than an individual to have all the information that is available to a group [13]. The heuristic delivers greater flexibility and much more adequate handling of group performance. The integration of AHP and ARAS techniques under PFSs provides means to DMs in utilization of the advantages for AHP in evaluating the problems linearly and benefits of ARAS in evaluating the experts' preference practically. This integration of techniques under PFSs leads to an accurate, simpler and more efficient algorithm in order to solve complex real case decision-making problems. This heuristic MCDM methodology is then validated on a case study for a NPD process. The contribution of the proposed chapter to the extant literature is the offering of a novel heuristic MCDM for the first time and applying it to a NPD problem to provide a successful selection procedure.

The flow of the chapter is as follows: the extant literature overview on the subject is given in Sect. 2. Section 3 presents the detailed description of the proposed heuristic methodology for NPD analysis. A case study is followed in Sect. 4 involving the proposed heuristic MCDM approach. The Sect. 5 provides the managerial implications of the study. The Conclusions section is delivered with some future suggestions at the end.

2 Literature Review

This section provides a comprehensive review of state of the art for NPD by examining online all databases within the scope of Google Scholar, Elsevier's Scopus, and Thomson Reuter's Web of Science for a period of the last 10 years. The selected publications are categorized to present a structured overview of ongoing research. As far as the author knows, there is no NPD study existing so far in a PF environment. The proposed research will be the first pioneering study that applies PF values with AHP and ARAS techniques. Since there is abundant amount of NPD studies in the literature, the following studies are narrowed down and aligned to our topic.

Having a novel NPD strategy means having a steady process in place to create a new product idea. Also, a necessary a rigorous analytical effort is needed to be undertaken in order to identify the NPD ideas. As noted above, this chapter aims to provide a heuristic MCDM method for NPD. Chen et al. [21] presented a MCDM method based on fuzzy DEMATEL and fuzzy ANP for NPD performance assessment in a lightning plant factory. Oliveira et al. [39] provide AHP and fuzzy TOPSIS based MCDM evaluation tool for NPD focused on a small and medium enterprise. Dursun and Arslan [26] also presented a fuzzy MCDM methodology for NPD process but based on quality function deployment approach and 2-tupple fuzzy linguistic by fuzzy COPRAS (COmplex PRoportional ASsessment) in a detergent manufacturer case. Dogu and Esra Albayrak [24] researched the marketing strategy in NPD by a MCDM method, namely Cognitive Map, under IF environment in technological device manufacturing company. Pourjavad and Peng [43] proposed an integrated MCDM approach based on the fuzzy AHP and fuzzy TOPSIS to evaluate NPD process for a pipe and fitting industry. Ayağ [5] is also proposed an integrated MCDM approach to concept evaluation in a NPD process but analytic network process (ANP) and TOPSIS techniques are the ones that are integrated. The validity of the proposed approach is presented on a real example of a hot runner system manufacturer. Ng and

Chuah [38] proposed an AHP based evaluation approach to evaluate the performances of alternatives in a NPD process and illustrate the validity of approach by a case study in an original equipment manufacturer of small electric appliance company. Yeh et al. [55] developed a hybrid MCDM approach in order to identify critical factors in NPD by utilizing DEMATEL and fuzzy AHP techniques. Chang [18] aimed to integrate ANP and TOPSIS to evaluate NPD project selection for a food industry. Lu et al. [35] proposed a MCDM evaluation approach using fuzzy hierarchical criteria GDM to assess NPD process in a garment industry.

The PFSs are also characterized by membership and non-membership degrees. In this set, the square sum of these two degrees is less than or equal to one but their sum is not required to be one [20, 25, 33, 53]. By utilizing this relaxed constraint condition, PFSs can not only utilize uncertain information but can also capture more ambiguous and vague information [19, 34, 58]. There are various MCDM methodologies and techniques utilized under PF environment, such as Choquet integral [42], TODIM [44], VIKOR [20], TOPSIS [33, 59], etc. The AHP approach under PFSs is utilized under interval valued setting for risk assessment by many researchers [27–29, 40, 50]. There is only a single research focuses on PF AHP method [9]. This research delivers and integration of PF AHP and PF COPRAS approaches to evaluate digital supply chain partners under a GDM environment for the first time in the literature. The proposed methodology in this paper has been using the main propositions of Buyukozkan and Gocer in utilization of the PF AHP approach.

Compared with the current MCDM approaches based on PFSs, no research has focused on the development of the proposed methodology, namely integrated AHP and ARAS techniques under PF environment. The review above shows that the NPD approaches usually focused on implementing NPD process in a type 1 fuzzy set environment and they are mostly in non-GDM settings. In order to fill this apparent gap, this study proposes a heuristic MCDM method.

In this chapter, the PF AHP approach and the PF ARAS approach are integrated to determine the most suitable NPD strategy. It must be emphasized that the proposed methodology is verified for the problem evaluation and the outcome obtained by the heuristic demonstrate the rank order of the NPD alternatives. To the best of author's knowledge, such a study has not yet been done in the literature in earlier publications, distinguishing itself by integrating AHP and ARAS techniques under PF environment with a GDM setting for a real case NPD selection problem.

3 A Heuristic Multi-criteria Methodology

The goal in this chapter is to distinguish the most desirable candidate in a set of NPD alternatives based on the assessments of DMs. Bearing in mind that NPD evaluations are characterized by high ambiguity, the heuristic MCDM methodology is applied to rank NPD alternatives. The following section briefly introduces the concept of PFSs and their operations.

3.1 Preliminaries

The concept of PFSs is initially proposed by Yager as an extension to IF sets to establish a distinctive theory in dealing with uncertainty and fuzziness by taking the membership degree and non-membership degree into consideration as pairs ($\mu_P(x) : X \to [0, 1], \upsilon_P(x) : X \to [0, 1]$) [51, 54].

X is a fixed nonempty universe and a PFS P in X is denoted as:

$$\mathrm{P} = \{\langle x, \mu_P(x), \upsilon_P(x)\rangle\}, \tag{1}$$

Subjected to:

$$0 \le (\mu_P(x))^2 + (\upsilon_P(x))^2 \le 1, \tag{2}$$

$$\pi_P(x) = \sqrt{1 - (\mu_P(x))^2 + (\upsilon_P(x))^2} \quad \forall x \ \mathrm{X}, \tag{3}$$

The PF arithmetic operations are presented by using two PF numbers, $p_1 = \left(\mu_{P_1}(x), \upsilon_{P_1}(x)\right)$ and $p_2 = \left(\mu_{P_2}(x), \upsilon_{P_2}(x)\right)$ and $\lambda > 0$, as follows [41, 59]:

$$p_1 \otimes p_2 = \left(\begin{array}{c}\mu_{P_1}(x) \cdot \mu_{P_2}(x), \\ \sqrt{\left(\upsilon_{P_1}(x)\right)^2 + \left(\upsilon_{P_2}(x)\right)^2 - \left(\upsilon_{P_1}(x)\right)^2 \cdot \left(\upsilon_{P_2}(x)\right)^2}\end{array}\right), \tag{4}$$

$$p_1 \oplus p_2 = \left(\begin{array}{c}\sqrt{\left(\mu_{P_1}(x)\right)^2 + \left(\mu_{P_2}(x)\right)^2 - \left(\mu_{P_1}(x)\right)^2 \cdot \left(\mu_{P_2}(x)\right)^2}, \\ \upsilon_{P_1}(x).\upsilon_{P_2}(x)\end{array}\right), \tag{5}$$

$$\frac{p_1}{p_2} = \left(\frac{\mu_{P_1}(x)}{\mu_{P_2}(x)}, \sqrt{\frac{\left(\upsilon_{P_1}(x)\right)^2 - \left(\upsilon_{P_2}(x)\right)^2}{1 - \left(\upsilon_{P_2}(x)\right)^2}}\right), \tag{6}$$

$$if\, \mu_{P_1}(x) \le \min\left\{\mu_{P_2}(x), \frac{\mu_{P_2}(x).\pi_{P_1}(x)}{\pi_{P_2}(x)}\right\}, \upsilon_{P_1}(x) \ge \upsilon_{P_2}(x)$$

$$p_1 \ominus p_2 = \left(\sqrt{\frac{\left(\mu_{P_1}(x)\right)^2 - \left(\mu_{P_2}(x)\right)^2}{1 - \left(\mu_{P_2}(x)\right)^2}}, \frac{\upsilon_{P_1}(x)}{\upsilon_{P_2}(x)}\right), \tag{7}$$

$$if\, \mu_{P_1}(x) \ge \mu_{P_2}(x), \upsilon_{P_1}(x) \le \min\left\{\upsilon_{P_2}(x), \frac{\upsilon_{P_2}(x) \cdot \pi_{P_1}(x)}{\pi_{P_2}(x)}\right\},$$

$$p_1^\lambda = \left(\mu_{P_1}(x)\right)^\lambda, \sqrt{1 - \left(1 - \left(\upsilon_{P_1}(x)\right)^2\right)^\lambda}, \tag{8}$$

$$\lambda p_1 = \left(\sqrt{1 - \left(1 - \left(\mu_{P_1}(x)\right)^2\right)^\lambda}, \left(\upsilon_{P_1}(x)\right)^\lambda\right), \tag{9}$$

$$p_1^C = \left(\upsilon_{P_1}(x), \mu_{P_1}(x)\right), \tag{10}$$

Pythagorean Fuzzy Weighted Geometric (PFWG) Aggregation [52, 54] operator is defined as follows:

$$PFWG = P_1^\lambda \otimes P_2^\lambda, \ldots, \otimes P_K^\lambda = \left\langle \frac{\prod_{k=1}^K \mu_{P_k}(x)^\lambda,}{\sqrt{1 - \prod_{k=1}^K \left(1 - \left(\upsilon_{P_k}(x)\right)^2\right)^\lambda}} \right\rangle \tag{11}$$

3.2 *Proposed Heuristic Approach*

The step-by-step presentation of the proposed methodology are provided in this section.

Step 1: Define decision criteria and available alternatives

The set of Alternatives ($A_i, i = 1, 2, \ldots, m$) are evaluated by the set of decision criteria ($C_j, j = 1, 2, \ldots, n$).

Step 2: Determine weights of DMs

The level of individual weights ($\lambda_k, \sum_{k=1}^K \lambda_k = 1$) in the set of DMs ($D_k, k = 1, 2, \ldots, K$) may vary since their experience, knowledge and responsibilities differ. In order to determine the weights of DMs, following procedure is applied:

- The importance of each DM is expressed in terms of linguistic terms.
- The PF linguistic scale in Table 1 is used to assess the weights, which are adapted from Cui et al. [22].
- The importance of each DM is aggregated with the PFWG aggregation operator.
- The weight of the kth DM is calculated by Eq. (12).

Table 1 Linguistic variables for assessment of DMs and alternatives

Preference		$(\mu_P(x)$,	$\upsilon_P(x))$
Extremely unimportant	EU	(0.15,	0.85)
Very unimportant	VU	(0.25,	0.75)
Unimportant	U	(0.35,	0.65)
Medium importance	MI	(0.50,	0.45)
Important	I	(0.65,	0.35)
Very important	VI	(0.75,	0.25)
Extremely important	EI	(0.85,	0.15)

Table 2 Linguistic variables for the importance of criteria

Linguistic variable		PFSs		Reciprocal	
		$(\mu_P(x),$	$v_P(x))$	$(\mu_P(x),$	$v_P(x))$
Equally important	EI	(0.07,	0.30)	(0.30,	0.07)
Intermediate	IV	(0.18,	0.49)	(0.49,	0.18)
Moderately more important	MI	(0.29,	0.60)	(0.60,	0.29)
Intermediate	IV2	(0.39,	0.65)	(0.65,	0.39)
Strongly more important	SI	(0.50,	0.67)	(0.67,	0.50)
Intermediate	IV3	(0.61,	0.66)	(0.66,	0.61)
Very strong importance	VSI	(0.71,	0.61)	(0.61,	0.71)
Intermediate	IV4	(0.82,	0.52)	(0.52,	0.82)
Extremely more important	EMI	(1.00,	0.00)	(0.00,	1.00)

$$\lambda_k = \frac{\frac{\sqrt{\mu_{P_j}(x) - \left(v_{P_j}(x)\right)^2}}{2}}{\sum_{k=1}^{K} \frac{\sqrt{\mu_{P_j}(x) - \left(v_{P_j}(x)\right)^2}}{2}}, \text{ where } \sum_{k=1}^{K} \lambda_k = 1 \tag{12}$$

Step 3: Define DMs' assessment on each factor

- DMs express their judgments in linguistic terms on each factor.
- The preference relation matrix ($\tilde{X}_k, k = 1, 2, \ldots, K$) for each DM is constructed by the linguistic scale in Table 1.
- The pairwise comparison matrix ($\tilde{P}_j^k, j = 1, 2, \ldots, n, k = 1, 2, \ldots, K$) of each criteria and sub-criteria for each DM is established by the linguistic scale in Table 2.

The PF AHP preference scale is constructed according to the conversion of consistency expressions to membership grades [1].

Step 4: Aggregate individual evaluations

- Establish GDM matrix of alternatives by the PFWG aggregation operator in Eq. (11).
- Establish GDM matrix of each DMs for the criteria evaluations to construct aggregated pairwise comparison matrix by the PFWG aggregation operator in Eq. (11).

Step 5: Check Consistency Ratio (CR)

By Eq. (13), CR is checked for the aggregated PF matrix. The RI (Random Index) is originated from Saaty [45] is presented in Table 3 up to fifteen criteria;

$$CR = \frac{RI - \frac{\sum \pi_P(x)_{ij}}{n}}{n - 1} \tag{13}$$

Here, elements' number of given matrix is represented by n. The hesitation value is represented by $\boldsymbol{\pi_P(x)_{ij}}$.

Table 3 Values of RI

n	RI	n	RI	n	RI
1	0.00	6	1.24	1.52	11
2	0.00	7	1.32	1.54	12
3	0.58	8	1.41	1.56	13
4	0.90	9	1.45	1.57	14
5	1.12	10	1.49	1.59	15

If the estimated CR value is less than or equal to 0.10, the matrix is seen as consistent. If it is not case, then the opinions of each DM should be collected once again.

Step 6: Construct criteria weights

PFWG aggregation operator in Eq. (11) is used to combine the PF judgment matrix. The PF AHP criteria weights ($\tilde{w}_j$) are constructed based on all expert opinions.

Step 7: Construct PF Optimal Decision Matrix ($\tilde{X}_{(Opt)}$)

A PF optimal decision matrix (see Eq. (14)) for $m + 1$ alternatives and n criteria is constructed.

$\tilde{x}_{0j}$ represents the optimal performance ratings of *jth* criterion.

The optimal values are constructed by Eqs. (15 and 16):

the higher value is the better is considered as a benefit criterion, then

$$\tilde{X}_{(Opt)} = \begin{bmatrix} \tilde{x}_{01} & \cdots & \tilde{x}_{11} & & \tilde{x}_{i1} & \cdots & \tilde{x}_{m1} \\ \vdots & \ddots & \vdots & \cdots & \vdots & \ddots & \vdots \\ \tilde{x}_{02} & \cdots & \tilde{x}_{12} & & \tilde{x}_{i2} & \cdots & \tilde{x}_{m2} \\ & \vdots & & \ddots & & \vdots & \\ \tilde{x}_{0j} & \cdots & \tilde{x}_{1j} & & \tilde{x}_{ij} & \cdots & \tilde{x}_{mj} \\ \vdots & \ddots & \vdots & \cdots & \vdots & \ddots & \vdots \\ \tilde{x}_{0n} & \cdots & \tilde{x}_{1n} & & \tilde{x}_{in} & \cdots & \tilde{x}_{mn} \end{bmatrix} \quad (14)$$

$$\left(\mu_{P_{ij}}\left(\tilde{x}_{0j}\right) = \max_i \mu_{P_{ij}}\left(\tilde{x}_{ij}\right), \upsilon_{P_{ij}}\left(\tilde{x}_{0j}\right) = \min_i \upsilon_{P_{ij}}\left(\tilde{x}_{ij}\right)\right) \quad (15)$$

the lower value is the better is considered as a cost criterion, then

$$\left(\mu_{P_{ij}}\left(\tilde{x}_{0j}\right) = \min_i \mu_{P_{ij}}\left(\tilde{x}_{ij}\right), \upsilon_{P_{ij}}\left(\tilde{x}_{0j}\right) = \max_i \upsilon_{P_{ij}}\left(\tilde{x}_{ij}\right)\right) \quad (16)$$

Step 8: Establish normalized decision matrix

Normalize the matrix by Eqs. (17) and (18), $\tilde{R} = \left[\tilde{r}_{ij}\right]_{m\times n}$ with $\tilde{r}_{ij} = (\mu_P(r), \upsilon_P(r))$.

$$\tilde{r}_{ij} = (\mu_P(r), \upsilon_P(r)) = \tilde{x}_{ij} = (\mu_P(x), \upsilon_P(x)), \quad \text{For benefit criteria} \tag{17}$$

$$\tilde{r}_{ij} = (\mu_P(r), \upsilon_P(r)) = \tilde{x}_{ij}^C = (\upsilon_P(x), \mu_P(x))^C, \text{For cost criteria} \tag{18}$$

Step 9: Establish weighted decision matrix
The weighted decision matrix $\tilde{\tilde{R}} = [\tilde{\tilde{r}}_{ij}]_{m \times n}$ is formed by Eq. (19).

$$\tilde{\tilde{R}} = \tilde{r}_{ij} \otimes \tilde{w}_j \tag{19}$$

Step 10: Determine Optimality Function Values
Optimality function values for all alternatives are calculated using Eq. (20) where the PF summation operator (see Eq. (5)) is applied. Here, $\tilde{f}_i$ represents the optimality function value for alternative i.

$$\tilde{f}_i = \oplus_{j=1}^{n} \tilde{\tilde{r}}_{ij} \qquad i = 0, 1, \ldots, m \tag{20}$$

Step 11: Determine Alternative Utility Degree
By the comparison of variant by Eq. (21), the alternative utility degree for alternative i ($\tilde{Q}_i$) is calculated where the PF division operator defined by Eq. (6) is used.

$$\tilde{Q}_i = \tilde{f}_i \oslash \tilde{f}_0 \tag{21}$$

Step 12: Rank Alternatives
The $\tilde{Q}_i$ values for $i = 1, \ldots, m$ are defuzzified [32] by Eq. (22).
The alternative that has the maximum alternative utility degree (Maximum Q_i) is considered as the best one.

$$Q_i = \frac{\sqrt{\mu_{P_j}(x) - \left(\upsilon_{P_j}(x)\right)^2}}{2}, \tag{22}$$

4 Case Study

Smartphones have altered the daily life of humans forever. The number of people accessing information via smartphones has been increasing with every passing minute. People use apps through smartphones to store and manage data, shop online, or socialize, etc. Therefore, NPD process in smartphone apps are analyzed as a case study for a Turkish utility apps developer. Following section presents the synopsis of the Developer Company and hierarchical criteria structure for assessing the NPD process in apps development.

4.1 Case Background and NPD Criteria Structuring for Apps

The proposed method is applied to a private Apps Developing Company located in Turkey. The identity of the company is unrevealed due to privacy and confidentiality reasons. The company will be addressed as the APPs Company, hereafter. The APPs Company is a young enterprise founded by classmates at collage university. It has been established in 2016 and based in Istanbul, Turkey, to come up with the solutions of Apple (iOS) and Android applications. The APPs Company has experienced in-house team and has been professionally developing apps for smartphones, tablets, wearables, and smart televisions. It provides tailored apps upon customer's demands and has a trustworthy experience in maintaining and developing all sorts of utility apps to meet latest industry standards. The competition in apps development is a fierce business and a major concern for the APPs Company, which it considers during NPD process and company strategy. Therefore, the assessment of the most suitable utility apps is a significant decision for the APPs Company.

A group of experts consisting of three DMs, an electrical engineer (software developer and co-founder), a top-level manager (software product and business development manager), and an academician (a scholar from a respectful Turkish university), are assigned to evaluate eight utility apps in a NPD strategy. The apps development process is an exceptional setting for NPD case, bearing in mind that there exists a noticeable difference between traditional production process and apps production. By the extant literature survey and experience of DMs, as well as evaluations of the executives of the APPs Company, the evaluation criteria are constructed to assess and rank the available apps alternative at hand. The findings of extant literature and the opinions of the DMs has led to identification of 5 main criteria and 16 sub-criteria. The necessity of benefit type and cost type attributes in a feasible manner during the criteria determination phase is stated as a requisite by the DMs. Therefore, the DMs have considered the number of criteria to be enough for the assessment process of the study. Table 4 displays the identified decision criteria and their simple description. Figure 1 presents the hierarchical structure of decision criteria in NPD process. As it is easily inferred from Table 4 and Fig. 1, third and fourth criteria are regarded as cost type of criteria while the rest are regarded as benefit type. The APPs Company aims to select the best candidate that addresses its needs and expectations. For this case, eight apps alternative are evaluated with which it has plans to work on. These alternatives are referred as $A_1, A_2, A_3, A_4, A_5, A_6, A_7$ *and* A_8 from now on.

Brief descriptions of the apps alternatives are narrated as follows:

$\boldsymbol{A}_1$: A translation app having translation features of more than 100 languages from all over the world with the instant camera option to take pictures of text for higher-quality translations.
$\boldsymbol{A}_2$: A cleaning and performance-boosting app having optimization features such as game booster, app cache cleaner, speed booster, and even battery saver, etc.
$\boldsymbol{A}_3$: A powerful barcode scanner app to use, import, create and share data in a matter of taps.

Table 4 NPD selection criteria and their description

Criteria	Description
Strengths (C_1)	This criterion measures how the creation of a new or different product that offers innovative new benefits to the end user [7, 8, 31]
Profitability (C_{11})	This sub criterion deals with cost-effectiveness in NPD process or share of market for a new or different product under the current competitive environment [8, 31]
Efficiency (C_{12})	This sub criterion evaluates the strategic direction to determine the improvements in efficiency of NPD process to reduce costs and time to reach to market [7, 8, 31]
Innovation (C_{13})	This sub criterion seeks to create, exchange, evolve, and apply a new or different product ideas, and thereby contribute to the success of NPD process [2, 3]
Strategic impact (C_2)	This criterion is the assessment of the wider strategic and business impacts of a new or different product proposals at the beginning of a NPD process [7, 8, 31]
Strategic value (C_{21})	This sub criterion assesses the value in NPD process that brings better a new or different product with faster and without too much effort to create value stream [8, 31]
Business impact (C_{22})	This sub criterion measures the impact that considers the nature of the business and the overall motivation in NPD [8, 31]
Total costs (C_3)	This criterion is one of the most significant one assessing the total expense in all operations [15–17]
Appraisal cost (C_{31})	This sub criterion measures the cost to acquire quality control and regulatory requirements [15–17]
Investment cost (C_{32})	This sub criterion measures the first cost of investment for related app [15–17]
Cost of service (C_{33})	This sub criterion is the cost of app to operate [15–17]
Follow-up cost (C_{34})	This sub criterion is found to affect the NPD process by measuring the follow-ups of app [15–17]
Risks (C_4)	This criterion assesses the risks to support the launching of a new or different product
Financial risk (C_{41})	This sub criterion deals with the risk in NPD process to make responses to the strategies, levels, and alteration in price [7, 8, 31]
Technical risk (C_{42})	This sub criterion includes the risk in NPD process that is required to get better targeted and more innovative products [7, 8, 16, 31]
Managerial risk (C_{43})	This sub criterion measures the risk that is the competitive reactions to NPD strategies [7, 8, 31]
Personnel risk (C_{44})	This sub criterion evaluates the risk that has the impact in NPD process on profits or market share [8, 31]

(continued)

Table 4 (continued)

Criteria	Description
Technical feasibility (C_5)	This criterion includes evaluating the ability of a new or different product to handle workloads adequately by ascertaining their needs [23, 37, 49]
Maintenance and support (C_{51})	This sub criterion includes the skills and requirements to maintain and support a new or different product, including troubleshooting, modifying, maintaining and enhancing legacy systems [23, 37, 49]
Platform capability (C_{52})	This sub criterion measures the ability to provide the required infrastructure to interface with all types of devices including smartphones, tablets, wearables, and smart televisions [23, 37, 49]
Operational feasibility (C_{53})	This sub criterion determines the reliability of response to the end user input at all stages of the development cycle to ensure better adaptation to the user requirements and market needs [23, 37, 49]

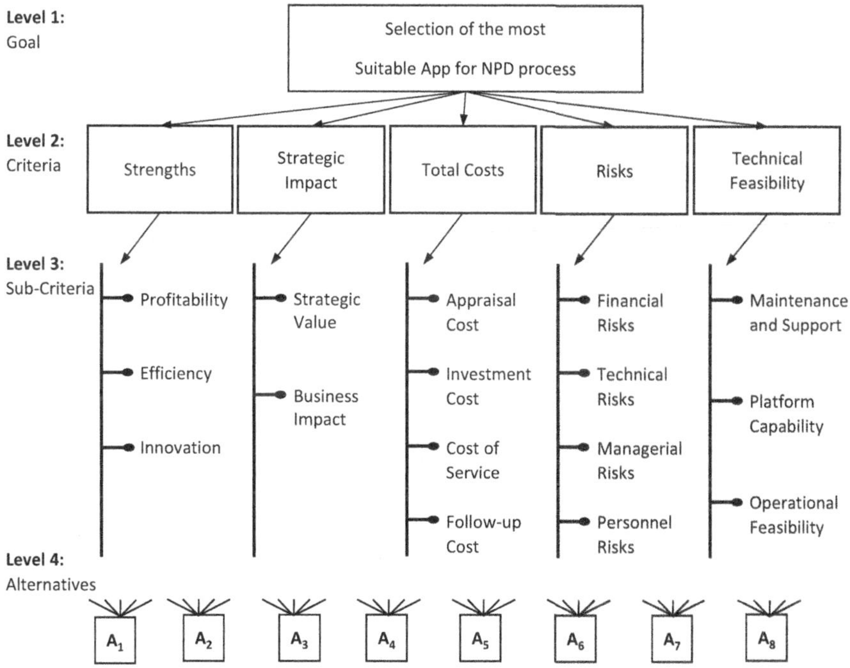

Fig. 1 Hierarchical structure of decision criteria in NPD process

A_4: A daily usage app having all in one combo of 15 tools that can prove to be quite useful to make users life easier such as flashlight, distance calculator, compass, and even more.
A_5: A security and privacy app to stop others from snooping into user's phone. It can easily hide videos and pictures in the device and can protect private memories easily.
A_6: A reminder app to take care of important errands having versatile and simple features to keep users life organized with to-do lists, reminders, and notes.
A_7: An excellent weather app that can decide on user's exact location by GPS tracking based on this information and sends precise weather alerts.
A_8: A smart authenticator app that verifies user's security stamp and gives only access to users even under device is not connected to data or Wi-Fi network.

4.2 Implementation of the Proposed Methodology

This subsection comprises the implementation of the proposed method by utilizing the NPD case study narrated in earlier subsection.

Step 1: As the overall goal is set to evaluate eight different app alternatives by five main and 16 sub-criteria.

Step 2: The priority of each DM is calculated by following the procedure in Step 2 of the methodology. The importance of DMs are gathered in terms of linguistic expressions by the preference scale in Table 1. DMs' opinions are aggregated with the PFWG aggregation operator. The level of influence of the each DM is computed by Eq. (11).

The linguistic expressions of DMs for their priorities are expressed in Table 5.

An example calculation for D_1 is presented below:

$$\lambda_1 = \frac{\frac{\sqrt{0.85-(0.15)^2}}{2}}{\left(\frac{\sqrt{0.85-(0.15)^2}}{2}\right)+\left(\frac{\sqrt{0.75-(0.25)^2}}{2}\right)+\left(\frac{\sqrt{0.65-(0.35)^2}}{2}\right)}$$
$$= 0.377,\ \lambda_2 = 0.337,\ \lambda_3 = 0.286$$

Table 5 The linguistic variables, corresponding PF values, and respective priority weights for Each DM

	D_1		D_2		D_3	
Linguistic terms	EI		VI		I	
	$(\mu_P(x),$	$\upsilon_P(x))$	$(\mu_P(x),$	$\upsilon_P(x))$	$(\mu_P(x),$	$\upsilon_P(x))$
PF value	(0.85,	0.15)	(0.75,	0.25)	(0.65,	0.35)
λ_k	0.377		0.337		0.286	

Step 3: Each DM gives their judgment on each factor in the form of linguistic terms. Table 6 presents the preference relation matrix ($\tilde{X}_k$) for each DM based on the linguistic scale in Table 1. Table 7 presents the pairwise comparison matrix ($\tilde{P}_j^k$) of each criteria and sub-criteria for each DM based on the linguistic scale in Table 2. Due to space requirements, the data is scaled to give a narrative of the core impression.

Step 4: The GDM matrix of alternatives is fused by the PFWG aggregation operator in Eq. (11). Table 8 presents the aggregated matrix of first alternative. Due to space requirements, a sample for the first alternative is disclosed. The GDM matrix of each DMs for the criteria evaluations is constructed to establish aggregated pairwise comparison matrix by the operator in Eq. (11). Table 9 presents the aggregated pairwise matrix.

Step 5: CR is checked for the aggregated PF matrix by utilizing Eq. (12). Since the calculated CR values for all matrices are smaller, or equal to 0.10, the judgment matrix is accepted as consistent.

Step 6: The PF AHP criteria weights ($\tilde{w}_j$) are constructed based on all expert opinions as displayed in Table 9. The defuzzified and respective ranking of each criterion is also given to better comprehend the assessment of each criterion. The defuzzification is done [32] by Eq. (16).

Step 7: A PF optimal decision matrix is constructed for each DM by Eqs. (15) and (16). The aggregated optimal decision matrix is displayed in Table 10.

Step 8: A normalization of decision matrix for each alternative based on each criteria is done by the formula in Eqs. (17) and (18) according to their characteristics of being whether cost or benefit criteria. The normalized values of first alternative is presented in Table 8 and normalized valued of optimal decision matrix is displayed in Table 10.

Step 9: The PF criteria values calculated in Step 6 are applied to get the weighed matrix by the formula in Eq. (19). The weighted matrix for first alternative is given in Table 8 and weighted matrix of optimal decision is presented in Table 10.

Step 10: The optimality function values are constructed by Eq. (20). The PF summation operator is utilized here to get the values. The optimality function value for each alternative is presented in Table 11.

Step 11: The alternative utility degree for each alternative is constructed by the comparison of variant using Eq. (15), where the PF division operator is used. Table 10 presents the corresponding alternative utility degree.

Step 12: The alternative that has the maximum alternative utility degree is considered as the best alternative of the evaluation. By defuzzifying the PF values, the ranking of alternatives is emerged. The best suitable alternative is calculated as second alternative '$\boldsymbol{A}_2$' while the sixth alternative '$\boldsymbol{A}_6$' is the last one in the ranking.

$\boldsymbol{A}_2$ denotes the best candidate, which means that the cleaning and performance boosting app is in first place with the maximum utility degree compared to other alternatives. In second place is the daily usage app '$\boldsymbol{A}_4$' and third place is the barcode scanner app '$\boldsymbol{A}_3$'. The worst-ranked NPD alternative is the reminder app '$\boldsymbol{A}_6$', which is in some way logical and expected outcome for the APPs Company.

Table 6 The preference relation matrix of alternatives

	D_1	D_2	D_3	D_1	D_2	D_3	D_1	D_2	D_3	D_1	D_2	D_3
	A_1			A_2			A_3			A_4		
C_{11}	MI	VU	VU	EI	MI	EI	I	U	VU	EU	I	VU
C_{12}	I	MI	MI	VI	I	EI	U	MI	I	I	MI	U
C_{13}	U	U	VI	MI	MI	MI	EI	I	EI	VU	U	MI
C_{21}	VU	U	U	MI	U	I	U	MI	MI	MI	VU	VI
C_{22}	MI	MI	U	I	MI	MI	MI	VI	VI	EI	EU	MI
C_{31}	U	MI	EI	MI	VI	MI	U	MI	EI	U	VI	EI
C_{32}	U	MI	MI	EI	VU	MI	EI	MI	I	VU	EU	EU
C_{33}	MI	I	U	I	EI	U	U	MI	U	MI	VI	EI
C_{34}	VU	MI	VU	MI	VI	VI	VI	EI	I	U	MI	VU
C_{41}	VU	MI	VI	MI	MI	I	MI	MI	EI	EI	EU	EI
C_{42}	MI	U	MI	VU	MI	VU	I	MI	MI	EU	I	VU
C_{43}	VU	U	U	I	I	EI	EI	VI	MI	VU	EI	MI
C_{44}	MI	EU	I	MI	EU	MI	U	EI	U	U	U	MI
C_{51}	MI	U	MI	MI	MI	MI	MI	MI	MI	U	U	U
C_{52}	MI	MI	U	VI	EI	U	EI	I	U	I	EI	U
C_{53}	EI	U	VU	U	MI	VI	MI	MI	EI	U	EI	I
	A_5			A_6			A_7			A_8		
C_{11}	I	MI	U	MI	I	MI	VI	VI	U	VU	VU	U
C_{12}	EU	I	VU	MI	EI	MI	MI	MI	EI	MI	U	MI
C_{13}	VU	U	MI	VU	I	EI	MI	U	VI	VU	EU	VI

(continued)

Table 6 (continued)

C_{21}	VU	EU	VI	MI	U	EI	MI	MI	U	U	VI	VI
C_{22}	M	EI	U	I	U	EI	I	EU	MI	EI	EU	MI
C_{31}	VI	MI	EI	MI	VI	MI	EI	VU	U	U	MI	VI
C_{32}	EU	VI	MI	U	MI	U	I	U	EI	I	EI	VI
C_{33}	VU	MI	U	I	EI	U	U	MI	EI	MI	U	MI
C_{34}	EU	EI	VU	MI	VI	EI	U	MI	VU	U	VU	U
C_{41}	EU	I	EI	MI	MI	EI	VI	VU	VU	VU	U	VU
C_{42}	EI	EU	VU	U	MI	I	MI	U	MI	MI	U	MI
C_{43}	EI	VU	MI	MI	MI	EI	EI	MI	MI	MI	MI	EU
C_{44}	U	U	EI	MI	VI	EI	I	I	EI	EI	MI	VI
C_{51}	U	U	I	EI	U	VI	EI	MI	VI	VU	I	U
C_{52}	MI	U	U	MI	MI	U	U	VI	I	EI	EI	VI
C_{53}	EI	U	I	I	U	I	EU	MI	MI	EU	U	U

Table 7 The pairwise comparison matrix of criteria

D_1	C_1	C_2	C_3	C_4	C_5		D_1	C_{11}	C_{12}	C_{13}	D_1	C_{21}	C_{22}
C_1	EI	MI	MI	1/MI	IV4		C_{11}	EI	IV4	SI	C_{21}	EI	VSI
C_2		EI	1/SI	IV2	MI		C_{12}		EI	1/SI	C_{22}		EI
C_3			EI	SI	1/IV		C_{13}			EI			
C_4				EI	MI								
C_5					EI								
D_2	C_1	C_2	C_3	C_4	C_5		D_2	C_{11}	C_{12}	C_{13}	D_2	C_{21}	C_{22}
C_1	EI	IV2	SI	1/VSI	IV2		C_{11}	EI	VSI	IV3	C_{21}	EI	IV4
C_2		EI	IV	VSI	IV		C_{12}		EI	IV4	C_{22}		EI
C_3			EI	IV2	1/IV		C_{13}			EI			
C_4				EI	SI								
C_5					EI								
D_3	C_1	C_2	C_3	C_4	C_5		D_3	C_{11}	C_{12}	C_{13}	D_3	C_{21}	C_{22}
C_1	EI	SI	IV	1/IV3	SI		C_{11}	EI	SI	VSI	C_{21}	EI	1/IV
C_2		EI	MI	IV4	IV		C_{12}		EI	IV4	C_{22}		EI
C_3			EI	VSI	MI		C_{13}			EI			
C_4				EI	VSI								
C_5					EI								
D_1	C_{31}	C_{32}	C_{33}	C_{34}	D_1	C_{41}	C_{42}	C_{43}	C_{44}	D_1	C_{51}	C_{52}	C_{53}
C_{31}	EI	VSI	IV3	1/IV	C_{41}	EI	MI	IV2	1/SI	C_{51}	EI	MI	MI
C_{32}		EI	1/VSI	IV	C_{42}		EI	1/VSI	IV2	C_{52}		EI	1/SI
C_{33}			EI	MI	C_{43}			EI	MI	C_{53}			EI

(continued)

Table 7 (continued)

D_1	C_1	C_2	C_3	C_4	C_5		D_1	C_{11}	C_{12}	C_{13}	D_1	C_{21}	C_{22}
C_{34}				EI	C_{44}				EI				
D_2	C_{31}	C_{32}	C_{33}	C_{34}	D_2	C_{41}	C_{42}	C_{43}	C_{44}	D_2	C_{51}	C_{52}	C_{53}
C_{31}	EI	SI	MI	1/VSI	C_{41}	EI	VSI	MI	1/IV4	C_{51}	EI	IV2	IV3
C_{32}		EI	IV2	VSI	C_{42}		EI	IV	IV4	C_{52}		EI	IV3
C_{33}			EI	IV4	C_{43}			EI	IV3	C_{53}			EI
C_{34}				EI	C_{44}				EI				
D_3	C_{31}	C_{32}	C_{33}	C_{34}	D_3	C_{41}	C_{42}	C_{43}	C_{44}	D_3	C_{51}	C_{52}	C_{53}
C_{31}	EI	IV3	VSI	1/IV	C_{41}	EI	MI	SI	1/MI	C_{51}	EI	MI	IV4
C_{32}		EI	IV4	SI	C_{42}		EI	MI	MI	C_{52}		EI	SI
C_{33}			EI	IV4	C_{43}			EI	SI	C_{53}			EI
C_{34}				EI	C_{44}				EI				

Table 8 The aggregated, normalized and weighted matrix of first alternative

	Aggregated		Normalized		Weighted	
	$(\mu_P(x),$	$v_P(x))$	$(\mu_P(x),$	$v_P(x))$	$(\mu_P(x),$	$v_P(x))$
C_{11}	0.325	0.672	0.325	0.672	0.014	0.894
C_{12}	0.552	0.416	0.552	0.416	0.024	0.841
C_{13}	0.435	0.580	0.435	0.58	0.018	0.886
C_{21}	0.308	0.693	0.308	0.693	0.014	0.846
C_{22}	0.451	0.522	0.451	0.522	0.018	0.810
C_{31}	0.509	0.501	0.501	0.509	0.019	0.831
C_{32}	0.437	0.542	0.542	0.437	0.019	0.826
C_{33}	0.493	0.499	0.499	0.493	0.021	0.826
C_{34}	0.316	0.682	0.682	0.316	0.021	0.793
C_{41}	0.432	0.578	0.578	0.432	0.019	0.870
C_{42}	0.443	0.533	0.533	0.443	0.018	0.859
C_{43}	0.308	0.693	0.693	0.308	0.028	0.820
C_{44}	0.359	0.652	0.652	0.359	0.028	0.844
C_{51}	0.443	0.533	0.443	0.533	0.015	0.819
C_{52}	0.451	0.522	0.451	0.522	0.020	0.789
C_{53}	0.444	0.591	0.444	0.591	0.019	0.837

Table 9 The criteria weights and respective ranking of each criterion

$\tilde{w}_j$	$(\mu_P(x),$	$v_P(x))$	Defuzzified	Normalized	Rank
C_{11}	0.043	0.797	−0.214	0.063	7
C_{12}	0.044	0.804	−0.218	0.063	6
C_{13}	0.040	0.822	−0.237	0.064	3
C_{21}	0.046	0.674	−0.120	0.058	16
C_{22}	0.039	0.726	−0.165	0.061	14
C_{31}	0.037	0.763	−0.194	0.062	10
C_{32}	0.035	0.780	−0.210	0.063	8
C_{33}	0.041	0.762	−0.189	0.062	11
C_{34}	0.031	0.767	−0.206	0.063	9
C_{41}	0.032	0.837	−0.260	0.066	1
C_{42}	0.033	0.821	−0.246	0.065	2
C_{43}	0.040	0.798	−0.219	0.064	5
C_{44}	0.042	0.818	−0.232	0.064	4
C_{51}	0.033	0.735	−0.179	0.061	12
C_{52}	0.045	0.694	−0.135	0.059	15
C_{53}	0.044	0.735	−0.166	0.061	13

Table 10 The aggregated, normalized and weighted matrix of optimal decision

	Aggregated		Normalized		Weighted	
	$(\mu_P(x),$	$\upsilon_P(x))$	$(\mu_P(x),$	$\upsilon_P(x))$	$(\mu_P(x),$	$\upsilon_P(x))$
C_{11}	0.711	0.295	0.711	0.295	0.031	0.817
C_{12}	0.741	0.269	0.741	0.269	0.033	0.819
C_{13}	0.777	0.24	0.777	0.24	0.031	0.833
C_{21}	0.478	0.515	0.478	0.515	0.022	0.774
C_{22}	0.644	0.343	0.644	0.343	0.025	0.764
C_{31}	0.509	0.501	0.501	0.509	0.019	0.831
C_{32}	0.182	0.819	0.819	0.182	0.029	0.788
C_{33}	0.348	0.648	0.648	0.348	0.027	0.795
C_{34}	0.311	0.719	0.719	0.311	0.022	0.793
C_{41}	0.404	0.643	0.643	0.404	0.021	0.866
C_{42}	0.284	0.731	0.731	0.284	0.024	0.837
C_{43}	0.308	0.693	0.693	0.308	0.028	0.820
C_{44}	0.333	0.664	0.664	0.333	0.028	0.840
C_{51}	0.500	0.450	0.500	0.450	0.017	0.796
C_{52}	0.629	0.415	0.629	0.415	0.028	0.755
C_{53}	0.584	0.393	0.584	0.393	0.026	0.782

Table 11 The optimality function value, alternative utility degree and respective rankings

	$\tilde{f}_i$		$\tilde{Q}_i$		Defuzzified	Rank
	$(\mu_P(x),$	$v_P(x))$	$(\mu_P(x),$	$v_P(x))$		
A_0	0.104	0.032				
A_1	0.080	0.058	0.770	0.048	0.438	6
A_2	0.087	0.051	0.837	0.040	0.457	1
A_3	0.081	0.058	0.784	0.049	0.441	3
A_4	0.082	0.061	0.793	0.052	0.444	2
A_5	0.081	0.063	0.778	0.054	0.440	4
A_6	0.078	0.060	0.749	0.050	0.431	8
A_7	0.080	0.057	0.770	0.048	0.438	5
A_8	0.079	0.066	0.764	0.058	0.435	7

5 Managerial Implications

The companies' profitability in today's competitive business environment depends heavily upon their ability to introduce new products. The digital NPD is, indeed, a cost generating, complicated and time-consuming process. The extensive steps of NPD such as assessing the market requirements, determining the type of app to build, deciding on the budget and, ultimately, the execution, can be discouraging sometimes. However, they all are equally significant. Neglecting any steps involved in the NPD processes can prejudice the others and can eventually create time delays and push up the cost value. Based on the identified lack of research and more general literature on apps development, this paper will contribute to NPD process, as well as to the MCDM literature. The one of the originalities of this chapter is the application of AHP technique. This technique has first been developed by Thomas L. Saaty more than forty years ago. AHP is still being used extensively and constructed to analyze decision-making problems using a pairwise comparison [10, 11]. PF AHP is a structured technique to evaluate MCDM problems by the help of PF values in the estimation of criteria weights. There have been a number of recent studies combining AHP under PF objective world environment [29, 30, 36]. The definite contribution of the proposed study is the utilization of AHP approach from the context of real world GDM problems by using PF values. Mohd and Abdullah [36] is the first study to combine AHP and PFSs in a conference proceeding. They offer a new scale under PF environment but the consistency of the pairwise comparison matrices are not checked in their paper and thus the legitimacy of their approach is not verified. In addition, this is a proceeding paper with several unexplained claims. Ilbahar et al. [30] also propose a new scale for AHP under PF environment. However, the Interval Valued PFSs is offered in their study rather than conventional PF values. Gul [28] and Gul and Ak [29] is also using the scale proposed by Ilbahar et al. [30]. This chapter utilizes the conversion of consistency expressions to membership grades [1] to construct PF AHP preference scale. By the proposed new pairwise comparison scale, the method gets an ability to check the inconsistencies in evaluation matrix. Thus, there is a strong discrepancy between the existing methods and the proposed approach. Besides, the ARAS technique is a robust MCDM approach, which encompasses implicit or explicit tradeoffs, usually having a fixed set of alternatives, and requires criteria comparisons [57]. The ARAS technique is proposed as an innovative approach that introduces a practical alternative, whose ratings could be used in expressing the DMs' judgments for the preferred actions of the criteria. The effectiveness of the AHP and ARAS integration is demonstrated in earlier studies using different objective world environments and has been successfully applied in several areas [12, 47, 48]. The key significance here is the effective utilization of PF objective world environment for this integration.

The analysis performed in the following subsections includes three stages, which determine the comparison, sensitivity and validity of the obtained outcome. The first stage consists of ranking of alternatives according to the various objective world environments. Crisp ARAS, type one Fuzzy ARAS, and IF ARAS is used to compare rankings with the proposed approach of PF ARAS. The next phase is the formation of 16 different scenarios in which the values of criteria weights are altered to simulate the ranking variations and Spearman's correlation checking is performed for all scenarios as the third stage.

5.1 Comparison of the Proposed Methodology Under Different Environments

The application of ARAS methodology under different objective world environments is employed using the same criteria weights. In Fig. 2, it can be observed that the best app alternative does not change the rank in any environments, i.e., in all four applications it is the best ranked alternative. In addition, the last ranking alternative also remains in its original positions while other alternatives vary. In summary, the comparison of the results obtained in Table 2 demonstrates the priorities of $\boldsymbol{A}_2$ over other alternatives. The performance boosting app is regarded as the most desirable by the proposed method, yet conversely, $\boldsymbol{A}_6$ is preferred as the last candidate.

5.2 Sensitivity Analysis for the Proposed Methodology

The ranks of alternatives throughout 16 scenarios are displayed in Fig. 3. The scenarios are formed in such a way that the criteria weights are altered one by one to a maximum weight of 'EMI' that is, PF values of [1.00, 0.00, 0.00], while the others remain unchanged in sequence. There are no scenarios encompasses identical ranks. The third and twelfth scenarios present greater differences in ranks than the initial scenario. This implies the increase in significance of the strategic value and managerial risk criteria significantly influences the rank. The major difference is in the eleventh and thirteenth scenarios when the criteria weights are altered. This implies that the risk criteria play a significant role in selecting the best suitable alternatives for the NPD process in apps development.

Figure 4 presents the Spearman's correlation coefficient through all the scenarios.

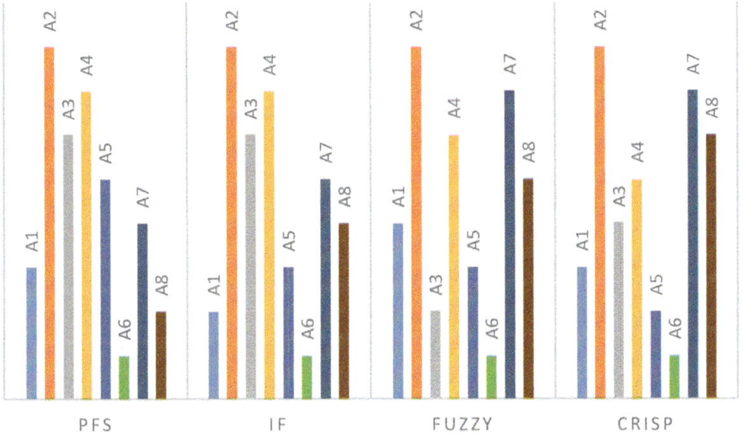

Fig. 2 Ranking of alternatives according to the various environments

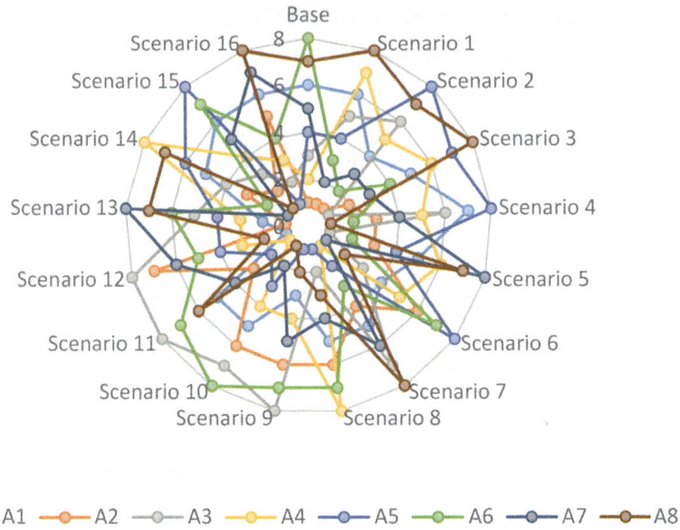

Fig. 3 Ranking order of alternatives through various scenarios of criteria variations

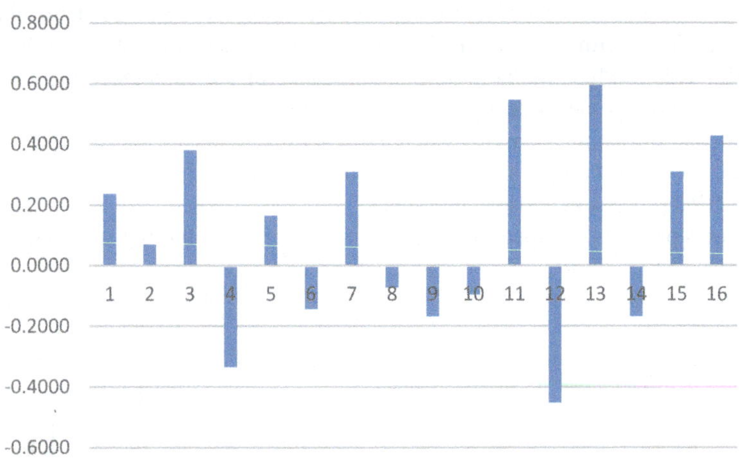

Fig. 4 The Spearman's correlation coefficient through various scenarios of criteria variations

This validates that the proposed approach is sensitive to changes of the criteria weights and that each factor plays a key role in assessing the NPD process for apps development. Spearman's correlation coefficient ranges in between 0.596 to −0.452, and shows the association between the variables would not be considered statistically significant, and the outcome of the proposed approach are considered stable. The results from this analysis have illustrated that each criterion variations results in a different path of ranks.

6 Conclusion

The NPD strategies has gaining a growing attention among researchers and practitioners in every passing year. The NPD process is considered as the leading task for organizations wishing to stay competitive in global markets. Still, the initiation of a new product is attached with risks and uncertainties. The NPD strategies can compensate these risks so that companies can adapt NPD processes more successfully. Thus, this chapter offers a systematic decision-making strategy to select the more rational alternative among NPD candidates. That is, a heuristic multi-criteria decision-making approach is proposed in this chapter to determine the best candidate of comparable new products alternatives.

The proposed heuristic approach is based on the integration of the AHP technique and the ARAS technique, under the PF environments of GDM setting. In this approach, the AHP is used to determine the weights of the evaluation criteria, whereas the ARAS is used to evaluate the alternatives and obtain the ranking order for each alternatives involved in the evaluation. The AHP technique is selected due to its substantial strength in determining criteria weights by its use of pairwise comparisons and hierarchical structure compared to other criteria weights determination techniques such as SWARA. Furthermore, the AHP technique could be more efficient compared to others when DMs are involved in the evaluation under GDM setting. The ARAS technique also has a very easy and comprehensive computational process, one of the reasons why it is chosen to evaluate the alternatives. Finally, the studied NPD case on the selection of app alternatives has confirmed the applicability of the proposed approach.

Compared to the extant literature, the contributions and several advantages of this research can be summarized as follows.

The PF environment has proven to be very useful in the application of a variety of MCDM technique. Whereas this environment has not yet been employed to the AHP technique and the ARAS technique, its applicability in these methods are validated here.

The relevant and available literature are analyzed to identify the selection criteria for NPD strategies in apps development. The proposed main and sub criteria are claimed to be the most relevant, which is verified by the DMs and the APPs Company, and cover the widest aspects in comparison to the extant literature.

Due to AHP's significant role in MCDM under uncertainty and its extensive application in wide variety of areas, this study validates that PF AHP can be successfully applied to decision making process related to the apps development problem, regarding the criteria defined in this study.

The involvement of PFS concept into MCDM techniques can result in much more reliable outcome due to the vagueness of information and the ambiguity of DMs. The integrated PFSs approach combines the PFSs theory with AHP can achieve more reliable weights.

One of the limitation in this study is the dependence to the experience of the DMs and the quality of the judgments from the DMs' assessment. The other one is the hierarchy among the main and sub criteria, which has not yet considered as mutual relationships, i.e., inter-dependence as in the case of ANP technique. However, the proposed approach has the benefit of allowing a quick resolution for NPD process. For the future studies, this integrated methodology could be applied to the other selection problems in variety of areas. Besides, the method can be extended into interval valued PF environments and its validity could be tested in uncertain MCDM environment. Furthermore, the development of the proposed approach with other techniques such as COPRAS, TOPSIS, and VIKOR is also promising for the analysis of the sensitivity of the method.

Acknowledgements The author gratefully acknowledges the experts' valuable contribution, which has been vital for the preparation of this study. This research has received the financial support of the Kahramanmaraş Sütçü İmam University.

References

1. Abdullah, L., Najib, L.: A new preference scale of intuitionistic fuzzy analytic hierarchy process in multi-criteria decision making problems. J. Intell. Fuzzy Syst. **26**(2), 1039–1049 (2014)
2. Achiche, S., Appio, F.P., McAloone, T.C., Di Minin, A.: Fuzzy decision support for tools selection in the core front end activities of new product development. Res. Eng. Design **24**(1), 1–18 (2013)
3. Akbariyeh, H., Seddigh, A.: The influence of strategic orientation on new product development: mediating role innovation, knowledge management and organisational capabilities: an empirical investigation. Int. J. Bus. Innov. Res. **13**(4), 502 (2017)
4. Atanassov, K.: Intuitionistic fuzzy sets. Fuzzy Sets Syst. **20**(1), 87–96 (1986)
5. Ayağ, Z.: An integrated approach to concept evaluation in a new product development. J. Intell. Manuf. **27**(5), 991–1005 (2016)
6. Bellman, R.E., Zadeh, L.A.: Decision-making in a fuzzy environment. Manag. Sci. **17**(4), B-141–B-164 (1970)
7. Büyüközkan, G., Arsenyan, J.: Collaborative product development: A literature overview. Prod. Plan. Control. **23**(1), 47–66 (2012)
8. Büyüközkan, G., Feyzıoğlu, O.: A fuzzy-logic-based decision-making approach for new product development. Int. J. Prod. Econ. **90**(1), 27–45 (2004)
9. Buyukozkan, G., Gocer, F.: A novel approach integrating AHP and COPRAS under Pythagorean fuzzy sets for digital supply chain partner selection. IEEE Trans. Eng. Manag. 1–18 (2019)
10. Büyüközkan, G., Gocer, F.: An intuitionistic fuzzy MCDM approach for effective hazardous waste management. In: Intelligence Systems in Environmental Management: Theory and Applications, vol. 113, pp. 21–40. Springer (2017a)
11. Büyüközkan, G., Göçer, F.: Application of a new combined intuitionistic fuzzy MCDM approach based on axiomatic design methodology for the supplier selection problem. Appl. Soft Comput **52**, 1222–38 (2017b)
12. Büyüközkan, G., Göçer, F.: An extension of ARAS methodology under interval valued intuitionistic fuzzy environment for digital supply chain. Appl. Soft Comput. **69**(C), 634–54 (2018a)
13. Büyüközkan, G., Göçer, F.: Smart medical device selection based on intuitionistic fuzzy Choquet integral. Soft Comput. (2018b)
14. Büyüközkan, G., Göçer, F., Feyzioğlu, O.: Cloud computing technology selection based on interval-valued intuitionistic fuzzy MCDM methods. Soft. Comput. **22**(15), 5091–5114 (2018)

15. Büyüközkan, G., Görener, A.: Evaluation of product development partners using an integrated AHP-VIKOR model. Kybernetes **44**(2), 220–237 (2015)
16. Büyüközkan, G., Güleryüz, S.: A new integrated intuitionistic fuzzy group decision making approach for product development partner selection. Comput. Ind. Eng. **102**, 383–395 (2016)
17. Büyüközkan, G., Güleryüz, S., Karpak, B.: A new combined IF-DEMATEL and IF-ANP approach for CRM partner evaluation. Int. J. Prod. Econ. **191**, 194–206 (2017)
18. Chang, K.-L.: Combined MCDM approaches for century-old Taiwanese food firm new product development project selection. Br. Food J. **115**(8), 1197–1210 (2013)
19. Chen, T.-Y.: A novel risk evaluation method of technological innovation using an inferior ratio-based assignment model in the face of complex uncertainty. Expert Syst. Appl. **95**, 333–350 (2018)
20. Chen, T.-Y.: Remoteness index-based Pythagorean fuzzy VIKOR methods with a generalized distance measure for multiple criteria decision analysis. Inf. Fusion **41**, 129–150 (2018)
21. Chen, W.-C., Lin, Y.-F., Liu, K.-P., Chang, H.-P., Wang, L.-Y., Tai, P.-H.: A complete MCDM model for NPD performance assessment in an LED-based lighting plant factory. Math. Probl. Eng. 1–24 (2018)
22. Cui, F.-B., You, X.-Y., Shi, H., Liu, H.-C.: Optimal siting of electric vehicle charging stations using Pythagorean fuzzy VIKOR approach. Math. Probl. Eng. 1–12 (2018)
23. Demirci, J.R., Bogen, D.L.: Feasibility and acceptability of a mobile app in an ecological momentary assessment of early breastfeeding. Matern. Child Nutr. **13**(3), e12342 (2017)
24. Dogu, E., Esra Albayrak, Y.: Criteria evaluation for pricing decisions in strategic marketing management using an intuitionistic cognitive map approach. Soft. Comput. **22**(15), 4989–5005 (2018)
25. Du, Y., Hou, F., Zafar, W., Qian, Yu., Zhai, Y.: A novel method for multiattribute decision making with interval-valued Pythagorean fuzzy linguistic information. Int. J. Intell. Syst. **32**(10), 1085–1112 (2017)
26. Dursun, M., Arslan, Ö.: An integrated decision framework for material selection procedure: a case study in a detergent manufacturer. Symmetry **10**(11), 657 (2018)
27. Gul, M., Guneri, A.F., Miralp Nasirli, S.: A fuzzy-based model for risk assessment of routes in oil transportation. Int. J. Environ. Sci. Technol. (2018)
28. Gul, M.: Application of Pythagorean fuzzy AHP and VIKOR methods in occupational health and safety risk assessment: the case of a gun and rifle barrel external surface oxidation and colouring unit. Int. J. Occup. Saf. Ergon. 1–14 (2018)
29. Gul, M., Ak, M.F.: A comparative outline for quantifying risk ratings in occupational health and safety risk assessment. J. Clean. Prod. **196**, 653–664 (2018)
30. Ilbahar, E., Karaşan, A., Cebi, S., Kahraman, C.: A novel approach to risk assessment for occupational health and safety using Pythagorean fuzzy AHP & fuzzy inference system. Saf. Sci. **103**, 124–136 (2018)
31. Kahraman, C., Büyüközkan, G., Ateş, N.Y.: A two phase multi-attribute decision-making approach for new product introduction. Inf. Sci. **177**(7), 1567–1582 (2007)
32. Kahraman, C., Oztaysi, B., Onar, S.C.: Multicriteria scoring methods using Pythagorean fuzzy sets. In: Advances in Fuzzy Logic and Technology (Proceedings of: EUSFLAT 2017), pp. 328–35. Springer (2018)
33. Liang, D., Xu, Z.: The new extension of TOPSIS method for multiple criteria decision making with hesitant Pythagorean fuzzy sets. Appl. Soft Comput. **60**, 167–179 (2017)
34. Liang, W., Zhang, X., Liu, M.: The maximizing deviation method based on interval-valued Pythagorean fuzzy weighted aggregating operator for multiple criteria group decision analysis. Discret. Dyn. Nat. Soc. 1–15 (2015)
35. Lu, J., Ma, J., Zhang, G., Zhu, Y., Zeng, X., Koehl, L.: Theme-based comprehensive evaluation in new product development using fuzzy hierarchical criteria group decision-making method. IEEE Trans. Industr. Electron. **58**(6), 2236–2246 (2011)
36. Mohd, W.R.W., Abdullah, L.: Pythagorean fuzzy analytic hierarchy process to multi-criteria decision making. In: AIP Conference Proceedings, vol. 1905, p. 040020 (2017)

37. Mylonas, A., Kastania, A., Gritzalis, D.: Delegate the smartphone user? Security awareness in smartphone platforms. Comput. Secur. **34**, 47–66 (2013)
38. Ng, C.Y., Chuah, K.B.: Evaluation of design alternatives' environmental performance using AHP and ER approaches. IEEE Syst. J. **8**(4), 1185–1192 (2014)
39. Oliveira, G.A., Tan, K.H., Guedes, B.T.: Lean and green approach: an evaluation tool for new product development focused on small and medium enterprises. Int. J. Prod. Econ. **205**, 62–73 (2018)
40. Ozdemir, Y., Gul, M.: Measuring development levels of NUTS-2 regions in Turkey based on capabilities approach and multi-criteria decision-making. Comput. Ind. Eng. **128**, 150–169 (2019)
41. Peng, X., Yang, Y.: Some results for Pythagorean fuzzy sets. Int. J. Intell. Syst. **30**(11), 1133–1160 (2015)
42. Peng, X., Yang, Y.: Pythagorean fuzzy Choquet integral based MABAC method for multiple attribute group decision making. Int. J. Intell. Syst. **31**(10), 989–1020 (2016)
43. Pourjavad, E., Peng, W.: An integrated fuzzy MCDM approach for risk evaluation of new product in a pipe industry. Int. J. Prod. Dev. **22**(3), 165 (2017)
44. Ren, P., Zeshui, X., Gou, X.: Pythagorean fuzzy TODIM approach to Multi-criteria decision making. Appl. Soft Comput. **42**, 246–259 (2016)
45. Saaty, T.L.: A scaling method for priorities in hierarchical structures. J. Math. Psychol. **15**(3), 234–281 (1977)
46. Saaty, T.L.: The Analytic Hierarchy Process: Planning, Priority Setting, Resource Allocation. McGraw, New York (1980)
47. Safaei Ghadikolaei, A., Esbouei, S.K.: Integrating FAHP and fuzzy ARAS for evaluating financial performance. *Boletim Da Sociedade Paranaense de Matemática* **32**(2), 163 (2014)
48. Sliogeriene, J., Turskis, Z., Streimikiene, D.: Analysis and choice of energy generation technologies: the multiple criteria assessment on the case study of Lithuania. Energy Procedia **32**, 11–20 (2013)
49. Smith, A., de Salas, K., Lewis, I., Schüz, B.: Developing smartphone apps for behavioural studies: the AlcoRisk app case study. J. Biomed. Inform. **72**, 108–119 (2017)
50. Tepe, S., Kaya, İ.: A fuzzy-based risk assessment model for evaluations of hazards with a real-case study. Hum. Ecol. Risk Assess. Int. J. 1–26 (2019)
51. Yager, R.R.: Pythagorean fuzzy subsets. In 2013 Joint IFSA World Congress and NAFIPS Annual Meeting (IFSA/NAFIPS), pp. 57–61. IEEE (2013)
52. Yager, R.R.: Pythagorean membership grades in multicriteria decision making. IEEE Trans. Fuzzy Syst. **22**(4), 958–965 (2014)
53. Yager, R.R.: Properties and applications of Pythagorean fuzzy sets. In: Studies in Fuzziness and Soft Computing, vol. 332, pp. 119–36 (2016)
54. Yager, R.R., Abbasov, A.M.: Pythagorean membership grades, complex numbers, and decision making. Int. J. Intell. Syst. **28**(5), 436–452 (2013)
55. Yeh, T.-M., Pai, F.-Y., Liao, C.-W.: Using a hybrid MCDM methodology to identify critical factors in new product development. Neural Comput. Appl. **24**(3–4), 957–971 (2014)
56. Zadeh, L.A.: Fuzzy sets. Inf. Control **8**(3), 338–353 (1965)
57. Zavadskas, E.K., Turskis, Z.: A new additive ratio assessment (ARAS) method in multicriteria decision-making. Technol. Econ. Dev. Econ. **16**(2), 159–172 (2010)
58. Zhang, X.: Multicriteria Pythagorean fuzzy decision analysis: a hierarchical QUALIFLEX approach with the closeness index-based ranking methods. Inf. Sci. **330**, 104–124 (2016)
59. Zhang, X., Xu, Z.: Extension of TOPSIS to multiple criteria decision making with Pythagorean fuzzy sets. Int. J. Intell. Syst. **29**(12), 1061–1078 (2014)

Application of Heuristics Algorithms on Product Development

Pavel Anselmo Alvarez

Abstract A relevant approach to increase success on products or services is based on customer oriented product design. In this case, the product will support expected requirements while meeting specific characteristic of attributes of particular type of products. Developing new products is an essential ability to compete in markets where important factors appear like globalization, customer demand and quality of products. Due that multi-criteria decision aids approaches are becoming relevant for the product design, thereby preference aggregation and disaggregation approaches complement existing methods in this area, since customer preferences need to be regarded and aggregated to customer preferences models concerning the target market. Besides that, customer needs to express their preference in an easy-to-answer format, taking into consideration they will not be surveyed again. This chapter is focused on describing a heuristic algorithm for product development, particularity focused on a case study of Agrifood products. The heuristic models customer preferences through the disaggregation of global preferences on corn toast products. The use of heuristics for disaggregating and modelling customer preferences has important implications for marketing managers concerning the development of new products, regarding they can be developed from consumer preferences expressed on attributes on known products.

1 Introduction

Adaptation for the market concerns consumer behavior, where considering consumer satisfaction allows obtaining competitive advantages [1]. The major focus of perceived quality is the consumers' satisfaction and related subjective perceptions of the product attributes [2]. Authors in [3] defined perceived quality as "consumer's evaluative and effective judgments concerning the overall excellence of a brand

P. A. Alvarez (✉)
Department of Economic and Management Sciences, Universidad Autónoma de Occidente, Culiacan, Mexico
e-mail: pavel.alvarez@uadeo.mx

C. Kahraman and S. Cebi (eds.), *Customer Oriented Product Design*, Studies in Systems, Decision and Control 279, https://doi.org/10.1007/978-3-030-42188-5_18

based on intrinsic as well as extrinsic attributes which are directly related to his or her satisfaction with that brand."

An important strategy in product development is to examine the target market to understand new demands or changes. Kotler and Armstrong [4] point out that new product's failure is related to ignorance or misinterpretation of market research, overestimation or market size or poor design. In [5], we can see a variety of approaches to develop forecast the expected market share of a product.

One popular approach for simulate market behavior on a new product introduction is using conjoint data [6]. After consumer preferences on various product attributes are estimated, a prediction of the hypothetical market can be performed [6]. In this way, the strategy for product development follows the estimation of the market to reduce uncertainty and focus the features on new product design.

Siskos et al. [7] present two directions on the analysis of consumer behavior and market trends: one toward the identification and determination of consumer general perceptions and attitudes and the other toward the examination of the existing interactions among consumers on different product criteria and the correlation among the attributes of the products.

Huber et al. [8] recommend the aggregation of heterogeneous individuals with different preferential models as an efficient way to represent the differential impact that certain features may have on specific brand. Following this recommendation, the prediction of the consumer's choice model should be performed at individual level. A consumer selection policy can be externalized by means of a set of reference products which he/she, the consumer, either has or can rank through simple questionnaires, familiar decision-making situations, and so on [7].

The aim of the chapter is to model in a simulation process the choice model of each consumer. Finding, adjacent preference information to construct a preferential model from a consumer with real and hypothetical data, the latter includes new products. In the rest of the chapter, a heuristic algorithm is presented to support the product development with the prediction of consumer's choice model. It includes some stage to find the consumer's choice model and predict in the market how well a new product can be accepted. A case of study is presented with agrifood products.

The rest of the chapter is organized as follows. In the following section, we present a literature review of heuristics implemented on new product development (NPD). Section 3 presents the Prediction of consumer's choice model, including a heuristic algorithm to elicitate consumer's preference with a preference disaggregation analysis as a model to analyze consumer behavior. The case study focused on corn toast is addressed in Sect. 4, simulating the computer behavior of the consumer in the introduction of a new product based on their preferences. Finally, concluding remarks and future directions are provided in Sect. 5.

2 Literature Review of New Product Development

The innovation process is an important activity for an organization to keep competitiveness. The new product development process is a very important stage. We can find very common to see designers/managers using several tools in a heuristic manner. The design of new products has important implications for marketing managers because a new concept can be tested before it enters the production stage. It is a common manner to take advantage of the information analyzed using data obtained from a market survey. The high predictive accuracy of the model may assist a firm in minimizing the uncertainty and risks associated with a new product launch [6].

Torrens [9] provided some insights to effectively apply design and validation methods and heuristics to the field of assistive technology product design, analyzing 20 case studies. In the study, the author identifies available research methods, design methods and heuristics, and define a critical pathway of method and heuristic application. The heuristics studied regards more to the process to ease the cognitive load to make design decisions, different for the heuristics to optimize objectives throughout finding feasible solutions from the operational research field.

A study using a genetic algorithm for a portfolio of product development projects is carried out by [10]. The study proposes two genetic algorithm approaches for scheduling project activities, minimizing the overall duration of the portfolio. In [11], authors study the impact of features involving online product reviews on information adoption by new product developers. The study suggests that helpfulness rating and the degree of referencing are positively associated with NPDs' information adoption, while the extremeness of product rating is negatively associated. Moreover, title attractiveness mitigates the negative relationship between the extremeness of product rating and information adoption.

Achiche et al. [12] analyzed the influencing factors along with data collection from managers. The study assessments try to suit product development teams in companies that are cost-focused and where the trade-offs between material, staff, and time to involve in core front end activities to hence the influence of their financial performances on the NPD process. The model is optimized by a genetic algorithm, applied to learn data obtained from five experienced managers, working for five different companies. The model reproduces the managers' estimations using the learned data sets.

A conjoint analysis heuristic-based in a utility function that explains market share is presented in [6]. It improves the value of market simulations in conjoint analysis. The approach deals with two issues: the lack of differential impact of attributes across alternatives and the absence of accounting for differential substitution across brands. One interesting aspect of the research is that the study is applied to real data from a market survey concerning milk. Authors carried out some simulation for different market scenarios, and the approach exhibits some theoretical properties that are necessary for the representation of consumer choice behavior.

Lockshin and Mueller [13] discussed three methods for setting the retail price for newly developed products: competitive comparison method, hedonic price analysis, and discrete choice analysis. The study discusses the types of new product development and the problems they create for setting the price. Varma et al. [14] proposed the Lagrangian decomposition heuristic for scheduling new product development projects in the pharmaceutical industry. In this problem, the effective use of their new product development capabilities should be carried out. The problem is about to schedule the testing, manufacturing, and process development activities on limited physical resources. The objective function is a weighted discounted cost expressed in terms of activity completion times.

Petrick and Echols [15] proposed a heuristic approach that requires firms concentrated on technological trajectories by combining technology road mapping, information technology and supply chain management to make more sustainable new product development decisions. Maravelias et al. [16] proposed a scheduling MILP model that addresses the scheduling of tests for pharmaceuticals and agrochemicals to optimize the overall costs. The model allows for the installation of new resources and handles the cost and duration of tests as functions of the type and amount of the resources assigned to each test. Authors avoid a single large-scale problem with a heuristic decomposition algorithm that relies on solving a reduced mixed-integer program that embeds the optimal schedules obtained for the individual products.

New product development with a real case was conducted in [17]. There is a firm implementing their approach via a Six Sigma project. It is stated that implementation saved several million dollars based on the schedules determined using simulation-based heuristic optimization. The authors developed a simulation-based framework for generating feasible solutions using a heuristic optimization approach. The heuristic optimization algorithms operate on a solution space consisting of the precedence relationships among activities in the new-product development process.

Maravelias and Grossmann [18] applied the simultaneous optimization of resource-constrained scheduling of testing tasks in new product development and design/planning of batch manufacturing facilities. A heuristic algorithm based on Lagrangean decomposition was developed to solve larger instances of this problem, obtaining optimal or near-optimal solutions, while being significantly faster than the full-space method.

A methodology and tool aimed at assisting in the reengineering of the processes and organization deployed in new product development were carried out by [19]. It was used process modeling and analysis techniques to the analysis of the operation of multifunctional project teams throughout the new product development process lifecycle. Wang [20] proposed a method that allows the construction of a schedule based on possibility theory to determine the satisfaction degrees of fuzzy temporal constraints. It is a methodology to schedule product development projects having imprecise temporal information.

3 Prediction of Consumer's Choice Model

The flow chart of Fig. 1 illustrates the process for applying a heuristic in the product

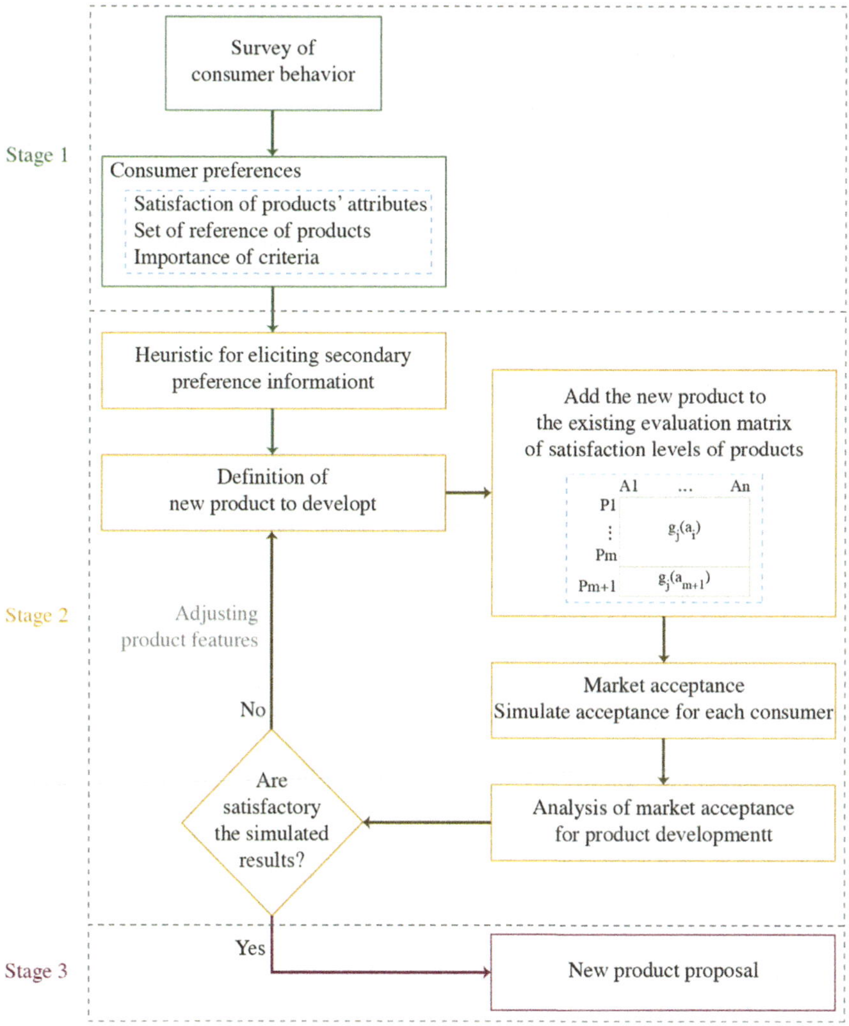

Fig. 1 Flow chart of heuristic algorithm for product development

development process. In Stage 1, the survey for gathering consumer preferences is carried out. The input information is stablished for applying the heuristic algorithm. In Stage 2, a heuristic algorithm is applied to find secondary preference information from consumer and then simulate the acceptance of a new product in the market. The heuristic includes the elicitation of secondary preference information of the consumers as weights and other parameters (disaggregated preferences) and secondary preference information is used to simulate the acceptance of new products on the market. Thereby, the prediction of the choice's model is performed with that preference information.

First, in product development, some strategy needs to be applied for new product design for setting up the features of the new product. Once the new product is defined, it is included in the evaluation matrix of each consumer in the format of satisfaction to the attributes of each product. Hence, the prediction of the choice's model is carried out, evaluating the products again with the second preference information inferred and acceptance of the new product is showed in a new ranking, where the position of the new product is analyzed. If the result is not satisfactory, some adjustments of the features of the new product are performed and the simulation is generated again until satisfactory result is reached (Stage 3).

4 Heuristic Algorithm to Elicitate Consumers' Preferences

In this section, a heuristic algorithm to elicitate secondary preference information from consumers is presented. It is a genetic algorithm which integrates an aggregation/disaggregation approach to elicitate secondary preference information. This preference information allows the evaluation and estimation at individual consumer level of the acceptance of a new product. When the new product is more accepted by a consumer, it will predominate the first places of her/his ranking, when the new product presents less acceptance, it will be shown in lower positions.

The preference disaggregation analysis regards overall preferences that for real problems could be infeasible, not just for the complexity of mathematical properties of the problem modeling, also because the inconsistency of data when decision-makers express their valued judgments, methods requires dealing with an inherent characteristic of multicriteria decision aids.

Figure 2 shows the scheme of the genetic algorithm-based heuristic for disaggregating consumer preference (secondary preference information). The heuristic elicitates the consumer's secondary preference as weights and other parameters. For the ELECTRE III method, weights (w), indifference (q), preference (p) and veto (v) threshold is derived by the heuristic. This secondary preference information allows aggregate the consumer preference as acceptance for the products, even if new products are added to the evaluation matrix. In this sense, the heuristic finds a consumer model as consistent as possible with her/his preference. The output of the heuristic is a set of parameters for each consumer to evaluate the same kind of product with different features. Thus, it is possible to simulate for the product development the acceptance of the new product in the market, when the scenery includes new products.

4.1 Aggregation Phase: ELECTRE III Method

At the aggregation phase the ELECTRE III method is applied. It constructs a fuzzy outranking relation expressed as $\sigma(a_i, a_l)$ [21]. ELECTRE III constructs three

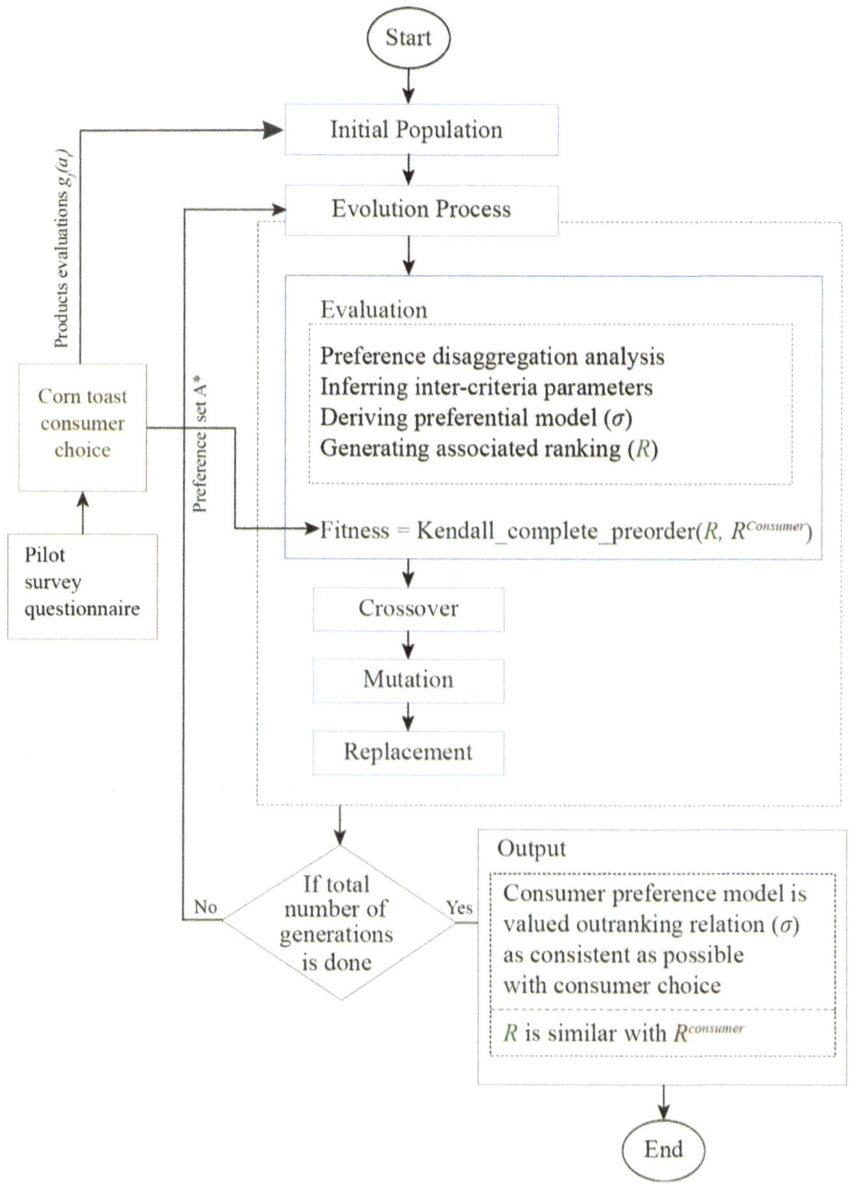

Fig. 2 General schema of genetic algorithm to disaggregate consumer preference

indices, the partial concordance index, comprehensive index, and discordant index, the last two are used to construct the fuzzy outranking relation. The discordance index $d_j(a_i, a_l)$ takes into account if a criterion is against with the assertion "a_i outranks a_l". The partial concordance index $c_j(a_i, a_l)$ evaluates if "action a_i is at least as good as action a_l" on the criterion g_i. The comprehensive concordance index $C(a_i, a_l)$ measures the performances on all criteria of the pair of alternatives $(a_i, a_l) \in A \times A$ to evaluate in what grade the criteria support the assertion "a_i outranks a_l". Then, the method estimates the credibility index $\sigma(a_i, a_l)$; $(0 \leq \sigma(a_i, a_l) \leq 1)$ which assesses the strength of the assertion that "a_i is at least as good as a_l", $a_i S a_l$.

$$\sigma(a_i, a_l) = \begin{cases} C(a_i, a_l) & if\ \bar{F}(a_i, a_l) = 0 \\ C(a_i, a_l) \times \prod\limits_{j \in \bar{F}(a_i, a_l)} \frac{1 - d_j(a_i, a_l)}{1 - C(a_i, a_l)} & if\ \bar{F}(a_i, a_l) \neq 0 \end{cases} \tag{1}$$

4.2 Exploitation Phase: Distillation Method

The distillation method is the ranking algorithm used for the exploitation phase to generate a complete or partial order from a fuzzy outranking relation. It is based on the degrees of credibility of each action generating a complete or partial preorder, resulting from the intersection of two complete preorders [22]. The distillation method contains two sub-processes, the ascending and descending distillation, each constructs a complete preorder, and the combination of the two complete preorders generate the final ranking. A general description of the distillation method is addressed in [23].

This algorithm needs the distillation threshold function $s(\lambda_k)$ as a piece of additional information to the distillation [24]. It acts as a threshold for adjustment in the research of pairs which are better than certain cut level.

$$s(\lambda_k)\alpha \cdot \lambda_k + \beta \tag{2}$$

where α and β are two thresholds to define the threshold function and $s(\lambda_k)$.

$$\lambda_0 = \max_{a_i, a_l \in A} \sigma(a_i, a_l) \tag{3}$$

For the next cut-off level, we need to find the maximum value lower than λ_0.

$$\lambda_1 = \max_{\sigma(a_i, a_l) < \lambda_k - s(\lambda_k)} \sigma(a_i, a_l) \tag{4}$$

With successive distillations, the cut-off level λ_1 is progressively reduced, which makes the condition weaker and it is much easier for a_i to be preferred than a_l [23].

Generating descending complete preorder

Step 1:

In the beginning, (3) represents the initial level to calculate the first cut-off level (4) to identify the best alternatives.

Step 2:

1.1 Find the highest degree of credibility (3) and then the crisp matrix with λ_k (4).
1.2 Calculation of power, weakness, and qualification using the crisp matrix.
1.3 Select the alternatives with a higher qualification, conforming the set D.

if |D| > 1 and λ_k <>0

- Repeat step 2

else

- A = A\D
- Add to descending total preorder
- $D = \emptyset$
- Return to Step 1

end

Ascending order requires the same procedure, instead, the best alternatives are positioned at last position on the ranking.

Intersection between descending and ascending preorders

a_i is strict preferred to a_l
if a_i is better rank in one order and a_i is at least as good as a_l in the other rank

a_i is indifferent to a_l
if the two actions belong to the same equivalence class in the two preorders

a_i is incomparable to a_l
if a_i is better ranked than a_l in one order and a_l is better ranked than a_i in the other order, or vice versa.

5 Preference Disaggregation Model for Consumer Behavior

The model presented in this section is based on [25], but it was first presented in [26]. It is designed to elicitate secondary customer preferential information at an individual level to evaluate products of the same kind with real or simulated information. This approach allows emulating the acceptance of new products as consumer behavior.

Due, the aggregation method is the ELECTRE III, one must specify the following parameters:

(i) The weight vector $W = (w_1, w_2, \ldots, w_n)$, $w_i \geq 0$ and $\sum_{i=1}^{n} w_i = 1$.

(ii) The vector of indifference thresholds $q = (q_1, q_2, \ldots, q_n)$, $q_i \geq 0$, $i = 1, 2, \ldots n$.
(iii) The vector of preference thresholds $p = (p_1, p_2, \ldots, p_n)$, $p_i \geq q_i$, $i = 1, 2, \ldots n$.
(iv) The vector of veto thresholds $v = (v_1, v_2, \ldots, v_n)$, $v_i \geq p_i$, $i = 1, 2, \ldots n$.

Problem formulation
The direct representation of the parameters regarding the preference, indifference and veto thresholds, does not allow an efficient search of the solution space. Therefore, an indirect representation scheme is employed, based on [25]. The above parameters are defined for each criterion j as follows:

$$g_j(b) - q_j = g_j(b) - p_j + y_j \qquad \forall j = 1, 2, \ldots, n \tag{5}$$

$$g_j(b) + p_j = g_j(b) - q_j + z_j \qquad \forall j = 1, 2, \ldots, n \tag{6}$$

$$g_j(b) - v_j = g_j(b) - p_j - x_j \qquad \forall j = 1, 2, \ldots, n \tag{7}$$

The parameters of the ELECTRE III method can be easily expressed in terms of the decision variables x_j, y_j, z_j and converted on parameters when needed.

$$q_j = \frac{z_j - y_j}{2} \tag{8}$$

$$p_j = \frac{z_j + y_j}{2} \tag{9}$$

$$v_j = x_j + \tfrac{1}{2} y_j + \tfrac{1}{2} z_j \tag{10}$$

The general form of the optimization problem can be expressed as follows:

$$\begin{aligned}
&Min \quad K(\vec{w}, \vec{x}, \vec{y}, \vec{z}, \vec{t}, \lambda) \\
&s.t. \\
&y_j + z_j \leq 2(x_j^* - x_{j*}) \quad j = 1, 2, \ldots, n \\
&w_1 + w_2 + \ldots + w_n = 1 \\
&\vec{z} \geq \vec{y} \\
&\vec{w}, \vec{x}, \vec{y}, \vec{z}, \vec{t} \geq \vec{0} \\
&t \in \{0, 1\}, \ \lambda \in (0, 1), \ \lambda > 0.5
\end{aligned} \tag{11}$$

where $x_j^* = \max_{i \in A^*}\{x_{ij}\}$, $x_{j*} = \min_{i \in A^*}\{x_{ij}\}$ and $(\vec{w}, \vec{x}, \vec{y}, \vec{z}, \vec{t})$ are the vectors of the decision variables w_j, x_j, y_j, z_j, t_j, $j = 1, 2, \ldots n$, and $K(\vec{w}, \vec{x}, \vec{y}, \vec{z}, \vec{t}, \lambda)$ denotes the ranking accuracy of the model corresponding to a specific solution $(\vec{w}, \vec{x}, \vec{y}, \vec{z}, \vec{t}, \lambda)$. To calculate the ranking accuracy, the elements x_j, y_j, z_j of the decision variables are transformed to the parameters of the ELECTRE III method q_j, p_j, v_j respectively with Eqs. (8), (9), (10), respectively.

Solution representation
$\tilde{P}$ is the representation of a potential solution. Let $\tilde{P} = p_1, p_2, \ldots, p_{5n+1}$ be the schematic representation of an individual's chromosome. $\tilde{P} \in \prod_{i=1}^{5n+1} C_i$, where C_i is the set of values that p_i can takes. It is obvious that the length of $\tilde{P}$ depends of the size of the number of criteria (attributes) n.

6 Case Study: Corn Toast

With interest in the economic development of the region, the product development process is inspired in the agricultural region capacities. There is in the region an enormous agricultural capacity and planting diversity. Because the new requirements on quality and nutritional values of food, the interest is to develop an Agrifood product. Hence a corn toast-based product is proposed for product development. There are many brands in the market of Mexico with corn toast product, but in the region, a few ones are in the market with high nutritional content. In this sense, the products included in the survey present related features with high nutritional content. The market study was performed with a pilot survey questionnaire in the small city, Culiacan, Mexico. The information acquired from the market was applied to test the heuristic for choice's model of the consumer at individual level.

6.1 Problem Description

As part of a product diversification strategy, one producer is working on product development. To reduce uncertainty of how the market will accept a new product, some simulation of the consumer acceptance is needed to understand the consumer behavior and the estimation of how the new product will be accepted in the market.

The problem is to elicitate secondary preferential information from consumer to evaluate products based on corn toast and then add new products a posteriori with a specific setting in its attributes. Hence, the first stage is to obtain the consumer satisfaction of current products in the market and then the preferential information of the consumer to simulate alternative scenarios.

6.2 Product's Attributes

The consumer behavior approach regards consumer perceptions and some behavior related to the market's products. When individual perception is concerned in market

Table 1 Ordering of attributes by consumers

		Attribute positions by consumer					
Code	Attribute	Con1	Con2	Con3	Con4	Con5	Con6
A1	Nutritional content	5	5	4	5	1	1
A2	Flavor	1	1	1	1	2	2
A3	Price	2	2	5	4	4	4
A4	Appearance	4	4	2	2	5	3
A5	Texture	3	3	3	3	3	5

studies, the products' attributes play an important element in the study because it is directly associated with the consumer's acceptance with some features of the products. When a product presents a combination of specific features, the consumer from a specific market niche could be attracted to that product.

In [27], authors created a rooibos sensory wheel by selecting 27 flavors, taste and mouthfeel attributes and grouping these terms together to form a logical overview of the sensory descriptors associated with rooibos. It seems that direct measurements of perceived intensities of target attributes are appearance, color, aroma, taste and texture [28]. In this sense, some attribute for consumer satisfaction evaluation were defined and used as decision criteria as listed below.

Nutritional content: The food nutritional content can be considered as a quality attribute; however, this attribute is perceived by the consumer for the survey as quality expectations in the nutritional content like healthy or convenient food for the consumer.

Flavor: The perceived perception resulting from stimulating a combination of the taste buds, the olfactory organs, and chemesthetic receptors within the oral cavity [29]. For corn toast product, the tastes are more related to the salty perception.

Price: The cost paid by the customer for the product.
Appearance: Related with the design, form and how the product looks.
Texture: It is the sensation or feels experienced with the corn toast.

Table 1 shows the order of importance of the attributes made by consumers. The major importance for the attribute is expressed with value 1; the minor importance is expressed with value 5. Using a weighted sum, we see the preference of attributes, where the flavor is the most important attribute, in second place appearance, third place texture, fourth place nutritional content and price in the last place (A2 > A4 > A5 > A1 > A3).

6.3 Products and Consumer' Preference Information

Six products are found in the local market that meets specification with added nutritional value for corn toast. Table 2 shows the name of the products. For this practical

Table 2 Ordering of product based corn toast

		Product positions					
Code	Product	Con1	Con2	Con3	Con4	Con5	Con6
P1	Trizalet	1	1	2	1	4	5
P2	Nopalia	5	5	4	6	2	1
P3	Blanmex	6	2	1	2	5	4
P4	Charras	2	3	3	5	3	6
P5	Nutri zuis	3	4	5	4	1	2
P6	Nutri well	4	6	6	3	1	3

Con it is the abbreviation used for consumer

illustration and test of the model, only information of six consumers from the survey is taken. In the survey, the consumer was asked to order the products regarding the most preferred in the highest position (value 1) and the lest preferred in the lowest position (5). We see some variation of position of the same product ranked by different consumers.

The consumer was asked to express her/his satisfaction level with the perceived experience with any product in each attribute. The highest satisfaction is expressed with value 1, the lowest satisfaction is expressed with value 5. Table 3 shows the evaluation tables, they correspond with the satisfaction level values expressed by consumers (satisfaction table).

6.4 Secondary Preference Information

The heuristic algorithm described in Fig. 2 was applied using the consumer preference information collected by the survey. As a result, secondary preference information was obtained, concerning parameters for the ELECTRE III method (see Table 4) and a preferential model of the consumer that reflects her/his choice's model (see Table 5).

The model implemented in the heuristic finds parameters as secondary preference information as consistent as possible to replicate the choice's model of the consumer. It looks for a set of parameters that generate a preferential model of the consumer as consistent as possible with their real preferences. It means that the preferential model from Table 5 should correspond with the ranking of the product as similar as possible with the order of products expressed by the consumer in the survey.

We see in Table 6 the comparison of the ranking generated by the heuristic with the ordering of products expressed by the consumer's choice. The predicted choice's model presents some differences with the consumer's choice. Those differences are identified in the low side of Table 6 as a pair of products. First, we will describe those differences and then we will explain why the heuristic is not finding the exact

Table 3 Evaluation of satisfaction level of products based on corn toast

Consumer 1						Consumer 2					
	A1	*A2*	*A3*	*A4*	*A5*		*A1*	*A2*	*A3*	*A4*	*A5*
P1	3	2	3	1	2	*P1*	1	1	3	2	2
P2	1	3	3	4	3	*P2*	3	3	2	4	4
P3	3	4	3	4	4	*P3*	1	2	3	2	2
P4	3	3	2	3	2	*P4*	2	3	3	3	3
P5	2	3	3	4	2	*P5*	1	3	3	2	3
P6	1	3	3	1	4	*P6*	1	2	2	1	1
Consumer 3						Consumer 4					
	A1	*A2*	*A3*	*A4*	*A5*		*A1*	*A2*	*A3*	*A4*	*A5*
P1	3	2	2	3	2	*P1*	2	2	2	2	1
P2	2	1	2	3	2	*P2*	3	2	2	3	1
P3	1	1	1	1	1	*P3*	2	2	2	3	2
P4	2	1	3	1	2	*P4*	1	2	2	2	3
P5	2	3	1	2	3	*P5*	1	1	1	1	3
P6	1	2	3	1	2	*P6*	1	1	3	1	1
Consumer 5						Consumer 6					
	A1	*A2*	*A3*	*A4*	*A5*		*A1*	*A2*	*A3*	*A4*	*A5*
P1	3	1	2	3	1	*P1*	5	4	2	2	4
P2	1	3	3	3	3	*P2*	1	1	4	3	1
P3	4	3	3	3	2	*P3*	5	3	2	3	3
P4	3	3	3	4	3	*P4*	2	2	3	2	2
P5	1	1	3	3	1	*P5*	1	1	4	2	1
P6	1	2	3	2	1	*P6*	1	1	4	1	1

Table 4 The inferred parameter with the heuristic

Consumer 1						Consumer 2					
	A1	*A2*	*A3*	*A4*	*A5*		*A1*	*A2*	*A3*	*A4*	*A5*
Consumer 3						Consumer 4					
	A1	*A2*	*A3*	*A4*	*A5*		*A1*	*A2*	*A3*	*A4*	*A5*
w	0.103	0.313	0.256	0.104	0.224	w	0.114	0.441	0.182	0.115	0.148
q	0	0	0	0	0	q	0	0	0	1	1
p	1	0	0	0	1	p	0	0	0	1	1
v	–	–	–	–	–	v	1	0	–	2	3
w	0.14	0.318	0.139	0.24	0.163	w	0.087	0.269	0.155	0.256	0.233
q	0	0	0	0	0	q	0	0	0	0	0
p	1	1	0	0	0	p	0	0	1	1	0
v	–	3	1	–	0	v	1	–	2	–	1
Consumer 5						Consumer 6					
	A1	*A2*	*A3*	*A4*	*A5*		*A1*	*A2*	*A3*	*A4*	*A5*
w	0.483	0.155	0.115	0.114	0.133	w	0.451	0.291	0.091	0.106	0.061
q	0	0	0	0	0	q	0	0	0	0	1
p	0	1	0	1	1	p	0	0	1	0	1
v	–	–	0	–	2	v	–	–	2	–	–

Table 5 Valued outranking relation of products

Consumer 1							Consumer 2						
	P1	*P2*	*P3*	*P4*	*P5*	*P6*		*P1*	*P2*	*P3*	*P4*	*P5*	*P6*
P1	1	0.897	1	0.744	0.897	0.897	*P1*	1	0.818	1	1	1	0.818
P2	0.359	1	1	0.416	0.776	0.896	*P2*	0	1	0	0	0	0
P3	0.359	0.36	1	0.103	0.36	0.48	*P3*	0	0.818	1	1	1	0.818
P4	0.583	0.897	1	1	0.897	0.793	*P4*	0	0.818	0	1	0	0
P5	0.583	0.897	1	0.64	1	0.793	*P5*	0	0.818	0	1	1	0
P6	0.463	0.776	1	0.52	0.776	1	*P6*	0	1	1	1	1	1
Consumer 3							Consumer 4						
	P1	*P2*	*P3*	*P4*	*P5*	*P6*		*P1*	*P2*	*P3*	*P4*	*P5*	*P6*
Consumer 5							Consumer 6						
	P1	*P2*	*P3*	*P4*	*P5*	*P6*		*P1*	*P2*	*P3*	*P4*	*P5*	*P6*
P1	1	0.542	0	0.302	0	0.62	*P1*	1	1	1	0	0	0
P2	1	1	0	0.76	0	0.62	*P2*	0	1	0	0	0	0
P3	1	1	1	1	1	1	*P3*	0	0	1	0	0	0
P4	0	0	0	1	0	0.86	*P4*	0	0	0	1	0.256	0
P5	0	0	0	0	1	0	*P5*	0	0	0	1	1	0
P6	0	0	0	0.682	0	1	*P6*	0.941	0.941	0.941	0.941	0.941	1
P1	1	0.517	1	1	0.517	0.403	*P1*	1	0.197	0.709	0.197	0.197	0.091
P2	0	1	0.867	1	0	0	*P2*	0	1	0	0.803	0.894	0.894
P3	0	0.517	1	0.517	0.229	0.115	*P3*	0.894	0.197	1	0.152	0.091	0.091
P4	0	0.403	0.753	1	0	0	*P4*	0.909	0.258	0.909	1	0.258	0.152
P5	0	1	1	1	1	0.886	*P5*	0	1	0	0.909	1	0.894
P6	0	1	1	1	0.845	1	*P6*	0	1	0	0.909	1	1

Table 6 Consistency of the outranking preference model with the consumer's choice

	Consumer 1		Consumer 2		Consumer 3		Consumer 4		Consumer 5		Consumer 6	
Product	Pref.	Heur.	Pref.	Heur.	Pref.	Heur.	Pref.	Heur.	Pref.	Heur.	Pref.	Heur.
P1	1	1	1	1	2	3	1	2	4	2	5	5
P2	5	5	5	6	4	2	6	6	2	3	1	3
P3	6	6	2	3	1	1	2	5	5	5	4	4
P4	2	2	3	5	3	5	5	4	3	4	6	6
P5	3	3	4	4	5	4	4	3	1	1	2	2
P6	4	4	6	2	6	6	3	1	1	1	3	1
Differences between consumers' orders and heuristic's order												
			(P2, P6) (P3, P6) (P4, P5) (P4, P6) (P5, P6)		(P1, P2) (P2, P4) (P4, P5)		(P1, P6) (P3, P4) (P3, P5) (P3, P6)		(P1, P2) (P1, P4)		(P2, P5) (P2, P6) (P5, P6)	

Pref. it is the abbreviation used for 'preference information'
Heur. it is the abbreviation used for 'heuristic'

consumer's choice (ordered products by the consumer). The choice's model of consumer 1 is predicted with high precision. The prediction for consumer 2 presents five pairs of products in opposite order. This is the lowest prediction of the consumer' choice. Prediction of consumer 3 and 6 show three differences, in consumer 4 four differences are shown and in consumer 5 two differences are presented.

Mainly, the difference between the ranking by the heuristic and the ranking by the consumers are due to the inconsistency of the consumer preference, evaluating the product (satisfaction table) and ordering the products. We are explaining this situation with the preference information of consumer 2. Based on the satisfaction level of the consumer 2, expressed on products of the Table 3, the product Nutri well (P6) presents the highest satisfaction on the majority of attributes, for attributes (A1, A2, A3, A4, A5) the corresponding values are (1, 1, 1, 2, 2). Other products evaluated with high satisfaction are P1 and P2 with values (1, 1, 2, 2, 3) and (1, 2, 2, 2, 3), respectively. Regarding the values of those products, it is supposed that at least P6 is ranked in the first places of the ordering. However, the consumer 6 allocated P6 in the last position of the ranking. This inconsistency is regarded in the consumer's preference model presenting a choice's model as consistent as possible with the consumer preference. However, the inconsistency of the consumer's choice is difficult to follow. The heuristic algorithm tries to predict the choice's model where P1 is the most preferred product, but it fails to try to place P6 in the last position because the satisfaction on the attributes of P6 is very high. Instead, P2 is located in the last position. Because of that, the main differences are presented in products P6 and P2. Hence, the more inconsistency expressed by the consumer's choice, the more differences the predicted choices model will show.

6.5 Simulating New Products in the Market

Concerning the orders of products generated by the consumers as a part of their preferences, it is feasible to establish an order of importance for the target of consumer. Applying a weighted sum, the next order is generated P1 > P5 > P3 > P4 > {P2, P6}. If a company is developing a new product, one approach concerns consumer behavior. In this sense, if the company wants to participate in the first places of the market, some similar attributes values need to be concerned from the more preferred product on the market. Assuming the P1 is one of the more preferred products by the market, the new product should share some features of P1.

Now, the simulation of a new product in the market will be performed. We consider two possible settings, each with specific features for the new product (NP). In each setting, satisfaction values are added for each consumer, the values are based on how the customer evaluated products with similar characteristic in the attributes. In Table 7, the values of attributes A1, A2, A3, A4 and A5 for each consumer are shown for new product 1 (NP1) and new product 2 (NP2). NP1 contains exactly the same features to P1. NP2 contain similar features to P4, but improving the attribute flavor (A2) and appearance (A4).

Table 7 Satisfaction level by the consumer for the new product

New product 1						New product 2					
	A1	*A2*	*A3*	*A4*	*A5*		*A1*	*A2*	*A3*	*A4*	*A5*
Con1	3	2	3	1	2	Con1	3	2	2	2	2
Con2	1	1	3	2	2	Con2	2	2	3	2	3
Con3	3	2	2	3	2	Con3	2	1	3	1	2
Con4	2	2	2	2	1	Con4	1	1	2	1	3
Con5	3	1	2	3	1	Con5	3	2	3	3	3
Con6	5	4	2	2	4	Con6	2	1	3	1	2

The gray cells are the modified attributes of the new product
Con it is the abbreviation used for the consumer

Table 8 Simulation of consumer preference with a new product 1

		Product positions					
Code	Product	Con1	Con2	Con3	Con4	Con5	Con6
P1	Trizalet	1	1	4	2	2	5
P2	Nopalia	4	6	2	5	3	4
P3	Blanmex	5	3	1	5	5	3
P4	Charras	2	5	3	4	4	2
P5	Nutri zuis	4	4	5	1	1	1
P6	Nutri well	3	2	6	3	1	1
NP1	New product	1	1	4	2	2	5

Table 9 Simulation of consumer preference with a new product 2

		Product positions					
Code	Product	Con1	Con2	Con3	Con4	Con5	Con6
P1	Trizalet	1	1	1	2	2	7
P2	Nopalia	4	6	6	5	3	4
P3	Blanmex	5	3	5	5	6	6
P4	Charras	2	5	7	6	5	5
P5	Nutri zuis	4	4	4	1	1	2
P6	Nutri well	3	2	2	4	1	1
NP2	New product	1	4	3	3	4	3

The prediction of choice's model for two settings of new product is described below. The simulation of the new product 1 in the market, allocates it in important positions in consumers' ranking (see Table 8). Due, the new product is based on one of the best product expressed by the consumers. The simulation of the new product 2 in the market allocates it in better positions than P4 (the original product in which the NP2 is based) (see Table 9). If we compare the NP2 with P4, it is easy to see the acceptance of the market. For consumer 1, consumer 2 and consumer 5, the NP2 roses one position in their corresponding ranking. For Consumer 4, NP2 roses four positions in the ranking. For consumer 4, NP2 roses three positions and for consumer 6 NP2 roses 2 positions.

The positions gained for NP2 for each consumer are the following. NP2 is allocated on position 1 tied with P1, for consumer 1. For consumer 2, NP2 is tied with P5. For consumer 3, it is allocated between P6 and P5, position 3. For consumer 4, NP2 is allocated in position 3, between P1 and P6. For consumer 5, NP2 is allocated in position 4, between P2 and P4. For consumer 6, NP2 is between P5 and P2, position 3.

The predictive level of the choice's model presents certain consistency with the consumer preference, due to the new product is evaluated according to the consumer's choice expressed in the real data of the market.

Some limitations of this research are related to some stages of product development. First, the heuristic algorithm aims to elicitate some adjacent preference information from consumer, in this sense, the new product design phase is not concerned to this study, but the simulation concerns how the new product will be accepted by market for each individual consumer. On the another hand, simulation of the market share is not included in the study, due, it is out of the scope of the heuristic algorithm. Differential substitution refers to the assumption that when a new product enters a market, it gains share mostly from similar products, rather than from the dissimilar ones [6]. Hence, the market share should be studied for a complete understanding of the new product on the market.

7 Concluding Remarks

The heuristic algorithm presented in this chapter aims to find the consumer's choice model to simulate the acceptance of the consumer on new products. Its development and implementation as support to product development; particularly in the simulation processes of new products added in the market. The heuristic address the problem of inferring consumer preference by preference disaggregation analysis and the problem of model consumer's preference expressed on product's attributes by ELECTRE III method. It shows some advantages for the product development because it allows to simulate market preference at individual level and predict the consumer's choice model even when the consumer is inconsistent with her/his preference and choices.

The heuristic algorithm still presents some limitation on product development, the new product design and market share. However, it shows certain support inferring adjacent preference information and aggregation consumer preference to rank products.

The originality of the research is aimed at the metaheuristic model and the kind of problem addressed in the new product development. Particularly, it supports the product design stage. Even scientific literature shows the importance and advances in new product development, it lacks on the new product design stage. The proposed model is applied in a real study case for corn toast, showing the advantage of simulating the consumer behavior and inferring parameters of ELECTRE model.

Some current research is addressing addressed the data from the market. In that sense, different to conjoint analysis approach can be designed and implemented to improve the knowledge in the analysis of consumer behavior to help organizations to be more competitive in new product trends.

References

1. Matsatsinis, N.F., Siskos, Y.: New product development methodology. In: Matsatsinis, N.F., Siskos, Y. (eds.) Intelligent Support Systems For Marketing Decisions. Kluwer Academic

Publisher, Norwell, MA (2003)
2. Gravin, D.A.: Competing On the eight dimensions of quality. Harvard Bus. Rev. **65**, 101–109 (1987)
3. Lee, D., Bae, S.: Effects of partitioned country of origin information on buyer assessment of binational products. In: Arnould, E.J., Scott, L.M., Provo, U.T. (eds.) NA—Advances in Consumer Research, vol. 26. pp. 344–351. Association for Consumer Research (1999)
4. Kotler, P., Armstrong, G.: Principles of Marketing, 12th edn. Prentice Hall, New Jersey (2008)
5. Fildes, R., Nikolopoulos, K., Crone, S.F., Syntetos, A.A.: Forecasting and operational research: a review. J. Oper. Res. Soc. **59**(9), 1150–1172 (2017). https://doi.org/10.1057/palgrave.jors.2602597
6. Tsafarakis, S., Grigoroudis, E., Matsatsinis, N.: Consumer choice behaviour and new product development: an integrated market simulation approach. J. Oper. Res. Soc. **62**(7), 1253–1267 (2011). https://doi.org/10.1057/jors.2010.70
7. Siskos, Y., Grigoroudis, E., Matsatsinis, N.F., Baourakis, G.: Disaggregation analysis in agricultural product consumer behavior. In: Pardalos, P.M., Siskos, Y., Zopounidis, C. (eds.) Advances in Multicriteria Analysis, vol. 5, pp. 185–202. Kluwer Academic Publishers (1995)
8. Huber, J., Orme, B., Miller, R.: Dealing with product similarity in conjoint simulations. In: Gustafsson, A., Herrmann, A., Huber, F. (eds.) Conjoint Measurement. Springer, Berlin, Heidelberg (2007)
9. Torrens, G.E.: The order and priority of research and design method application within an assistive technology new product development process: a summative content analysis of 20 case studies. Disabil. Rehabil.-Assist. Technol. **13**(1), 66–77 (2018). https://doi.org/10.1080/17483107.2017.1280547
10. Yassine, A.A., Mostafa, O., Browning, T.R.: Scheduling multiple, resource-constrained, iterative, product development projects with genetic algorithms. Comput. Ind. Eng. **107**, 39–56 (2017). https://doi.org/10.1016/j.cie.2017.03.001
11. Lee, K.Y., Yang, S.B.: The role of online product reviews on information adoption of new product development professionals. Internet Res. **25**(3), 435–452 (2015). https://doi.org/10.1108/IntR-11-2013-0238
12. Achiche, S., Appio, F.P., McAloone, T.C., Di Minin, A.: Fuzzy decision support for tools selection in the core front end activities of new product development. Res. Eng. Des. **24**(1), 1–18 (2013). https://doi.org/10.1007/s00163-012-0130-4
13. Lockshin, L., Mueller, S.: Pricing for new product development. In: Jaeger, S.R., MacFie, H. (eds.) Consumer-Driven Innovation in Food and Personal Care Products. Woodhead Publishing in Food Science Technology and Nutrition, vol. 195, pp. 303–331. (2010)
14. Varma, V.A., Uzsoy, R., Pekny, J., Blau, G.: Lagrangian heuristics for scheduling new product development projects in the pharmaceutical industry. J. Heuristics **13**(5), 403–433 (2007). https://doi.org/10.1007/s10732-007-9016-4
15. Petrick, I.J., Echols, A.E.: Technology roadmapping in review: a tool for making sustainable new product development decisions. Technol. Forecast. Soc. Change **71**(1–2), 81–100 (2004). https://doi.org/10.1016/s0040-1625(03)00064-7
16. Maravelias, C.T., Grossmann, I.E.: Optimal resource investment and scheduling of tests for new product development. Comput. Chem. Eng. **28**(6–7), 1021–1038 (2004). https://doi.org/10.1016/j.compchemeng.2003.09.019
17. Bassett, M.H., Gardner, L.L., Steele, K.: Dow AgroSciences uses simulation-based optimization to schedule the new-product development process. Interfaces **34**(6), 426–437 (2004). https://doi.org/10.1287/inte.1040.0099
18. Maravelias, C.T., Grossmann, I.E.: Simultaneous planning for new product development and batch manufacturing facilities. Ind. Eng. Chem. Res. **40**(26), 6147–6164 (2001). https://doi.org/10.1021/ie010301x
19. Haque, B., Pawar, K.S., Barson, R.J.: Analysing organisational issues in concurrent new product development. Int. J. Prod. Econ. **67**(2), 169–182 (2000). https://doi.org/10.1016/s0925-5273(00)00017-7

20. Wang, J.R.: A fuzzy set approach to activity scheduling for product development. J. Oper. Res. Soc. **50**(12), 1217–1228 (1999)
21. Roy, B.: The Outranking Approach and the Foundations of ELECTRE Methods. In: Bana e Costa, C.A. (ed.) Reading in Multiple Criteria Decision Aid, pp. 155–183. Springer, Berlin (1990)
22. Dias, J., José, F., Roy, B.: The Software Electre III–IV. Lamsade (2006)
23. Giannoulis, C., Ishizaka, A.: A Web-based decision support system with ELECTRE III for a personalised ranking of British universities. Decis. Support Syst. **48**(3), 488–497 (2010). https://doi.org/10.1016/j.dss.2009.06.008
24. Dias, J., Figueira, J.R., Roy, B.: The software ELECTRE III–IV: methodology and user manual (Version 3.X). Université Paris-Dauphine LAMSADE, Paris (2006)
25. Doumpos, M., Marinakis, Y., Marinaki, M., Zopounidis, C.: An evolutionary approach to construction of outranking models for multicriteria classification: the case of the ELECTRE TRI method. Eur. J. Oper. Res. **199**(2), 496–505 (2009). https://doi.org/10.1016/j.ejor.2008.11.035
26. Alvarez, P.A., Leyva López, J.C., López Parra, P.: A new disaggregation preference method for new products design. In: Martínez, L., Karre, E.E. (ed.) 13th International FLINS Conference, Belfast, Northern Ireland, UK 2018. Computer Engineering and Information Science. World Scientific
27. Koch, I.S., Muller, M., Joubert, E., van der Rijst, M., Næs, T.: Sensory characterization of rooibos tea and the development of a rooibos sensory wheel and lexicon. Food Res. Int. **46**(1), 217–228 (2012). https://doi.org/10.1016/j.foodres.2011.11.028
28. Bleibaum, R.N., Stone, H., Tan, T., Labreche, S., Saint-Martin, E., Isz, S.: Comparison of sensory and consumer results with electronic nose and tongue sensors for apple juices. Food Qual. Prefer. **13**(6), 409–422 (2002). https://doi.org/10.1016/s0950-3293(02)00017-4
29. ASTM-E253-18a: Standard Terminology Relating to Sensory Evaluation of Materials and Products. ASTM International (2018). https://doi.org/10.1520/e0253-18a

Warehouse Location Design Using AS/RS Technologies: An Interval Valued Intuitionistic Fuzzy AHP Approach

Cengiz Kahraman, Başar Öztayşi and Sezi Cevik Onar

Abstract An automated storage and retrieval system (AS/RS) is a type of warehouse automation technology specifically designed to buffer, store, and retrieve product and inventory on demand. AS/RS technology varies substantially, and can consist of shuttles, cranes, carousels, vertical lift modules (VLMs), micro-loads, mini-loads, unit-loads, or other systems. Design of a warehouse location involving AS/RS technologies is a multicriteria decision making problem with several criteria under vague and imprecise environment. In this chapter, two different warehouse location design alternatives involving AS/RS technologies are evaluated by using an interval valued intuitionistic fuzzy analytic hierarchy process (AHP) approach.

1 Introduction

Warehouse design is important for the overall efficiency of a company. Inefficient warehouse designs lead to financial damages. In order to improve the warehouse design and inventory management processes automated storage and retrieval technologies (AS/RS technologies) can be used. AS/RS technologies are automated stock management technologies that are used in various areas including production lines, logistics and storage centers. These technologies often involve automated machines that can move up and down through single or multiple storage areas for carrying and storing items. AS/RS technologies provide efficiency in space usage. They increase storage capacities and therefore very beneficial for long term planning. Inventory management and control processes can be improved by using AS/RS technologies. They provide an agile system that decreases the response time for locating, storing and retrieving materials or products. The continuous information that is provided by these systems decreases the inventory shortage and enhance the rotation of the stoked items. They deliver flexibility in the design of the storage place and easily coordinate with other inventory management technologies such as automated guided vehicles.

C. Kahraman (✉) · B. Öztayşi · S. C. Onar
Department of Industrial Engineering, Istanbul Technical University, 34367 Macka, Istanbul, Turkey
e-mail: kahramanc@itu.edu.tr

C. Kahraman and S. Cebi (eds.), *Customer Oriented Product Design*, Studies in Systems, Decision and Control 279, https://doi.org/10.1007/978-3-030-42188-5_19

The costs are minimized through reduced energy and heating costs and minimal scrap due to automation.

There are five types of AS/RS technologies, namely, unit load AS/RS, miniload AS/RS, person-on-board AS/RS, deep-lane AS/RS and automated item retrieval systems. Unit load AS/RS technologies are usually used to manage inventories that are in standard-size containers. Miniload AS/RS systems are used to manage small size inventories usually stored in bins or drawers. In the person-on-board AS/RS technologies, there is a human operator that is responsible for individual items in a bin or drawer. These systems are usually used for managing inventories which are less than unit load. Deep-lane AS/RS technologies are used to manage large quantities of inventory where ten items can be stored in the same rack. The automated item retrieval systems are used to manage individual items or small inventories where the inventories are stored in lanes.

Selecting the appropriate AS/RS technology and designing the warehouses based on the selected technology is critical for the long term success of a company. Not only financial criteria but also the reliability criteria should be taken into account for selecting the appropriate AS/RS technology [1]. Yet, evaluating these criteria is hard and often involves imprecision. Fuzzy multi-criteria decision making techniques are appropriate for dealing with decision problems where multiple uncertain dimensions are involved.

In order to better deal with the fuzziness, the fuzzy sets theory has been extended and extensions of fuzzy sets have been introduced. Intuitionistic fuzzy sets (IFS), introduced by Atanassov [2], define membership and non-membership values for dealing with vagueness. The sum of membership and non-membership values in the IFS do not have to be one that enables representing the hesitancy in a system. This representation of fuzziness brings flexibility to the decision makers. Therefore, intuitionistic fuzzy sets have been widely used in the literature for various problems such as quality management, energy management, manufacturing management, economic analysis, engineering economics [3–7]. In this study, interval valued intuitionistic fuzzy AHP method is used to evaluate AS/RS technologies for designing a warehouse. This interval valued intuitionistic fuzzy AHP approach enables dealing with conflicting criteria and makes it possible to handle uncertainty in the evaluation of AS/RS technologies.

The rest of the chapter is organized as follows. Section 2 briefly explains the different AS/RS Technologies. The preliminaries of the intuitionistic fuzzy sets are given in Sect. 3. The utilized methodology, interval valued intuitionistic fuzzy AHP is given in Sect. 4. Section 5 explains the warehouse location design problem using AS/RS Technologies. Section 6 contains a real case study for the evaluation of location design alternatives. The last section concludes and gives further suggestions.

2 AS/RS Technologies

AS/RS refers to a variety of means under computer control for automatically depositing and retrieving loads from defined storage locations [8]. Although similar technologies exist for more than 40 years, until 2000s AS/RS term has only been used as crane and aisle applications. Today AS/RS are widely used in modern distribution centers as they provide fast, accurate and efficient handling of materials, 24 h a day. In its most basic form, an AS/RS consists of storage racks where products are stored and retrieved automatically. This broad definition brings a wide range of possible AS/RS implementations [9]. In their extended literature review [10] provide a classification for various AS/RS system options (Fig. 1).

One of the most commonly adopted AS/RS is the unit load, which has in each aisle one crane that cannot leave its designated aisle. In this systems, product handling in this case is accomplished by unit load (e.g., full pallet quantities) only; no people are involved to handle individual products. The racks in the basic version are stationary and single-deep, which means that every load is directly accessible by the crane. Unit Load AS/RS systems are used to store and retrieve loads which are stored in standard-sized containers. Storage and retrieval systems are computer controlled and are designed to handle the standard-sized containers. In most common case, a mechanical clamp wich is located on the storage and retrieval machine handles the load. On the other hand, vacuum or a magnet-based mechanism can also be utilized for handling the containers. Unit load AS/RS systems are generally designed for unit loads more than 200 kilos (Fig. 2).

Mini Load AS/RS systems are designed to handle small loads such as individual parts, tools, and supplies that are contained in bins or drawers in the warehouse. MiniLoad systems perform best in cases where the availability of space is limited. They are also feasible for cases where the volume is too low. In such systems the

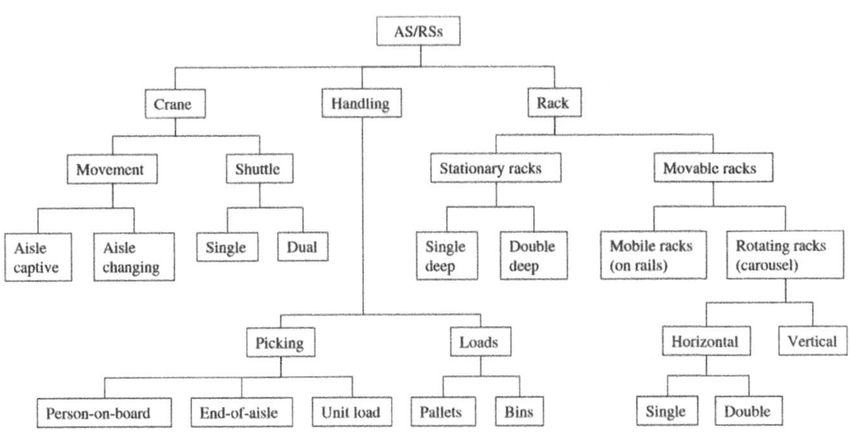

Fig. 1 Implementation options of AS/RS systems [10]

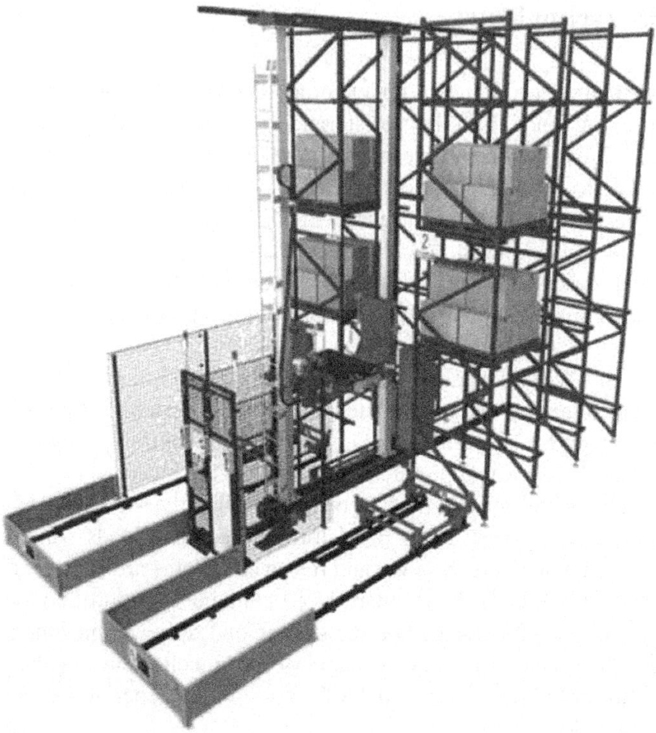

Fig. 2 Unit Load AS/RS Technology

AS/RS drops off the retrieved unit loads at a workstation and a picker at this workstation takes the required amount of products from the mini load after which the AS/RS moves the remainder of the load back into the storage rack (Fig. 3).

Alternatively, the picking operation can be integrated with the AS/RS system. When the system is designed so that a person can ride along, the system is entitled "person-onboard" AS/RS. In such systems the load is not automatically retrieved from its location the person on the board pick the items. Person on board ASRS system allow storage of items in low quantities. A human operator rides the carriage to pick up individual items form a bin or drawer. The system gives the operator the chance to select the items and place them in a module. The items are then carried by the system to the end of the aisle or to a conveyor to reach its destination (Fig. 4).

Deep-Lane AR/SR systems are high density unit load storage system that is appropriate for storing large quantities of stock. In such systems the loads are stored in multi deep storage with up to 10 items in a single rack, one load behind the next. Each rack is designed for flow-through, with input and output on the opposite side. Machine is used on the entry side of the rack for input load and loads are retrieved from other side by an S/R type machine. Storage in the racks may occur single or double deep. In a double-deep rack, each rack location has space for two unit loads;

Fig. 3 Mini Load AS/RS Technology

one load is stored in front of the other load. A load can only be put into or retrieved from the second position if there is no load in the first position. Double-deep storage might be beneficial if the variety of loads is relatively low and the turnover rate of these loads is high [11]. Modifications to the crane may be required to be able to store and retrieve loads from both positions. Carousel systems (horizontal or vertical, single or double) are suitable for storing small and medium-sized products at different levels. A crane is used to store and retrieve items from the rotating carousel. The lower and upper part of a double carousel can rotate independently of each other (Fig. 5).

Final type of AS/RS is called autonomous vehicle storage and retrieval systems. This system separates horizontal and vertical travel. Vehicles travel horizontally over rails through aisles, while lifts are used to transfer loads vertically. This system is designed for retrieval of individual items or small product cartoons. The items are stored in lanes rather than bins or drawers. When an item is retrieved from the front it is delivered to the pickup station by pushing it from its lane or propping onto a conveyor. The supply of items in each lane is periodically replenished and thus permitting first-in/First-out inventory rotation. After moving itself to the correct lane, the picking head activates (Fig. 6).

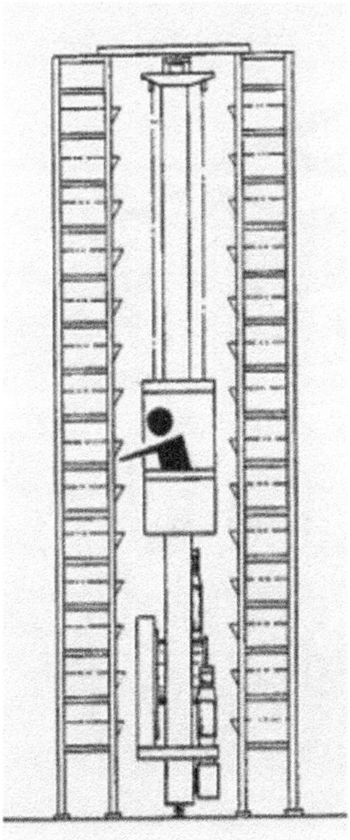

Fig. 4 Person on board AS/RS technology

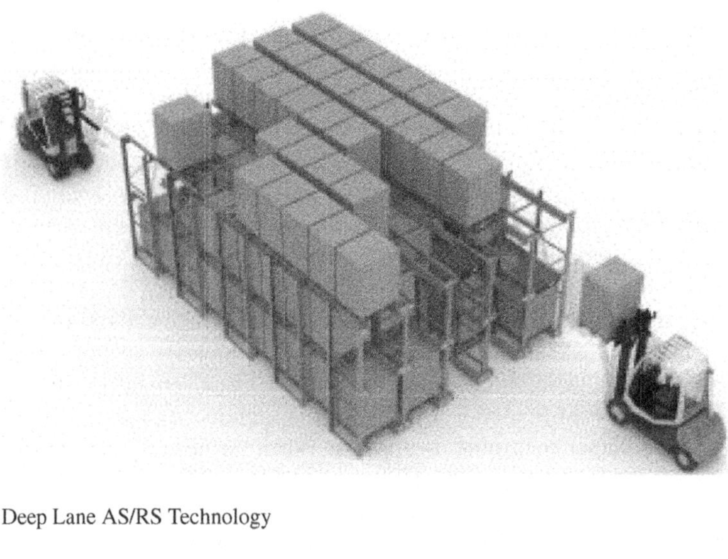

Fig. 5 Deep Lane AS/RS Technology

Fig. 6 Autonomous vehicle storage and retrieval systems

3 Preliminaries: Intuitionistic Fuzzy Sets

Atanassov [2] defines an intuitionistic fuzzy set $\tilde{I}$ as in Eqs. (1) and (2).

$$\tilde{I} = \left\{\langle x, \mu_{\tilde{I}}(x), v_{\tilde{I}}(x)\rangle; x \in X\right\}, \tag{1}$$

$$0 \leq \mu_{\tilde{I}}(x) + v_{\tilde{I}}(v) \leq 1, \tag{2}$$

where $X \neq \emptyset$ be a given set, $\tilde{I}$ is an intuitionistic fuzzy set in X, $\mu_{I} \rightarrow [0, 1]$ is the membership function and $v_{\tilde{I}} \rightarrow [0, 1]$ is the non-membership function. Atanassov and Gargov [12] enlarged the intuitionistic fuzzy sets to interval valued intuitionistic fuzzy sets (IVIF) and defined an IVIF sets $\widetilde{II}$ over X as in Eqs. (3) and (4).

$$\widetilde{II} = \left\{< x, \tilde{\mu}_{\widetilde{II}}(x), \tilde{v}_{\widetilde{II}}(x) > | x \in X\right\}, \tag{3}$$

$$0 \leq sup\tilde{\mu}_{\widetilde{II}}(x) + \sup \tilde{v}_{\widetilde{II}}(x) \leq 1, \forall x \in X \tag{4}$$

where $C \subseteq [0, 1]$ is the set of all closed subintervals of the interval with the membership $\tilde{\mu}_{\widetilde{II}} \rightarrow C \subseteq [0, 1]$ and non-membership function $\tilde{v}_{\widetilde{II}}(x) \rightarrow C \subseteq [0, 1]$.

IVIF sets $\widetilde{II}$ has the membership values with the lower point of closed interval $\tilde{\mu}_{\widetilde{II}L}(x)$ and the upper point of closed interval $\tilde{\mu}_{\widetilde{II}U}(x)$ and has the non-membership values with the lower point of closed interval $\tilde{v}_{\widetilde{II}L}(x)$ and the upper point of closed interval $\tilde{v}_{\widetilde{II}U}(x)$. $\widetilde{II}$ can be denoted as in Eqs. (5) and (6).

$$\widetilde{II} = \left\{ < x, \left[\tilde{\mu}_{\widetilde{II}L}(x), \tilde{\mu}_{\widetilde{II}U}(x) \right], [\tilde{v}_{\widetilde{II}L}(x), \tilde{v}_{\widetilde{II}U}(x)] > | x \in X \right\} \tag{5}$$

$$0 \leq \tilde{\mu}_{\widetilde{II}U}(x) + \tilde{v}_{\widetilde{II}U}(x) \leq 1, \tilde{\mu}_{\widetilde{II}L}(x) \geq 0, \tilde{v}_{\widetilde{IIL}}(x) \geq 0 \tag{6}$$

The operations on two IVIF sets $\widetilde{II}_1$ and $\widetilde{II}_2$ are defined as in Eqs. (7–9) [12].

$$\widetilde{II}_1 \oplus \widetilde{II}_2 = \left(\left[\mu_1^- + \mu_2^- - \mu_1^- \mu_2^-, \mu_1^+ + \mu_2^+ - \mu_1^+ \mu_2^+ \right] . \left[v_1^- v_2^-, v_1^+ v_2^+ \right] \right) \tag{7}$$

$$\widetilde{II}_1 \otimes \widetilde{II}_2 = \left(\left[\mu_1^- \mu_2^-, \mu_1^+ \mu_2^+ \right], \left[v_1^- + v_2^- - v_1^- v_2^-, v_1^+ + v_2^+ - v_1^+ v_2^+ \right] \right) \tag{8}$$

$$\lambda \widetilde{II}_1 = \left(\left[1 - \left(1 - \mu_1^-\right)^\lambda, 1 - \left(1 - \mu_1^+\right)^\lambda \right], \left[\left(v_1^-\right)^\lambda, \left(v_1^+\right)^\lambda \right] \right) \tag{9}$$

4 IVIF-AHP Method

The multicriteria decision making methodology used in this paper is based on [13] work. They propose an approach for multi-criteria decision making (MCDM) problems with interval-valued intuitionistic fuzzy preference relations (IVIFPRs). Let $\tilde{R} = \left(\tilde{r}_{ij}\right)_{n \times n} = \left(\left[\mu_{ij}^-, \mu_{ij}^+ \right], \left[v_{ij}^-, v_{ij}^+ \right] \right)_{n \times n}$ be an interval-valued intuitionistic judgment matrix as in Eq. (10).

$$\tilde{R} = \begin{bmatrix} \left(\left[\mu_{11}^-, \mu_{11}^+ \right], \left[v_{11}^-, v_{11}^+ \right] \right) & \cdots & \left(\left[\mu_{1n}^-, \mu_{1n}^+ \right], \left[v_{1n}^-, v_{1n}^+ \right] \right) \\ \vdots & \ddots & \vdots \\ \left(\left[\mu_{n1}^-, \mu_{n1}^+ \right], \left[v_{n1}^-, v_{n1}^+ \right] \right) & \cdots & \left(\left[\mu_{nn}^-, \mu_{nn}^+ \right], \left[v_{nn}^-, v_{nn}^+ \right] \right) \end{bmatrix} \tag{10}$$

The reciprocal value of $\left(\left[\mu_{ij}^-, \mu_{ij}^+ \right], \left[v_{ij}^-, v_{ij}^+ \right] \right)$ in $\tilde{R}$ is $\left(\left[v_{ji}^-, v_{ji}^+ \right], \left[\mu_{ji}^-, \mu_{ji}^+ \right] \right)$. For instance, the reciprocal value of $\left(\left[\mu_{n1}^-, \mu_{n1}^+ \right], \left[v_{n1}^-, v_{n1}^+ \right] \right)$ is $\left(\left[v_{1n}^-, v_{1n}^+ \right], \left[\mu_{1n}^-, \mu_{1n}^+ \right] \right)$.

The score judgment matrix of $\tilde{R}$ is represented by the matrix $\tilde{S} = \left(\tilde{s}_{ij}\right)_{n \times n} = \left[\mu_{ij}^- - v_{ij}^+, \mu_{ij}^+ - v_{ij}^- \right]$ as in Eq. (11).

$$\tilde{S} = \begin{bmatrix} \left[\mu_{11}^{-} - v_{11}^{+}, \mu_{11}^{+} - v_{11}^{-}\right] & \cdots & \left[\mu_{1n}^{-} - v_{1n}^{+}, \mu_{1n}^{+} - v_{1n}^{-}\right] \\ \vdots & \ddots & \vdots \\ \left[\mu_{n1}^{-} - v_{n1}^{+}, \mu_{n1}^{+} - v_{n1}^{-}\right] & \cdots & \left[\mu_{nn}^{-} - v_{nn}^{+}, \mu_{nn}^{+} - v_{nn}^{-}\right] \end{bmatrix} \quad (11)$$

The interval multiplicative matrix $\tilde{A} = (\tilde{a}_{ij})_{n \times n} = \left[10^{\left(\mu_{ij}^{-} - v_{ij}^{+}\right)}, 10^{\left(\mu_{ij}^{+} - v_{ij}^{-}\right)}\right]$ is given as in Eq. (12).

$$\tilde{A} = \begin{bmatrix} \left[10^{\left(\mu_{11}^{-} - v_{11}^{+}\right)}, 10^{\left(\mu_{1n}^{+} - v_{1n}^{-}\right)}\right] & \cdots & \left[10^{\left(\mu_{ij}^{-} - v_{ij}^{+}\right)}, 10^{\left(\mu_{ij}^{+} - v_{ij}^{-}\right)}\right] \\ \vdots & \ddots & \vdots \\ \left[10^{\left(\mu_{n1}^{-} - v_{n1}^{+}\right)}, 10^{\left(\mu_{n1}^{+} - v_{n1}^{-}\right)}\right] & \cdots & \left[10^{\left(\mu_{nn}^{-} - v_{nn}^{+}\right)}, 10^{\left(\mu_{nn}^{+} - v_{nn}^{-}\right)}\right] \end{bmatrix}. \quad (12)$$

The steps of the fuzzy multicriteria decision making method are summarized in the following:

Step 1: Assign the interval-valued intuitionistic preferences for each pair of alternatives and obtain the interval-valued intuitionistic preference relation $\tilde{R} = (\tilde{r}_{ij})_{n \times n}$.

Step 2: Calculate the score judgment matrix $\tilde{S} = (\tilde{s}_{ij})_{n \times n}$ and the interval multiplicative matrix $\tilde{A} = (\tilde{a}_{ij})_{n \times n}$.

Step 3: Determine the priority vector of the interval multiplicative matrix $\tilde{A} = (\tilde{a}_{ij})_{n \times n}$ by calculating the $\tilde{w}_i$ interval for each criteria using Eq. (13).

$$\tilde{w}_i = \left[\frac{\sum_{j=1}^{n} \tilde{a}_{ij}^{-}}{\sum_{i=1}^{n} \sum_{j=1}^{n} \tilde{a}_{ij}^{+}}, \frac{\sum_{j=1}^{n} \tilde{a}_{ij}^{+}}{\sum_{i=1}^{n} \sum_{j=1}^{n} \tilde{a}_{ij}^{-}}\right] \quad (13)$$

Step 4: Construct the possibility degree matrix $P = (p_{ij})_{m \times n}$ by comparing the obtained weights in Step 3. To do this, use Eq. (14):

$$P(w_i \geq w_j) = \frac{min\left\{L_{w_i} + L_{w_j}, max\left(w_i^{+} - w_j^{-}, 0\right)\right\}}{L_{w_i} + L_{w_j}} \quad (14)$$

where $L_{w_i} = w_i^{+} - w_i^{-}$ and $L_{w_j} = w_j^{+} - w_j^{-}$ and $p_{ij} \geq 0$, $p_{ij} + p_{ji} = 1$, $p_{ii} = 1/2$.

Step 5: Prioritize the $P = (p_{ij})_{m \times n}$ by Eq. (15):

$$w_i = \frac{1}{n}\left[\sum_{j=1}^{n} p_{ij} + \frac{n}{2} - 1\right] \quad (15)$$

Step 6: Normalize the weights vector obtained in Step 5.

Table 1 Linguistic scale and its corresponding IVIFS

Linguistic terms	Membership	Non-membership
Absolutely Low (AL)	([0, 0.2], [0.5, 0.8])	([0.5, 0.8], [0, 0.2])
Very Low (VL)	([0.1, 0.3], [0.4, 0.7])	([0.4, 0.7], [0.1, 0.3])
Low (L)	([0.2, 0.4], [0.3, 0.6])	([0.3, 0.6], [0.2, 0.4])
Medium Low (ML)	([0.3, 0.5], [0.2, 0.5])	([0.2, 0.5], [0.3, 0.5])
Equal (E)	([0.4, 0.6], [0.2, 0.4])	([0.2, 0.4], [0.4, 0.6])
Medium High (MH)	([0.5, 0.7], [0.1, 0.3])	([0.1, 0.3], [0.5, 0.7])
High (H)	([0.6, 0.8], [0, 0.2])	([0, 0.2], [0.6, 0.8])
Very High (VH)	([0.7, 0.9], [0, 0.1])	([0, 0.1], [0.7, 0.9])
Absolutely High (AH)	([0.8, 1.0], [0, 0])	([0, 0], [0.8, 1.0])

The scale given in Table 1 is used to assign the interval-valued intuitionistic preferences.

For exactly equal, we assign EE = ([0.5, 0.5], [0.5, 0.5]).

5 Warehouse Location Design with AS/RS Technologies

Studies on AS/RS system design provide some characteristic of the design and operation of a warehouse system design. Bozer and White [14] underline the most important factors related to layout and design of AS/RS such as load size determination, individual storage space dimensions, estimation of storage spaces, number of machines to be used, size parameters of the machines, cycle time for a unit load. Cycle time is a criterion that will be considered in the application section.

The design process starts with determining the load size based on the flow in the warehouse. The movement frequency of parts, tools, fixtures, pallets, and other supplies express the overall work flow in the ware house. The work flow is determined by variety and volume of part types and the type of production system. The width, length and height of rack structure of the AS/RS aisle are related to the unit load dimensions. The dimensions of the unit load with proper clearances provide the individual storage space dimensions and account for the size of supporting beams in the rack structure. Storing unique items of unusual and complex shape are excluded from the AS/RS design. Weight of the unit load is also considered in the structural design.

Individual storage space Dimensions is another important issue in AS/RS technology. Let l, b and h be the length, width and height of the unit load. The length (L), width (W) and height (H) of the rack structure of the AS/RS aisle are related to the unit load dimensions and number of compartments.

Estimating Storage Spaces should be implemented before AS/RS technology is purchased. Storage policies are used to determine the number of storage spaces in AS/RS. In dedicated storage policy, a particular set of storage slots is allocated to a specific product. Hence the sum of the maximum inventory levels for the entire products match with the number of slots required to store the product. In case of randomized storage policy, any compartment in the storage aisle is equally probable to be selected for transaction. Likewise, there is equal chance of each unit of particular product to be retrieved when a retrieval operation is performed. Thus, in a long run, maximum of the aggregate inventory level of all the products is taken into account for determining the storage space number.

The system throughput refers to the number of loads to be stored and number of loads to be retrieved per hour. It is defined as hourly rate of S/R transactions (number of loads stored and number of loads retrieved) that an automated storage system can perform. A dual command cycle is used to increase the throughput, since it reduces travel time per transaction. Factors that influence the throughput include: 1. Speed of S/R machine, 2. Mix of single- and dual-cycle transactions, 3. Percent utilization of the storage racks, 4. Arrangement of stored items, 5. AS/RS control system speed, 6. Speed and efficiency of the material- handling equipment such as AGVs, conveyors, and forklifts used to move loads to the input and removal loads from the output.

The size parameters are crucial for the performance of AS/RS technologies. Therefore, it is necessary to estimate size parameters for the AS/RS selection decision. The main size parameters are system length, system width and the system height. For estimating the size parameters, the number of rows and bays that are used in the storage process should be determined. The bay width, system width, rack length, system length, bay depth, the length of single storage space and centre to centre rack support width are the parameters that are used in the calculations. For instance, the bay width is calculated by using length of storage space and centre to centre rack support width.

The single and dual command cycle times are one of the most important criteria for AS/RS technology selection. The single command cycle time refers to the cycle time during storage or retrieval operations. The machines pick up times, the travel time to the storage area, the deposit time, return time to the station after the loading are considered under single command storage or retrieval cycle time. In the calculation of Dual command cycle time, time spent by the S/R machine from the P/D station to the storage area for putting down the load, S/R machine's travel time for recovering the load and the traveling back time to the station for depositing the load is considered.

The utilization ratio is an important determinant that shoes the performance of S/R technologies. The utilization percentage depends on the single command cycle time and dual command cycle times. The cost of the S/R technologies is affected by the utilization ratio.

A suitable AS/RS technology satisfying the above necessities should be selected for warehouses. In the following section, two warehouse layout design alternatives are evaluated based on interval-valued intuitionistic fuzzy AHP.

6 Evaluation of Warehouse Layout Design Alternatives

XYZ company plans to use an AS/SR technology in their warehouse. Location design of their warehouses involves the alternatives presented in Fig. 7. Warehouse design is composed of a vertical machine center, a turning center, a control center, a horizontal machine center, a washing machine, an AGV, and an AS/RS system. In the first design, the AS/RS is circled by AGV including the whole parts of the warehouse system. In the second alternative, AS/RS technology is again circled by AGV but surrounded by all other parts. The considered criteria in the selection of design alternatives are the *cycle time* (necessary duration to complete and to store a product), *Cost* (total expenses to complete and to store a product), *Space dimensions for the AS/RS system*, *Complexity of the system*, and *Flexibility* to adopt new circumstances.

Table 2 presents the linguistic evaluations for the pairwise comparisons of the criteria.

Table 3 shows the corresponding numerical assessments for the linguistic terms given in Table 2. These are the interval-valued intuitionistic preference relations.

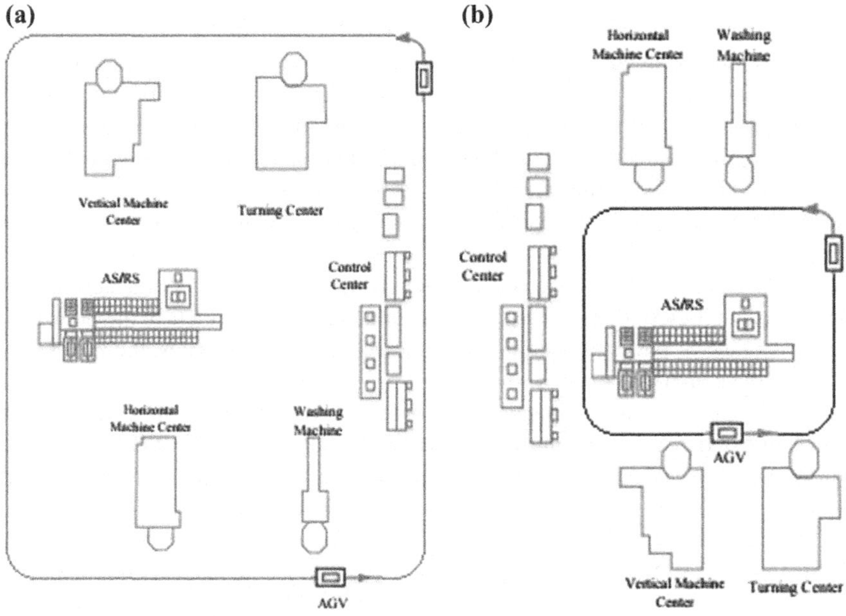

Fig. 7 Warehouse location alternatives involving AS/RS technology

Table 2 Linguistic evaluations for comparisons of the criteria

Criteria	C1	C2	C3	C4	C5
C1: Cycle Time	EE	MH	H	VL	VL
C2: Cost		EE	AE	VL	VL
C3: Required Area			EE	AL	AL
C4: Complexity				EE	AE
C5: Flexibility					EE

Table 3 Interval-valued intuitionistic preference relations

	C1	C2	C3
C1: Cycle Time	([0.5, 0.5], [0.5, 0.5])	([0.5, 0.7], [0.1, 0.3])	([0.6, 0.8], [0, 0.2])
C2: Cost	([0, 0.2], [0.6, 0.8])	([0.5, 0.5], [0.5, 0.5])	([0.4, 0.6], [0.2, 0.4])
C3: Space Dimension	([0.3, 0.6], [0.2, 0.4])	([0.4, 0.7], [0.1, 0.3])	([0.5, 0.5], [0.5, 0.5])
C4: Complexity	([0.2, 0.5], [0.3, 0.5])	([0.3, 0.6], [0.2, 0.4])	([0.1, 0.3], [0.5, 0.7])
C5: Flexibility	([0.3, 0.6], [0.2, 0.4])	([0.4, 0.7], [0.1, 0.3])	([0.2, 0.4], [0.4, 0.6])
	C4	C5	
C1: Cycle Time	([0.1, 0.3], [0.4, 0.7])	([0.1, 0.3], [0.4, 0.7])	
C2: Cost	([0.1, 0.3], [0.4, 0.7])	([0.1, 0.3], [0.4, 0.7])	
C3: Space Dimension	([0, 0.2], [0.5, 0.8])	([0, 0.2], [0.5, 0.8])	
C4: Complexity	([0.5, 0.5], [0.5, 0.5])	([0.4, 0.6], [0.2, 0.4])	
C5: Flexibility	([0.2, 0.5], [0.3, 0.5])	([0.5, 0.5], [0.5, 0.5])	

Table 4 presents the score judgment matrix.

Table 5 gives the interval multiplicative matrix and the summation of the lower and upper interval values.

Table 6 presents the interval valued weights of the criteria. It is the priority vector of the interval multiplicative matrix.

Table 7 shows the possibility degree matrix.

Table 8 presents the prioritized possibility degree matrix and the normalized criteria weights.

Table 9 summarizes the weights of the alternatives with respect to each criterion.

Table 10 presents the calculation of overall priorities of the alternatives.

Based on the overall priorities on Table 10, Design 1 is the preferred alternative.

Table 4 Score judgment matrix

	C1		C2		C3		C4		C5	
C1: Cycle Time	0	0	0.2	0.6	0.4	0.8	− 0.6	− 0.1	− 0.6	− 0.1
C2: Cost	− 0.6	− 0.2	0	0	0	0.4	− 0.6	− 0.1	− 0.6	− 0.1
C3: Space Dimension	− 0.8	− 0.4	− 0.4	0	0	0	− 0.8	− 0.3	− 0.8	− 0.3
C4: Complexity	0.1	0.6	0.1	0.6	0.3	0.8	0	0	0	0.4
C5: Flexibility	0.1	0.6	0.1	0.6	0.3	0.8	− 0.4	0	0	0

Table 5 Interval multiplicative matrix

	C1		C2		C3		C4		C5			
	–	+	–	+	–	+	–	+	–	+	TOTAL –	TOTAL +
C1: Cycle Time	1.00	1.00	1.58	3.98	2.51	6.31	0.25	0.79	0.25	0.79	5.60	12.88
C2: Cost	0.25	0.63	1.00	1.00	1.00	2.51	0.25	0.79	0.25	0.79	2.75	5.73
C3: Space Dimension	0.16	0.40	0.40	1.00	1.00	1.00	0.16	0.50	0.16	0.50	1.87	3.40
C4: Complexity	1.26	3.98	1.26	3.98	2.00	6.31	1.00	1.00	1.00	2.51	6.51	17.78
C5: Flexibility	1.26	3.98	1.26	3.98	2.00	6.31	0.40	1.00	1.00	1.00	5.91	16.27
Total											22.65	56.07

Table 6 Priority vector of the interval multiplicative matrix

Interval valued criteria weights	
w1	[0.1, 0.569]
w2	[0.049, 0.253]
w3	[0.033, 0.15]
w4	[0.116, 0.785]
w5	[0.105, 0.718]

Table 7 Possibility degree matrix

$p(C1 \geq C2) = 0.772$	$p(C3 \geq C1) = 0.086$
$p(C2 \geq C1) = 0.228$	$p(C1 \geq C3) = 0.914$
$p(C4 \geq C1) = 0.602$	$p(C1 \geq C5) = 0.428$
$p(C1 \geq C4) = 0.398$	$p(C2 \geq C3) = 0.685$
$p(C5 \geq C1) = 0.572$	$p(C3 \geq C2) = 0.315$
$p(C2 \geq C4) = 0.157$	$p(C2 \geq C5) = 0.181$
$p(C4 \geq C2) = 0.843$	$p(C5 \geq C2) = 0.819$
$p(C3 \geq C4) = 0.043$	$p(C4 \geq C3) = 0.957$
$p(C3 \geq C5) = 0.061$	$p(C5 \geq C3) = 0.939$
$p(C4 \geq C5) = 0.530$	$p(C5 \geq C4) = 0.470$

Table 8 Prioritized possibility degree matrix and the normalized criteria weights

Criteria weights	Possibility degrees	Normalized criteria weights
w1	0.902	0.226
w2	0.650	0.163
w3	0.501	0.125
w4	0.987	0.247
w5	0.960	0.240
TOTAL	4	1

7 Conclusions

Automatic storage and retrieval systems are automated systems for placement and retrieval of loads from a specific location. This system is widely used in manufacturing hubs, distribution centers and warehouses. It can be beneficial where there is high volume movement, storage density has to be high due to space constraints; accuracy is important. It enhances inventory control and tracking. AS/RS provides a greater flexibility, easily replaceable thus decreased downtime, labor cost decrease, space utilization increase. They can operate in matrix of storage containers and removes personnel from difficult working conditions for example in cold storage. However,

Table 9 Weights of the alternatives with respect to criteria

Criteria	Design1	Design2	Weight
Design1	EE	VL	0.425
Design2		EE	0.575
Criteria	Design1	Design2	Weight
Design1	EE	L	0.463
Design2		EE	0.537
Criteria	Design1	Design2	Weight
Design1	EE	MH	0.594
Design2		EE	0.406
Criteria	Design1	Design2	Weight
Design1	EE	H	0.616
Design2		EE	0.384
Criteria	Design1	Design2	Weight
Design1	EE	ML	0.513
Design2		EE	0.487

Table 10 Overall priorities of the alternatives

Alternatives	Criteria weights					Overall priorities
	C1	C2	C3	C4	C5	
	0.226	0.163	0.125	0.247	0.24	
Design 1	0.425	0.463	0.594	0.616	0.513	0.52
Design 2	0.575	0.537	0.406	0.384	0.487	0.48

they require skilled technicians and staff and high installation cost which mean high investment.

Selection of an appropriate design for the warehouse location when AS/RS technology is used is a multicriteria decision making problem under uncertainty based on linguistic terms rather than exact numerical evaluations. The vagueness and impreciseness in this problem have been handled by interval valued intuitionistic fuzzy sets. The multicriteria decision making method has been selected as AHP. Two warehouse location alternatives involving two AS/RS alternatives have been evaluated successfully.

For further research, we suggest the other extensions to be integrated by AHP method such as hesitant sets [15] hesitant fuzzy linguistic term sets [16], fuzzy inference system [17], Pythagorean fuzzy sets [7], picture fuzzy sets, spherical fuzzy sets, or q-rung orthopair fuzzy sets in order to evaluate AS/RS technology investments in warehouses. Other fuzzy multicriteria decision models using axiomatic design can be used [18–22]. Their results can be compared with this work.

References

1. Bolturk, E.: AS/RS technology selection using spherical fuzzy TOPSIS and neutrosophic TOPSIS. In: Intelligent and Fuzzy Techniques in Big Data Analytics and Decision Making, pp 969–976. Springer, Cham (2020)
2. Atanassov, K.: Intuitionistic fuzzy sets. Fuzzy Sets Syst. **20**, 87–96 (1986)
3. Kahraman, C., Cevik Onar, S., Oztaysi, B.: A comparison of wind energy investment alternatives using interval-valued intuitionistic fuzzy benefit/cost analysis. Sustainability **8**(2), 118 (2016)
4. Kahraman, C., Çevik Onar, S., Öztayşi, B.: Engineering economic analyses using intuitionistic and hesitant fuzzy sets. J. Intell. Fuzzy Syst. **29**(3), 1151–1168 (2015)
5. Kahraman, C., Cevik Onar, S., Cebi, S., Oztaysi, B.: Extension of information axiom from ordinary to intuitionistic fuzzy sets: an application to search algorithm selection. Comput. Ind. Eng. **105**, 348–361 (2017)
6. Kahraman, C., Cevik Onar, S., Oztaysi, B.: B2C marketplace prioritization using hesitant fuzzy linguistic AHP. Int. J. Fuzzy Syst. **20**(7), 2202–2215 (2018)
7. Cevik Onar, S., Oztaysi, B., Kahraman, C.: Multicriteria evaluation of cloud service providers using pythagorean fuzzy TOPSIS. J. Mult.-Valued Logic Soft Comput. **30**, 263–283 (2018)
8. MHI (2019), Materials handling Institute, Glossary http://www.mhi.org/glossary?fq=title-first-letter%3Aa-c&page=2&sort=title-sort%20asc%2Ccompany-sort%20asc
9. Gagliardi, J.P., Renaud, J., Ruiz, A.: Models for automated storage and retrieval systems: a literature review. Int. J. Prod. Res. **50**(24), 7110–7125 (2012)
10. Roodbergen, K.J., Vis, I.F.A.: A survey of literature on automated storage and retrieval systems. Eur. J. Oper. Res. **194**, 343–362 (2009)
11. Tompkins, J.A., White, J.A., Bozer, Y.A., Tanchoco, J.M.A.: Facilities Planning, 3rd edn. Wiley (2003)
12. Atanassov, G.Gargov: Interval valued intuitionistic fuzzy sets. Fuzzy Sets Syst. **31**(3), 343–349 (1989)
13. Wu, J., Huang, H., Cao, Q.W.: Research on AHP with interval-valued intuitionistic fuzzy sets and its application in multi-criteria decision making problems. Appl. Math. Model. **37**(24), 9898–9906 (2013)
14. Bozer, Y.A., White, J.A.: Optimum designs of automated storage/retrieval systems. In: TIMS/ORSA Joint National Meeting (1980)
15. Estrella, F.J., Cevik Onar, S., Rodríguez, R.M., Oztaysi, B., Martinez, L., Kahraman, C.: Selecting firms in university technoparks: a hesitant linguistic fuzzy TOPSIS model for heterogeneous contexts. J. Intell. Fuzzy Syst. **33**(2), 1155–1172 (2017)
16. Kahraman, C., Oztaysi, B., Cevik Onar, S.: A multicriteria supplier selection model using hesitant fuzzy linguistic term sets. In: Proceedings of the 11th International FLINS Conference (FLINS 2014) Decision Making and Soft Computing (2014)
17. Behret, H., Öztayşi, B., Kahraman, C.: A Fuzzy inference system for supply chain risk management. In: Practical Applications of Intelligent Systems, 429–438. Springer, Berlin (2011)
18. Kaya, I., Öztayşi, B., Kahraman, C.: A two-phased fuzzy multicriteria selection among public transportation investments for policy-making and risk governance. Int. J. Uncertain. Fuzziness Knowl.-Based Syst. **20**(01), 31–48 (2012)
19. Kahraman, C., Cebi, S., Cevik Onar, S., Oztaysi, B.: A novel trapezoidal intuitionistic fuzzy information axiom approach: An application to multicriteria landfill site selection. Eng. Appl. Artif. Intell. **67**, 157–172 (2018)
20. Kahraman, C., Oztayşi, B., Cevik Onar, S.: An integrated intuitionistic fuzzy AHP and TOPSIS approach to evaluation of outsource manufacturers. J. Intell. Syst. **3**(3), 197–204 (2017)

21. Kahraman, C., Oztaysi, B., Cevik Onar, S., Dogan, O.: Intuitionistic fuzzy originated interval type-2 fuzzy AHP: an application to damless hydroelectric power plants. Int. J. Anal. Hierarchy Process **10**(2), 266–292 (2018)
22. Kahraman, C., Parchami, A., Cevik Onar, S., Oztaysi, B.: Process capability analysis using intuitionistic fuzzy sets. J. Intell. Fuzzy Syst. **32**(3), 1659–1671 (2017)

Part VI
Customer Oriented Product Design Using Other Methods

Customer Oriented Product Design with Conjoint Analysis

Selçuk Alp and Ersoy Öz

Abstract Conjoint analysis is a technique used to measure consumer preferences related to the production of a product or service. This analysis can be used to decide if it is possible for the so-called product to retain in the market and compete with other products. In the study, the factors to be used in conjoint analysis were determined as RAM, Graphic Card, Processor Model, Storage Type, Screen Size and Operating System. This chapter has revealed the attributes that academic staff working on data analysis at Yildiz Technical University attaches the highest importance to when choosing a laptop. Sincsatisfaction with product dee all of the cards identified within the context of research would be complicated, 18 cards were prepared using orthogonal design. Apart from the 18 cards prepared, 3 simulation cards were also created. Results of the analysis reveal that the computer processor ranks first and other attributes are ranked as memory capacity (RAM), operating system, storage type, graphics card, and screen size respectively.

1 Introduction

Conjoint analysis is made up of a combination of the words "consider" and "jointly", which means joint effect or thinking jointly. Conjoint analysis is a dependency technique developed for the evaluation of a product or services. It is primarily used in design and market researches. It is also frequently used in assigning the weights of different attributes and determining their levels [1].

Conjoint analysis is a technique used to measure consumer preferences related to the production of a product or service. Thus, attributes of a product or service that are more important to consumers are determined. With conjoint analysis, the consumers' buying behavior and the most preferred features (attributes) of a product or service are determined before the development of a product or updating an existing product.

S. Alp (✉)
Department of Industrial Engineering, Yildiz Technical University, Istanbul, Turkey
e-mail: alp@yildiz.edu.tr

E. Öz
Department of Statistics, Yildiz Technical University, Istanbul, Turkey

C. Kahraman and S. Cebi (eds.), *Customer Oriented Product Design*, Studies in Systems, Decision and Control 279, https://doi.org/10.1007/978-3-030-42188-5_20

By doing so, this analysis can be used to decide if it is possible for the so-called product to retain in the market and compete with other products.

The advantage of conjoint analysis over other analyses is that it provides numerical and easily interpretable results based on non-numerical attributes and variables as well as these results pave the way for future.

The rest of the chapter is organized as follows. Section 2 presents the definitions on product design and literature review. Section 3 gives the details of conjoint analysis. Section 4 includes an application and Sect. 5 concludes the chapter.

2 Product Design

2.1 Customer-Oriented Product Design

Production is a business function that is under the influence of physical factors such as materials, facilities, and machines as well as other factors that are difficult to measure such as labor, management, economic events, consumer expectations, education, and social structure. Although production management has the basic activities grouped under the pre-planning, planning and control stages, product design is a pre-planning activity. Product design is a set of activities aimed at clearly determining the physical properties and functions of the product to be produced by the enterprise ([2], 52–53).

Determining the product structure is a typical engineering function. At this stage, a detailed drawing or specification of dimensions, weights, colors, and other features is prepared. That a company develops a new product or makes innovations that improve the quality of an existing product can help the enterprise gain a significant share of the market.

Customer orientation, starting with the step of recognizing the customer and continuing with the efforts to satisfy them, is the process where the managers and all units of the enterprise understand the importance of customers and act accordingly. In other words, the company's ability and willingness to identify, analyze, understand, and meet customer needs can be defined as customer orientation [3].

Customer dissatisfaction occurs in cases where the designed product does not meet customer expectations or fit for the intended use of customers. The enterprises, for this reason, employ various methods and models for their products in order to meet the expectations and demands of their customers today.

Conjoint analysis is a multivariate statistical analysis technique used to understand how the features developed for products or services will be taken.

2.2 *Customer-Oriented Product Design with Conjoint Analysis*

Conjoint analysis is a technique used to determine which attributes are important in product preferences and which combinations made up of attributes and levels are most preferred.

Conjoint analysis is based on the premise that consumers evaluate the value by combining the separate amounts of value provided by each attribute of a real or virtual object. In this case, consumers make the best choice estimates with the help of objects created by the combination of attributes [4].

Objectives of conjoint analysis are;

- to determine the relative importance levels of the variables that consumers attach importance to during the selection process,
- to calculate the proportional significance values based on the utility coefficients obtained,
- to estimate brand market shares,
- to determine the most preferred product combination, and
- to divide the market into sections based on the similarities in the attribute levels preferred by consumers.

Advantages;

- It ensures obtaining a hypothetical series of goods or services,
- Results can be easily interpreted thanks to the valuable information provided by conjoint analysis before the production,
- Classifier, sequencer, spaced and proportional scale variables can be used in the analysis.

Disadvantages;

- All attributes and levels must be known before the survey,
- Analysis might become complicated in case attributes and levels are numerous.

2.3 *Literature Review on Customer-Oriented Product Design*

Conjoint Analysis was used to determine the product and/or service characteristics of customer requirements with Quality Function Deployment (QFD) in the studies conducted by Moody K et al. [5], Lin et al. [6], Üçler et al. [7], and Akbaba [8].

Chen and Chuang [9] proposed a new model that determined the most suitable features for a new product to be developed using the Kano model and they conducted an empirical study on Mobile Phones.

Lin et al. [10] suggested a new approach using AHP and TOPSIS methods to help identify customer requirements and design characteristics and to ensure effective evaluation of the final design solution.

Fuller et al. [11] discussed the issue of identifying inputs for new product development for online customers. Kaulio [12] analyzed customer, consumer and user involvement in product development in terms of total quality management.

2.4 Literature Review on Conjoint Analysis

Tunc and Corba [13] used conjoint analysis to determine the most appropriate combination of factors influencing patients' choice of health facilities.

The study conducted by Arslan [14] aimed to determine the most suitable investment instrument by conjoint analysis method. In that context, variables such as stock, gold, foreign exchange, real estate investment instruments as well as the amount of invested money, risk level, profitability ratio and maturity period were used. As a result of the analysis, foreign exchange has been found to be the investment instrument with the highest benefit function.

The study conducted by Karaarslan and Altuntaş [15] using conjoint analysis primarily aims to determine the orientation of young individuals' job preferences when the factors affecting their career choices are evaluated together. As the research results suggest, young individuals prefer to do a career in high income, dynamic and challenging jobs where they can do their own professions. Income has been found to be the most important factor affecting career choice.

Cacace et al. [16] used conjoint analysis to determine the Occupational Health services in small-scale enterprises. As a result of the study, it has been determined that those enterprises prefer services offered at affordable prices in cooperation with social insurance providers.

Yavuz and Çemrek [17] used conjoint analysis to determine the housing preferences of healthcare professionals. Zardari and Cordery [18], in their study, first determined the attributes and levels of attributes for water allocation problem. Having analyzed the attributes and levels with conjoint analysis, they found that the most important attribute was "annual net farm income". Bridges et al. [19] tried to determine which strategies should be given priority for controlling liver cancer in Asia by using conjoint analysis. Lee et al. [20] employed conjoint analysis in their study to evaluate the acceptability of HIV vaccine. Fisher et al. [21] used conjoint analysis to analyze the decision-making models of emergency service nurses. The results of the study indicate that nurses rely on patients' future health status, functional status and family input when making critical health decisions for patients. Miletic and Misetic [22] made a research for the design of the campus by taking the students' opinions about the restructuring of the campus. The analysis included six attributes consisting of transportation, architecture, accommodation, campus-neighborhood relationship, accommodation, and dominant campus opinion. As a result of the analysis, transportation organization has been found to be the most important criterion for the participants. Şen and Çemrek (2004) conducted a study on the preferences of private courses using conjoint analysis [23]. Gürbüz and Kaygısız [24] conducted a research on the transportation sector using conjoint analysis. They tried to determine to what extent the individuals give importance to the attributes.

The purpose of the study conducted by Ness and Gerhardy [25] was to analyze the relationship between the quality and freshness attributes of eggs and the characteristics of consumers by using conjoint analysis that is a statistical method.

3 Conjoint Analysis

Conjoint analysis is a multivariate statistical analysis technique used to figure out how the features (attributes) developed for products or services will be taken. What distinguishes conjoint analysis from many other analyses is that it offers the opportunity to compare qualities (attributes) quantitatively. Through this analysis, the relationship between the attributes that cannot be quantified for a specific product or service and the levels of these attributes and the significance levels of these attributes are transformed into quantitative data.

Conjoint analysis is based on the customer interests' shifting from various attributes and levels to some other preferences in alternative choices. The participants evaluate the attribute levels that make up alternative choices based on some rules to form an overall view [26].

The early studies on conjoint analysis began in the 1920s. In 1964, R. Duncan Luce and John W. Tukey [27] highlighted the importance of the "Conjoint Measurement". The first consumer-oriented study with conjoint analysis was carried out by Paul Green and Vithala R. Rao in 1971 [28]. The study conducted by Gren and Srinivasan using conjoint analysis in 1978 has been cited by researchers as a resource since then.

Conjoint analysis model as a simple dependent model can be formulated as follows:

$$Y_i = X_1 + X_2 + X_3 + \cdots + X_p \tag{1}$$

Variable Y herein might have been measured by classifier, sequencer or spaced (non-metrical, metric) scale, while variables $X_1 + X_2 + X_3 + \cdots + X_p$ might have been measured by classifier and sequencer (non-metrical) scale [29, 30].

The application steps of Conjoint Analysis are as follows:

Step 1. The problem to be evaluated with conjoint analysis is determined. The features (attributes) and their levels related to the problem are defined. The characteristic of a product is called an attribute which consists of various levels. The design of a well-structured conjoint analysis depends on the correct definition of attributes and levels [31]. At this stage, the researcher should determine the most important attributes and their levels which are considered to have effects on consumers' preferences of a product or service. Great attention must be paid to identifying all variables that might be effective in determining consumers' preferences regarding the attributes and levels of a product and service [17].

Step 2. The selection model is determined. At this stage, the researcher chooses one of the models that are vector model, ideal point model, part worth function model, and mixed model. Thus, the mathematical function used to determine the direction of the relationship between the attributes/levels and the selection of attributes is selected [32].

$p = 1, 2, \ldots, t$ stands for the set of t attributes or factors that have been selected. Say y_{ip} stand for the level of the pth attribute for the jth stimulus. It is firstly considered that the case where y_{ip} is naturally a continuous variable.

In ***the vector model*** preference,

$$s_j = \sum\nolimits_{p=1}^{t} w_p y_{ip}, \tag{2}$$

gives the preference s_i for the jth stimulus as the model posits; the $\{w_p\}$ stand for the individual's weights for the t attributes [33].

The ideal-point model suggests that there is a negative relationship between the preference s_j and the squared (weighted) distance d_j^2 of the location $\{y_{ip}\}$ of the jth stimulus from the individual's ideal point $\{x_p\}$. Based on the Formula, d_j^2 is defined as

$$d_j^2 = \sum\nolimits_{p=1}^{t} w_p (y_{ip} - x_p)^2. \tag{3}$$

As ***the part worth function model*** suggests,

$$s_j = \sum\nolimits_{p=1}^{t} f_p(y_{ip}), \tag{4}$$

where f_p is the function that stands for the part worth different levels of y_{ip} for the pth attribute.

The features of the three models mentioned above can be combined to formulate **a mixed model** as shown below:

$$s_j = \sum\nolimits_{q=1}^{T} v_q z_{jq}, \tag{5}$$

T, herein, stands for the total number of pseudo-attributes (also the total number of estimated parameters) and the $\{z_{jq}\}$ are defined from the y_{ip}.

Step 3. One of the three profile presentation methods is selected in order to collect the evaluations of the respondents [34, 35]: these methods are the trade-off method where two factors are evaluated at a time; the full profile method where all factors are evaluated altogether; and the pairwise combination comparison method emerging in the form of a mixture of the two other approaches.

The trade-off method is performed by evaluating two attributes at a time. In this method, the respondents sort all possible combinations of the two attributes in order of preference [36]. However, this method comes with some disadvantages such as showing similar tables to the respondents over and over and the difficulties in placing the tables in order of importance [17].

The full profile method is performed by using one level of each attribute defined within the context of research [36].

The Pairwise Combination Method involves preferring one of the profiles formed using the sorting scale over the other. Unlike the full profile method, profiles do not include each attribute, and if the number of attributes is excessive, profiles are created with several attributes at a time for simplification. This method is especially used in specific methods such as adaptive conjoint analysis [4].

Step 4. Test combinations are prepared depending on the profile selection methods. This is done by choosing either a full factorial design or a fractional factorial design. The full factorial design consists of all possible combinations of levels and the respondents are provided with all possible situations [4]. The fractional factorial design is used to reduce the number of combinations obtained with the full factorial design. Here, the reduction process is carried out by making sure that there is no relationship between the levels of attributes (that is they are orthogonal) and they have the same number and are in balance [4]. Orthogonal design is a special case of the fractional factorial design method in which only main effects are taken into consideration and the effects of levels are ignored assuming that the preferences of attributes and levels are independent of each other. In the studies using orthogonal design, the number of attributes and levels under 7 attributes is 3 or 4, while the number of profile cards is designed as 16–18 [37–39].

Step 5. The test combinations designed in Step 4 are evaluated by the respondents based on the sorting technique. As a result of the evaluations, the partial utility coefficients of the attributes created for the product or service are calculated, which allows the researcher to determine which attribute is more prioritized for the product or service [31]. The results of the conjoint analysis are obtained by interpreting the partial benefit values calculated based on the answers given by each participant and the proportional significance values calculated accordingly ([37], p. 50).

4 Application

30 academic staff, who attaches importance to computer preferences and work on data analysis at Yildiz Technical University, were considered as participants in order to analyze the use of conjoint analysis in customer-oriented product design. The respondents were interviewed face-to-face and they were asked to sort the cards containing laptop features, which will be explained below, in order of preference. The conjoint analysis has been performed based on the answers given by respondents.

4.1 Data Set

The number of attributes and levels in Conjoint Analysis is important because it affects the statistical adequacy and reliability of the results. The number of attributes in a typical Conjoint Analysis is generally 6–7 [40]. Given the assumption that the interpretation of the data would be difficult as well as the respondents would have difficulty ranking the cards presented to them with the increase in the number of attributes and therefore reliable data cannot be obtained, and taking the studies on product preferences in the literature into consideration, RAM, Graphic Card, Processor, Storage Type, Screen Size and Operating System have been considered as attributes in this chapter. Brand and quality are of the significant factors affecting consumer preferences. However, these variables were not included in the study in order to eliminate their effects on consumer preferences based on the assumption that the significance weights of these attributes would be high for consumers if included in the study.

Table 1 shows the attributes and levels to be used in Conjoint Analysis that aims to determine the factors affecting consumers' laptop preferences.

Three different methods are used for data collection in conjoint analysis. These are as follows: a) two attributes simultaneously method; b) component comparison method; and c) full profile method. In the current study, the full profile method was used by asking the respondents to sort the selection cards from the most preferred (1 - First) to the least preferred.

Table 1 Attributes and Levels

Attribute	Levels	Number of levels	Direction of relationship
RAM	Less than 8 GB 8–12 GB More than 12 GB	3	Linear more
Graphic card	Integrated(Shared) Dedicated	2	Discrete
Processor model	IntelCore_I5 IntelCore_I5 IntelCore_I5	3	Linear more
Storage type	HDD SSD SSD + HDD	3	Discrete
Screen size	Less than 14 in. 14–15 in. More than 15 in.	3	Linear more
Operating system	NonOp MacOs Windows	3	Discrete

The numbers of levels of the variables in this chapter are 3, 2, 3, 3, 3, and 3 respectively. Thus, the total number of selection cards including all combinations of levels was calculated as 486 (3*2*3*3*3*3).

Since all of the 486 cards identified within the context of research would be complicated, difficult and time-consuming to be sorted by the respondents, 18 cards were prepared using orthogonal design. Apart from the 18 cards prepared, 3 simulation cards were also created.

The respondents were asked to sort the 21 cards (given in Appendix 1) from the most preferred to the least preferred, which means the full profile method. Each of the 21 cards included all the relevant factors and different combinations of factor levels were generated.

4.2 Analysis

Pearson's **R** statistics, the ratio of the model's suitability to individuals' preferences, was found 0.993 ($p = 0.0000$). The model can be said to be significant based on the p value obtained.

As a result of the conjoint analysis, the attributes that are important to respondents are determined with importance values as shown in Table 2. The order of laptop preferences related to the attributes is as follows: Processor Model (38,303%), memory capacity (RAM) (16,341%), Operating System (14,189%), storage type (13,704%), graphics card (10,691%) and Screen Size (6,173%) respectively.

Examining the partial utility coefficients of the levels of these attributes, it is seen that the respondents prefer a laptop whose operating system is IntelCore_I9 (10.938) or IntelCore_I7 (7.292) and RAM is more than 12 GB. The partial utility coefficients of all levels of the attributes are shown in Table 3.

Considering the partial utility coefficients of the attributes, it is seen that the most important level for Processor Model is InterCore_i9 with 10.938, and the levels InterCore_i7 and InterCore_i5 have a positive high utility level with 7,292 and 3,646 utility coefficients.

Table 2 Importance Values

Attributes	Importance level (%)
RAM	16.341
Graphic card	10.691
Processor model	38.303
Storage type	13.704
Screen size	6.173
Operating system	14.189

Table 3 Utilities

Attributes	Levels	Utility estimate	Std. Error
GP	Integrated(Shared)	−0.781	0.46
	Dedicated	0.781	0.146
Storage type	SSD	0.429	0.195
	HDD	−1.060	0.195
	SSD + HDD	0.631	0.195
Operating system	NonOp	−0.810	0.195
	Windows	0.762	0.195
	MacOs	0.048	0.195
RAM	<8 GB	1.616	0.169
	8–12 GB	3.232	0.338
	>12 GB	4.848	0.507
Processor model	IntelCore_I5	3.646	0.169
	IntelCore_I7	7.292	0.338
	IntelCore_I9	10.938	0.507
Screen size	<14 in.	0.420	0.169
	14–15 in.	0.839	0.338
	>15 in.	1.259	0.507
(Constant)		−1.603	0.603

With respect to the partial utility coefficients of memory capacity (RAM), it is seen that the utility coefficients of all three levels are positive (4,848 for ">12"; 3,232 for "8–12"; 1,616 for "<8") and are relatively high.

Even though the utility coefficients of Operating System are low, it is seen that the values of Windows (0,762) and MacOS (0,048) are positive, while NonOp, which represents the absence of an operating system, has a negative value (−0,810). This points out that there is a negative opinion about purchasing laptops without an operating system.

Examining the storage type, it is seen that the utility coefficients of SSD + HDD (0,631) and SSD (0,429) are positive, while the utility coefficient of HDD (−1,060) is negative. This shows that the respondents are reluctant to buy a laptop with HDD-type storage.

Regarding the partial utility coefficients of the attribute Graphics card, it is seen that Dedicated (0781) has a positive value, while Integrated(Shared) (−0,781) has a negative value. It can be concluded that the respondents would not be willing to buy a laptop with an Integrated(Shared)-level graphics card.

When it comes to the levels of screen size, it has been found that the utility coefficients increase (0,420, 0,839, 1,259) as the screen size increases ("<14", "14–15", ">15"), which means the respondents prefer larger screens.

The score values calculated based on the attribute levels of the cards that were asked to the respondents are given in Appendix 2. The most preferred cards are ">12" for RAM, "Dedicated" for Graphic Card, "Intel Core I9" for Processor Model, "SSD" for Storage type, ">15" for Screen Size, and the laptop with no operating system. Appendix 2 can be examined for other orders of preferences.

5 Discussion and Conclusion

This study has revealed the attributes that academic staff working on data analysis at Yildiz Technical University attaches the highest importance to when choosing a laptop. In the conversations with the respondents at the card-sorting stage, it has been seen that most of the academic staff had their own laptops but they wanted to change these for their academic studies.

Having looked at the research results, it is seen that the computer processor ranks first and other attributes are ranked as memory capacity (RAM), operating system, storage type, graphics card, and screen size respectively.

In general, offering laptops with a high processor model and RAM, screen size as large as possible and a storage type that is not HDD would be appropriate for a laptop company that considers academic staff working in the field of data analysis as target customers. The presence of an operating system on the computer whose graphics card is Dedicated will increase the attractiveness of the laptop. It would be more appropriate for the companies to market suitable computers with appropriate price and features considering the level-utility balance.

The enterprises need to benefit from marketing departments when developing existing products or designing new products that are expected to meet the changing preferences and demands of consumers. Marketing departments have to carry out various market researches while meeting these needs. These researches are only possible with scientific studies. The length, cost, efficiency, and reliability of scientific studies are very important for enterprises. Conjoint analysis is, therefore, a reliable guide for marketing departments.

The number of participants surveyed can be increased so that more accurate results can be obtained. Moreover, different groups can be surveyed, and differences and similarities between groups can be evaluated.

Appendix 1

18 Cards and Simulation Cards

Card	RAM	GraphicCard	Processor	Storage type	Screen size	Operating system
1	<8	Integrated(Shared)	IntelCore_I7	HDD	>15	NonOp
2	>12	Integrated(Shared)	IntelCore_I5	HDD	>15	MacOs
3	>12	Integrated(Shared)	IntelCore_I5	SSD + HDD	14–15	NonOp
4	<8	Integrated(Shared)	IntelCore_I9	HDD	14–15	Windows
5	>12	Integrated(Shared)	IntelCore_I9	HDD	<14	Windows
6	<8	Dedicated	IntelCore_I9	SSD + HDD	<14	MacOs
7	<8	Dedicated	IntelCore_I9	SSD + HDD	14–15	Windows
8	>12	Dedicated	IntelCore_I9	SSD	>15	NonOp
9	<8	Integrated(Shared)	IntelCore_I5	SSD	<14	NonOp
10	>12	Integrated(Shared)	IntelCore_I5	HDD	>15	NonOp
11	8–12	Integrated(Shared)	IntelCore_I9	SSD + HDD	14–15	NonOp
12	8–12	Integrated(Shared)	IntelCore_I9	SSD	>15	MacOs
13	>12	Integrated(Shared)	IntelCore_I7	SSD + HDD	<14	MacOs
14	>12	Dedicated	IntelCore_I7	SSD	14–15	Windows
15	8–12	Integrated(Shared)	IntelCore_I7	SSD + HDD	>15	Windows
16	8–12	Integrated(Shared)	IntelCore_I5	SSD	<14	Windows
17	<8	Dedicated	IntelCore_I5	SSD + HDD	>15	Windows
18	8–12	Dedicated	IntelCore_I5	HDD	14–15	MacOs
19	8–12	Dedicated	IntelCore_I7	HDD	<14	NonOp
20	8–12	Integrated(Shared)	IntelCore_I9	SSD	<14	NonOp
21	<8	Integrated(Shared)	IntelCore_I7	SSD	14–15	MacOs

Appendix 2

Score Values of Cards

Card	RAM	GraphicCard	Processor	Storage type	Screen size	Operating system	Score
8	>12	Dedicated	IntelCore_I9	SSD	>15	NonOp	17.445
7	<8	Dedicated	IntelCore_I9	SSD + HDD	14–15	Windows	15.567
5	>12	Integrated(Shared)	IntelCore_I9	HDD	<14	Windows	15.127
12	8–12	Integrated(Shared)	IntelCore_I9	SSD	>15	MacOs	15.125
14	>12	Dedicated	IntelCore_I7	SSD	14–15	Windows	14.951
6	<8	Dedicated	IntelCore_I9	SSD + HDD	<14	MacOs	14.434
11	8–12	Integrated(Shared)	IntelCore_I9	SSD + HDD	14–15	NonOp	14.049
20	8–12	Integrated(Shared)	IntelCore_I9	SSD	<14	NonOp	13.428
13	>12	Integrated(Shared)	IntelCore_I7	SSD + HDD	<14	MacOs	12.458
15	8–12	Integrated(Shared)	IntelCore_I7	SSD + HDD	>15	Windows	12.395
4	<8	Integrated(Shared)	IntelCore_I9	HDD	14–15	Windows	12.314
19	8–12	Dedicated	IntelCore_I7	HDD	<14	NonOp	9.855
21	<8	Integrated(Shared)	IntelCore_I7	SSD	14–15	MacOs	9.443
17	<8	Dedicated	IntelCore_I5	SSD + HDD	>15	Windows	8.695
3	>12	Integrated(Shared)	IntelCore_I5	SSD + HDD	14–15	NonOp	8.373
2	>12	Integrated(Shared)	IntelCore_I5	HDD	>15	MacOs	7.96
16	8–12	Integrated(Shared)	IntelCore_I5	SSD	<14	Windows	7.708
1	<8	Integrated(Shared)	IntelCore_I7	HDD	>15	NonOp	7.516
18	8–12	Dedicated	IntelCore_I5	HDD	14–15	MacOs	7.486
10	>12	Integrated(Shared)	IntelCore_I5	HDD	>15	NonOp	7.102
9	<8	Integrated(Shared)	IntelCore_I5	SSD	<14	NonOp	4.52

References

1. Rencher, A.C.: Methods of Multivariate Analysis. Wiley, New York (2002)
2. Kobu, B.: Üretim Yönetimi, Geliştirilmiş ve Değiştirilmiş 9. Avcıol Basım Yayım, Baskı, İstanbul (1999)
3. Matsuo, M.: Customer orientation, conflict, and innovativeness in Japanese sales departments. J. Bus. Res. **59**(2), 242–250 (2006)
4. Hair, J.J.F., Black, W.C., Babin, B.J., Anderson, R.E.: Multivariate Data Analysis. Pearson, New Jersey (2006)
5. Moody, K.A., LAL, R.R., PANDEY, V.: Analysis of customer oriented product development with quality function deployment. Int. J. Mech. Eng. Technol. **8**(5), 270–279 (2017)
6. Lin, M.C., Wang, C.C., Chen, T.Z.: A strategy for managing customer-oriented product design. Concur. Eng. Res. Appl. **14**(3), 231–244 (2006)
7. Üçler, Ç., Vayvay, Ö., Çobanoğlu, E.: Customer-focused product development and a case study in turkish refrigerator market. İstanbul Ticaret Üniversitesi Fen Bilimleri Dergisi **5**(10), 81–97 (2006)
8. AKBABA, A.: The quality function deployment (QFD) approach in customer focused service production: an application study for hospitality industry, Anatolia: Turiz. Araştırmaları Derg. **16**(1), 59–81 (2005)
9. Chen, C.C., CHUANG, M.C.: integrating the kano model into a robust design approach to enhance customer satisfaction with product design. Int. J. Prod. Econ. **114**(2), 667–681 (2008)
10. Lin, M.C., Wang, C.C., Chen, T.Z., Chang, A.A.: Using AHP and TOPSIS apprioaches in customer-driven product design process. Comput. Ind. **59**, 17–31 (2008)
11. Fuller, J., Bartl, M., Ernst, H.,Mühlbacher, H.: Community based innovation: How to integrate members of virtual communities into new product development. Electronic. Commer. Res. **6**(1), 57–73 (2006)
12. Kaulio, M.A.: Customer, consumer and user involvement in product development: a framework and a review of selected methods. Total Qual. Manag. **9**(1), 141–149 (1998)
13. Tunc, T., Corba, B. Ş.: Determining the optimal combinations of the factors affecting the choice of health facility. Int. J. Arts Commer. **6**(5), 37–50 (2017)
14. Arslan, H.: Konjoint Analizi ile En Uygun Yatırım Aracının Belirlenmesi. İstanbul ticaret Üniversitesi Sosyal Bilimler Dergisi **29**(2), 305–317 (2016)
15. Karaarslan, M.H., Altuntaş, B.: Kariyer Tercihini Etkileyen Faktör Düzeylerinin Öneminin Konjoint Analizi ile Belirlenmesi. İnsan ve Toplum Bilimleri Araştırmaları Dergisi **5**(7), 1972–1988 (2016)
16. Cacace, M., Franz, I., Ratz, D.: Using conjoint analysis to elicit preferences for occupational health services in small and microenterprises. Athens J. Health **1**(4), 237–254 (2014)
17. Yavuz, S., Çemrek, F.: The determination of residential preferences of healthcare workers through conjoint analysis. Atatürk Üniversitesi Sosyal Bilimler Enstitüsü Dergisi **17**(2), 379–396 (2013)
18. Zardari, N.H., Cordery, I.: Determining irrigators preferences for water allocation criteria using conjoint analysis. J. Water Resour. Prot. **4**, 249–255 (2012)
19. Bridges, J.F.P., Dong, L., Gallego, G., Blauvelt, B.M., Joy, S.M., Pawlik, T.M.: Prioritizing strategies for comprehensive liver cancer control in Asia: a conjoint analysis. BMC Health Serv. Res. **12**(376), 1–12 (2012)
20. Lee, S.J., Newman, P.A., Comulada, W.S., Cunningham, W.E., Duan, N.: Use of conjoint analysis to assess HIV vaccine acceptability: feasibility of an innovation in the assessment of consumer health-care preferences. Int. J. STD AIDS **23**, 235–241 (2012)
21. Fisher, K., Orkin, F., Frazer, C.: Utilizing conjoint analysis to explicate health care decision making by emergency department nurses: a feasibility study. Appl. Nurs. Res. **23**(1), 30–35 (2010)
22. Miletic, M., Misetić, A.: University campus borongaj in zagreb; application of conjoint analysis. Prost.: Sch. J. Arch. Urban Plan. **18–2**(40), 412–423 (2010)

23. Şen, H., Çemrek, F.: Konjoint Analizi ile Özel Dersane Tercihine Yönelik Bir Uygulama. Osmangazi Üniversitesi Sosyal Bilimler Dergisi, 5(2) (2004)
24. Gürbüz, H., Kaygisiz, Z.: Conjoint analysis and a practice in arrival sector. Süleyman Demirel Ünivesitesi İktisadi ve İdari Bilimler Fakültesi Dergisi **9**(1), 139–148 (2004)
25. Ness, M.R., Gerhardy, H.: Consumer preferences for quality and freshness attributes of eggs. Br. Food J. **96**(3), 26–34 (1994)
26. Katoshevski, R., Timmermans, H.: Using conjoint analysis to formulate user-centred guidelines for urban design: the example of new residential development in Israel. J. Urban Des. **6**(1), 37–53 (2001)
27. Luce, R.D., Tukey, J.W.: Simultaneous conjoint measurement: a new type of fundamental measurement. J. Math. Psychol. **1**(1), 1–27 (1964)
28. Green, P., Rao, V.: Conjoint measurement for quantifying judgmental data. J. Mark. Res. **8**(3), 355–363 (1971). https://doi.org/10.2307/3149575
29. Sönmez, H.: The usage of conjoint analysis in marketing research and an application. PhD thesis. Anadolu University Graduate School of Natural and Applied Science Statistics Program (2001)
30. Sönmez, H.: Müşteri tercihleri için konjoint analizi uygulaması: ev bilgisayarı nasıl seçilir. Anadolu Üniversitesi Sosyal Bilimler Dergisi **2006**(2), 185–196 (2006)
31. Aydin, N., Yalçin, E.: Choice based conjoint analysis and an application. Dumlupinar Univ. J. Soc. Sci. **50**, 26–48 (2016)
32. Erdoğan, C.: Tüketicinin Otomobil Tercihinin Konjoint Analizi İle Belirlenmesi. Yüksek Lisans Tezi. Gazi Üniversitesi Fen Bilimleri Enstitüsü, Ankara (2006)
33. Green, P.E., Srinivasan, V.: Conjoint analysis in consumer research :issues and outlook. J. Consum. Res. **5**, 103–123 (1978)
34. Green, P.E., Srinivasan, V.: Conjoint analysis in marketing: new developments with implications for research and practice. J. Mark. **54**, 3–19 (1990)
35. Rao, R.V.: Applied Conjoint Analysis. Springer, New York (2014)
36. ORME, B.: Formulating attributes and levels in conjoint analysis. Sawtooth Software Inc (2002)
37. BAŞARAN, Z.K.: Çok Değişkenli İstatistiksel Analiz Tekniklerinden Bulanık Konjoint Analizi ve Çay İşletmelerine Uygulanması. Yüksek Lisans Tezi. İstanbul: Marmara Üniversitesi Sosyal Bilimler Enstitüsü (2010)
38. Yalniz, A., ve Bilen, L.: Kasko Sigortalarında Konjoint Analizi ile Tüketici Tercihi. Hazine Dergisi 8, 53–70 (1997)
39. Tatlidil, H.: Konjoint Analizi. Ders Notları, Hacettepe Üniversitesi, İstatistik Bölümü, Ankara (1995)
40. Cengiz, M., Girginer, N.: Determining consumers' preferences with using conjoint analysis: case of refrigerator. Eskişehir Osman. Univ. J. Econ. Adm. Sci. **7**(1), 269–290 (2012)

Assessing Customer Preferences on a New Design of Refrigerator Using Conjoint Analysis

Nezir Aydin and Sukran Seker

Abstract Home appliances become necessities to continue a daily life of human beings. Since, home appliances are not just used to conduct daily life works but also to design or create a stylish appearance within the houses, they are purchased by almost all residents and, thus, demand increases. On the other hand, new entrance of retailers and producers to the market increases, simultaneously. Therefore, to stay influential in the market, marketing departments of the companies need to compete in a competitive rivalry. Within this context, marketers analyze the market using different methods to satisfy customers' needs and expectations. Conjoint Analysis (CA) is one of the most important method to determine the importance of product or service attributes to customers, and the utilities that are delivered from these attributes to customers. In this chapter, the applicability of CA is experienced in home appliances industry by designing a new refrigerator. Obtained results are found to be reliable and satisfactory to retain competitive in the market, and the method may be used as a guide to determine the best design of home appliances (refrigerator) considering customer expectations, needs, and perceptions.

1 Introduction

Home appliances are electrical, electronic, or mechanical devices that are used by humankind to accommodate their daily life. Several types of home appliances are available for different types of needs such as cooling, heating places, preserving products, or cleaning clothes, rooms, walls, etc. Each of these devices are produced for different needs of people. Some major appliances including refrigerator, oven, washing machine, and vacuum cleaner etc., are must to exist in each house. Since the existence of such appliances is a must in each household the marketing of these devices becomes a strategic issue for producers [8].

N. Aydin (✉) · S. Seker
Department of Industrial Engineering, Yildiz Technical University, Istanbul, Turkey
e-mail: nzraydin@yildiz.edu.tr

S. Seker
e-mail: sseker@yildiz.edu.tr

C. Kahraman and S. Cebi (eds.), *Customer Oriented Product Design*, Studies in Systems, Decision and Control 279, https://doi.org/10.1007/978-3-030-42188-5_21

The competitiveness in the marketing of these devices increases every day because of several reasons, such as the entrance of new players to the market, change in customer needs and expectations, improvements in the technology, new designs etc. Therefore, marketing departments need to spend more effort to consider all factors to increase their market share. Besides common factors, such as quality, price, and brand name of the product, some other factors, i.e. product design, appearance, and size, are considered as product specific factors. Before launching a new product to the market besides advertising and promotions, companies need to conduct surveys to determine customer expectations and needs in terms of advantages, properties, and designs provided by new technologies to reach a high profitable growth. It is highly significant that companies give intensive importance to market research and resource allocations, so that they observe changes in consumers' needs, habits, and desires. Since catching or understanding these issues are significant, several and different marketing strategies and techniques are used by the marketing departments of the companies [1, 26].

One of these important methods are statistical techniques. The statistical techniques used in marketing research make businesses more successful in determining consumer preferences. Further, Conjoint Analysis (CA) is a widespread advanced statistical technique that has been used in recent years. The word Conjoint is the combination of the words CONside and JOINT, and its first applications started in 1920s in automotive, but it is applied to mobile technologies, information technologies, retail, and health sectors so far. CA is one of the multivariate analysis techniques preferred by researchers to determine the characteristics of consumer behaviors. The most important feature that distinguishes CA from many other statistical techniques is that it provides the opportunity to compare the qualitatively properties in a quantitative way [30].

Statistical techniques are the main tools for data mining, machine learning and intelligent decision making. As a statistical method, CA analysis have been used in several intelligent decision making processes. Furthermore, in the last decade to accomplish better performances of CA, machine learning and data mining have been used in CA [21]. Thus CA can be considered as an intelligent technique, since it utilizes data mining and machine learning processes in determining the appropriate attributes to foresee customer expectations.

In the application of CA, the combination of attributes and their levels are evaluated by correspondents. They select the combination that provides them the highest utility and they states whether they will purchase or not to purchase the selected new design of hypothetical product [17].

Basically, CA searches for the answers of the following questions:

- What changes companies can make on their products or services,
- What points they need to stand in the product development studies,
- How the current and potential customers in the market can react to these changes,
- How the general tendency in the market will affect the products and services of the company,

- What are the possible effects of the competitors' new products and services on consumers,
- How pricing will reflect the market i.e.

In this chapter the implementation of CA is tested for evaluating and prioritizing the customers' wishes and expectations on the selection of new designs of a refrigeration. By the application, the attitudes of Turkish home appliances' consumers in purchasing refrigerator are investigated. They are asked to select one of the provided combinations of a new refrigerator design which satisfy their needs and provides the highest utility.

This chapter is organized as follows: In Sect. 2, related literature to home appliances and CA are provided. In Sect. 3, the basic concepts on the method-CA are presented. In Sect. 4, an industrial application on the basis of a new refrigerator design is conducted to show the applicability of CA. In Sect. 5, results and discussions are carried out. Conclusions and limitations are provided in the last section.

2 Literature Review

The main devices of the modern life of human beings are home appliances. More sophisticated devices and home appliances are started to use in a daily life of human beings, in the last century [32]. Actually, human beings are in search of making their life more expedient and easier, in returns their lifestyle is started to be formed by the home appliances used. These home appliances are used to complete functions, as cleaning, preventing, cooking, and ironing etc., and are used in almost all places in a house, such as kitchens, bedrooms, living rooms, and bathrooms. They are not just used as devices to complete works but also used as decoration, which shows style of the owners. For this purpose the design and color of these appliances are important issues for the household. Even appliances are used for different purposes, i.e., refrigerator for keeping foods fresh, oven to cook, TVs and music players for entertainment etc., their accordance is important to create a beautiful ambiance. Some of the main appliances that are purchased, in the modern life, can be summarized as, refrigerator, TV, music player, air conditioner, iron, fan, vacuum cleaner, oven, and stove, and many more devices can be counted. Since a vast number of appliances are purchased and needed to invest a good amount of money, prices of them become another important issue for the household. Further, improvements in technology make appliances more complex and user friendly, especially with the invention of newer technologies such as Internet of Things (IoT). With these improvements, household thinks of replacing even existing and fine functioning appliances too. Even only these three criteria (appearance, price, and new technology-properties of appliance) creates a high tradeoff and uncertainty for the household in the decision making process. Thus, household's preferences change based on their expectations and house designing styles.

Therefore, merchandisers would like to know what properties, such as, color, design, technology etc., are important or preferable to customers in order to design their products in accordance with the consumers' preferences [14]. Within this sense, several methodologies are proposed to measure customer preferences in marketing industry. One of the techniques proposed in the literature is CA, which is developed to search customer preferences [20] and used as market search by [11]. Later, CA became one of the common methodologies to assess consumers' expectations and preferences in marketing studies [12, 13, 28]. In CA, responders evaluate suggested alternatives to determine their expectations and preferences [23]. Suggested alternatives usually are determined based on product properties, such as color, size, etc., and level of the property, such as height or length in size. The main goal of CA is to determine responders' behaviors in making decisions when there is a trade-off, for instance purchasing a higher volume refrigerator with higher energy consumption or lesser volume refrigerator with lower energy consumption, between suggested alternatives. CA is, mainly, a questionnaire method, which asks responders to assess suggested products. Once responders assessed the alternatives, these assessments are decomposed to determine the level of the properties of alternatives and thus CA is called a "decompositional" technique [29].

While determining consumer preferences on a product, which aims to get the most out of it [22], designing a new service or device is objected considering the selected attributes. Eliciting the attributes that delivers more benefits to the design is very important taking into account a close-fitting to the desires and preferences of the customers [7, 19]. When a customer states his/her preferences on a product s/he also values the attributes related to the combination selected [3]. CA allows creating a set of conjectural products via conjoining some levels of elicited attributes. Created conjectural products are provided to the respondents to assess and deliver their preferences [27].

CA method is applied in a vast number of areas to design products or services considering consumer preferences. For instance, Vick and Scott [33] applied CA in health economics, Dauda and Lee [6] in finance, Gamble et al. [9] in marketing, Chocarro et al. [4] and Hailu et al. [15] in food industry, Karniouchina et al. [16] in operations management, Wu et al. [34] in vehicle industry, Daghan and Akkoyunlu [5] in online learning, Prasad [25] in banking-credit cards, Oyatoye et al. [24] in computer selection, and many more.

3 Method—Conjoint Analysis

Considering product or service as a marketing strategy, multiple decisions require to be taken into account together. These decisions which are made in uncertain environment without knowing uncertain competitive responses of competitors, are with

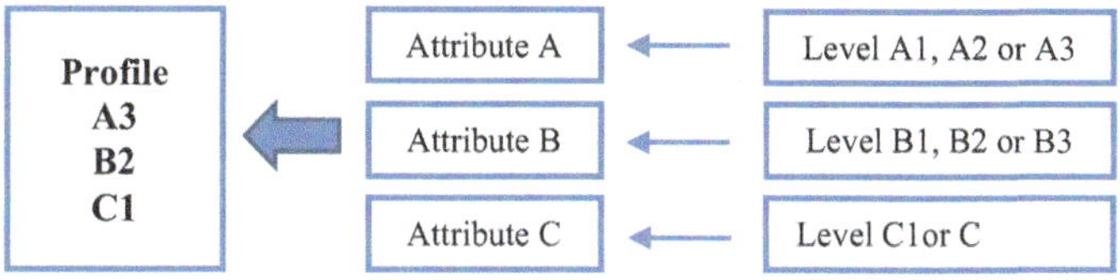

Fig. 1 Relationship between designs, attributes and their levels [28]

respect to determining product characteristics, positioning, communication, distribution and pricing decisions for certain target group. The competitiveness of an enterprise is provided to analyze customers' behavior such as how to select among competitive alternatives and how to respond to new product development while making decisions [11].

In the process of decision making for strategic decisions, marketing experts employ CA that is known as multivariate technique to evaluate customer needs and expectations on the basis of customers' utility functions. CA is a methodology in which a decision maker must choose a number of options that vary simultaneously from a range of attributes at the same time. The objective is to assess customers' trade-offs among multi-attribute alternatives for satisfying customer value during new product design. Accordingly, CA is used to construct a model which includes specific experiments among customers by implementing decision-making processes. The method has the ability to determine which product properties generate the most value, while it allows to reveal how customers can react to different products [17].

CA is applied by setting up and conducting specific tests among customers to reveal customers opinions towards a new product design. To implement a CA, potential customers are expected to make decisions about the characteristics of product that affect purchasing decisions instead of evaluating each property separately. Performing this analysis provides optimal product propositions of product or service [11, 28]. Marketing experts use CA which is one of the best methods to investigate and analyze customer needs and expectations (Fig. 1).

The basic CA is based on the utility model which determines the relationship between attribute levels and consumer utility. The utility model is expressed as follows [18]:

$$U(X) = \sum_{i=1}^{m} \sum_{j=1}^{k_i} a_{ij} X_{ij} \tag{1}$$

m shows the number of product attributes. $j = 1, 2, \ldots k_i$ represent the levels of attribute i. The total utility of product mix is expressed as $U(X)$. Accordingly, a_{ij} shows the partial utility value of attribute i with respect to level j. X_{ij} is a dummy variable ($X_{ij} = 1$ when X is exists, otherwise $X_{ij} = 0$). This equation shows the common basic utility model. As a result, the importance I_i of attribute I is shown as the utility range which is the largest difference in the partial values.

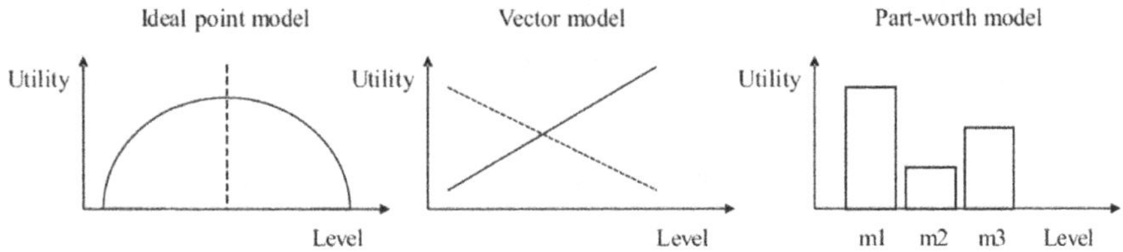

Fig. 2 Utility models [29]

$$I_i = \{\max(a_{ij}) - \min(a_{ij})\} \text{ (for all the levels of attribute i)} \quad (2)$$

The relative importance W_i of attribute i is obtained from standardizes calculation of I_i.

$$W_i = \frac{I_i}{\sum_{i=1}^{m} I_i} \quad (3)$$

The objective of CA is to determine the utility value of each attribute's level taking into account rating of each attribute's level for each product identified by the decision makers. The utility values of model are calculated by dummy variable regression. Furthermore, decision makers choose the alternative to maximize utility so that conjoint methods define the preference structure of consumers on the basis of the assessment of the product's attributes.

The utility model conducts the relationship between attribute levels and respondent utility. Figure 2 illustrates three different utility models.

An **ideal point model** suggest that an attribute has an 'optimal' level that can maximizes consumer utility. A **vector model**, accepts that higher (or lower) levels of an attribute can arise consumer utility. The **part-worth model** does not cover any a priori assumptions regarding the utility caused by specific levels. Thus, the part-worth model is the most flexible approach [29].

4 Industrial Application

In this section a company, which operates in the market of home appliances sector of Turkey, is analyzed for market searches of a refrigerator. The main goal of this section is to determine attributes that affects customers' preferences and behaviors in selecting the refrigerator. For this purpose questionnaires are developed to gather data (evaluations) from customers. The questionnaires are constructed by taking into account refrigerator design examples as shown in Fig. 3 (a spurious example) and considering literature reviews, expert opinions, and target customers groups. The questionnaires include refrigerator attributes and their levels.

Fig. 3 Refrigerator design examples [31]

The attributes that are determined are color, price, volume, energy consumption, product life, service reliability, and freezer type. The attributes and their levels are presented in Table 1.

The sample demographics of respondents are shown in Tables 2, 3, and 4.

Lastly, respondents selected the refrigerator that the combination of attributes and their levels provides the highest utility to them, and later stated whether they

Table 1 Attributes and their levels

Attributes	Level of attributes		
Color	White	Black	Gray
Price (₺)	<4000	4000–7000	>7000
Volume (Lt)	300	500	700
Energy consumption	A+	A++	A+++
Product life (year)	5–10	11–15	16–20
Service reliability	Low	Medium	High
Freezer type	Low frost	No frost	

Table 2 Gender

Gender	Frequency
Man	52
Woman	146

Table 3 Age

Age	18–25	26–35	36–50	>50
Frequency	35	66	58	39

Table 4 Education level

Education	Primary-high school	Undergraduate	Post graduate
Frequency	65	72	61

will purchase or not to purchase the selected refrigerator. In their selection process respondents faced with the tradeoff between attributes and their levels. Since all respondents made a choice, non-selection option was not added to the alternatives.

Once the attributes and their levels are determined by the participations of experts and customers, ideal combinations of attributes are constructed using orthogonal design in SPSS [10]. The orthogonal design applied in this study was created with the value of 1,000,000 for randomization. Thus the required number of combinations is created. Considering the selected attributes and their levels 33 combinations are generated as orthogonal design.

A specific code syntax, which is provided as follows and available in the conjoint.sps, is used to assess the data collected:

```
CONJOINT PLAN='File specification'
  /DATA=' File specification '
  /SEQUENCE=PREF1 TO PREF31
  /SUBJECT=ID
  /FACTORS=Color (LINEAR)
  Price (LINEAR )
  Volume (LINEAR)
  Energy (DISCRETE)
  Life (DISCRETE)
  Service (DISCRETE)
  Freezer (DISCRETE)
  /PRINT=SUMMARYONLY.
```

5 Results and Discussions

The data collected is analyzed in SPSS using syntax provided in Sect. 4. According to the results of CA, attribute levels, utility scores and attributes' importance scores are obtained. The utility scores, which behave as coefficient in regression analysis, delivers numerical degrees on the preferences of each attribute level. The higher utility scores specify more desired preferences. The ranking of utility scores from the highest to the lowest shows the measure of preferable or vital attributes to overall favorites. Also, in CA, calculated values of Pearson's R and Kendall's tau shows the correlation between the observed and estimated preferences. The relative importance of attributes provided in Table 5 and Fig. 4 shows the importance level of each attribute among all attributes. According to the relative importance gathered by CA, the customer tendency to choose a refrigerator with the highest "service reliability" and "product life", which means these two attributes are leading factors that affect customers' choice. "Service reliability" and "product life" relative importance's are obtained as 17,949 and 17,947, respectively, as the two most effective attributes.

As the third most important attribute that effects customer choice is determined as "energy consumption" with 17,171 relative importance score. Actually comparing relative importance's of top three attributes there is not much difference between them. One can say that in choosing a refrigerator, customers value "service reliability", product life" and "energy consumption" the most.

As anticipated, an inverse relationship between price and its utility is obtained, which means higher prices cause lower utility values. Furthermore, as expected,

Table 5 Relative importance of attributes

Attribute	Level	Utility	Standard error	Relative importance
Color	White	−0.758	0.958	%15.520
	Black	0.236	0.958	
	Gray	0.522	0.958	
Price	<4000	−1.350	0.830	%14.878
	4000–7000	−2.700	1.660	
	>7000	−4.050	2.489	
Volume	300 LT	1.144	0.830	%10.410
	500 LT	2.289	1.660	
	700 LT	3.433	2.489	
Energy consumption	A+	1.965	0.830	%17.171
	A++	3.931	1.660	
	A+++	5.896	2.489	
Product life	5–10	−0.497	0.958	%17.947
	11–15	−0.633	0.958	
	16–20	1.131	0.958	
Service reliability	Low	−0.031	0.958	%17.949
	Medium	−1.289	0.958	
	High	1.319	0.958	
Freezer type	Low frost	−0.246	0.719	%6.125
	No frost	0.246	0.719	

Pearson's R = 0.693 Kendals Tau = 0.4654

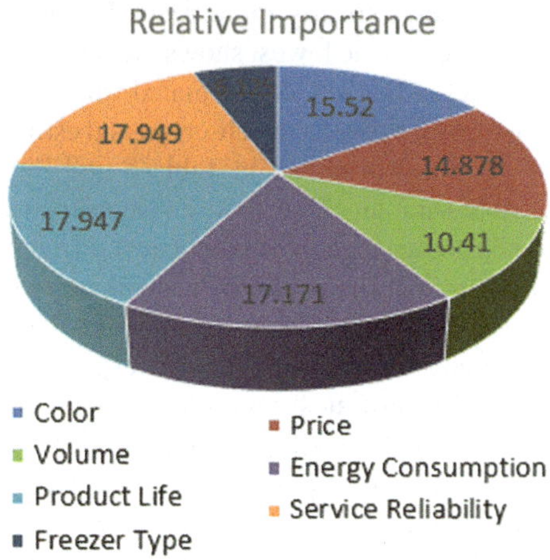

Fig. 4 Relative importance of attributes

"A+++" as a level of "energy consumption" has the highest utility score. Considering the results, manufacturers or retailers may want to give highest priority to "A+++" as a marketing strategy to gain competitiveness in the market. Besides "A+++", considering "16–20" as a level in "product life" and "high" as a level in "service reliability" are other two important attributes and their levels that should be insisted on while establishing marketing strategies. In Table 5, Pearson's R and Kendall's tau values provide correlation measures between the observed and estimated preferences [2]. The values obtained show that the observed and estimated utility ranking scores are correlated.

6 Conclusions

The selection of home appliances, which are used in a daily life of a modern humankind, requires maximum care and attention since they affect almost all activities and make these activities more efficient and comfortable. In today's competitive environment and fast changes, especially in technology, brings a high challenge for companies to satisfy customers' desires. Instead of satisfying all desires that are requested by the customers, fulfillment the most attributes that affect their choices, which they value the most, may be more profitable and effective. To accomplish this goal marketing specialists perform CA, which is one of the most efficient and effective method to use. CA is applied to build and employ specified experiments on consumers to construct a decision making process. In this context CA applied to numerous numbers of marketing research projects. CA is one of the multivariate analysis techniques to determine the importance of product or service attributes to customers, and the utilities that are delivered from these attributes to consumers.

The aim of this chapter is to analyze the applicability of CA for assessing and ranking the customers' desires and expectations. Thus, the behaviors of Turkish home appliances' customers in purchasing refrigerator are analyzed. To do this, we first identified the most important attributes and their related attribute levels for a refrigerator. Then, we conducted CA on nearly 200 customers. Considering the selected attributes and their levels 33 combinations are generated as orthogonal design. They are asked to select one of the provided combinations which satisfy their needs and provides them the highest utility. According to the relative importance obtained by CA, the customers tend to choose a refrigerator with higher "service reliability" and "product life", and less "energy consumption".

With the implementation of this study, it is clearly seen that CA is beneficial to identify, analyze, evaluate and interpret the consumer preferences in purchasing home appliances, and to point out the most critical issues for companies who want to develop and grow in home appliance sector.

The limitation of the study is the sample size, which consists of only 200 consumers, that does not represent the whole country's tendency. Further, increasing sample size will give a chance to analyze responses based on customers' age, sex, occupation, and education levels.

References

1. Awan, A., Rehman, A.: Impact of customer satisfaction on brand loyalty: an empirical analysis of home appliances in Pakistan. Br. J. Mark. Stud. **2**(8), 18–32 (2014). ISSN 2053-4043 (Print), ISSN 2053-4051
2. Backhaus, K., Erichson, B., Plinke, W., Weiber, R.: Multivariate Analysemethoden. Springer, Berlin, Heidelberg (2016)
3. Bouza, C.N.: Técnicas para el estudio de la cantidad total en el análisis conjuntual: modelos decisionales. Investig. Oper. **33**(1), 67–84 (2013)
4. Chocarro, R., Cortiñas, M., Elorz, M.: The impact of product category knowledge on consumer use of extrinsic cues—a study involving agrifood products. Food Qual. Prefer. **20**(3), 176–186 (2009)
5. Daghan, G., Akkoyunlu, B.: An examination through conjoint analysis of the preferences of students concerning online learning environments according to their learning styles. Int. Educ. Stud. **5**(4), 122–138 (2012)
6. Dauda, S.Y., Lee, J.: Technology adoption: a conjoint analysis of consumers' preference on future online banking services. Inf. Syst. **53**, 1–15 (2015)
7. Duarte, C.M., Olivares, A.I., Puyana, J.R.M., De Paolini, M.: El uso del análisis conjunto en el diseño de nuevos sistemas de transporte: el caso del auto tren del Distrito Cultural Universitario en Guadalajara. Contexto. Rev. Fac. Arquit. Univ. Autónoma Nuevo León **8**(8), 29–49 (2014)
8. Fernandez, V.: Forecasting home appliances Jeles. J. Int. Consum. Mark. **12**(2), 39–61 (2000). https://doi.org/10.1300/J046v12n02_04
9. Gamble, J., Jaeger, S.R., Harker, F.R.: Preferences in pear appearance and response to novelty among Australian and New Zealand consumers. Postharvest Biol. Technol. **41**(1), 38–47 (2006)
10. Gelmar, G., Reyner, P., Rodobaldo, M., Alexender, S.: Conjoint analysis tool for determining attributes on désigning a graduate degree program. J. Int. Bus. Res. Mark. **1**(5) (2016)
11. Green, P.E., Rao, V.R.: Conjoint measurement-for quantifying judgmental data. J. Mark. Res. **8**(3), 355–363 (1971)

12. Green, P.E., Krieger, A.M., Wind, Y.: Thirty years of conjoint analysis: reflections and prospects. Interfaces **31**, S56–S73 (2001)
13. Gustafsson, A., Herrmann, A., Huber, F. (eds.): Conjoint Measurement: Methods and Applications. Springer Science & Business Media (2013)
14. Haaijer, M.: Modeling conjoint choice experiments with the probit model. Doctoral dissertation, University of Groningen (1999)
15. Hailu, G., Boecker, A., Henson, S., Cranfield, J.: Consumer valuation of functional foods and nutraceuticals in Canada. A conjoint study using probiotics. Appetite **52**(2), 257–265 (2009)
16. Karniouchina, E.V., Moore, W.L., van der Rhee, B., Verma, R.: Issues in the use of ratings-based versus choice-based conjoint analysis in operations management research. Eur. J. Oper. Res. **197**(1), 340–348 (2009)
17. Kotri, A.: Analyzing Customer Value Using Conjoint Analysis: The Example of a Packaging Company. Tartu University Press (2006). ISSN 1406-5967, ISBN 9949-11-479-9
18. Li, B., Wang, L., Li, B.: Demand attributes and market segmentation: an evaluation of refrigerator purchase behavior in rural China. Int. J. China Mark. **1**(2), 2011 (2011)
19. Lopes, S.D.F., Boubeta, A.R., Prado, E.P., Mallou, J.V.: Efecto del orden de presentación de los atributos sobre los resultados del Análisis Conjunto. Metod. Encuestas **11**(1), 103–119 (2009)
20. Luce, R.D., Tukey, J.W.: Simultaneous conjoint measurement: a new type of fundamental measurement. J. Math. Psychol. **1**(1), 1–27 (1964)
21. Maldonado, S., Montoya, R., Weber, R.: Advanced conjoint analysis using feature selection via support vector machines. Eur. J. Oper. Res. **241**(2), 564–574 (2015)
22. Mora, M., Infante, R., Espinoza, J., Predieri, S.: Actitudes y preferencias de consumidores chilenos e italianos hacia los damascos. Econ. Agrar. Recur. Nat. **10**(1), 83–96 (2006)
23. Oppewal, H., Vriens, M.: Measuring perceived service quality using integrated conjoint experiments. Int. J. Bank Mark. **18**(4), 154–169 (2000)
24. Oyatoye, E.O., Otike-Obaro, A.E., Golda Nkeiruka, E.: Using Conjoint Analysis to Study the Factors Important to University Students in Nigeria When They Select a Laptop Computer. Peer reviewed, University of Lagos, Nigeria (2016)
25. Prasad, S.: Changing preferences of Indian customers' towards combinations of services offered through credit cards: a conjoint analysis (2013)
26. Prinzie, A., Poel, D.: Predicting home-appliance acquisition sequences: Markov/Markov for discrimination and survival analysis for modeling sequential information in NPTB models. Decis. Support Syst. **44**(2007), 28–45 (2007)
27. Rahimi, S., Fallahnezhad, M.S., Owlia, M.S., Abooie, M.H.: Investigation of customer priorities for machine made carpet through conjoint and cluster analysis (case study in Yazd, Iran). Int. J. Bus. Dev. Stud. **6**(1), 83–98 (2014)
28. Ramírez-Hurtado, J.M.: Measuring preferences: from conjoint analysis to integrated conjoint experiments. Rev. Métodos Cuantitativos Econ. Empres. **9**, 28–43 (2010)
29. Steiner, M., Meißner, M.: A user's guide to the galaxy of conjoint analysis and compositional preference measurement. Mark. ZFP **40**(2), 3–25 (2018)
30. Turanlı, M., Cengiz, D., Işık, M.: Konjoint Analizi İle Gazete Tercihlerini Etkileyen Faktörlerin Belirlenmesi, İstanbul Üniversitesi, İktisat Fakültesi. Ekonometri Ve İstatistik Dergisi, Ekonometri ve İstatistik Sayı **19**, 1–26 (2013)
31. Tuvie (2019). http://www.tuvie.com/modern-nord-fridge-design-by-olga-kalugina/. Accessed 27 Oct 2019
32. Uma, K., Sasikala, P.: Study on consumer's perceptions and buying behaviour towards selected home appliances products in Madurai district. Indian J. Appl. Res. **4**(4), 87 (2014)
33. Vick, S., Scott, A.: Agency in health care. Examining patients' preferences for attributes of the doctor–patient relationship. J. Health Econ. **17**(5), 587–605 (1998)
34. Wu, W.Y., Liao, Y.K., Chatwuthikrai, A.: Applying conjoint analysis to evaluate consumer preferences toward subcompact cars. Expert Syst. Appl. **41**(6), 2782–2792 (2014)

A Literature Review on Fuzzy FMEA and an Application on Infant Car Seat Design Using Spherical Fuzzy Sets

Elif Haktanir and Cengiz Kahraman

Abstract Failure mode and effects analysis (FMEA) is a structured approach to determine the potential failures that may arise during the implementation of a process. Because of the vagueness and impreciseness in the definition of severity, occurrence probability, and detectability factors, fuzzy set-based techniques can be used in FMEA. Spherical fuzzy sets are a recently developed extension of ordinary fuzzy sets integrating the principles of neutrosophic sets and Pythagorean fuzzy sets. In this chapter, we summarize the literature on fuzzy FMEA and propose a spherical fuzzy FMEA model with an illustrative application on infant car seats design.

1 Introduction

Failure mode and effects analysis (FMEA) method has been developed as a formal step-by-step procedure to evaluate the priorities of potential failure modes systematically. The easiness of its application has brought the method popularity among the other safety and reliability analysis tools for product and process development/design. FMEA has been applied in many different areas since its introduction to the literature but mostly on aerospace, automotive and healthcare industries [1–3].

Although FMEA is a well-accepted and widely used safety analysis tool, the traditional FMEA method is still being criticized by having several drawbacks. One of the inadequacies of classical FMEA is about how the method achieves the risk ranking. In traditional FMEA, risk priority number (RPN) is calculated to evaluate the risk level of the products or the processes which is the multiplication of the three components of FMEA: Severity (S), Occurrence (O), and Detectability (D). Severity

E. Haktanir · C. Kahraman (✉)
Department of Industrial Engineering, Istanbul Technical University, 34367 Besiktas, Istanbul, Turkey
e-mail: kahramanc@itu.edu.tr

E. Haktanir
e-mail: elif.haktanir@altinbas.edu.tr

E. Haktanir
Department of Industrial Engineering, Altınbas University, 34217 Bagcilar, Istanbul, Turkey

C. Kahraman and S. Cebi (eds.), *Customer Oriented Product Design*, Studies in Systems, Decision and Control 279, https://doi.org/10.1007/978-3-030-42188-5_22

is the degree of resulting severity when a potential failure occurs. Occurrence in the FMEA method can be defined as the probability that a potential failure will occur. Detectability is the difficulty of the failure to be detected. Traditional FMEA generally uses 1–10 scale to measure those three parameters. The main disadvantage of the method is believed to be the possible identical RPN values as the multiplication result while the risk implication of the parameters is totally different. To overcome these shortcomings, usage of fuzzy logic has been proposed to deal with ambiguous and imprecise data.

Fuzziness has been often handled in the literature from different points of view. The first significant works on fuzzy logic have been proposed by using ordinary fuzzy sets. Several extensions of ordinary fuzzy sets have been developed by various researchers in the literature. Figure 1 shows the most frequently used extensions of ordinary fuzzy sets together with their researchers.

In the following, these extensions are briefly explained. Then, a comprehensive literature review on their usage on FMEA is presented.

Zadeh [4] introduced the fuzzy set theory to the literature. The first type of fuzzy sets is ordinary fuzzy sets, which are represented by only a membership degree for an x value as given in Eq. (1).

$$\tilde{A} = \{(x, \mu(x))|x \in X\} \tag{1}$$

where X represents the universe, $0 \le \mu(x) \le 1$, and the non-membership degree equals $1 - \mu(x)$.

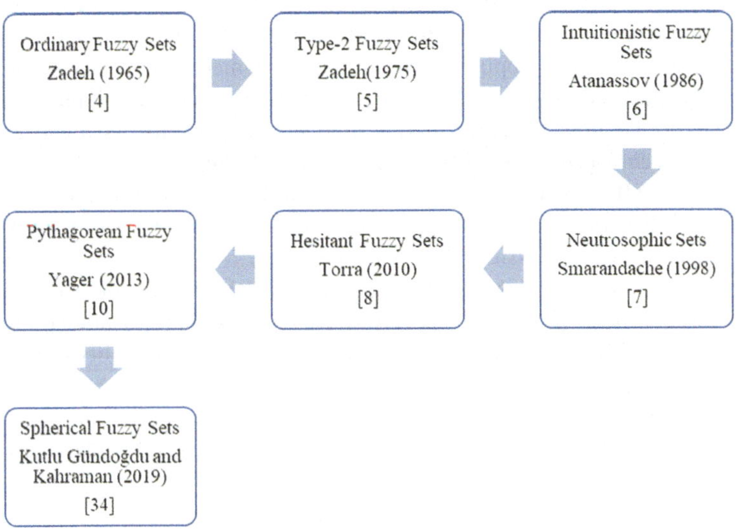

Fig. 1 Extensions of fuzzy sets

Zadeh [5] developed Type-2 fuzzy sets since ordinary fuzzy sets (Type-1 fuzzy sets) were criticized by some researchers. A type-2 fuzzy set $\tilde{\tilde{A}}$ in the universe of discourse X which can be represented by a Type-2 membership function $\mu_{\tilde{\tilde{A}}}$ is given as:

$$\tilde{\tilde{A}} = \left\{ \left((x, u), \mu_{\tilde{\tilde{A}}}(x, u) \right) \middle| \forall x \in X,\ \forall u \in J_x \subseteq [0, 1], 0 \leq \mu_{\tilde{\tilde{A}}}(x, u) \leq 1 \right\} \quad (2)$$

where J_x denotes an interval [0, 1]. Moreover, $\tilde{\tilde{A}}$ represented by:

$$\tilde{\tilde{A}} = \int_{x \in X} \int_{u \in J_x} \mu_{\tilde{\tilde{A}}}(x, u)/(x, u)$$

where $\int\int$ dotes union over all admissible x and u.

Intuitionistic fuzzy sets (IFSs) are a generalization of Zadeh's ordinary fuzzy sets proposed by Atanassov [6]. These sets involve the degrees of membership and non-membership together with experts' hesitancy. Let U be a universe of discourse. An IFS $\tilde{A}$ is an object having the form,

$$\tilde{A} = \left\{ u, \left(\mu_{\tilde{A}}(u), v_{\tilde{A}}(u) \right) \middle| u \in U \right\} \quad (3)$$

where the function $\mu_{\tilde{A}} : U \to [0, 1]$, $v_{\tilde{A}}(u) : U \to [0, 1]$ are the degree of membership, non-membership of u to $\tilde{A}$, respectively, satisfying that $0 \leq \mu_{\tilde{A}}(u) + v_{\tilde{A}}(u) \leq 1$. For any IFS $\tilde{A}$ and $u \in U$, $\pi_{\tilde{A}} = 1 - \mu_{\tilde{A}}(u) - v_{\tilde{A}}(u)$ is called degree of hesitancy of u to $\tilde{A}$.

Smarandache [7] generalized the IFSs to the neutrosophic sets. Let U be a universe of discourse. Neutrosophic set $\tilde{A}$ in U is an object having the form,

$$\tilde{A} = \left\{ u, \left(T_{\tilde{A}}(u), I_{\tilde{A}}(u), F_{\tilde{A}}(u) \right) \middle| u \in U \right\} \quad (4)$$

where, $T_{\tilde{A}}$ is the truth-membership function, $I_{\tilde{A}}$ is the indeterminacy-membership function and $F_{\tilde{A}}$ is the falsity-membership function. They are assigned independently, and their sum can be $0 \leq T_{\tilde{A}}(u) + I_{\tilde{A}}(u) + F_{\tilde{A}}(u) \leq 3$.

Hesitant fuzzy sets initially developed by Torra [8] are the extensions of ordinary fuzzy sets which handle the situations where a set of values are possible for the membership of a single element. Let X be a fixed set, a hesitant fuzzy set on X is in terms of a function that when applied to X returns a subset of [0, 1]. Mathematical expression for hesitant fuzzy sets is as follows.

$$E = \{\langle x, h_E(x)\rangle | x \in X\}, \quad (5)$$

where $h_E(x)$ shows that possible membership degrees of the element $x \in X$ to the set E.

Atanassov [9] proposed intuitionistic Type-2 fuzzy sets (IFS2), which they are later called Pythagorean fuzzy sets (PFSs) by Yager [10]. In PFSs, the sum of membership degree and non-membership degree can be greater than 1 but their squared sum should be less than or equal to 1. Let U be a universe of discourse, a PFS $\tilde{P}$ is an object having the form,

$$\tilde{P} = \left\{ u, \left(\mu_{\tilde{P}}(u), \nu_{\tilde{P}}(u) \right) \middle| u \in U \right\} \tag{6}$$

where the function $\mu_{\tilde{P}} : U \rightarrow [0, 1]$, $\nu_{\tilde{P}}(u) : U \rightarrow [0, 1]$ are the degree of membership, non-membership of u to P, respectively, satisfying that $0 \leq \mu_{\tilde{P}}^2(u) + \upsilon_{\tilde{P}}^2(u) \leq 1$. For any PFS $\tilde{A}$ and $u \in U$, $\pi_{\tilde{p}} = (1 - \mu_{\tilde{P}}^2(u) - \nu_{\tilde{P}}^2(u))^{1/2}$ is called degree of hesitancy of u to $\tilde{p}$.

Kutlu Gündoğdu and Kahraman [11] have recently introduced the spherical fuzzy sets (SFSs). The SFSs allow the hesitancy of a decision maker to be assigned independently from membership and non-membership degrees. The advantage of the SFSs is that they gather the positive sides of other fuzzy sets extensions in a unique theory. The preliminaries of SFSs are presented in detail in Sect. 3.

The rest of this study is organized as follows. Section 2 presents a literature review on FMEA using ordinary fuzzy sets and extended fuzzy sets. Section 3 gives the preliminaries of single-valued spherical fuzzy sets (SVSFs) and interval-valued spherical fuzzy sets (IVSFs). Section 4 presents the proposed SVSF and IVSF FMEA methods. Section 5 illustrates the proposed models with an application on infant car seat design. Finally, Sect. 6 concludes the paper with future directions.

2 Literature Review

There are various fuzzy FMEA studies in the literature on different areas. These ordinary fuzzy FMEA studies are briefly summarized in Table 1. We have selected the publications have more than 100 citations and presented them based on the rank of citations.

Figure 2 illustrates the publication frequencies on fuzzy FMEA based on years. There is a significant increase in the number of publications on fuzzy FMEA after 2013. Figure 3 presents the number of fuzzy FMEA publications with respect to countries. The first three leading countries are China, Iran, and Turkey. Figure 4 shows the classification of fuzzy FMEA publications based on subject areas. The most published subject areas related to fuzzy FMEA are *Engineering, Computer Science,* and *Business, Management and Accounting*.

Although fuzzy FMEA has been studied by many researchers in the literature, there is still a gap for the integration of the fuzzy set extensions with FMEA. The existing publications of FMEA using extended fuzzy sets are reviewed below.

Chang and Cheng [12] proposed an efficient and comprehensive algorithm to evaluate the risk of failure. Their approach integrated intuitionistic fuzzy sets with

Table 1 Ordinary fuzzy FMEA studies

Author (year)	Integrated method with FMEA	Application area	Citation rate
Pillay and Wang (2003) [20]	Grey relation theory	Marine industry-fishing vessel	333
Wang et al. (2009) [21]	Alpha-level sets, linear programming models, and centroid defuzzification method	Numerical example	292
Xu et al. (2002) [22]	–	Diesel engine's turbocharger system	258
Kutlu and Ekmekçioğlu (2012) [23]	TOPSIS and AHP	Automotive industry	223
Sharma et al. (2005) [24]	–	Paper industry-hydraulic system	181
Chin et al. (2009) [25]	Evidential reasoning (ER) approach	Fishing vessel	171
Yang et al. (2008) [26]	Fuzzy rule-based Bayesian reasoning (FuRBaR) approach and fuzzy rule-based evidential reasoning method	Tandem offloading operation	166
Peláez and Bowles (1996) [27]	Fuzzy Cognitive Maps (FCM)	Water tank leveling system	150
Liu et al. (2012) [28]	Extended VIKOR method	General anesthesia process	149
Liu et al. (2011) [29]	Fuzzy evidential reasoning (FER) approach and grey theory	Numerical example (adapted from Wang et al. 2009).	144
Tay and Lim (2006) [30]	Guided rules reduction system (GRRS)	Semiconductor manufacturing process	142
Chang et al. (1999) [31]	Grey theory	Numerical example	139
Abdelgawad and Fayek (2010) [32]	Fuzzy AHP	Pipeline project	111
Hu et al. (2009) [33]	Fuzzy AHP	Electronic manufacturer in Taiwan	108
Guimarães and Lapa (2007) [34]	Fuzzy inference system (FIS)	Four-loop pressurized water reactor (PWR) containment cooling system (CCS)	104
Chin (2008) [35]	–	Permanent magnet direct current (PMDC) micro-motor development project for printer carriage drive application	101

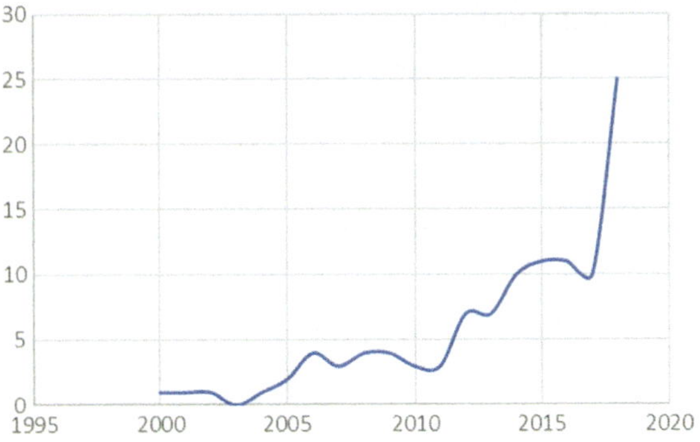

Fig. 2 Publication frequencies on fuzzy FMEA based on years

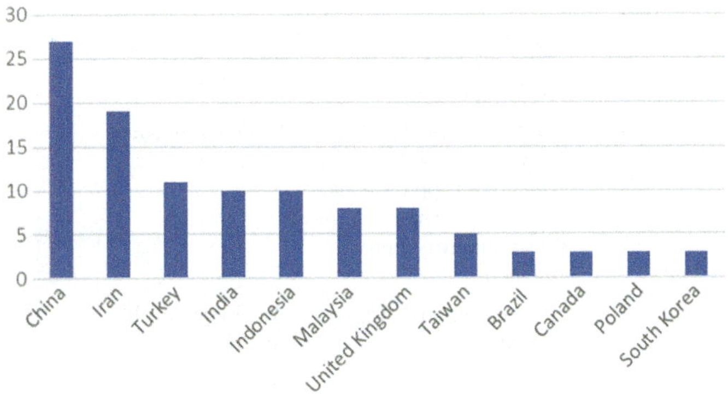

Fig. 3 Publications on fuzzy FMEA with respect to countries

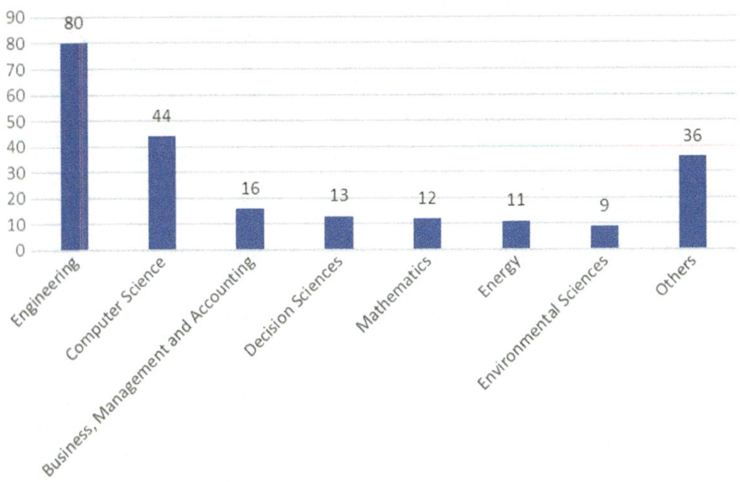

Fig. 4 Classification of fuzzy FMEA publications based on subject areas

the decision-making trial and evaluation laboratory (DEMATEL) approach on risk assessment. The approach they proposed resolves some of the shortcomings of the conventional RPN method.

Since traditional fuzzy set could not permit the membership with a set of possible values Geng and Zhang [13] used the hesitant fuzzy linguistic term and semantic-based computation method to express and deal with the evaluation information in FMEA and DEMATEL. They applied the rough set idea to transfer the hesitant fuzzy linguistic term set into interval-valued number to cope with the hard-computing problem brought by inconsistent granularity among different hesitant fuzzy linguistic terms. They used the hesitant fuzzy VIKOR method evaluate the severity, frequency and detection of failure modes. They presented a case study of failure assessment for a crane to illustrate the effectiveness of the method they proposed.

Akyuz and Celik [14] developed a quantitative risk-based approach integrating interval type-2 fuzzy sets with FMEA to perform a comprehensive risk analysis. The IT2FSs deal with ambiguity and vagueness of linguistic assessment of decision-makers while the FMEA method provides a robust risk analyzing tool with relevant control options. Linguistic evaluations in risk assessment are transformed into useful information in terms of enhancing safety and pollution prevention in maritime industry. They showed the proposed approach, potential failures and effects with a real shipboard oil spill case application.

Sayyadi Tooranloo et al. [15] evaluated knowledge management failure factors in an intuitionistic fuzzy environment as a case study in Khuzestan Oil and Gas Company. They identified a series of failure factors affecting knowledge management in organizations based on a review of the related literature and similar studies. They determined 16 failure factors in the implementation of knowledge management based on interviews with company experts. They investigated the knowledge management failure factors in Khuzestan Oil and Gas Company according to the specified factors as well as the integration of multiple criteria decision-making techniques in an intuitionistic fuzzy environment. The results indicated that lack of management commitment and leadership is the most important factor affecting the failure of knowledge management in the given company.

Wang et al. [16] proposed an improved FMEA method based on intuitionistic fuzzy multiplicative MOORA (IF-MULTIMOORA). They used the linguistic variables represented by intuitionistic fuzzy numbers to consider the evaluation information of failure modes and risk factor weights. They utilized the intuitionistic fuzzy weighted average operator to aggregate the experts' evaluation information of different priority levels. They developed a comprehensive weighting method based on expert evaluation method and deviation maximization model to determine risk factor weights. They introduced the intuitionistic fuzzy aggregation operators and improved Hamming distance to improve the MULTIMOORA approach, and then determined the risk ranking of failure modes. They provided a case study of railway line maintenance system to illustrate the validity and rationality of the proposed method.

Mirghafoori et al. [17] conducted their study to address the quality of the electronic library services with FMEA in the intuitionistic-fuzzy environment in the Library of

Yazd University. They identified20 failure modes in Electronic library service quality after reviewing the literature and interviews with library and information science (LIS) experts. They assessed students the priority and importance of each failure modes by entropy based on FMEA model in the intuitionistic-fuzzy environment after they designed and distributed questionnaires among LIS experts.

Mete [18] proposed a FMEA-based AHP-MOORA integrated method under Pythagorean fuzzy sets to assess occupational risks in a natural gas pipeline construction project. To show the feasibility and effectiveness of the integrated method, they presented a case study for concrete coating process of natural gas pipeline project. They also presented a comparative study, correlation analysis, and sensitivity analysis to approve the novel risk assessment approach. They showed that the integrated approach they proposed gives more reasonable results for evaluating the occupational risks in project of pipeline construction using the advantage of Pythagorean fuzzy sets that reflects uncertainty in more suitable way.

3 Preliminaries of Single Valued and Interval Valued Spherical Fuzzy Sets

Intuitionistic and Pythagorean fuzzy membership functions are composed of membership, non-membership and hesitancy parameters, which can be calculated by $\pi_{\tilde{I}} = 1 - \mu - \vartheta$ and $\pi_{\tilde{P}} = (1 - \mu_{\tilde{P}}^2(u) - \nu_{\tilde{P}}^2(u))^{1/2}$, respectively. Neutrosophic membership functions are also defined by three parameters truthiness, falsity and indeterminacy, whose sum can be between 0 and 3, and the value of each is between 0 and 1 independently. Kutlu Gündoğdu and Kahraman [11] proposed SFSs as an extension of IFSs and neutrosophic sets. In SFSs, while the squared sum of membership, non-membership and hesitancy parameters can be between 0 and 1, each of them can be defined between 0 and 1 independently. In the following, some definitions of SFSs are presented.

Definition 1 SVSF $\tilde{A}_S$ of the universe of discourse U is given by

$$\tilde{A}_S = \left\{ u, \left(\mu_{\tilde{A}_S}(u), \nu_{\tilde{A}_S}(u), \pi_{\tilde{A}_S}(u)\right) \middle| u \in U \right\} \tag{7}$$

where $\mu_{\tilde{A}_S} : U \to [0, 1]$, $\nu_{\tilde{A}_S}(u) : U \to [0, 1]$, $\pi_{\tilde{A}_S} : U \to [0, 1]$ and

$$0 \le \mu_{\tilde{A}_S}^2(u) + \nu_{\tilde{A}_S}^2(u) + \pi_{\tilde{A}_S}^2(u) \le 1, \forall u \in U \tag{8}$$

For each u, the numbers $\mu_{\tilde{A}_S}(u)$, $\nu_{\tilde{A}_S}(u)$ and $\pi_{\tilde{A}_S}(u)$ are the degree of membership, non-membership and hesitancy of u to $\tilde{A}_S$, respectively.

Definition 2 The basic arithmetic operations for SVSFSs are defined as follows where $\lambda > 0$.

$$\tilde{A}_S \oplus \tilde{B}_S = \left\langle \begin{array}{c} \left(\mu^2_{\tilde{A}_S} + \mu^2_{\tilde{B}_S} - \mu^2_{\tilde{A}_S}\mu^2_{\tilde{B}_S}\right)^{\frac{1}{2}}, v_{\tilde{A}_S} v_{\tilde{B}_S}, \\ \left(\left(1 - \mu^2_{\tilde{B}_S}\right)\pi^2_{\tilde{A}_S} + \left(1 - \mu^2_{\tilde{A}_S}\right)\pi^2_{\tilde{B}_S} - \pi^2_{\tilde{A}_S}\pi^2_{\tilde{B}_S}\right)^{1/2} \end{array} \right\rangle \tag{9}$$

$$\tilde{A}_S \otimes \tilde{B}_S = \left\langle \begin{array}{c} \mu_{\tilde{A}_S}\mu_{\tilde{B}_S}, \left(v^2_{\tilde{A}_S} + v^2_{\tilde{B}_S} - v^2_{\tilde{A}_S} v^2_{\tilde{B}_S}\right)^{\frac{1}{2}}, \\ \left(\left(1 - v^2_{\tilde{B}_S}\right)\pi^2_{\tilde{A}_S} + \left(1 - v^2_{\tilde{A}_S}\right)\pi^2_{\tilde{B}_S} - \pi^2_{\tilde{A}_S}\pi^2_{\tilde{B}_S}\right)^{1/2} \end{array} \right\rangle \tag{10}$$

$$\lambda \cdot \tilde{A}_S = \left\langle \left(1 - \left(1 - \mu^2_{\tilde{A}_S}\right)^{\lambda}\right)^{1/2}, v^{\lambda}_{\tilde{A}_S}, \left(\left(1 - \mu^2_{\tilde{A}_S}\right)^{\lambda} - \left(1 - \mu^2_{\tilde{A}_S} - \pi^2_{\tilde{A}_S}\right)^{\lambda}\right)^{1/2} \right\rangle \tag{11}$$

$$\tilde{A}^{\lambda}_S = \left\langle \mu^{\lambda}_{\tilde{A}_S}, \left(1 - \left(1 - v^2_{\tilde{A}_S}\right)^{\lambda}\right)^{1/2}, \left(\left(1 - v^2_{\tilde{A}_S}\right)^{\lambda} - \left(1 - v^2_{\tilde{A}_S} - \pi^2_{\tilde{A}_S}\right)^{\lambda}\right)^{1/2} \right\rangle \tag{12}$$

Definition 3 Spherical Weighted Geometric Mean (SWGM) with respect to, $w = (w_1, w_2 \ldots, w_n)$; $w_i \in [0, 1]$; $\sum_{i=1}^{n} w_i = 1$, is defined as;

$$SWGM_w(A_1, \ldots\ldots, A_n) = A_{S1}^{w1} + A_{S2}^{w2} + \cdots + A_{Sn}^{wn} = \left\{ \begin{array}{c} \prod_{i=1}^{n} \mu_{A_{Si}}^{w_i}, \left[1 - \prod_{i=1}^{n} (1 - v^2_{A_{Si}})^{w_i}\right]^{\frac{1}{2}}, \\ \left[\prod_{i=1}^{n} (1 - v^2_{A_{Si}})^{w_i} - \prod_{i=1}^{n} (1 - v^2_{A_{Si}} - \pi^2_{A_{Si}})^{w_i}\right]^{1/2} \end{array} \right\} \tag{13}$$

Definition 4 Score function for sorting SVSFSs is defined by Eq. (14);

$$Score(A_S) = \left(\mu_{A_S} - \pi_{A_S}\right)^2 - \left(v_{A_S} - \pi_{A_S}\right)^2 \tag{14}$$

IVSFSs, some definitions and arithmetic operations are developed by Kutlu Gündoğdu and Kahraman [19].

Definition 5 IVSF $\tilde{A}_{IVS}$ of the universe of discourse U is given by

$$\tilde{A}_{IVS} = \{u, ([\mu_L(u), \mu_U(u)], [v_L(u), v_U(u)], [\pi_L(u), \pi_U(u)]) | u \in U\} \tag{15}$$

where

$$0 \leq \mu^2_U(u) + v^2_U(u) + \pi^2_U(u) \leq 1 \forall u \in U \tag{16}$$

For each u, the numbers $\mu_U(u)$, $v_U(u)$ and $\pi_U(u)$ are the upper degrees of membership, non-membership and hesitancy of u to $\tilde{A}_{IVS}$, respectively.

Definition 6 Some operations are defined over the IVSFS as below where $\lambda > 0$.

$$\tilde{A}_{IVS} \oplus \tilde{B}_{IVS} =$$

$$\left\{\begin{array}{c}\left[\begin{array}{c}\left(\left(\mu^L_{\tilde{A}_{IVS}}\right)^2+\left(\mu^L_{\tilde{B}_{IVS}}\right)^2-\left(\mu^L_{\tilde{A}_{IVS}}\right)^2\left(\mu^L_{\tilde{B}_{IVS}}\right)^2\right)^{\frac{1}{2}}\\ \left(\left(\mu^U_{\tilde{A}_{IVS}}\right)^2+\left(\mu^U_{\tilde{B}_{IVS}}\right)^2-\left(\mu^U_{\tilde{A}_{IVS}}\right)^2\left(\mu^U_{\tilde{B}_{IVS}}\right)^2\right)^{\frac{1}{2}}\end{array}\right],\left[v^L_{\tilde{A}_{IVS}}v^L_{\tilde{B}_{IVS}},v^U_{\tilde{A}_{IVS}}v^U_{\tilde{B}_{IVS}}\right],\\ \left[\begin{array}{c}\left(\left(1-\left(\mu^L_{\tilde{A}_{IVS}}\right)^2\right)\left(\pi^L_{\tilde{A}_{IVS}}\right)^2+\left(1-\left(\mu^L_{\tilde{B}_{IVS}}\right)^2\right)\left(\pi^L_{\tilde{B}_{IVS}}\right)^2-\left(\pi^L_{\tilde{A}_{IVS}}\right)^2\left(\pi^L_{\tilde{B}_{IVS}}\right)^2\right)^{\frac{1}{2}},\\ \left(\left(1-\left(\mu^U_{\tilde{A}_{IVS}}\right)^2\right)\left(\pi^U_{\tilde{A}_{IVS}}\right)^2+\left(1-\left(\mu^U_{\tilde{B}_{IVS}}\right)^2\right)\left(\pi^U_{\tilde{B}_{IVS}}\right)^2-\left(\pi^U_{\tilde{A}_{IVS}}\right)^2\left(\pi^U_{\tilde{B}_{IVS}}\right)^2\right)^{\frac{1}{2}}\end{array}\right]\end{array}\right\} \quad (17)$$

$\tilde{A}_{IVS}\otimes\tilde{B}_{IVS}=$

$$\left\{\begin{array}{c}\left[\mu^L_{\tilde{A}_{IVS}}\mu^L_{\tilde{B}_{IVS}},\mu^U_{\tilde{A}_{IVS}}\mu^U_{\tilde{B}_{IVS}}\right],\left[\begin{array}{c}\left(\left(v^L_{\tilde{A}_{IVS}}\right)^2+\left(v^L_{\tilde{B}_{IVS}}\right)^2-\left(v^L_{\tilde{B}_{IVS}}\right)^2\left(v^L_{\tilde{B}_{IVS}}\right)^2\right)^{\frac{1}{2}},\\ \left(\left(v^U_{\tilde{A}_{IVS}}\right)^2+\left(v^U_{\tilde{B}_{IVS}}\right)^2-\left(v^U_{\tilde{A}_{IVS}}\right)^2\left(v^U_{\tilde{B}_{IVS}}\right)^2\right)^{\frac{1}{2}}\end{array}\right],\\ \left[\begin{array}{c}\left(\left(1-\left(v^L_{\tilde{B}_{IVS}}\right)^2\right)\left(\pi^L_{\tilde{A}_{IVS}}\right)^2+\left(1-\left(v^L_{\tilde{A}_{IVS}}\right)^2\right)\left(\pi^L_{\tilde{B}_{IVS}}\right)^2-\left(\pi^L_{\tilde{A}_{IVS}}\right)^2\left(\pi^L_{\tilde{B}_{IVS}}\right)^2\right)^{\frac{1}{2}},\\ \left(\left(1-\left(r^U_{\tilde{B}_{IVS}}\right)^2\right)\left(\pi^U_{\tilde{A}_{IVS}}\right)^2+\left(1-\left(v^U_{\tilde{A}_{IVS}}\right)^2\right)\left(\pi^U_{\tilde{B}_{IVS}}\right)^2-\left(\pi^U_{\tilde{A}_{IVS}}\right)^2\left(\pi^U_{\tilde{B}_{IVS}}\right)^2\right)^{\frac{1}{2}}\end{array}\right]\end{array}\right\} \quad (18)$$

$\lambda\cdot\tilde{A}_{IVS}=$

$$\left\{\begin{array}{c}\left[\left(1-\left(1-\left(\mu^L_{\tilde{A}_{IVS}}\right)^2\right)^{\lambda}\right)^{\frac{1}{2}},\left(1-\left(1-\left(\mu^U_{\tilde{A}_{IVS}}\right)^2\right)^{\lambda}\right)^{\frac{1}{2}}\right],\left[\left(v^L_{\tilde{A}_{IVS}}\right)^{\lambda},\left(v^U_{\tilde{A}_{IVS}}\right)^{\lambda}\right],\\ \left[\begin{array}{c}\left(\left(1-\left(\mu^L_{\tilde{A}_{IVS}}\right)^2\right)^{\lambda}-\left(1-\left(\mu^L_{\tilde{A}_{IVS}}\right)^2-\left(\pi^L_{\tilde{A}_{IVS}}\right)^2\right)^{\lambda}\right)^{\frac{1}{2}},\\ \left(\left(1-\left(\mu^U_{\tilde{A}_{IVS}}\right)^2\right)^{\lambda}-\left(1-\left(\mu^U_{\tilde{A}_{IVS}}\right)^2-\left(\pi^U_{\tilde{A}_{IVS}}\right)^2\right)^{\lambda}\right)^{\frac{1}{2}}\end{array}\right]\end{array}\right\} \quad (19)$$

$\tilde{A}^{\lambda}_{IVS}=$

$$\left\{\begin{array}{c}\left[\left(\mu^L_{\tilde{A}_{IVS}}\right)^{\lambda},\left(\mu^U_{\tilde{A}_{IVS}}\right)^{\lambda}\right],\left[\left(1-\left(1-\left(v^L_{\tilde{A}_{IVS}}\right)^2\right)^{\lambda}\right)^{\frac{1}{2}},\left(1-\left(1-\left(v^U_{\tilde{A}_{IVS}}\right)^2\right)^{\lambda}\right)^{\frac{1}{2}}\right],\\ \left[\begin{array}{c}\left(\left(1-\left(v^L_{\tilde{A}_{IVS}}\right)^2\right)^{\lambda}-\left(1-\left(v^L_{\tilde{A}_{IVS}}\right)^2-\left(\pi^L_{\tilde{A}_{IVS}}\right)^2\right)^{\lambda}\right)^{\frac{1}{2}},\\ \left(\left(1-\left(v^U_{\tilde{A}_{IVS}}\right)^2\right)^{\lambda}-\left(1-\left(v^U_{\tilde{A}_{IVS}}\right)^2-\left(\pi^U_{\tilde{A}_{IVS}}\right)^2\right)^{\lambda}\right)^{\frac{1}{2}}\end{array}\right]\end{array}\right\} \quad (20)$$

Definition 7 Let $\tilde{A}_j=\left[\mu^L_{\tilde{A}_{IVS_j}},\mu^U_{\tilde{A}_{IVS_j}}\right],\left[v^L_{\tilde{A}_{IVS_j}},v^U_{\tilde{A}_{IVS_j}}\right],\left[\pi^L_{\tilde{A}_{IVS_j}},\pi^U_{\tilde{A}_{IVS_j}}\right]$ be a collection of IVS Weighted Arithmetic Mean (IVSWAM) with respect to, $w_j=(w_1,w_2\ldots,w_n)$; $w_j\in[0,1]$ and $\sum_{j=1}^{n}w_j=1$, IVSWAM is defined as;

$$IVSWAM_w\left(\tilde{A}_1,\tilde{A}_2\ldots,\tilde{A}_n\right)=w_1\cdot\tilde{A}_1\oplus W_2\cdot\tilde{A}_2\oplus\cdots\oplus W_n\cdot\tilde{A}_n=$$

$$\left\{\begin{array}{c}\left[\left(1-\prod_{j=1}^{n}\left(1-\left(\mu^{L}_{\tilde{A}_{IVS_j}}\right)^2\right)^{w_j}\right)^{\frac{1}{2}},\left(1-\prod_{j=1}^{n}\left(1-\left(\mu^{U}_{\tilde{A}_{IVS_j}}\right)^2\right)^{w_j}\right)^{\frac{1}{2}}\right],\\ \left[\prod_{j=1}^{n}\left(v^{L}_{\tilde{A}_{IVS_j}}\right)^{w_j},\prod_{j=1}^{n}\left(v^{U}_{\tilde{A}_{IVS_j}}\right)^{w_j}\right]\\ \left[\begin{array}{c}\left(\prod_{j=1}^{n}\left(1-\left(\mu^{L}_{\tilde{A}_{IVS_j}}\right)^2\right)^{N/j}-\prod_{j=1}^{n}\left(1-\left(\mu^{L}_{\tilde{A}_{IVS_j}}\right)^2-\left(\pi^{L}_{IVS_j}\right)^2\right)^{w_j}\right)^{\frac{1}{2}},\\ \left(\prod_{j=1}^{n}\left(1-\left(\mu^{U}_{\tilde{A}_{IVS_j}}\right)^2\right)^{w_j}-\prod_{j=1}^{n}\left(1-\left(\mu^{U}_{\tilde{A}_{IVS_j}}\right)^2-\left(\pi^{U}_{\tilde{A}_{IVS_j}}\right)^2\right)^{w_j}\right)^{\frac{1}{2}}\end{array}\right]\end{array}\right\} \tag{21}$$

Definition 8 Score function of sorting IVSFSs is defined by;

$$Score(A_{IVS})=\frac{\left(\mu^{L}_{\tilde{A}_{IVS}}\right)^2+\left(\mu^{U}_{\tilde{A}_{IVS}}\right)^2-\left(v^{L}_{\tilde{A}_{IVS}}\right)^2-\left(v^{U}_{\tilde{A}_{IVS}}\right)^2-\left(\frac{\pi^{L}_{\tilde{A}_{IVS}}}{2}\right)^2-\left(\frac{\pi^{U}_{\tilde{A}_{IVS}}}{2}\right)^2}{2} \tag{22}$$

4 Single Valued and Interval Valued Spherical Fuzzy FMEA Methods

In this section, SVSF and IVSF FMEA methods will be presented. Firstly, SVSF FMEA is defined as follows. Three components of FMEA Severity (S), Probability (P), and Detectability (D) are assumed to be SVSF numbers: $\tilde{S}_s$, $\tilde{O}_s$, and $\tilde{D}_s$. The SVSF Risk Priority Number $(\widetilde{RPN}_s)$ is then calculated as in Eq. (23).

$$\widetilde{RPN}_s=\tilde{S}_s\otimes\tilde{O}_s\otimes\tilde{D}_s \tag{23}$$

The multiplication of the first two elements gives Eq. (24).

$$\tilde{S}_s\otimes\tilde{O}_s=\left\langle\mu_{\tilde{S}_s}\mu_{\tilde{O}_s},\left(v^2_{\tilde{S}_s}+v^2_{\tilde{O}_s}-v^2_{\tilde{S}_s}v^2_{\tilde{O}_s}\right)^{1/2},\left(\left(1-v^2_{\tilde{O}_s}\right)\pi^2_{\tilde{S}_s}+\left(1-v^2_{\tilde{S}_s}\right)\pi^2_{\tilde{O}_s}-\pi^2_{\tilde{S}_S}\pi^2_{\tilde{O}_S}\right)^{1/2}\right\rangle \tag{24}$$

Let $\left(v^2_{\tilde{S}_s}+v^2_{\tilde{O}_s}-v^2_{\tilde{S}_s}v^2_{\tilde{O}_s}\right)^{1/2}$ is represented by $\tilde{R}_s$ and $\left(\left(1-v^2_{\tilde{O}_s}\right)\pi^2_{\tilde{S}_s}+\left(1-v^2_{\tilde{S}_s}\right)\pi^2_{\tilde{O}_s}-\pi^2_{\tilde{S}_s}\pi^2_{\tilde{O}_s}\right)$ is represented by $\tilde{T}_s$. Then, $\widetilde{RPN}_s$ can be calculated as in Eq. (25).

$$\tilde{S}_s \otimes \tilde{O}_s \otimes \tilde{D}_s = \left\langle \begin{array}{c} \mu_{\tilde{S}_s}\mu_{\tilde{O}_s}\mu_{\tilde{D}_s}, \left(v_{\tilde{R}_s}^2 + v_{\tilde{D}_s}^2 - v_{\tilde{R}_s}^2 v_{\tilde{D}_s}^2\right)^{\frac{1}{2}}, \\ \left(\left(1 - v_{\tilde{D}_s}^2\right)\pi_{\tilde{T}_s}^2 + \left(1 - v_{\tilde{T}_s}^2\right)\pi_{\tilde{D}_s}^2 - \pi_{\tilde{T}_s}^2\pi_{\tilde{D}_s}^2\right)^{\frac{1}{2}} \end{array} \right\rangle. \quad (25)$$

SVSF scales are given by Tables 2, 3, and 4. Table 2 presents the SVSF scale for severity. Table 3 presents the SVSF scale for occurrence while Table 4 shows the SVSF scale for detectability.

The most important possible failure mode is the mode having the largest $\widetilde{RPN}_s$. The score function defined by Eq. (8) can be used to compare the $\widetilde{RPN}_s$.

Table 2 SVSF scale for severity

SVSF value	Severity	Meaning
(0.1, 0.9, 0.1)	Extremely low	Unreasonable to expect that the minor nature of this failure would have any real effect on the system performance. Customer probably will not even notice the failure
(0.2, 0.8, 0.2)	Very low	Failure causes only a very low customer attention. Customer will probably only notice a very low deterioration on the system performance
(0.3, 0.7, 0.3)	Low	Failure causes only a low customer attention. Customer will probably only notice a low deterioration on the system performance
(0.4, 0.6, 0.4)	Slightly low	Failure causes only a slightly low customer attention. Customer will probably only notice a slightly low deterioration on the system performance
(0.5, 0.5, 0.5)	Moderate	Failure causes some customer dissatisfaction. Customer is made uncomfortable or is annoyed by the failure. Customer will notice some system performance deterioration
(0.6, 0.4, 0.4)	Slightly high	Failure causes slightly high customer dissatisfaction. Customer is made uncomfortable or is annoyed by the failure. Customer will notice some system performance deterioration
(0.7, 0.3, 0.3)	High	High degree of customer dissatisfaction due to the nature of the failure such as an inoperable vehicle or an inoperable convenience subsystem. Failure does not involve safety or noncompliance with government regulations
(0.8, 0.2, 0.2)	Very high	Very high degree of customer dissatisfaction due to the nature of the failure such as an inoperable vehicle or an inoperable convenience subsystem. Failure does not involve safety or noncompliance with government regulations
(0.9, 0.1, 0.1)	Very very high	Very very high ranking when a potential failure mode affects safe vehicle operation and/or involves noncompliance with government regulations
(1.0, 0.0, 0.0)	Extremely high	Extremely high ranking when a potential failure mode affects safe vehicle operation and/or involves noncompliance with government regulations

Table 3 SVSF scale for occurrence

SVSF value	Occurrence	Meaning
(0.1, 0.9, 0.1)	Extremely low	Failure is extremely unlikely
(0.2, 0.8, 0.2)	Very low	Failure is very unlikely
(0.3, 0.7, 0.3)	Low	Relatively few failures
(0.4, 0.6, 0.4)	Slightly low	Relatively slightly few failures
(0.5, 0.5, 0.5)	Moderate	Maybe few failures
(0.6, 0.4, 0.4)	Slightly high	Occasional failures
(0.7, 0.3, 0.3)	High	Repeated failures
(0.8, 0.2, 0.2)	Very high	Highly repeated failures
(0.9, 0.1, 0.1)	Very very high	Failure is almost inevitable
(1.0, 0.0, 0.0)	Extremely high	Failure is inevitable

Table 4 SVSF scale for detectability

SVSF value	Detectability	Meaning
(0.1, 0.9, 0.1)	Extremely high	Program will detect a potential design weakness
(0.2, 0.8, 0.2)	Very very high	Program will almost certainly detect a potential design weakness
(0.3, 0.7, 0.3)	Very high	Program has a very good chance of detecting a potential design weakness
(0.4, 0.6, 0.4)	High	Program has a good chance of detecting a potential design weakness
(0.5, 0.5, 0.5)	Slightly high	Program may very probably detect a potential design weakness
(0.6, 0.4, 0.4)	Moderate	Program may detect a potential design weakness
(0.7, 0.3, 0.3)	Slightly low	Program is not very likely to detect a potential design weakness
(0.8, 0.2, 0.2)	Low	Program is not likely to detect a potential design weakness
(0.9, 0.1, 0.1)	Very low	Program probably will not detect a potential design weakness
(1.0, 0.0, 0.0)	Non-detection	Program cannot detect a potential design weakness or there is no design verification program

IVSF FMEA is defined as follows. Three components of FMEA Severity (S), Probability (P), and Detectability (D) are assumed to be IVSF numbers: $\tilde{S}_{IVS}$, $\tilde{O}_{IVS}$, and $\tilde{D}_{IVS}$. The IVSF Risk Priority Number $(\widetilde{RPN}_{IVS})$ is then calculated as in Eq. (26).

$$\widetilde{RPN}_{IVS} = \tilde{S}_{IVS} \otimes \tilde{O}_{IVS} \otimes \tilde{D}_{IVS} \tag{26}$$

Table 5 IVSF scale for severity

IVSF value	Severity
([0.00, 0.05], [0.90, 0.95], [0.10, 0.15])	Extremely low
([0.10, 0.15], [0.80, 0.85], [0.20, 0.25])	Very low
([0.20, 0.25], [0.70, 0.75], [0.30, 0.35])	Low
([0.30, 0.35], [0.60, 0.65], [0.40, 0.45])	Slightly low
([0.40, 0.45], [0.50, 0.55], [0.50, 0.55])	Moderate
([0.50, 0.55], [0.40, 0.45], [0.40, 0.45])	Slightly high
([0.60, 0.65], [0.30, 0.35], [0.30, 0.35])	High
([0.70, 0.75], [0.20, 0.25], [0.20, 0.25])	Very high
([0.80, 0.85], [0.10, 0.15], [0.10, 0.15])	Very very high
([0.90, 0.95], [0.00, 0.05], [0.00, 0.05])	Extremely high

$$\tilde{S}_{IVS} \otimes \tilde{O}_{IVS} = \left\langle \begin{array}{c} \mu_{\tilde{S}_{IVS}} \mu_{\tilde{O}_{IVS}}, \left(v^2_{\tilde{S}_{IVS}} + v^2_{\tilde{O}_{IVS}} - v^2_{\tilde{S}_{IVS}} v^2_{\tilde{O}_{IVS}}\right)^{\frac{1}{2}}, \\ \left(\left(1 - v^2_{\tilde{O}_{IVS}}\right)\pi^2_{\tilde{S}_{IVS}} + \left(1 - v^2_{\tilde{S}_{IVS}}\right)\pi^2_{\tilde{O}_{IVS}} - \pi^2_{\tilde{S}_{IVS}} \pi^2_{\tilde{O}_{IVS}}\right)^{1/2} \end{array} \right\rangle \quad (27)$$

Let $\left(v^2_{\tilde{S}_{IVS}} + v^2_{\tilde{O}_{IVS}} - v^2_{\tilde{S}_{IVS}} v^2_{\tilde{O}_{IVS}}\right)^{1/2}$ is represented by $\tilde{R}_{IVS}$ and $\left(\left(1 - v^2_{\tilde{O}_{IVS}}\right)\pi^2_{\tilde{S}_{IVS}} + \left(1 - v^2_{\tilde{S}_{IVS}}\right)\pi^2_{\tilde{O}_{IVS}} - \pi^2_{\tilde{S}_{IVS}} \pi^2_{\tilde{O}_{IVS}}\right)$ is represented by $\tilde{T}_{IVS}$. Then, $\widetilde{RPN}_{IVS}$ can be calculated as in Eq. (28).

$$\tilde{S}_{IVS} \otimes \tilde{O}_{IVS} \otimes \tilde{D}_{IVS} = \left\langle \begin{array}{c} \mu_{\tilde{S}_{IVS}} \mu_{\tilde{O}_{IVS}} \mu_{\tilde{D}_{IVS}}, \left(v^2_{\tilde{R}_{IVS}} + v^2_{\tilde{D}_{IVS}} - v^2_{\tilde{R}_{IVS}} v^2_{\tilde{D}_{IVS}}\right)^{\frac{1}{2}}, \\ \left(\left(1 - v^2_{\tilde{D}_{IVS}}\right)\pi^2_{\tilde{T}_{IVS}} + \left(1 - v^2_{\tilde{T}_{IVS}}\right)\pi^2_{\tilde{D}_{IVS}} - \pi^2_{\tilde{T}_{IVS}} \pi^2_{\tilde{D}_{IVS}}\right)^{\frac{1}{2}} \end{array} \right\rangle \quad (28)$$

Tables 5, 6, and 7 present the IVSF scales for severity, occurrence, and detectabil-

Table 6 IVSF scale for occurrence

IVSF value	Occurrence
([0.00, 0.05], [0.90, 0.95], [0.10, 0.15])	Extremely low
([0.10, 0.15], [0.80, 0.85], [0.20, 0.25])	Very low
([0.20, 0.25], [0.70, 0.75], [0.30, 0.35])	Low
([0.30, 0.35], [0.60, 0.65], [0.40, 0.45])	Slightly low
([0.40, 0.45], [0.50, 0.55], [0.50, 0.55])	Moderate
([0.50, 0.55], [0.40, 0.45], [0.40, 0.45])	Slightly high
([0.60, 0.65], [0.30, 0.35], [0.30, 0.35])	High
([0.70, 0.75], [0.20, 0.25], [0.20, 0.25])	Very high
([0.80, 0.85], [0.10, 0.15], [0.10, 0.15])	Very very high
([0.90, 0.95], [0.00, 0.05], [0.00, 0.05])	Extremely high

Table 7 IVSF scale for detectability

IVSF value	Detectability
([0.00, 0.05], [0.90, 0.95], [0.10, 0.15])	Extremely high
([0.10, 0.15], [0.80, 0.85], [0.20, 0.25])	Very very high
([0.20, 0.25], [0.70, 0.75], [0.30, 0.35])	Very high
([0.30, 0.35], [0.60, 0.65], [0.40, 0.45])	High
([0.40, 0.45], [0.50, 0.55], [0.50, 0.55])	Slightly high
([0.50, 0.55], [0.40, 0.45], [0.40, 0.45])	Moderate
([0.60, 0.65], [0.30, 0.35], [0.30, 0.35])	Slightly low
([0.70, 0.75], [0.20, 0.25], [0.20, 0.25])	Low
([0.80, 0.85], [0.10, 0.15], [0.10, 0.15])	Very low
([0.90, 0.95], [0.00, 0.05], [0.00, 0.05])	Non-detection

ity, respectively where they have the same meanings as defined for SVSF scales given in Tables 2, 3, and 4.

The most important possible failure mode is the mode having the largest $\widetilde{RPN}_{IVS}$. The score function defined by Eq. (22) can be used to compare the $\widetilde{RPN}_{IVS}$.

5 An Illustrative Application on Infant Car Seat Design

FMEA has four different types: System FMEA, Design FMEA, Process FMEA, and Service FMEA. Design FMEA is a method being used to identify the potential risks in a new or changed design of a product/service. This section illustrates the proposed methods on an infant car seat design application.

Suppose that an infant car seat manufacturer in Turkey plans to determine the priorities of potential failures for the design phase of their new product. The determined potential failures of the design process are *infant cannot physically fit into the seat, infant "sinks into" the seat too much, natural freq. ranges overlap (occupied seat vs. infant sensitivity), not held tight enough during cornering, seatbelt does not fit, inadequate vibration isolation*, and *seat is ugly, lacking style*. The proposed SVSF FMEA method is illustrated below.

Table 8 presents the SVSF assessments of three experts for the $\tilde{S}_s$, $\tilde{O}_s$, and $\tilde{D}_s$ values of the potential failure modes. Table 9 shows the aggregated SVSF values of $\tilde{S}_s$, $\tilde{O}_s$ and $\tilde{D}_s$ (by using Eq. (13)), the calculated $\widetilde{RPN}_S$ (by using Eq. (23)) and the corresponding importance scores (by using Eq. (14)). Weights of the experts are taken as 0.4, 0.3 and 0.3 respectively depending on their experience levels.

The results indicate that the potential failure mode with the highest priority is *infant "sinks into" the seat too much*. The same problem will be solved by using IVSF FMEA as follows (Tables 10 and 11).

The results indicate that the potential failure mode with the highest priority is *inadequate vibration isolation*.

Table 8 SVSF assessments of three experts for the potential failure modes

Potential failure modes	Experts	$\tilde{S}_s$	$\tilde{O}_s$	$\tilde{D}_s$
Infant cannot physically fit into the seat	E_1	(0.7, 0.3, 0.3)	(0.5, 0.5, 0.5)	(0.4, 0.6, 0.4)
	E_2	(0.6, 0.4, 0.4)	(0.5, 0.5, 0.5)	(0.6, 0.4, 0.4)
	E_3	(0.7, 0.3, 0.3)	(0.5, 0.5, 0.5)	(0.5, 0.5, 0.5)
Infant "sinks into" the seat too much	E_1	(0.5, 0.5, 0.5)	(0.6, 0.4, 0.4)	(0.5, 0.5, 0.5)
	E_2	(0.4, 0.6, 0.4)	(0.7, 0.3, 0.3)	(0.7, 0.3, 0.3)
	E_3	(0.6, 0.4, 0.4)	(0.5, 0.5, 0.5)	(0.6, 0.4, 0.4)
Natural freq. ranges overlap (occupied seat vs. infant sensitivity)	E_1	(0.7, 0.3, 0.3)	(0.2, 0.8, 0.2)	(0.4, 0.6, 0.4)
	E_2	(0.8, 0.2, 0.2)	(0.3, 0.7, 0.3)	(0.5, 0.5, 0.5)
	E_3	(0.9, 0.1, 0.1)	(0.2, 0.8, 0.2)	(0.4, 0.6, 0.4)
Not held tight enough during cornering	E_1	(0.9, 0.1, 0.1)	(0.1, 0.9, 0.1)	(0.2, 0.8, 0.2)
	E_2	(0.9, 0.1, 0.1)	(0.1, 0.9, 0.1)	(0.3, 0.7, 0.3)
	E_3	(1.0, 0.0, 0.0)	(0.2, 0.8, 0.2)	(0.1, 0.9, 0.1)
Seatbelt does not fit	E_1	(0.8, 0.2, 0.2)	(0.2, 0.8, 0.2)	(0.3, 0.7, 0.3)
	E_2	(0.6, 0.4, 0.4)	(0.3, 0.7, 0.3)	(0.3, 0.7, 0.3)
	E_3	(0.7, 0.3, 0.3)	(0.2, 0.8, 0.2)	(0.3, 0.7, 0.3)
Inadequate vibration isolation	E_1	(0.6, 0.4, 0.4)	(0.4, 0.6, 0.4)	(0.7, 0.3, 0.3)
	E_2	(0.6, 0.4, 0.4)	(0.5, 0.5, 0.5)	(0.8, 0.2, 0.2)
	E_3	(0.6, 0.4, 0.4)	(0.6, 0.4, 0.4)	(0.9, 0.1, 0.1)
Seat is ugly, lacking style	E_1	(0.7, 0.3, 0.3)	(0.2, 0.8, 0.2)	(0.8, 0.2, 0.2)
	E_2	(0.6, 0.4, 0.4)	(0.2, 0.8, 0.2)	(0.9, 0.1, 0.1)
	E_3	(0.5, 0.5, 0.5)	(0.3, 0.7, 0.3)	(1.0, 0.0, 0.0)

Table 9 Aggregated SVSF expert assessments, $\widetilde{RPN}_S$, and importance scores

Potential failure modes	Aggregated $\tilde{S}_s$	Aggregated $\tilde{O}_s$	Aggregated $\tilde{D}_s$	$\widetilde{RPN}_S$	Importance score
Infant cannot physically fit into the seat	(0.6684, 0.3341, 0.3364)	(0.5000, 0.5000, 0.5000)	(0.4830, 0.5213, 0.4345)	(0.1614, 0.7175, 0.5255)	0.0957
Infant "sinks into" the seat too much	(0.4939, 0.5100, 0.4452)	(0.5950, 0.4103, 0.4192)	(0.5842, 0.4216, 0.4316)	(0.1717, 0.7029, 0.5292)	0.0976
Natural freq. ranges overlap (occupied seat vs. infant sensitivity)	(0.7857, 0.2272, 0.2301)	(0.2259, 0.7748, 0.2290)	(0.4277, 0.5734, 0.4304)	(0.0759, 0.8635, 0.3215)	−0.2235

(continued)

Table 9 (continued)

Potential failure modes	Aggregated $\tilde{S}_s$	Aggregated $\tilde{O}_s$	Aggregated $\tilde{D}_s$	$\widetilde{RPN}_S$	Importance score
Not held tight enough during cornering	(0.9289, 0.0837, 0.0839)	(0.1231, 0.8774, 0.1274)	(0.1835, 0.8186, 0.1943)	(0.0210, 0.9616, 0.1174)	−0.7033
Seatbelt does not fit	(0.7050, 0.3039, 0.3097)	(0.2259, 0.7748, 0.2290)	(0.3000, 0.7000, 0.3000)	(0.0478, 0.9028, 0.2581)	−0.3713
Inadequate vibration isolation	(0.6000, 0.4000, 0.4000)	(0.4830, 0.5213, 0.4345)	(0.7857, 0.2272, 0.2301)	(0.2277, 0.6480, 0.5017)	0.0537
Seat is ugly, lacking style	(0.6042, 0.4021, 0.4124)	(0.2259, 0.7748, 0.2290)	(0.8862, 0.1384, 0.1396)	(0.1209, 0.8194, 0.3249)	−0.2029

Table 10 IVSF assessments of three experts for the potential failure modes

Potential failure modes	Experts	$\tilde{S}_s$	$\tilde{O}_s$	$\tilde{D}_s$
Infant cannot physically fit into the seat	E_1	([0.60, 0.65], [0.30, 0.35], [0.30, 0.35])	([0.40, 0.45], [0.50, 0.55], [0.50, 0.55])	([0.30, 0.35], [0.60, 0.65], [0.40, 0.45])
	E_2	([0.50, 0.55], [0.40, 0.45], [0.40, 0.45])	([0.40, 0.45], [0.50, 0.55], [0.50, 0.55])	([0.50, 0.55], [0.40, 0.45], [0.40, 0.45])
	E_3	([0.60, 0.65], [0.30, 0.35], [0.30, 0.35])	([0.40, 0.45], [0.50, 0.55], [0.50, 0.55])	([0.40, 0.45], [0.50, 0.55], [0.50, 0.55])
Infant "sinks into" the seat too much	E_1	([0.40, 0.45], [0.50, 0.55], [0.50, 0.55])	([0.50, 0.55], [0.40, 0.45], [0.40, 0.45])	([0.40, 0.45], [0.50, 0.55], [0.50, 0.55])
	E_2	([0.30, 0.35], [0.60, 0.65], [0.40, 0.45])	([0.60, 0.65], [0.30, 0.35], [0.30, 0.35])	([0.60, 0.65], [0.30, 0.35], [0.30, 0.35])
	E_3	([0.50, 0.55], [0.40, 0.45], [0.40, 0.45])	([0.40, 0.45], [0.50, 0.55], [0.50, 0.55])	([0.50, 0.55], [0.40, 0.45], [0.40, 0.45])
Natural freq. ranges overlap (occupied seat vs. infant sensitivity)	E_1	([0.60, 0.65], [0.30, 0.35], [0.30, 0.35])	([0.10, 0.15], [0.80, 0.85], [0.20, 0.25])	([0.30, 0.35], [0.60, 0.65], [0.40, 0.45])
	E_2	([0.70, 0.75], [0.20, 0.25], [0.20, 0.25])	([0.20, 0.25], [0.70, 0.75], [0.30, 0.35])	([0.40, 0.45], [0.50, 0.55], [0.50, 0.55])

(continued)

Table 10 (continued)

Potential failure modes	Experts	$\tilde{S}_s$	$\tilde{O}_s$	$\tilde{D}_s$
	E_3	([0.80, 0.85], [0.10, 0.15], [0.10, 0.15])	([0.10, 0.15], [0.80, 0.85], [0.20, 0.25])	([0.30, 0.35], [0.60, 0.65], [0.40, 0.45])
Not held tight enough during cornering	E_1	([0.80, 0.85], [0.10, 0.15], [0.10, 0.15])	([0.00, 0.05], [0.90, 0.95], [0.10, 0.15])	([0.10, 0.15], [0.80, 0.85], [0.20, 0.25])
	E_2	([0.80, 0.85], [0.10, 0.15], [0.10, 0.15])	([0.00, 0.05], [0.90, 0.95], [0.10, 0.15])	([0.20, 0.25], [0.70, 0.75], [0.30, 0.35])
	E_3	([0.00, 0.05], [0.90, 0.95], [0.10, 0.15])	([0.10, 0.15], [0.80, 0.85], [0.20, 0.25])	([0.00, 0.05], [0.90, 0.95], [0.10, 0.15])
Seatbelt does not fit	E_1	([0.70, 0.75], [0.20, 0.25], [0.20, 0.25])	([0.10, 0.15], [0.80, 0.85], [0.20, 0.25])	([0.20, 0.25], [0.70, 0.75], [0.30, 0.35])
	E_2	([0.50, 0.55], [0.40, 0.45], [0.40, 0.45])	([0.20, 0.25], [0.70, 0.75], [0.30, 0.35])	([0.20, 0.25], [0.70, 0.75], [0.30, 0.35])
	E_3	([0.60, 0.65], [0.30, 0.35], [0.30, 0.35])	([0.10, 0.15], [0.80, 0.85], [0.20, 0.25])	([0.20, 0.25], [0.70, 0.75], [0.30, 0.35])
Inadequate vibration isolation	E_1	([0.50, 0.55], [0.40, 0.45], [0.40, 0.45])	([0.30, 0.35], [0.60, 0.65], [0.40, 0.45])	([0.60, 0.65], [0.30, 0.35], [0.30, 0.35])
	E_2	([0.50, 0.55], [0.40, 0.45], [0.40, 0.45])	([0.40, 0.45], [0.50, 0.55], [0.50, 0.55])	([0.70, 0.75], [0.20, 0.25], [0.20, 0.25])
	E_3	([0.50, 0.55], [0.40, 0.45], [0.40, 0.45])	([0.50, 0.55], [0.40, 0.45], [0.40, 0.45])	([0.80, 0.85], [0.10, 0.15], [0.10, 0.15])
Seat is ugly, lacking style	E_1	([0.60, 0.65], [0.30, 0.35], [0.30, 0.35])	([0.10, 0.15], [0.80, 0.85], [0.20, 0.25])	([0.70, 0.75], [0.20, 0.25], [0.20, 0.25])
	E_2	([0.50, 0.55], [0.40, 0.45], [0.40, 0.45])	([0.10, 0.15], [0.80, 0.85], [0.20, 0.25])	([0.80, 0.85], [0.10, 0.15], [0.10, 0.15])
	E_3	([0.40, 0.45], [0.50, 0.55], [0.50, 0.55])	([0.20, 0.25], [0.70, 0.75], [0.30, 0.35])	([0.90, 0.95], [0.00, 0.05], [0.00, 0.05])

Table 11 Aggregated IVSF expert assessments, $\widetilde{RPN}_{IVS}$, and importance scores

Potential failure modes	Aggregated $\tilde{S}_{IVS}$	Aggregated $\tilde{O}_{IVS}$	Aggregated $\tilde{D}_{IVS}$	$\widetilde{RPN}_{IVS}$	Importance score
Infant cannot physically fit into the seat	([0.5734, 0.6236], [0.3270, 0.3774], [0.3309, 0.3801])	([0.4000, 0.4500], [0.5000, 0.5500], [0.5000, 0.5500])	([0.4021, 0.4517], [0.5030, 0.5537], [0.4346, 0.4853])	([0.0922, 0.1268], [0.7069, 0.7650], [0.5307, 0.5346])	−0.6011
Infant "sinks into" the seat too much	([0.4103, 0.4600], [0.4939, 0.5445], [0.4445, 0.4951])	([0.5100, 0.5602], [0.3923, 0.4432], [0.4027, 0.4511])	([0.5017, 0.5519], [0.4012, 0.4522], [0.4131, 0.4612])	([0.1050, 0.1422], [0.6806, 0.7418], [0.5305, 0.5380])	−0.5625
Natural freq. ranges overlap (occupied seat vs. infant sensitivity)	([0.7049, 0.7573], [0.1911, 0.2454], [0.2120, 0.2561])	([0.1382, 0.1862], [0.7686, 0.8187], [0.2359, 0.2856])	([0.3341, 0.3838], [0.5681, 0.6182], [0.4365, 0.4875])	([0.0325, 0.0541], [0.8561, 0.8992], [0.3301, 0.3302])	−0.7960
Not held tight enough during cornering	([0.8383, 0.8929], [0.0000, 0.1079], [0.0762, 0.1146])	([0.0549, 0.0924], [0.8688, 0.9188], [0.1384, 0.1867])	([0.1270, 0.1697], [0.7962, 0.8465], [0.2170, 0.2654])	([0.0058, 0.0140], [0.9541, 0.9779], [0.1345, 0.1361])	−0.9378
Seatbelt does not fit	([0.6222, 0.6731], [0.2781, 0.3299], [0.2923, 0.3391])	([0.1382, 0.1862], [0.7686, 0.8187], [0.2359, 0.2856])	([0.2000, 0.2500], [0.7000, 0.7500], [0.3000, 0.3500])	([0.0172, 0.0313], [0.8986, 0.9335], [0.2612, 0.2605])	−0.8558
Inadequate vibration isolation	([0.5000, 0.5500], [0.4000, 0.4500], [0.4000, 0.4500])	([0.4021, 0.4517], [0.5030, 0.5537], [0.4346, 0.4853])	([0.7049, 0.7573], [0.1911, 0.2454], [0.2120, 0.2561])	([0.1417, 0.1882], [0.6288, 0.6930], [0.5059, 0.5306])	−0.4773
Seat is ugly, lacking style	([0.5213, 0.5716], [0.3812, 0.4322], [0.3932, 0.4412])	([0.1382, 0.1862], [0.7686, 0.8187], [0.2359, 0.2856])	([0.8114, 0.8699], [0.0000, 0.1323], [0.1173, 0.1483])	([0.0585, 0.0926], [0.8063, 0.8582], [0.3250, 0.3406])	−0.7151

6 Conclusion

Product development should be based on a successful design process which should consider the consumers' requirements and the possible failures that may exist within the design of a product or a process. FMEA is one of the techniques dealing with the potential failures that may cause the process to end with a loss rather than a benefit. Fuzzy sets approaches let linguistic assessments to be incorporated into the numerical models making the evaluation process flexible and more informative. Fuzzy FMEA enables fuzzy numbers to be employed in the FMEA model providing flexible assessments rather than evaluations based on exact numbers. Besides, SFSs involve more information than ordinary fuzzy sets since SFSs take care of the hesitancy of experts in the calculation process. For further research, we suggest severity, occurrence and detectability parameters to be weighted based on the type of the problem. These weights can be also fuzzy numbers in order to incorporate another flexible element to the FMEA process.

References

1. Papadopoulos, Y., Parker, D., Grante, C.: Automating the failure modes and effects analysis of safety critical systems. In: Proceedings of IEEE International Symposium on High Assurance Systems Engineering, vol. 8, pp. 310–311 (2004)
2. Shebl, N.A., Franklin, B.D., Barber, N.: Failure mode and effects analysis outputs: are they valid? BMC Health Serv. Res. **12**(1) (2012)
3. Liu, H.C., Fan, X.J., Li, P., Chen, Y.Z.: Evaluating the risk of failure modes with extended MULTIMOORA method under fuzzy environment. Eng. Appl. Artif. Intell. **34**, 168–177 (2014)
4. Zadeh, L.A.: Fuzzy sets. Inf. Control. 338–353 (1965)
5. Zadeh, L.A.: The concept of a linguistic variable and its application to approximate reasoning. Inf. Sci. **8**, 199–249 (1975)
6. Atanassov, K.T.: Intuitionistic fuzzy sets. Fuzzy Sets Syst. **20**(1), 87–96 (1986)
7. Smarandache, F.: Neutrosophic set-a generalization of the intuitionistic fuzzy set. Int. J. Pure Appl. Math. 287 (2003)
8. Torra, V.: Hesitant fuzzy sets. Int. J. Intell. Syst. 529–539 (2010)
9. Atanassov, K.T.: Intuitionistic Fuzzy Sets Theory and Applications. Springer (1999)
10. Yager, R.: Pythagorean fuzzy subsets. In: Joint IFSA World Congress and NAFIPS Annual Meeting, Edmonton, Canada, pp. 57–61 (2013)
11. Kutlu Gündoğdu, F., Kahraman, C.: Spherical fuzzy sets and spherical fuzzy TOPSIS method. J. Intell. Fuzzy Syst. **36**(1), 337–352 (2019)
12. Chang, K., Cheng, C.: A risk assessment methodology using intuitionistic fuzzy set in FMEA. Int. J. Syst. Sci. **41**(12), 1457–1471 (2010)
13. Geng, X., Zhang, Y.: Improved FMEA approach for risk evaluation based on hesitant fuzzy set. Comput. Integr. Manuf. Syst. CIMS **23**(2), 340–348 (2017)
14. Akyuz, E., Celik, E.: A quantitative risk analysis by using interval type-2 fuzzy FMEA approach: the case of oil spill. Marit. Policy Manag. **45**(8), 979–994 (2018)
15. Sayyadi Tooranloo, H., Ayatollah, A.S., Alboghobish, S.: Evaluating knowledge management failure factors using intuitionistic fuzzy FMEA approach. Knowl. Inf. Syst. **57**(1), 183–205 (2018)
16. Wang, R., Zhu, J., Li, Y.: Improved FMEA method for risk evaluation based on intuitionistic fuzzy MULTIMOORA. Comput. Integr. Manuf. Syst. CIMS **24**(2), 290–301 (2018)

17. Mirghafoori, S.H., Izadi, M.R., Daei, A.: Analysis of the barriers affecting the quality of electronic services of libraries by VIKOR, FMEA and entropy combined approach in an intuitionistic-fuzzy environment. J. Intell. Fuzzy Syst. **34**(4), 2441–2451 (2018)
18. Mete, S.: Assessing occupational risks in pipeline construction using FMEA-based AHP-MOORA integrated approach under Pythagorean fuzzy environment. Hum. Ecol. Risk Assess. **25**(7), 1645–1660 (2019)
19. Kutlu Gündoğdu, F., Kahraman, C.: A novel fuzzy TOPSIS method using emerging interval-valued spherical fuzzy sets. Eng. Appl. Artif. Intell. **85**, 307–323 (2019)
20. Pillay, A., Wang, J.: Modified failure mode and effects analysis using approximate reasoning. Reliab. Eng. Syst. Saf. **79**(1), 69–85 (2003)
21. Wang, Y., Chin, K., Poon, G.K.K., Yang, J.: Risk evaluation in failure mode and effects analysis using fuzzy weighted geometric mean. Expert. Syst. Appl. **36**(2 Part 1), 1195–1207 (2009)
22. Xu, K., Tang, L.C., Xie, M., Ho, S.L., Zhu, M.L.: Fuzzy assessment of FMEA for engine systems. Reliab. Eng. Syst. Saf. **75**(1), 17–29 (2002)
23. Kutlu, A.C., Ekmekçioğlu, M.: Fuzzy failure modes and effects analysis by using fuzzy TOPSIS-based fuzzy AHP. Expert Syst. Appl. **39**(1), 61–67 (2012)
24. Sharma, R.K., Kumar, D., Kumar, P.: Systematic failure mode effect analysis (FMEA) using fuzzy linguistic modelling. Int. J. Qual. Reliab. Manag. **22**(9), 986–1004 (2005)
25. Chin, K., Wang, Y., Ka Kwai Poon, G., Yang, J.: Failure mode and effects analysis using a group-based evidential reasoning approach. Comput. Oper. Res. **36**(6), 1768–1779 (2009)
26. Yang, Z., Bonsall, S., Wang, J.: Fuzzy rule-based Bayesian reasoning approach for prioritization of failures in FMEA. IEEE Trans. Reliab. **57**(3), 517–528 (2008)
27. Peláez, C.E., Bowles, J.B.: Using fuzzy cognitive maps as a system model for failure modes and effects analysis. Inf. Sci. **88**(1–4), 177–199 (1996)
28. Liu, H., Liu, L., Liu, N., Mao, L.: Risk evaluation in failure mode and effects analysis with extended VIKOR method under fuzzy environment. Expert Syst. Appl. **39**(17), 12926–12934 (2012)
29. Liu, H., Liu, L., Bian, Q., Lin, Q., Dong, N., Xu, P.: Failure mode and effects analysis using fuzzy evidential reasoning approach and grey theory. Expert Syst. Appl. **38**(4), 4403–4415 (2011)
30. Tay, K.M., Lim, C.P.: Fuzzy FMEA with a guided rules reduction system for prioritization of failures. Int. J. Qual. Reliab. Manag. **23**(8), 1047–1066 (2006)
31. Chang, C., Wei, C., Lee, Y.: Failure mode and effects analysis using fuzzy method and grey theory. Kybernetes **28**(8–9), 1072–1080 (1999)
32. Abdelgawad, M., Fayek, A.R.: Risk management in the construction industry using combined fuzzy FMEA and fuzzy AHP. J. Constr. Eng. Manag. **136**(9), 1028–1036 (2010)
33. Hu, A.H., Hsu, C., Kuo, T., Wu, W.: Risk evaluation of green components to hazardous substance using FMEA and FAHP. Expert. Syst. Appl. **36**(3 Part 2), 7142–7147 (2009)
34. Guimarães, A.C.F., Lapa, C.M.F.: Fuzzy inference to risk assessment on nuclear engineering systems. Appl. Soft Comput. J. **7**(1), 17–28 (2007)
35. Chin, K., Chan, A., Yang, J.: Development of a fuzzy FMEA based product design system. Int. J. Adv. Manuf. Technol. **36**(7–8), 633–649 (2008)

Examination of Factors Affecting Customer Oriented Product Design in Automotive Industry

Cigdem Kadaifci and Irem Ucal Sari

Abstract Existing competitive environment leads manufacturers to focus on customer expectations more than ever. One of the reasons behind this focus is that the design concepts which are flexible to customer needs add incredible value to the product. In this chapter, customer oriented product design for an automobile is examined according to design concepts and their causal relationships using fuzzy cognitive mapping (FCM) to help automotive industry to be customer focused. Individual assessments are used in the analysis rather than aggregate them to obtain common customer's preferences. It is found that the relations and the steady states of the concepts are differentiated among the customers. The results of the study can be used as a road map for automotive industry for customer oriented product design.

1 Introduction

Examining the production technologies evolving from mechanization to mass production, automation through information technologies, mass customization and finally individualization (i.e. personalization) demonstrate that changing technologies and customer profiles lead organizations to evolve from being production oriented to being product oriented. Nowadays, customer oriented product design is more important for the producers to be able to satisfy demanding customer needs and in this way, increase their compatibility in the market place. To survive in a highly dynamic environment and sustain their position in the market, the companies need to develop capabilities especially on the product development process with reference to customer requirements. As Ulrich and Eppinger [47] indicated in their book, the economic success of companies is directly related with how they are able to describe the exact customer requirements and how they reflect these requirements to the newly designed products as quickly as possible [47]. Note that, customer oriented design

C. Kadaifci (✉) · I. U. Sari
Management Faculty, Industrial Engineering Department, Istanbul Technical University, Istanbul, Turkey
e-mail: kadaifci@itu.edu.tr

C. Kahraman and S. Cebi (eds.), *Customer Oriented Product Design*, Studies in Systems, Decision and Control 279, https://doi.org/10.1007/978-3-030-42188-5_23

will induce customer satisfaction, which, in turn, provides a long-term cash flow for the future of a company [31].

Product development is basically a process starts with identification of the opportunities (i.e. ideas for a new product) and right after a comprehensive planning phase, customer needs are identified in order to experience the use environment of the product, to perform competitive benchmarking, and as a result, to establish the specifications of the product to be designed [47]. The customer oriented product development, on the other hand, is based on the capability of producing only what customers want with a market push approach by including customers into the design process and by implementing innovative technologies [15]. With the increased expectations of the customers and the ability to reach technology easier than before, developing a customer oriented product, which has exact features that are expected by the customer, creates a considerable value in today's competitive environment. Especially for the high value products such as automobiles, a variety of options attracts customers and increases brand value in the sector. Most of the customers are willing to pay additional amounts for specific configurations, emerge from the need of the automotive industry to differentiate their products, such as digital driving mirrors, parking cameras and remote lock systems. Due to focusing only on the technology and innovation results in similar products, companies in this industry aim to provide both high-tech and customer appealing products [33]. Also, it is important to understand customer's willingness to pay on different modules and their variations.

The effects of different modules for the automotive industry are investigated in lots of papers in various ways. Most of the researchers focused on some parts of the automobiles such as seats, steering wheel, and dashboard design separately and try to find the added value to the product for different modifications of that component. He et al. [17] analyzed automobile seat design in the user's vision. Ren et al. [39] examined automobile dashboard design using Kansei engineering. Elarusi et al. [13] studied optimum design of car seat climate control system. Ibragimova [19] analyzed steering wheel cover design. A method for designing user-centric products in automotive industry is described by Mohammed et al. [33] considering ISO 9241: the human centered design process (the standard used to be known as ISO 13407). This proposed method is actually a process oriented approach aims to integrate customers into each step of the user-centric product design process. A case study examines whether Volvo Car Group is customer oriented or not through a designed survey and tries to find an answer to a particular question: is Volvo Car Group consistent with its vision "Designed Around You" [4]. The transformation from mass customization to personalization is examined from Industry 4.0 era perspective and the BMW case is provided to demonstrate how this era contributes to the design of personalized products through emerging technologies [22].

In fact, before focusing on a specific configuration, the customer response for different configurations should be examined. Also, it is important to reveal causal relationships among these configurations from the point of view of customer satisfaction. Therefore, in this chapter, automotive design concepts are examined to find out which configurations have a significant influence on customers' preferences considering the long-term behavior of the system by using fuzzy cognitive mapping

(FCM). FCM is an essential tool for examining the causal relationships among several concepts. It helps decision makers to find out long term states of concepts using their interrelationships.

The rest of the chapter is constructed as follows: In Sect. 2, a detailed literature review on customer oriented product design and FCM is given. Then the product design concepts in the automotive industry are determined in Sect. 3. After providing the basics of the method in Sect. 4, automobile design parameters are examined using FCM in the application part (Sect. 5). At the end, the results of the study are discussed in the conclusion.

2 Literature Review

Ellis and Jacobs [14] expanded traditional four marketing utilities; time, place, possession and form to include other activities such as image creation, confidence building, and product customization which may lead to modifying the predisposition and perception of customers. Although it is mentioned first in the late 1970s, product customization was not very popular for a long time due to the priorities of the producers and limited customer expectations. Sioukas [41] introduced customer oriented product design terminology to the literature to emphasize the importance of involving customers in the design phase of a product. After a while in 2005, a design alternative generation procedure is proposed to improve product development and design efficiency by linking user requirements with design characteristics that integrates a rule-based system and a design prototyping approach [26]. At the same time, a matrix approach is introduced to generate feasible design alternatives for a particular set of customer needs that utilizes congruent matrix multiplication operator, relational database system, and component design strategy [27]. Since then different methods such as quality function deployment [30, 32, 34, 35, 49], Kansei Engineering [20, 44], adaptive clustering analysis [10], fuzzy inference process [16], morphological analysis [28], multiple regression analysis [29], and fuzzy rule-based systems [52] are utilized to integrate customer requirements and expectations in the product design phase of new product development process.

Besides these methods, it is found that FCM is a widely-used tool to identify relationships between product design parameters. Efe [12] proposed a fuzzy cognitive mapping approach to calculate correlations among design requirements of dishwashers. Townsend and Urbanic [46] utilized this method for the redesign of an assembly production system. Kumar et al. [25] proposed a structural hierarchy model to analyze the changing decision making model of consumers in the digital market. Hu et al. [18] proposed a hybrid method using genetic algorithm and FCM to analyze customers' dynamic requirements in product life-cycle management. Chen [9] integrated fuzzy decision tree and FCM in engineering design perspectives to optimize complex interactions and experiential design systems with ambiguous relationships. Uzochukwu et al. [48] developed a product sustainment simulator using

FCM. Storto [42] used a framework based on FCM to describe the technical problem-solving activity that occurred in new product development in the food machinery industry. Ahmadi et al. [1] developed a new approach to manage interrelated tasks in the introduction of a new enterprise resource planning system using FCM and the fuzzy analytical hierarchy process. Chen [8] designed an autonomous agent-based tracing system that utilizes the internet of things architecture and combines FCM and fuzzy rule method for product usage life cycle of an agriculture food product. Dong et al. [11] developed a model to form the synesthesia between customer and designer to increase the requirement transmission accuracy and help design tasks and validate the method with a mobile phone case study. Chen [7] designed an intelligent agent-based tracing system for a backward design process of product usage life cycle that utilizes the internet of things architecture and FCM. Cheah et al. [6] used FCM constructor to generate Bayesian belief networks to achieve design knowledge from domain experts systematically. Jie et al. [21] proposed a method of mapping customer's Kansei cognition to product shape features to address the problem of how to use customer cognitive knowledge to support product shape design.

3 Product Design in the Automotive Industry

The automotive industry is growing day by day with an increase in living standards. The product design in the automotive industry is critical in various aspects. Due to the value of the product is high in this industry; customers want to achieve their requirements more than usual products. There are lots of modular tools that a customer can offer for an automobile. This diversity in the applicable tools and high variety in their designs make the design of an automobile one of the best examples of customer oriented product design.

Customer preferences for automobile design are studied by various researchers. Khalid and Helander [23] classified customer needs on a product as prestige, utility, aesthetics, functionality, pleasure and usability. In that paper, holistic design (overall size, overall design, overall shape and color combination), functional design (information display, size of the button, arrangement) and product styling (form, color, size) are used as design parameters to define customer needs in automobile design. Taha et al. [43] defined user preferences for automobiles as good design, multi-functionality, advanced technology, good brand, user-friendliness, environmentally friendliness, and good performance. Nepal et al. [36] defined fuel economy, engine power, external styling, front-seat room, engine quietness, interior quietness, seat comfort, air conditioner, vehicle stability, smooth-riding, instrument panel appearance, cup holders and interior styling as customer satisfaction attributes for an automotive. You et al. [54] analyzed customer preferences for interior parts of an automobile. Crash pad, transmission gear, steering wheel, shift knob, metal gray inlay, audio panel, and wood grain inlay are analyzed with the design parameters such as embossing shape, embossing size, brightness, softness, hue, material type, color uniformity, shininess, saturation, and surface slipperiness.

4 Fuzzy Cognitive Mapping

Fuzzy Cognitive Mapping (FCM), proposed by Bart Kosko [24], is an extension of the classical Cognitive Mapping (CM) method [3] and used for modeling complex systems, specifically where uncertainty and fuzziness exist [2]. In CM, concepts are represented by nodes and causal relationships among these concepts are represented by edges either as an increase (+1) or as a decrease (−1) [40]. In other words, for the relationship between two concepts, three possible values can be indicated by an expert: no relationship, positive relationship, or negative relationship. On the other hand, in FCM, the degrees of relationships being positive or negative can be assigned. Even though they are proposed to reflect causality in the system, CM can be considered as a picture of a system representing the expert's way of thinking, while FCM helps the analyst to discover the dynamic system behavior throughout the time.

Concepts are represented by C_i, $i = 1, 2, \ldots, N$, where N is the total number of concepts, and weighted arcs (w_{ij}) show the causal relationships [38]. It is possible to represent FCM either by a graph or by a relationship matrix called adjacency matrix. In Fig. 1, a sample FCM with five nodes is given along with its matrix representation in Table 1.

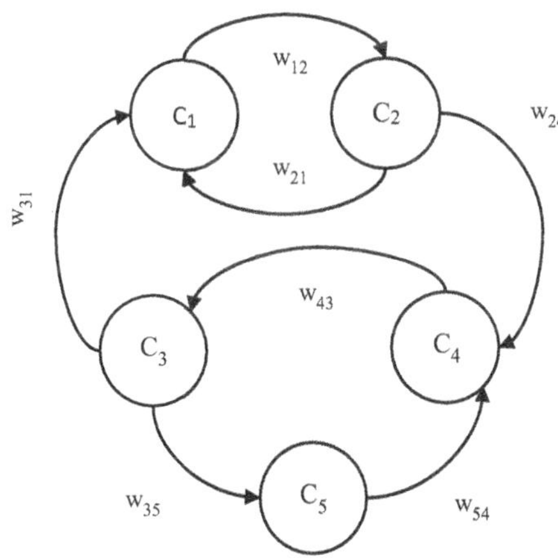

Fig. 1 A sample FCM (graphical representation)

Table 1 A sample FCM (matrix representation)

	C_1	C_2	C_3	C_4	C_5
C_1	0	w_{12}	0	0	0
C_2	w_{21}	0	0	w_{24}	0
C_3	w_{31}	0	0	0	w_{35}
C_4	0	0	w_{43}	0	0
C_5	0	0	0	w_{54}	0

In FCM, the diagonal entries of the adjacency matrix are zero, because a concept is not allowed to affect itself. The causal relationships, w_{ij}, take values based on the following rules [37]:

- $w_{ij} > 0$ indicates a causal increase
- $w_{ij} < 0$ indicates causal decrease
- $w_{ij} = 0$ indicates no causality

FCM is a simulation process, triggered by the initial conditions of the selected concepts [40] and ends when the system reaches an equilibrium, shows a limit cycle, or chaotic behavior [45, 50]. Each iteration produces vectors representing the concept values at a particular time and the final output of the process is a steady state vector consisting of the future values of the concepts. The mathematical representation of the simulation process is given below in Eq. (1):

$$A_i^{(k+1)} = f\left(A_i^{(k)} + \sum_{\substack{j=1, \\ j \neq i}}^{n} A_j^{(k)} w_{ji}\right) \tag{1}$$

Here, $A_i^{(k+1)}$ denotes the concept value at step $(k + 1)$, $A_i^{(k)}$ denotes the concept value at step (k), w_{ji} denotes the weight of the relationship between C_j and C_i [37]. f, a threshold function, reduces the concept values in a comparable range [5]. The common functions used in FCM applications are bivalent, trivalent, hyperbolic tangent, and sigmoid functions (see: [45].

5 Application

In this chapter, it is decided to examine the design concepts of an automobile in physical, technological, and comfort-oriented aspects. The physical design includes six design concepts which are overall size, overall shape, color, bumper shape, baggage shape and textile of seats. Multimedia control panel, locking system, gear technology, the quietness of the engine, and cruise system are taken as the technological design concepts. In comfort-oriented design, seat comfort, cup holders, steering wheel and arrangement of the control panel are selected. Besides the changing preferences of customers considering these fifteen concepts, each of them may have a different effect on the aesthetics and performance related perceptions. In order to reveal the effects of fifteen concepts on customer perceptions, aesthetics, prestige, pleasure, and usability are included in the analysis. The list of these concepts is given in Table 2.

The causal relationships among these concepts are assessed by four customers who have their own automobiles. They are asked to assess the strength of the existing relationships by indicating the causality as linguistic variables in the interval [−1,1], defined as "very very high", "very high", "high", "medium", "low", very low", and "very very low". The membership functions, μ, of the fuzzy sets characterizing these linguistic terms are given in Fig. 2 [51].

Table 2 The list of concepts

	Concepts	Group
C1	Overall size	Physical design concepts
C2	Overall shape	
C3	Color	
C4	Bumper shape	
C5	Baggage shape	
C6	Textile of seats	
C7	Multimedia control panel	Technological design concepts
C8	Locking system	
C9	Gear technology	
C10	Quietness of the engine	
C11	Cruise system	
C12	Seat comfort	Comfort-oriented design concepts
C13	Cup holders	
C14	Steering wheel	
C15	Arrangement of the control panel	
C16	Aesthetics	Customer preferences
C17	Prestige	
C18	Pleasure	
C19	Usability	

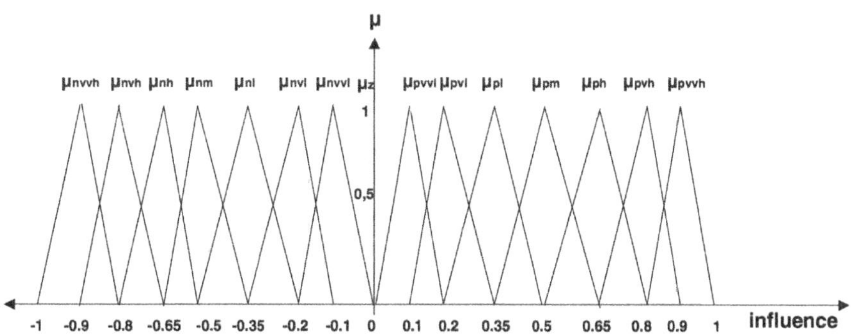

Fig. 2 Membership functions of linguistic variables

The linguistic assessments of each customer for the existing causal relationships are transformed into crisp values using an appropriate defuzzification technique. In this study, the graded mean integration approach in Eq. (2) is used [53].

$$P(\widetilde{A}) = \frac{a_1 + 4a_2 + a_3}{6} \tag{2}$$

After defuzzification, the adjacency matrix of each customer is obtained. The matrices are given in Table 3 and the corresponding maps are provided in Fig. 3. Obviously, both the existence and the strength of the relationships vary according to the customers. While the first one constructs the most complex map, the fourth one indicates there is nearly the least number of direct causal relationships among these concepts comparing to other customers. These differences reflect the changing customer preferences for automobiles.

In order to trigger FCM simulations, an initial vector, A0 = [1, 0, 0, 0, 0, 0, 1, 0, 0, 0, 0, 1, 0, 0, 0, 0, 0, 0, 0], where "Overall Size" from physical design group, "Multimedia Control Panel" from technological design group, and "Seat Comfort" from comfort-oriented design group are activated, is formed. In other words, one concept from each group is activated to observe the behavior of the systems constructed by different customers. Keeping initial vectors same provides comparable steady state values, because the only parameter changes in the FCM formula is the adjacency matrix for each customer. Equation (1) is run until each customer's system reaches equilibrium and the steady state vectors (i.e. future concept values) are obtained. Note that, the sigmoid function is used as a threshold function to reduce the concept values into a normalized range through iterations.

The steady state values of each customer are given in Table 4. The behavior of the systems during time is also represented in Fig. 4 for the fourth customer's assessments as an example.

According to the results, the first and the second customer's preference related concepts will be active in the long run. "Pleasure" and "Usability" have the highest value for the last two customers. While "Overall Shape" is the only common concept that will be active in four systems, "Bumper Shape" has the highest value according to the first customer. Considering technological and comfort-oriented design groups, it can be concluded that the ranking of the concepts is quite different for the customers. For instance, "Cruise System" has the highest value in the technological design group in the second customer's system. On the other hand, for the first customer, "Quietness of the Engine" and "Locking System" have the highest values, respectively. Interestingly, based on the third and fourth customers' assessments, the simulation yields equal concept values for the technological design group.

"Arrangement of the Control Panel" has the highest value among comfort-oriented design concepts. For the first customer, it will be better to indicate that "Steering Wheel" will matter compared to other customers.

6 Conclusion

Product design is a critical stage in manufacturing that differentiates the companies in competitive market places. With the changing and increasing customer expectations, companies have to focus on customer oriented product design to take customers'

Table 3 The causal relationship assessments of each customer

Customer 1

	C1	C2	C3	C4	C5	C6	C7	C8	C9	C10	C11	C12	C13	C14	C15	C16	C17	C18	C19
C1	0.00	0.79	0.00	0.65	0.79	0.00	0.00	0.10	0.50	0.65	0.00	0.00	0.00	0.35	0.35	0.79	0.79	0.79	0.79
C2	0.00	0.00	0.00	0.65	0.50	0.00	0.00	0.00	0.00	0.00	0.00	0.00	0.00	0.00	0.00	0.79	0.79	0.21	0.50
C3	0.00	0.00	0.00	0.00	0.00	0.50	0.00	0.00	0.00	0.00	0.00	0.00	0.00	0.00	0.00	0.79	0.65	0.35	0.21
C4	0.00	0.21	0.00	0.00	0.00	0.00	0.00	0.00	0.00	0.00	0.00	0.00	0.00	0.00	0.00	0.79	0.90	0.50	0.50
C5	0.50	0.65	0.00	0.50	0.00	0.00	0.00	0.50	0.00	0.00	0.00	0.21	0.00	0.00	0.00	0.65	0.35	0.50	0.79
C6	0.00	0.00	0.50	0.00	0.00	0.00	0.00	0.00	0.00	0.00	0.00	0.79	0.10	0.10	0.00	0.79	0.79	0.79	0.79
C7	0.00	0.00	0.00	0.00	0.00	0.00	0.00	0.00	0.00	0.00	0.00	0.00	0.00	0.50	0.79	0.50	0.50	0.79	0.90
C8	0.00	0.00	0.00	0.00	0.00	0.00	0.00	0.00	0.00	0.00	0.00	0.00	0.00	0.00	0.21	0.00	0.35	0.35	0.50
C9	0.10	0.00	0.00	0.00	0.00	0.00	0.00	0.00	0.00	0.65	0.79	0.21	0.00	0.50	0.65	0.00	0.00	0.90	0.90
C10	0.00	0.00	0.00	0.00	0.00	0.00	0.00	0.00	0.00	0.00	0.00	0.00	0.00	0.00	0.00	0.00	0.10	0.65	0.10
C11	0.00	0.00	0.00	0.00	0.00	0.00	0.00	0.00	0.00	0.00	0.00	0.00	0.00	0.50	0.50	0.00	0.50	0.65	0.79
C12	0.00	0.00	0.00	0.00	0.00	0.50	0.00	0.00	0.00	0.00	0.00	0.00	0.00	0.00	0.00	0.00	0.50	0.79	0.79
C13	0.00	0.00	0.00	0.00	0.00	0.00	0.00	0.00	0.00	0.00	0.00	0.00	0.00	0.00	0.50	0.35	0.10	0.65	0.79
C14	0.00	0.00	0.00	0.00	0.00	0.00	0.00	0.00	0.00	0.00	0.00	0.00	0.00	0.00	0.79	0.79	0.65	0.79	0.79
C15	0.00	0.00	0.00	0.00	0.00	0.00	0.50	0.35	0.10	0.00	0.00	0.00	0.35	0.35	0.00	0.50	0.35	0.79	0.79
C16	0.00	0.00	0.00	0.00	0.00	0.00	0.00	0.00	0.00	0.00	0.00	0.00	0.00	0.00	0.00	0.00	0.79	0.65	0.00
C17	0.00	0.00	0.00	0.00	0.00	0.00	0.00	0.00	0.00	0.00	0.00	0.00	0.00	0.00	0.00	0.00	0.00	0.50	0.00
C18	0.00	0.00	0.00	0.00	0.00	0.00	0.00	0.00	0.00	0.00	0.00	0.00	0.00	0.00	0.00	0.00	0.00	0.00	0.00
C19	0.00	0.00	0.00	0.00	0.00	0.00	0.00	0.00	0.00	0.00	0.00	0.00	0.00	0.00	0.00	0.10	0.00	0.79	0.00

(continued)

Table 3 (continued)

Customer 2

	C1	C2	C3	C4	C5	C6	C7	C8	C9	C10	C11	C12	C13	C14	C15	C16	C17	C18	C19
C1	0.00	0.21	0.00	0.50	0.50	0.00	0.50	0.00	0.00	0.00	0.00	0.50	0.00	0.00	0.50	0.65	0.65	0.50	0.50
C2	0.35	0.00	0.00	0.50	0.50	0.00	0.00	0.00	0.00	0.00	0.00	0.00	0.00	0.00	0.00	0.90	0.90	0.50	0.35
C3	0.00	0.00	0.00	0.00	0.00	0.00	0.00	0.00	0.00	0.00	0.00	0.00	0.00	0.00	0.00	0.79	0.65	0.00	0.00
C4	0.50	0.65	0.00	0.00	0.00	0.00	0.00	0.00	0.00	0.00	0.00	0.00	0.00	0.00	0.00	0.79	0.65	0.00	0.00
C5	0.65	0.65	0.00	0.00	0.00	0.00	0.00	0.00	0.00	0.00	0.00	0.00	0.00	0.00	0.00	0.65	0.35	0.00	0.65
C6	0.00	0.00	0.00	0.00	0.00	0.79	0.00	0.00	0.00	0.00	0.00	0.35	0.00	0.00	0.00	0.50	0.35	0.50	0.10
C7	0.00	0.00	0.00	0.00	0.00	0.00	0.00	0.00	0.00	0.00	0.00	0.00	0.00	0.00	0.50	0.21	0.65	0.65	0.90
C8	0.00	0.00	0.00	0.00	0.00	0.00	0.00	0.00	0.00	0.00	0.00	0.00	0.00	0.00	0.00	0.00	0.65	0.65	0.79
C9	0.00	0.00	0.00	0.00	0.00	0.00	0.00	0.00	0.00	0.79	0.90	0.00	0.00	0.00	0.00	0.00	0.65	0.90	0.90
C10	0.00	0.00	0.00	0.00	0.00	0.00	0.00	0.00	0.00	0.00	0.00	0.00	0.00	0.00	0.00	0.00	0.35	0.90	0.65
C11	0.00	0.00	0.00	0.00	0.00	0.00	0.00	0.00	0.00	0.00	0.00	0.00	0.00	0.00	0.35	0.00	0.90	0.90	0.90
C12	0.00	0.00	0.00	0.00	0.00	0.00	0.00	0.00	0.00	0.00	0.00	0.00	0.00	0.00	0.00	0.00	0.00	0.90	0.90
C13	0.00	0.00	0.00	0.00	0.00	0.00	0.00	0.00	0.00	0.00	0.00	0.00	0.00	0.00	0.00	0.35	0.35	0.79	0.90
C14	0.00	0.00	0.00	0.00	0.00	0.00	0.00	0.00	0.00	0.00	0.00	0.00	0.00	0.00	0.00	0.00	0.00	0.35	0.50
C15	0.00	0.00	0.00	0.00	0.00	0.00	0.00	0.00	0.00	0.00	0.00	0.00	0.00	0.00	0.00	0.90	0.65	0.90	0.90
C16	0.00	0.00	0.00	0.00	0.00	0.00	0.00	0.00	0.00	0.00	0.00	0.00	0.00	0.00	0.00	0.00	0.90	0.65	0.00
C17	0.00	0.00	0.00	0.00	0.00	0.00	0.00	0.00	0.00	0.00	0.00	0.00	0.00	0.00	0.00	0.00	0.00	0.35	0.00
C18	0.00	0.00	0.00	0.00	0.00	0.00	0.00	0.00	0.00	0.00	0.00	0.00	0.00	0.00	0.00	0.00	0.00	0.00	0.00
C19	0.00	0.00	0.00	0.00	0.00	0.00	0.00	0.00	0.00	0.00	0.00	0.00	0.00	0.00	0.00	0.00	0.00	0.79	0.00

(continued)

Table 3 (continued)

Customer 3

	C1	C2	C3	C4	C5	C6	C7	C8	C9	C10	C11	C12	C13	C14	C15	C16	C17	C18	C19
C1	0.00	0.50	0.00	0.00	0.00	0.00	0.00	0.00	0.00	0.00	0.00	0.00	0.00	0.00	0.50	0.79	0.35	0.10	0.90
C2	0.00	0.00	0.00	0.00	0.00	0.00	0.00	0.00	0.00	0.00	0.00	0.00	0.00	0.00	0.00	0.90	0.10	0.90	0.00
C3	0.00	0.00	0.00	0.00	0.00	0.00	0.00	0.00	0.00	0.00	0.00	0.00	0.00	0.00	0.00	0.65	0.00	0.65	0.00
C4	0.10	0.65	0.00	0.00	0.00	0.00	0.00	0.00	0.00	0.00	0.00	0.00	0.00	0.00	0.00	0.79	0.00	0.50	0.00
C5	0.00	0.65	0.00	0.00	0.00	0.00	0.00	0.00	0.00	0.00	0.00	0.00	0.00	0.00	0.00	0.79	0.00	0.35	0.90
C6	0.00	0.00	0.00	0.00	0.00	0.00	0.00	0.00	0.00	0.00	0.00	0.50	0.00	0.00	0.00	0.10	0.10	0.50	0.00
C7	0.00	0.50	0.00	0.00	0.00	0.00	0.00	0.00	0.00	0.00	0.00	0.00	0.00	0.00	0.79	0.00	0.00	0.50	0.50
C8	0.00	0.00	0.00	0.00	0.00	0.00	0.00	0.00	0.00	0.00	0.00	0.00	0.00	0.00	0.00	0.00	0.00	0.00	0.50
C9	0.00	0.00	0.00	0.00	0.00	0.00	0.00	0.00	0.00	0.00	0.00	0.00	0.00	0.00	0.00	0.00	0.00	0.90	0.90
C10	0.00	0.00	0.00	0.00	0.00	0.00	0.00	0.00	0.00	0.00	0.00	0.00	0.00	0.00	0.00	0.00	0.00	0.21	0.21
C11	0.00	0.00	0.00	0.00	0.00	0.00	0.00	0.00	0.00	0.00	0.00	0.00	0.00	0.00	0.35	0.00	0.35	0.79	0.90
C12	0.00	0.00	0.00	0.00	0.00	0.00	0.00	0.00	0.00	0.00	0.00	0.00	0.00	0.00	0.00	0.00	0.90	0.79	0.90
C13	0.00	0.00	0.00	0.00	0.00	0.00	0.00	0.00	0.00	0.00	0.00	0.00	0.00	0.00	0.00	0.00	0.00	0.50	0.79
C14	0.00	0.50	0.00	0.00	0.00	0.00	0.00	0.00	0.00	0.00	0.00	0.00	0.00	0.00	0.00	0.00	0.00	0.00	0.90
C15	0.00	0.21	0.00	0.00	0.00	0.00	0.00	0.00	0.00	0.00	0.00	0.00	0.00	0.00	0.00	0.65	0.00	0.00	0.65
C16	0.00	0.00	0.00	0.00	0.00	0.00	0.00	0.00	0.00	0.00	0.00	0.00	0.00	0.00	0.00	0.00	0.65	0.65	0.00
C17	0.00	0.00	0.00	0.00	0.00	0.00	0.00	0.00	0.00	0.00	0.00	0.00	0.00	0.00	0.00	0.00	0.00	0.21	0.00
C18	0.00	0.00	0.00	0.00	0.00	0.00	0.00	0.00	0.00	0.00	0.00	0.00	0.00	0.00	0.00	0.00	0.00	0.00	0.00
C19	0.00	0.00	0.00	0.00	0.00	0.00	0.00	0.00	0.00	0.00	0.00	0.00	0.00	0.00	0.00	0.00	0.00	0.79	0.00

(continued)

Table 3 (continued)

Customer 4

	C1	C2	C3	C4	C5	C6	C7	C8	C9	C10	C11	C12	C13	C14	C15	C16	C17	C18	C19
C1	0.00	0.79	0.00	0.00	0.10	0.00	0.00	0.00	0.00	0.00	0.00	0.00	0.00	0.00	0.65	0.79	0.35	0.21	0.50
C2	0.00	0.00	0.00	0.00	0.00	0.00	0.00	0.00	0.00	0.00	0.00	0.00	0.00	0.00	0.00	0.90	0.65	0.90	0.00
C3	0.00	0.00	0.00	0.00	0.00	0.00	0.00	0.00	0.00	0.00	0.00	0.00	0.00	0.00	0.00	0.00	0.00	0.79	0.00
C4	0.10	0.50	0.00	0.00	0.00	0.00	0.00	0.00	0.00	0.00	0.00	0.00	0.00	0.00	0.00	0.00	0.00	0.00	0.00
C5	0.00	0.00	0.00	0.00	0.00	0.00	0.00	0.00	0.00	0.00	0.00	0.00	0.00	0.00	0.00	0.00	0.00	0.35	0.00
C6	0.00	0.00	0.00	0.00	0.00	0.00	0.00	0.00	0.00	0.00	0.00	0.50	0.00	0.00	0.00	0.21	0.10	0.50	0.00
C7	0.00	0.50	0.00	0.00	0.00	0.00	0.00	0.00	0.00	0.00	0.00	0.00	0.00	0.00	0.00	0.35	0.10	0.65	0.00
C8	0.00	0.00	0.00	0.00	0.00	0.00	0.00	0.00	0.00	0.00	0.00	0.00	0.00	0.00	0.00	0.00	0.00	0.35	0.00
C9	0.00	0.00	0.00	0.00	0.00	0.00	0.00	0.00	0.00	0.00	0.00	0.00	0.00	0.00	0.00	0.00	0.00	0.00	0.90
C10	0.00	0.00	0.00	0.00	0.00	0.00	0.00	0.00	0.00	0.00	0.00	0.00	0.00	0.00	0.00	0.00	0.00	0.00	0.65
C11	0.00	0.00	0.00	0.00	0.00	0.00	0.00	0.00	0.00	0.00	0.00	0.00	0.00	0.00	0.00	0.00	0.00	0.00	0.35
C12	0.00	0.00	0.00	0.00	0.00	0.00	0.00	0.00	0.00	0.00	0.00	0.00	0.00	0.00	0.00	0.00	0.00	0.00	0.90
C13	0.00	0.00	0.00	0.00	0.00	0.00	0.00	0.00	0.00	0.00	0.00	0.00	0.00	0.00	0.00	0.00	0.00	0.10	0.00
C14	0.00	0.35	0.00	0.00	0.00	0.00	0.00	0.00	0.00	0.00	0.00	0.00	0.00	0.00	0.00	0.00	0.00	0.00	0.90
C15	0.00	0.21	0.00	0.00	0.00	0.00	0.00	0.00	0.00	0.00	0.00	0.00	0.00	0.00	0.00	0.65	0.00	0.00	0.00
C16	0.00	0.00	0.00	0.00	0.00	0.00	0.00	0.00	0.00	0.00	0.00	0.00	0.00	0.00	0.00	0.00	0.79	0.90	0.00
C17	0.00	0.00	0.00	0.00	0.00	0.00	0.00	0.00	0.00	0.00	0.00	0.00	0.00	0.00	0.00	0.00	0.00	0.21	0.00
C18	0.00	0.00	0.00	0.00	0.00	0.00	0.00	0.00	0.00	0.00	0.00	0.00	0.00	0.00	0.00	0.00	0.00	0.00	0.00
C19	0.00	0.00	0.00	0.00	0.00	0.00	0.00	0.00	0.00	0.00	0.00	0.00	0.00	0.00	0.00	0.00	0.00	0.90	0.00

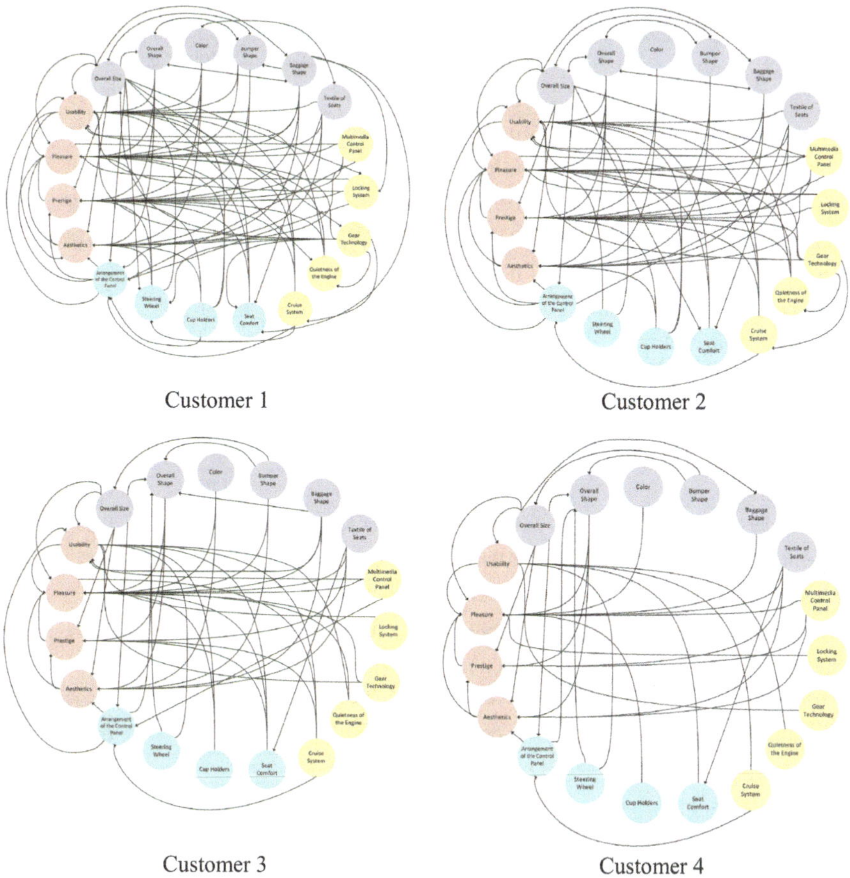

Fig. 3 Individual maps of customers

interest. Most of the design concepts of a product interact with each other. To understand the dynamics of design concepts long term effects of these interactions should be analyzed.

Fuzzy cognitive mapping is useful to identify the causal reasoning among design concepts. In the literature, it is seen that most of the papers use aggregate judgments in the determination of causal relationships. In the customer oriented product design, individual preferences become crucial. Therefore, in this chapter, fuzzy cognitive mapping is applied for each customer separately to discover the differentiation in steady state values of design concepts. The results of the study indicate that the causal relationships change with the diversity in customer preferences which confirms that the customer oriented product design should be flexible to satisfy each customer.

In this chapter, it is found that overall shape, arrangement of the control panel, pleasure and usability design concepts should be flexible in the automotive industry due to they have the highest values for all of the customers. Besides that, locking system, bumper shape, engine quietness, cruise system, seat comfort, steering wheel,

Table 4 The steady state values

	Customer 1	Customer 2	Customer 3	Customer 4
Overall size	0.786	0.900	0.678	0.678
Overall shape	0.908	0.900	0.952	0.925
Color	0.766	0.659	0.659	0.659
Bumper shape	0.921	0.852	0.659	0.659
Baggage shape	0.876	0.852	0.659	0.678
Textile of seats	0.840	0.810	0.659	0.659
Multimedia control panel	0.781	0.772	0.659	0.659
Locking system	0.846	0.659	0.659	0.659
Gear technology	0.781	0.659	0.659	0.659
Quietness of the engine	0.868	0.787	0.659	0.659
Cruise system	0.806	0.801	0.659	0.659
Seat comfort	0.867	0.826	0.746	0.746
Cup holders	0.768	0.659	0.659	0.659
Steering wheel	0.944	0.659	0.659	0.659
Arrangement of the control panel	0.984	0.880	0.877	0.770
Aesthetics	0.999	0.997	0.989	0.961
Prestige	1.000	1.000	0.947	0.935
Pleasure	1.000	1.000	0.999	0.997
Usability	1.000	1.000	0.999	0.979

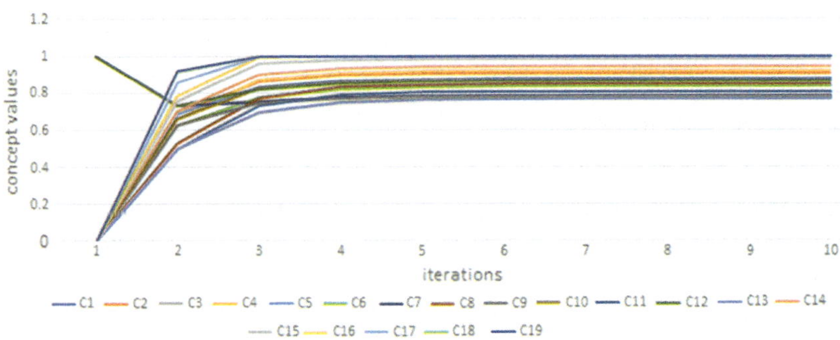

Fig. 4 The behavior of system during time (Customer 4)

aesthetics and prestige should also be flexible to be customer oriented according to the results of the study. For further researches, it is suggested to analyze customer preferences on product design concepts which differentiate among customers using multi-criteria decision making tools to determine options for each flexible design concept. This approach can also be considered as a tool to examine the changing preferences of customers in other industries such as fashion, electronics, durable goods, construction, etc.

References

1. Ahmadi, S., Yeh, C.H., Papageorgiou, E.I., Martin, R.: An FCM–FAHP approach for managing readiness-relevant activities for ERP implementation. Comput. Ind. Eng. **88**, 501–517 (2015)
2. Ahmadi, S., Forouzideh, N., Alizadeh, S., Papageorgiou, E.: Learning fuzzy cognitive maps using imperialist competitive algorithm. Neural Comput. Appl. **26**(6), 1333–1354 (2015)
3. Axelrod, R.: The analysis of cognitive maps. Struct. Decis. 55–73 (1976)
4. Blom, J., Josefsson, A.: (2013). Customer-oriented product development: an explorative study within a car manufacturer. Master Thesis. Department of Product and Production Development, Chalmers University of Technology, Gothenburg, Sweden
5. Bueno, S., Salmeron, J.L.: Benchmarking main activation functions in fuzzy cognitive maps. Expert Syst. Appl. **36**(3), 5221–5229 (2009)
6. Cheah, W.P., Kim, Y.S., Kim, K.Y., Yang, H.J.: Systematic causal knowledge acquisition using FCM constructor for product design decision support. Expert Syst. Appl. **38**(12), 15316–15331 (2011)
7. Chen, R.Y.: Intelligent IoT-based tracing system for backward design using FCM and fuzzy rule. In: 2013 Fourth Global Congress on Intelligent Systems, pp. 229–233. IEEE (2013)
8. Chen, R.Y.: Autonomous tracing system for backward design in food supply chain. Food Control **51**, 70–84 (2015)
9. Chen, R.Y.: Fuzzy dual experience-based design evaluation model for integrating engineering design into customer responses. Int. J. Interact. Design Manuf. (IJIDeM) **10**(4), 439–458 (2016)
10. Dai, W H.: Customer oriented product design by adaptive clustering analysis. In Applied Mechanics and Materials, vol. 26, pp. 690–693. Trans Tech Publications (2010)
11. Dong, Y., Guo, G., Liu, W.: Requirement elicitation in product emotional design using fuzzy clustering and fuzzy cognitive model. Proc. Inst. Mech. Eng. B J. Eng. Manuf. **228**(1), 156–160 (2014)
12. Efe, B.: Fuzzy cognitive map based quality function deployment approach for dishwasher machine selection. Appl. Soft Comput. **83**, 105660 (2019)
13. Elarusi, A., Attar, A., Lee, H.: Optimal design of a thermoelectric cooling/heating system for car seat climate control (CSCC). J. Electron. Mater. **46**(4), 1984–1995 (2017)
14. Ellis, D.S., Jacobs, L.W.: Marketing utilities: a new look. J. Acad. Mark. Sci. **5**(1–2), 21–26 (1977)
15. Fornasiero, R., Carpanzano, E.: Advances in customer-oriented manufacturing and value chain management. Int. J. Comput. Integr. Manuf. **30**(7), 677–679 (2017)
16. Gologlu, C., Mizrak, C.: An integrated fuzzy logic approach to customer-oriented product design. J. Eng. Des. **22**(2), 113–127 (2011)
17. He, C., Sun, X., Xiao, W.: Research on evaluation of CMF based on vision in automobile seat design. In: International Conference on Applied Human Factors and Ergonomics, pp. 146–158. Springer, Cham (2019)
18. Hu, J., Fang, J., Qiu, R., Wu, Y.: Customer requirement analysis of engineering project safety management intelligent system. ICCREM 2018: Innovative Technology and Intelligent Construction, pp. 252–262. American Society of Civil Engineers, Reston, VA (2018)
19. Ibragimova, E., Vermeeren, A., Vink, P., Mueller, N., Verboom, L.: The smart steering wheel cover design: a case study of industrial-academic collaboration in human-computer interaction. In: International Conference on HCI in Business, pp. 688–698. Springer, Cham (2015)
20. İlhan, E.Ü., Çelik, H.: Perceived quality and craftsmanship—common terminology proposal for automotive world with guidance of Kansei engineering. In: Advances in Affective and Pleasurable Design, pp. 391–399. Springer, Cham (2017)

21. Jie, Y., Yu, Y., Chuan, Z., Xiaolei, W., Qiang, Z.: Customer's kansei cognition model in product shape design. J. Comput. Aided Design Comput. Graph. **22**(3), 538–544 (2010)
22. Kabasakal, I., Keskin, F. D., Ventura, K., Soyuer, H.: From mass customization to product personalization in automotive industry: potentials of industry 4.0. J. Manage. Market. Logistics **4**(3), 244–250 (2017)
23. Khalid, H.M., Helander, M.G.: A framework for affective customer needs in product design. Theor. Issues Ergon. Sci. **5**(1), 27–42 (2004)
24. Kosko, B.: Fuzzy cognitive maps. Int. J. Man Mach. Stud. **24**(1), 65–75 (1986)
25. Kumar, A., Mangla, S.K., Luthra, S., Rana, N.P., Dwivedi, Y.K.: Predicting changing pattern: building model for consumer decision making in digital market. J. Enterp. Inf. Manage. **31**(5), 674–703 (2018)
26. Lin, M.C., Chen, L.A.: An integrated prototyping and knowledge representation procedure for customer-oriented product design. Int. J. Prod. Dev. **2**(4), 353–370 (2005)
27. Lin, M.C., Chen, L.A.: A matrix approach to the customer-oriented product design. Concurrent Eng. **13**(2), 95–109 (2005)
28. Lin, M.C., Chen, M.S., Lin, C.C., Lin, C.P.: A customer-oriented decision-making procedure for the design of electronic viewfinder interchangeable lens cameras. In: Concurrent Engineering Approaches for Sustainable Product Development in a Multi-disciplinary Environment, pp. 969–980. Springer, London (2013)
29. Lin, M.C., Lin, Y.H., Liu, S.F., Chen, M.S.: An integrated analysis of customer requirements for bicycle leisure activity functional clothing design. In: Advances in Ergonomics in Design, pp. 391–403. Springer, Cham (2016)
30. Lin, M.C., Wang, C.C., Chen, T.C.: A strategy for managing customer-oriented product design. Concurrent Eng. **14**(3), 231–244 (2006)
31. Matzler, K., Hinterhuber, H.H., Bailom, F., Sauerwein, E.: How to delight your customers. J. Prod. Brand Manage. **5**(2), 6–18 (1996)
32. Miao, Y., Liu, Y., Chen, Y., Zhou, J., Ji, P.: Two uncertain chance-constrained programming models to setting target levels of design attributes in quality function deployment. Inf. Sci. **415**, 156–170 (2017)
33. Mohammed, A., Hussain, M.I., Zain, Z.B.: Application of user centered design for customer requirement in design process for automotive manufacturing organizations. IOSR J. Mech. Civil Eng. **14**(2), 27–31 (2017)
34. Nahm, Y.E.: New competitive priority rating method of customer requirements for customer-oriented product design. Int. J. Precision Eng. Manuf. **14**(8), 1377–1385 (2013)
35. Nahm, Y.E.: A novel approach to prioritize customer requirements in QFD based on customer satisfaction function for customer-oriented product design. J. Mech. Sci. Technol. **27**(12), 3765–3777 (2013)
36. Nepal, B., Yadav, O.P., Murat, A.: A fuzzy-AHP approach to prioritization of CS attributes in target planning for automotive product development. Expert Syst. Appl. **37**(10), 6775–6786 (2010)
37. Papageorgiou, E.I., Markinos, A., Gemptos, T.: Application of fuzzy cognitive maps for cotton yield management in precision farming. Expert Syst. Appl. **36**(10), 12399–12413 (2009)
38. Papakostas, G.A., Boutalis, Y.S., Koulouriotis, D.E., Mertzios, B.G.: Fuzzy cognitive maps for pattern recognition applications. Int. J. Pattern Recogn. Artif. Intell. **22**(08), 1461–1486 (2008)
39. Ren, H., Tan, Y., Zhang, N.: Research on form design of automotive dashboard based on Kansei engineering. In: IOP conference series: materials science and engineering, vol. 573, No. 1, p. 012090. IOP Publishing (2019)
40. Salmeron, J.L., Mansouri, T., Moghadam, M.R.S., Mardani, A.: Learning fuzzy cognitive maps with modified asexual reproduction optimisation algorithm. Knowl. Based Syst. **163**, 723–735 (2019)
41. Sioukas, A.V.: User involvement for effective customization: an empirical study on voice networks. IEEE Trans. Eng. Manage. **42**(1), 39–49 (1995)
42. Storto, C.L.: Small group problem-solving and knowledge creation: a fuzzy cognitive maps investigation. Adv. Sci. Lett. **22**(5–6), 1282–1286 (2016)

43. Taha, Z., Alli, H., Rashid, A., Hanim, S.: Users' requirements and preferences in the success of a new product: a case study of automotive design. In: Applied Mechanics and Materials, vol. 215, pp. 484–488. Trans Tech Publications (2012)
44. Tama, I.P., Azlia, W., Hardiningtyas, D.: Development of customer oriented product design using Kansei engineering and Kano model: Case study of ceramic souvenir. Procedia Manuf. **4**, 328–335 (2015)
45. Tsadiras, A.K.: Comparing the inference capabilities of binary, trivalent and sigmoid fuzzy cognitive maps. Inf. Sci. **178**(20), 3880–3894 (2008)
46. Townsend, V., Urbanic, J.: Fuzzy cognitive modeling with users for design system analysis. Res. Eng. Des. 1–29 (2019)
47. Ulrich, K.T., Eppinger, S.D.: Product Design and Development, 5th edn. McGraw-Hill (2012)
48. Uzochukwu, B.M., Udoka, S.J., Balogun, F.: Development and implementation of product sustainment simulator utilizing fuzzy cognitive map (FCM). Benchmarking: Int. J. **23**(2), 425–442 (2016)
49. Wang, F., Li, H., Liu, A., Zhang, X.: Hybrid customer requirements rating method for customer-oriented product design using QFD. J. Syst. Eng. Electron. **26**(3), 533–543 (2015)
50. Wu, K., Liu, J., Chi, Y.: Wavelet fuzzy cognitive maps. Neurocomputing **232**, 94–103 (2017)
51. Xirogiannis, G., Glykas, M., Staikouras, C.: Fuzzy cognitive maps in banking business process performance measurement. In: Fuzzy Cognitive Maps, pp. 161–200. Springer, Berlin, Heidelberg (2010)
52. Xu, Y., Chen, G., Zheng, J.: An integrated solution—KAGFM for mass customization in customer-oriented product design under cloud manufacturing environment. Int. J. Adv. Manuf. Technol. **84**(1–4), 85–101 (2016)
53. Yong, D.: Plant location selection based on fuzzy TOPSIS. Int. J. Adv. Manuf. Technol. **28**, 839–844 (2006)
54. You, H., Ryu, T., Oh, K., Yun, M.H., Kim, K.J.: Development of customer satisfaction models for automotive interior materials. Int. J. Ind. Ergon. **36**(4), 323–330 (2006)

Baby Cradle Design with Kansei Knowledge Mining Based on Rough Set Theory

Esra Akgül, Mihrimah Özmen, Emel Kizilkaya Aydoğan and Cem Sinanoğlu

Abstract Today, producing consumer-oriented products are a key to commercial success. The successful companies have to aim to capture consumer's psychological perception. Kansei Engineering is a type of methodology to determine consumer's perception on product elements for using new product development. Kansei evaluation dataset were significantly developed such as uncertainty reason techniques including rough set, fuzzy set and neural networks. In literature, the rough set theory has been widely used approach to reduce the complexity of the knowledge database. In this chapter, to extract more accurate design parameters and Kansei knowledge, rough set rule based mining was applied for Kansei engineering research to baby cradle design. The proposed design methodology results show that obtained rules are consistent with customer expectations.

1 Introduction

The new product design or product development is quite important for companies especially with the rapidly changing market requirements in today's competitive market. The understanding of customer needs plays vital role a new product design or product development. Companies must analyze customer needs well to maintain their market position and deal with their competitors.

Nowadays, customers' purchasing decision depends not only on the product features but also on the feelings that the product makes in people. To be successful in the design process of the new product, companies must capture the voice of the customer. And so, they can improve the attractiveness of the product.

The customer needs on the product depend on emotions or feelings except for economic and technological concerns. Emotion is known as an intense and intentional state which involves relationship between the person and the stimulus [1]. The

E. Akgül · C. Sinanoğlu
Department of Industrial Design Engineering, Erciyes University, Kayseri, Turkey

M. Özmen · E. K. Aydoğan (✉)
Department of Industrial Engineering, Erciyes University, Kayseri, Turkey
e-mail: ekaydogan@erciyes.edu.tr

C. Kahraman and S. Cebi (eds.), *Customer Oriented Product Design*, Studies in Systems, Decision and Control 279, https://doi.org/10.1007/978-3-030-42188-5_24

customers are affected by the connotation and denotation of product concepts which are related to product element properties such as aesthetics, design, technological etc. [2] To capture customer voice or customer feeling on product, a lots of methodology such as Kano models [3], quality function deployment (QFD) [4], Conjoint Analysis [5] are used product design stage. Another methodology is the Kansei Engineering.

Kansei that is known emotion, feeling and affective in the literature is a Japanese term which evokes senses, feelings, impressions [6]. Kansei Engineering is a type of product design methodology which translates the consumer's feeling for a product into design elements [7]. This methodology is introduced by Mitsuo Nagamachi from Hirosima University in 1970s. Kansei engineering aims to understand emotions related to product elements and to produce customer-oriented successful product. Firstly, it is successfully used to develop Mazda Motor Corporation for developing the Miata model sports car which symbolized "human-machine unity" [8, 9].

A general model of Kansei Engineering is showed in Fig. 1. According to the general model, the methodology starts with choice of product domain. After choosing product, the semantic space to measure a product's affective response and product's physical structure to describe a product's affective response are defined. The obtained results synthesize with different statistical methods such as multi-regression-analysis, fuzzy logics or neural networks [10]. Then, the test of validity is done and the model builds for the product.

Kansei Engineering methodology is separated into different eight types to use solving design problems related to customer tendencies. The first type, Type I Kansei Engineering means Category Classification from zero level concept to nth level concept. Type II is known as Kansei Engineering System that uses the computer system. Type III is described as Hybrid Kansei Engineering System which utilizes a mathematical model to reason the appropriate ergonomic design. Type IV means Kansei

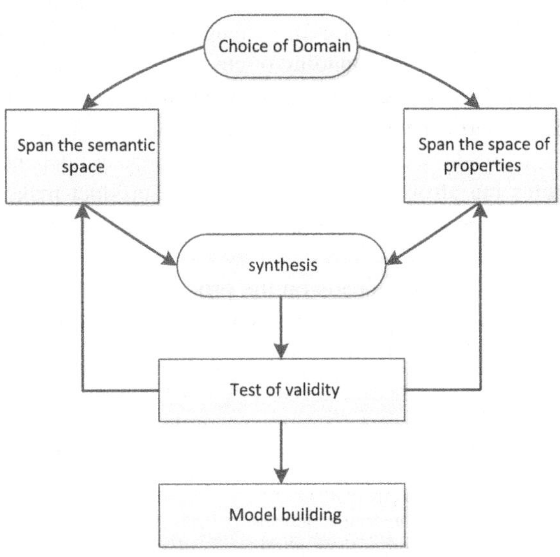

Fig. 1 A general model of Kansei engineering [21]

Engineering Modelling that implements more than Expert Systems with mathematical models. Type V is known as Virtual Kansei Engineering that is extended type of Kansei Engineering Systems or techniques. Type VI is a Collaborative Kansei Engineering Designing that utilizes a mutual Kansei database and collaborate through a network to develop a new product design. Type VII is Concurrent Kansei Engineering which perform from different departments and lastly, Type VIII is Rough Sets Kansei Engineering to dealing with ambiguous and uncertain Kansei data [11].

Kansei engineering is used different design area from car, home appliances to fabric, furniture, pen, ceramics etc. To determine relation of between Kansei data and product elements, different algorithms and statistical models can be applied. Data mining is also one of the knowledge discovery processes. Rough set theory is a mathematical theory of approximation of concepts, and various applications of rough set theory in Kansei engineering, in particular, applications of extracting decision rules are also widely studied. Mori presented that rough set approach based on Kansei data for car design solution [12]. Nishino et al. showed that extracting the decision rules of manufacturing conditions from much ambiguous data between coffee flavor evaluations such as taste and aroma with rough set method [13]. Nishino et al. proposed that the identification of customers' needs and the clarification of effects between design parameters to develop customer-oriented product by using rough set model for car floor carpet [14]. Zhou et al. presented that rough set based k-optimal rule mining approach from the customer domain to the designer domain for truck cab interior design [15]. Shieh et al. examine that toothbrush data for exploring the relationship between form and color during a product evaluation [16].

In this chapter, rough set based Kansei Engineering approach is used to explore the essential elements of the new customer-oriented baby cradles. The rest of this chapter is organized as follows. In Sect. 2, rough set classification system is introduced. In Sect. 3, the proposed new baby cradle design framework and a pilot example is presented. Preliminary practical implications and more future challenges are discussed in Sect. 4.

2 Rough Set Classification System

Rough set theory, introduced by Pawlak [17], is an effective tool to manage uncertain and vague problems or data related to indiscernibility relations and classification, information systems, attribute dependence and approximation accuracy, reduct and core attribute sets, and decision rules.

Rough set theory applies the unclear relation and data pattern comparison based on the concept of an information system with indiscernible data, where the data is uncertain or inconsistent. The data is grouped into classes called elementary sets. Feature/attribute selection is crucial in data processing that consists of relevant (or maybe irrelevant) object patterns, but it may be redundant in data pattern recognition.

Our study uses LEM2 (Learning by Examples-Module 2) algorithm for the rule induction. The LEM2 method which was also implemented in ROSE2 [18], is the generation of a minimal set of rules covering all objects. The pseudo code of the standard LEM2 algorithm is as follows [19]:

```
Algorithm LEM2(X)
// X represents a set of objects representing lower/upper
// approximation of the concept selected.
// t represents an attribute-value pair.
// The algorithm returns a single local covering L of X.
{
  G := X;
  L := ∅;
  while (G ≠ ∅) do
  {
    T := ∅;
    T(G) := {t\ [t] ∩ G ≠ ∅};
    while ((T = ∅)or(not([T] ⊆ X)))
    {
      Select a pair t ∈ T(G) such that |[t] ∩ G| is maximum.
      if a tie occurs arbitrarily select any one pair.
      T := T ∪ {t};
      if ([t] ∩ G ≠ ∅)
        {
          G := [t] ∩ G;
          T(G):= {t\[t] ∩ G ≠ ∅; t ∈ T(G)}
          T(G) := T(G) − T;
        } }
          for each t in T do
          { if ((T − {t}) ≠ ∅) then
               if ([T − {t}] ⊆ X) then
                T := T − {t};
          }
          L := L ∪ {T};
          G := X − ∪T∈L[T];
      }
      for each T ∈ L do
          if (∪P∈L−{T}[P] = X) then
              L := L − {T};}
```

3 The Proposed Method—The New Baby Cradle Design

When customers decide to buy something, they establish a relation with the product and are affected. The manufacturers have to catch customers' feeling and reflect on their products this feeling for successful in the market. Similar products also emerge in baby cradles like many products in the market. The baby cradle is chosen for the applied of customer-oriented design process based on Kansei Engineering. A baby cradle is a kind of furniture product used to put the baby to sleep and produced by wood, textiles or metals.

The aim of the case study is to applied to rough set theory and to create design rules between the most desirable design elements from a set of selected products and the customers' Kansei preferences. The proposed methodology is given in Fig. 2.

Stage 1. Span the Space of Baby Cradle Design Properties:

The most preferred baby cradle design sample space was spanned and then the design properties were selected from the market. The researchers and designers extract the

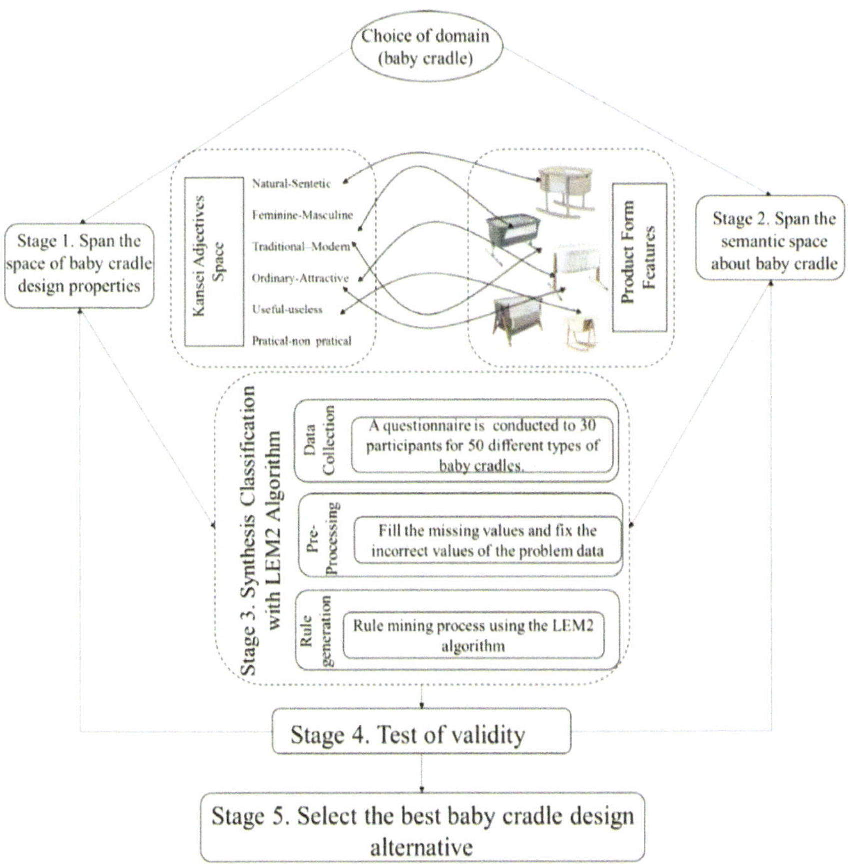

Fig. 2 The proposed methodology for baby cradle

design elements of the baby cradle for this chapter. According to this, the baby cradle has basic four product elements such as materials of stroller, materials of leg, the shape of stroller and shape of the leg. The morphologic analysis of baby cradle is shown Table 1.

Stage 2. Span the Semantic Space About Baby Cradle:

Magazines, book, web site and catalog are spanned to determine customer perceptions for baby cradle design. As semantic elements for Kansei evaluation, different 100 Kansei adjectives are collected for this study. From these Kansei adjectives six of them are chosen with a group of academicians, designers and engineers. They are given in Table 2.

Stage 3. Synthesis Classification with LEM2 Algorithm

The sub stages of stage 3 are detailed below:

Stage 3.1. Data Collection

A questionnaire is prepared using the pictures of the baby cradle for evaluating baby cradle with semantic differential scale. Totally 15 male (25–40 years) and 15 female

Table 1 Morphologic analysis of baby cradle

Items	Elements			
I_1:Materials of stroller	 1:Denim	 2:Wool	 3:Quilted	4:Mesh
I_2:Materials of Leg	 1:Wood	 2:Metal	 3:Wood&Metal	
I_3:Style of stroller	 1:Rectangle	 2:Oval	 3:Rounded rectangle	
I_4:Shape of leg	 1	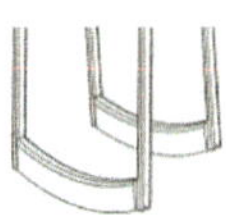 2	 3	
	 4	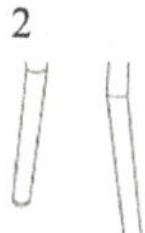 5	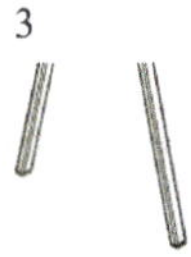 6	

Table 2 Kansei adjectives for baby cradle

	Kansei adjective
K_1	Natural–synthetic
K_2	Feminine–masculine
K_3	Traditional–modern
K_4	Ordinary–attractive
K_5	Useful–useless
K_6	Practical–non practical

(24–38 years) volunteer subjects have evaluated 50 different baby cradle designs. Six pairwise adjectives including Natural-Synthetic, Feminine-Masculine, Traditional–Modern, Ordinary-Attractive and Useful-Useless, Practical-Non-practical were used for semantic differential (SD) evaluations. Each adjective word is evaluated using a 7-point semantic scale.

The data obtained from the survey with 30 participants were analyzed for 50 different types of baby cradles. The mean values for 6 Kansei word pairs were calculated for each of the baby cradles.

Stage 3.2. Pre-processing

The survey dataset is preprocessed to obtain design rules. For this purpose, traditional preprocessing methods such as data cleaning, transformation of attributes, and data partitioning are applied. The dataset records containing a missing value are not discarded. The missing value in continuous attributes is with the most frequent value of the attribute.

Stage 3.3. Rule Generation

Lastly, the pre-processed Kansei data and cradle design parameters are analyzed by using the LEM2 algorithm for creating rules about design and customers' perception.

According to results of between the Kansei data and design parameters of baby cradle, the nine different design rules are generated. These are given in Table 3 as below:

For example: In Rule 1 If Baby cradle's stroller material is ***Denim*** and Baby cradle's Leg's material is ***Wood***, then this baby cradle is perceived as ***Natural*** by customers. In Rule 6 If Baby cradle's stroller material is ***Quilted*** and Baby cradle has 6th type Leg's shape, then this baby cradle is perceived neither ***Traditional*** nor ***Modern*** by customers.

Stage 4. Test of Validity

The design rule validity is tested with accuracy, which is a statistical measure that is obtained by use of the number of TP, TN, FP and FN according to the confusion matrix Table 4. Generally, accuracy is the most commonly used metric in classification and is formulated as given in Eq. 1. [20].

Table 3 The obtained design rules

R^1	If I_1 is 1 and I_2 is 1, then Class is K_1: Natural
R^2	If I_1 is 1 and I_3 is 1, then Class is K_4: Ordinary
R^3	If I_4 is 6 then Class is K_4: Ordinary-Attractive
R^4	If I_1 is 1 and I_4 is 2, then Class is K_6: Practical
R^5	If I_1 is 2 and I_4 is 2, then Class is K_2: Feminine
R^6	If I_1 is 3 and I_4 is 5, then Class is K_3: Traditional–Modern
R^7	If I_2 is 2 and I_4 is 4, then Class is K_6: Practical
R^8	If I_1 is 1 and I_3 is 3, then Class is K_2: Masculine
R^9	If I_3 is 2 and I_4 is 3, then Class is K_5: Useful

Table 4 Confusion matrix of classification problem

	Predicted positive class	Predicted negative class
Positive class	TP	FN
Negative class	FP	TN

$$\mathrm{Acc} = \frac{\mathrm{TP} + \mathrm{TN}}{\mathrm{TP} + \mathrm{FN} + \mathrm{FP} + \mathrm{TN}} \tag{1}$$

These nine design rules predicted 90% accuracy whereas it means the rules are covered 90% of the customer questionnaire answers. Besides obtained design rules are discussed with expert group and also they are confirmed the rules are consistent with customer perception.

Stage 5. Select the Best Baby Cradle Design Alternative

In previous steps, we obtain design rules for baby cradle and we have confirmed these rules by a group of academicians, designers and engineers. According to these rules and experts' opinions, from alternatives a baby cradle design is selected which best fits the customers' needs and feelings. For producc natural, practical and attractive products we prefer denim material for stroller, Wood leg, oval style stroller and type 2 shape of leg. The selected baby cradle design morphological properties and one of the alternative prototype are given in Table 5 as below:

4 Conclusion

Baby cradle is one of the most useful baby furniture in terms of parents' perception product. Many furniture companies are attempting to know the customer needs and perceptions for making favorite products. Kansei engineering is one of the techniques used to increase customer satisfaction. On the other hand, it is difficult to evaluate customers' preferences on Kansei parameters of products because of uncertainty, and imprecision of the consumer. In this chapter, we proposed the integration of rough

Table 5 The design morphological properties and alternative baby cradle prototype

Items	I_1:Materials of stroller	I_2:Materials of Leg	I_3:Style of stroller	I_4:Shape of leg
Elements			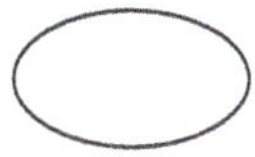	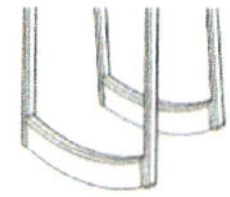
	1:Denim	3:Wood&Metal	2:Oval	2
Alternative prototype				
	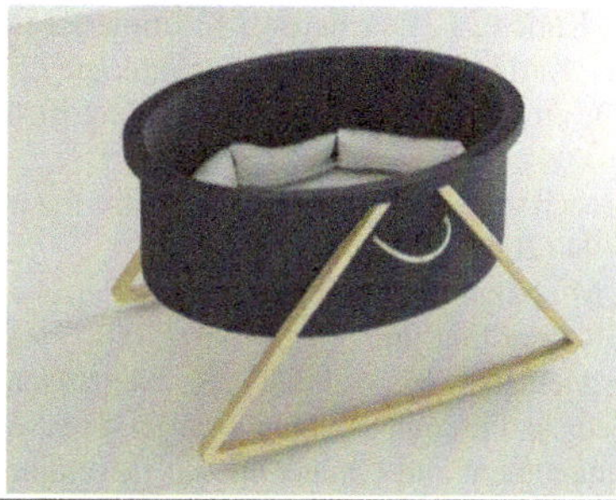			

sets theory and Kansei engineering methodology in order to find the relationships between the Kansei (emotions) and the product parameters. This proposed Kansei engineering approach is based on the rough set theory for developing customer-oriented products. It aims at representing the baby cradle preferences of the customers emotional requirement by means of simple decision rules. The findings from this study would increase customer satisfaction and help product designers. In future research, comparisons can be conducted between rough kansei engineering approach and fuzzy Kansei engineering approach. Furthermore, meta-heuristic algorithms can be applied to the model to compare with the results.

References

1. Lu, W., Petiot, J.F.: Affective design of products using an audio-based protocol: application to eyeglass frame. Int. J. Ind. Ergon. **44**(3), 383–394 (2014)
2. Petiot, J.F., Yannou, B.: Measuring consumer perceptions for a better comprehension, specification and assessment of product semantics. Int. J. Ind. Ergon **33**(6), 507–525 (2004)
3. Kano, N., Seraku, N., Takahashi, F., Tsuji, S.: Attractive quality and must-be quality. J Jpn. Soc. Qual. Cont. **41**, 39–48 (1984)
4. Cohen, L.: Quality Function Deployment: How to Make QFD Work for You. Prentice Hall (1995)
5. Luce, R.D., Turkey, J.W.: Simultaneous conjoint measurement: a new type of fundamental measurement. J. Math. Psych. **1**(1), 1–27 (1964)
6. Schütte, S., Eklund, J.: Design of rocker switches for work-vehicles—an application of Kansei engineering. Appl. Ergon. **36**(5), 557–567 (2005)

7. Nagamachi, M.: Kansei engineering: a new ergonomic consumer-oriented technology for product development. Int. J. Ind. Ergon. **15**(1), 3–11 (1995)
8. Nagamachi, M., Tachikawa, M., Imanishi, N., Ishizawa, T., Yano, S.: A successful statistical procedure on Kansei engineering products. In: 11th QMOD Conference. Quality Management and Organizational Development Attaining Sustainability from Organizational Excellence to Sustainable Excellence; 20–22 August; 2008 in Helsingborg; Sweden (No. 033, pp. 987–995). Linköping University Electronic Press (2008)
9. Tama, I.P., Azlia, W., Hardiningtyas, D.: Development of customer oriented product design using Kansei engineering and Kano model: Case study of ceramic souvenir. Procedia Man. **4**, 328–335 (2015)
10. Häfner, P., Ommeln, M., Katičić, J., Ovtcharova, J.: Immersive Kansei Engineering: A New Method and Its Potentials. Karlsruher Institut für Technologie (2018)
11. Lokman, A.M.: Design & emotion: the Kansei engineering methodology. Malays. J. Comp. **1**(1), 1–11 (2010)
12. Morı, N.: Rough set approach to product design solution for the purposed Kansei. Bull. Japan. Soc. Sci. Des. **48**(6), 85–94 (2002)
13. Nishino, T., Nagamachi, M., Sakawa, M.: Acquisition of kansei decision rules of coffee flavor using rough set method. Kansei Eng. Int. **5**(4), 41–50 (2006)
14. Nishino, T., Satsuta, R., Uematsu, M., Sugihara, S., Nagamachi, M.: Identification of customers' latent Kansei needs and product design by rough set based approach. In: 11th QMOD Conference. Quality Management and Organizational Development Attaining Sustainability from Organizational Excellence to Sustainable Excellence; Linköping University Electronic Press, vol. 033 Sweden, Helsingborg, pp. 341–350 (2008)
15. Zhou, F., Jiao, R.J., Schaefer, D., Chen, S.: Rough set based rule mining for affective design. In: DS 58-9: Proceedings of ICED 09, the 17th international conference on engineering design, vol. 9, Human Behavior in Design, Palo Alto, CA, USA, pp. 245–254 (2009)
16. Shieh, M.D., Yeh, Y.E., Huang, C.L.: Eliciting design knowledge from affective responses using rough sets and Kansei engineering system. J Ambient. Intell. Hum.Ized Comp. **7**(1), 107–120 (2016)
17. Pawlak, Z.: Rough sets. Int. J. Comp. Inform. Sci. **11**(5), 341–356 (1982)
18. Predki, B., Slowinski, R., Stefanowski, J., Susmaga, R., Wilk, S.Z.: ROSE—software implementation of the rough set theory. In: Polkowski, L., Skowron, A. (eds.) Rough design of waiting areas 407 Sets and Current Trends in Compter: Lecture Notes in Artificial Intelligence, vol. 1424, Springer, Berlin (1998)
19. Sabu, M.K., Raju, G.: Rule induction using Rough Set Theory—an application in agriculture. In: 2011 International Conference on Computer, Communication and Electrical Technology (ICCCET). IEEE, pp. 45–49 (2011)
20. Aydogan, E.K., Ozmen, M., Delice, Y.: CBR-PSO: cost based rough particle swarm optimization approach for high-dimensional imbalanced problems. Neural Comp. App. 1–19 (2018)
21. Schütte, S. (2002). Designing feelings into products: integrating Kansei engineering methodology in product development. Thesis, Department Mechanical Engineering, Linkopings University, Linkoping, Sweden

Index

C. Kahraman and S. Cebi (eds.), *Customer Oriented Product Design*, Studies in Systems, Decision and Control 279, https://doi.org/10.1007/978-3-030-42188-5

T

U

V

W

The manufacturer's authorised representative in the EU is Springer Nature Customer Service Centre GmbH, Europaplatz 3, 69115 Heidelberg, Germany. If you have any concerns regarding our products, please contact ProductSafety@springernature.com

Printed and bound by CPI Group (UK) Ltd, Croydon, CR0 4YY
22/09/2026
02221048-0010